FACILITY LAYOUT AND LOCATION

PRENTICE-HALL INTERNATIONAL SERIES IN INDUSTRIAL AND SYSTEMS ENGINEERING

W. J. Fabrycky and J. H. Mize, Editors

BLANCHARD *Logistics: Engineering and Management*
FABRYCKY, GHARE, AND TORGERSEN *Industrial Operations Research*
FRANCIS AND WHITE *Facility Layout and Location: An Analytical Approach*
GOTTFRIED AND WEISMAN *Introduction to Optimization Theory*
KIRKPATRICK *Introductory Statistics and Probability for Engineering, Science, and Technology*
MIZE, WHITE, AND BROOKS *Operations Planning and Control*
OSTWALD *Cost Estimating for Engineering and Management*
SIVAZLIAN AND STANFEL *Analysis of Systems in Operations Research*
SIVAZLIAN AND STANFEL *Optimization Techniques in Operations Research*
THUESEN, FABRYCKY, AND THUESEN *Engineering Economy*, 4/E
WHITEHOUSE *Systems Analysis and Design Using Network Techniques*

FACILITY LAYOUT AND LOCATION

AN ANALYTICAL APPROACH

RICHARD L. FRANCIS

Department of Industrial and Systems Engineering
University of Florida, Gainesville

JOHN A. WHITE

School of Industrial and Systems Engineering
Georgia Institute of Technology, Atlanta

PRENTICE-HALL, INC., *Englewood Cliffs, New Jersey*

Library of Congress Cataloging in Publication Data

Francis, Richard L
Facility layout and location.

(Prentice-Hall international series in industrial and systems engineering)

Includes bibliographies.

1. Plant layout. 2. Factories—Location.
3. Operations research. I. White, John A joint author. II. Title.

TS178.F7 658.2'1 73-18455

ISBN 0-13-299149-7

To the Memory of

David F. Baker

10 9 8

Prentice-Hall International, Inc., *London*
Prentice-Hall of Australia, Pty. Ltd., *Sydney*
Prentice-Hall of Canada, Ltd., *Toronto*
Prentice-Hall of India Private Limited, *New Delhi*
Prentice-Hall of Japan, Inc., *Tokyo*

CONTENTS

PREFACE

This book is written primarily for those college students who have an interest in facility layout and location problems. Students majoring in industrial engineering, management science, operations research, systems engineering, transportation science, and urban and regional design are logical candidates for a course which provides an analytical treatment of the subject of facility layout and location. Previous acquaintance with the subject of facility layout and location would be beneficial, but is not necessary. However, we do assume that the students have had an introductory course in linear programming or have been exposed to the subject in an introductory operations research course.

Facility layout and location is an area which offers considerable potential for the application of the analytical approaches of operations research. Students with diverse backgrounds and interests can find a number of interesting facility layout and location problems to be solved in a variety of contexts. A number of applied and theoretical problems remain to be solved.

The subject matter presented in this book has been taught in both a senior-level undergraduate course and a beginning graduate course taken by industrial engineering, systems engineering, and operations research students. A two-course sequence can be taught using this text, supplemented in the second course with selected reference papers from professional journals.

The objectives of this book are

1. To provide the facilities analyst with new techniques, approaches, and philosophies for the solution of facility layout and location problems.
2. To stimulate interest in facility layout and location problems within a wide variety of academic disciplines.
3. To provide an opportunity for a shift in the emphasis on qualitative–quantitative aspects of facility layout and location in college-level instruction on the subject.
4. To provide a classification of the rapidly expanding body of literature on facility layout and location problems and attempt to treat a selected portion of the literature in a unified manner.

The treatment of facility layout and location provided in this book is intentionally different from that found in other facility layout and location texts. We feel that, with the development of new research results in the area, the time has come when it is appropriate to take a more analytical approach to facility layout and location problems in a text than is commonly taken.

Since 1960 over 500 papers have been published in the area of facility layout and location. Since this book is not a survey text, a number of the facilities problems discussed in these papers have necessarily been omitted. In considering the tradeoff between unity of the presentation and breadth of coverage, we chose to provide an in-depth treatment of a small number of related problems.

Depending on the orientation of the course, the subject matter in this text can be covered in a variety of ways. We have found it convenient to cover Chapters 1, 2, 3, 4, 5, and 6 (omitting proofs) in an undergraduate course and Chapters 1, parts of 4, 5, 6, 7, 8, 9, and 10 in a graduate course (including proofs). If only one course is to be taught from the book, we recommend Chapters 1, 3, 4, 5, 6, 9 and the heuristic procedures in Chapter 8, with Chapter 2 assigned as supplementary reading. In the latter course, the coverage of the proofs would depend upon the backgrounds of the students. To facilitate the selection of appropriate sections, the more advanced material is indicated with footnotes. The designated sections can be omitted without loss of continuity in the presentation.

A large number of persons have contributed to the development of this text. We are deeply indebted to the counsel provided by Dr. David F. Baker of The Ohio State University, to whose memory this book is dedicated. Additionally, the support and encouragement provided by the Department of Industrial Engineering and Operations Research at Virginia Polytechnic Institute and State University, and the Departments of Industrial and Systems Engineering at both The Ohio State University and the University of Florida are gratefully acknowledged. Dr. P. M. Dearing, Dr. James W. Eyster, Dr. Stephen D. Roberts, and Dr. Kerry E. Kilpatrick read early drafts of the manuscript and provided helpful observations. To the many undergraduate and graduate students who assisted us in this effort, a special acknowledgment is given. Our appreciation is also expressed to Dr. Salah E. Elmaghraby for encouraging us to write the manuscript.

Finally, we wish to express our appreciation to the editors of *AIIE Transactions*, *Naval Research Logistics Quarterly*, and *Operations Research* for allowing us to include in the text material previously published by us in the indicated journals.

RICHARD L. FRANCIS
JOHN A. WHITE

A Journey through Facility Layout and Location

Facility layout and location is the subject of this book.
I wonder what it's about? I think I'll take a look.
Analytical approaches are taken in this text.
As I read the Preface, I wonder what is next?

Ten little chapters are waiting to be read.
I'm reading the Introduction and like what is said.
Facilities location, I've got it on my mind.
And can hardly wait to see the other nine.

The plant layout problem is the subject of Chapter Two.
Very little is presented here that is altogether new.
It's quite apparent the material fills a need.
Now there remain only eight more to read.

Computerized layout programs are discussed in Chapter Three.
Such programs can be useful; that is plain to see.
In developing layout designs, they are very deft.
And if you've been counting, seven now are left.

Single facility location is treated in Chapter Four.
And mathematical approaches come walking through the door.
The variety of applications seems quite a mix.
And more are bound to come in the remaining six.

Multiple facilities are considered in Chapter Five.
Analytic approaches continue, but I manage to survive.
Mathematical modeling dominates the game.
And the five to be read are probably the same.

Chapter Six is about discrete layout and location.
And introduced is a lot of new notation.
I've gone this far, I think I'll go some more.
After all, the number left is now no more than four.

Chapter Seven presents continuous layout design.
And from behind a cloud, the sun begins to shine.
This material is finally making a lot of sense to me.
And I can hardly wait to read the other three.

Quadratic assignment problems are discussed in Chapter Eight.
This material has application, or so the authors state.
I'm going to apply the material. It's something I want to do.
I wonder what's coming up in the remaining two?

Minimax problems are the subject of Chapter Nine.
I'm finally catching on and handling the material fine.
In practice, minimax objectives do occur a lot.
Now, one chapter remains to complete this simple plot.

Plant location and covering problems are presented last.
The reason for this choice is the way the die was cast.
About facilities layout and location, ten chapters I have read.
And I'm sure you will agree, it's time to go to bed!

FACILITY LAYOUT AND LOCATION

INTRODUCTION

1.1 Background

This is a book dealing with the subject of facility layout and location. Since you have elected to read this far, it is quite likely that either (1) you wonder what facility layout and location is all about and hope to satisfy your curiosity, (2) you are interested in facility layout and location and wish to see what this book has to offer on the subject, or (3) you are a student taking a course for which this book is the required text, and you have been assigned the first chapter to read (even though the professor will begin lecturing from some later chapter). So be it.

If you are of the first type, let us shorten your search by stating that facility layout and location is about the location of facilities, as well as the determination of the configuration of certain types of facilities. The latter activity is called facility layout. Traditionally, the subject of this book is referred to as *plant layout*. The more general term of facility layout and location is used here, as opposed to plant layout and location, for a number of reasons. One reason is that we wish to avoid the possibility of librarians categorizing this book as a botany or horticulture text. Certainly, facility has a very broad connotation. However, it is felt that this book will be of

1

only limited use to botanists and horticulturists. Another reason is that we wish to emphasize the fact that facility layout and location problems include plant layout problems as a subclass.

If you are a person interested in facility layout and location and are interested in contrasting the contents of this book with that of others, we believe you may be surprised at the departure from tradition employed in this book. Hopefully, after recovering from your surprise, you will judge the contents of this book to be a valuable contribution toward the solution of facility layout and location problems.

Finally, if you are in the third category, a student, we have endeavored to provide you with a book that will stimulate your interest in, and provide you with some knowledge of, the subject. Facility layout and location is an area offering considerable opportunities for persons trained in the analysis of such problems.

Facility layout and location problems have been the subject of analysis for centuries. As an illustration, a special case of a location problem treated in Chapter 4 was solved as early as the seventeenth century. Even though facility layout and location problems have received considerable attention over the years, it was not until the emergence of the interest in operations research that the subject received renewed attention in a number of disciplines. Today, we find that there exists a strong interdisciplinary interest in facility layout and location.

Economists, operations researchers, urban planners, management scientists, architects, regional scientists, home economists, and engineers from several disciplines have discovered a commonality of interest in their concern for the location and layout of facilities. Each tends to bring to the subject a different interpretation of the problem and different approaches for its solution. Indeed, each defines "facility" in a different way. As mentioned earlier, the term can be given a very general interpretation.

As evidence of the interdisciplinary interest in facility layout and location problems, consider the number of professional journals that publish location analysis or facility-design research papers. Typical of the journals that publish such papers are

AIIE Transactions
International Journal of Production Research
The Logistics and Transportation Review
Naval Research Logistics Quarterly
Operations Research
Operational Research Quarterly
Management Science
SIAM Review
SIAM Journal of Applied Mathematics
Journal of Regional Science
Journal of Farm Economics
Geographical Analysis
International Economic Review
Econometrica
Journal of the American Institute of Planners
Mathematical Programming
Networks
Transportation Science

The journals cited provide an indication of the variety of disciplines having an interest in the subject matter treated in this text. We shall attempt to further motivate the interdisciplinary appeal of the subject in the examples and exercises at the end of the chapters.

In comparison with other facility layout and location texts, we take an analytical approach to the facility location and layout problems treated. Mathematical optimization models are developed and solutions obtained. We provide an in-depth treatment of a relatively limited number of aspects of the facility layout and location problem, as compared with the traditional textbook approach, which examines a broad range of problems in a less analytical fashion. Whereas the traditional approach relies heavily on intuition and engineering judgment, we bring analysis to bear on a smaller set of problems. The solutions obtained through analysis serve as aids in decision making. Factors not included in the analysis should be considered, along with the analytical results, in reaching final decisions concerning the layout and location of facilities.

In some cases the solutions obtained from analysis will appear impractical from an operational viewpoint. Such solutions should be interpreted as *benchmarks* against which operationally accepted solutions are compared. Thus, the models presented should be viewed as design tools, just as many models from other disciplines are viewed.

Even though heavy emphasis is given in this book to analytical approaches, you should not overlook the value of the qualitative aspects of facility layout and location. It should be realized that the analytical approach yields a solution to the *model*, but not necessarily the *problem*. There remain a number of nonquantifiable questions, which must be considered in making the transition from model to problem. On the other hand, it is equally important that you not dismiss the analytical approach in favor of a completely qualitative approach. For there are quantitative aspects of the facility layout and location problem that cannot be reckoned with accurately through intuition alone.

1.2 Model Classification

Reference has been made to the development and use of models in solving facility layout and location problems. Perhaps it is worthwhile to review one classification of models and contrast the approach taken in the past with that taken in this book with regard to the analysis of facility layout and location problems.

Models can be classified as iconic, analog, and symbolic. Iconic models are scalar representations of objects in that they look like the objects represented. Iconic models may be two- or three-dimensional representations and closely maintain the visual effects of the situation under study. Analog models substitute one property for another. After the problem is solved in the substituted state, the solution is translated back to the original dimensions or properties. Symbolic or mathematical models are an abstract representation of a system.

Prior to the development of interest in analytical approaches, facility layout and location problems were solved primarily using iconic models. Templates and scale models were maneuvered on a floor plan of the facility until a number of alternative solutions were obtained. The generation of these alternatives was largely dependent upon the subjective criteria of the analyst. The alternatives were judged on the basis of visual effects, as well as other qualitative objectives. In some cases, checklists and rules of thumb were employed in an attempt to reduce the degree of subjectivity involved. Gradually, such approaches have been replaced by a greater degree of reliance on quantitative analysis.

In this book, our concern is with the development of symbolic or mathematical models. There are basically two types of mathematical models: descriptive and prescriptive (normative). In the former, the model is used to *describe* the behavior of the system. Normally, as an example, queueing models are descriptive. Normative models are used to *prescribe* a course of action that, in some sense, is optimal. Thus, normative models require a measure of effectiveness against which alternative solutions can be judged. Mathematical programming models are examples of typical normative models.

1.3 Criterion Selection

The facility layout and location models developed in this text are normative models. For this reason the problem of selecting an appropriate criterion must be considered. The choice of the criterion to be used in choosing the "best" solution from among several alternative solutions is not an easy one in the case of facility location and design problems.

Perhaps the most common approach is to employ the criterion of minimizing some function of distance traveled. The choice of this approach is justified on the basis that material-handling costs are minimized by minimizing distance traveled. The degree of emphasis given to this criterion has been criticized by Vollmann and Buffa [16]. They point out that most facility location and layout models developed in the past have been based on the assumptions that

1. Cost and flow data exist for definitionally unknown conditions.
2. Material-handling costs are linear, incremental, and assignable to specific activities.
3. Material-handling cost is the only significant factor.
4. Flow data are deterministic.
5. No interaction exists between the facility location problem and other system problems.

Vollmann and Buffa raise many valid criticisms concerning facility location and layout research efforts. Indeed, their paper should be required reading for the student seriously interested in the subject of facility location and design. Their discussion, unfortunately, raises more questions than it answers. The discussion does emphasize the necessity for giving serious thought to the selection of an appropriate criterion. The analyst should not blindly assume that minimizing total distance traveled is an appropriate objective. As an absurd illustration, with such an objective, locating a school such that 500 students must each travel 10 miles to school is as desirable as locating the school such that 499 students travel zero miles and one student travels 5,000 miles.

Instead of the criterion of minimizing total distance traveled, one may wish to minimize the maximum distance traveled. Such a solution is called a minimax location. In the case of the location of a school, a minimax location would guarantee that the largest distance traveled by a student be as small as possible. Such an objective is encountered in the design of stadiums and theaters, as well as the location of fire stations, hospitals, and civil-defense units.

A number of models developed in subsequent chapters are based on the objective of minimizing material-handling cost. However, the definition of material-handling cost is expanded to include those costs which vary incrementally as a function of the distances between the components of the system under study. Such a definition includes the cost of the movement of men, materials, equipment, and information.

The two distance measures commonly used in our subsequent discussion are rectilinear and Euclidean distances. The rectilinear distance between the coordinate locations (x_A, y_A) and (x_B, y_B) equals $|x_A - x_B| + |y_A - y_B|$. The Euclidean distance, or straight-line distance, equals $[(x_A - x_B)^2 + (y_A - y_B)^2]^{1/2}$. Material-handling cost will be related to these distance measures in later discussions.

1.4 Cost Elements

A number of costs are required in the various models developed in this text. More specifically, incremental costs are required. As an illustration, suppose that trucks are used to perform all material handling between stores in a city. A new store is to be located in the city, and there are two possible locations, *P* and *Q*, for the new store. If the new store is located at *P*, there will be 50 miles of travel per day between the new store and other existing stores; if located at *Q*, there will be 70 miles of travel per day. If the number of trucks and employees required for material handling is the same whether the new store is located at *P* or *Q*, the only incremental costs are those resulting from the additional travel of 20 miles per day, plus the costs of site preparation and constructing the store at each site. If the new store is actually an existing store that is to be relocated, relocation costs must be considered.

In some cases the choice of location affects the type of material-handling equipment to be used. For example, suppose that there exists a machine at location A. A new machine, B, can be located at either X or Y. If located at X, a roller conveyor will be installed between A and X, and Euclidean distance will be the appropriate distance measure. If located at Y, a lift truck will transport items between A and Y, and rectilinear distance will be the appropriate distance metric. The incremental costs to be considered are

1. Cost of purchasing, installing, and maintaining a roller conveyor if located at X.
2. Cost of operating the lift truck if located at Y.
3. Cost of in-process inventory differences resulting from different material-handling systems.
4. Other costs produced by the differences in material-handling systems.

In the case of lift trucks, the incremental annual cost of item movement between machines A and B is typically the product of the incremental cost per mile, number of round trips between A and B per year, and round-trip distance per trip between A and B. If a conveyor connects machines A and B, the incremental annual cost is the incremental cost of owning, operating, and maintaining a conveyor of sufficient length to connect A and B and of sufficient capacity to meet the yearly material-handling requirements. Normally, this cost is assumed to be a linear function of conveyor length.

1.5 Model Validation

In addition to an appropriately selected criterion, the validation of the model is a very important step in the analysis of facility location and design problems. By validation we mean the process of verifying that the model does indeed accurately represent the physical system under study. One measure of the validity of our models is whether or not they lead to reliable predictions of the system's performance and subsequent improvements in the system. Normally, it is quite difficult to validate a facility location and design model.

In practice, model validation usually consists of a verification of the assumptions of the model. In a number of cases this approach is justified. However, as will be seen in later discussion, the model can be relatively insensitive to a number of the assumptions. Consequently, it does not necessarily follow that a model based on inaccurate assumptions will yield inaccurate solutions. The true test of the validity of the model is whether or not it serves as an aid in obtaining solutions that are better than would otherwise be obtained.

There exist tradeoffs between the degree of realism in the model, its ease of manipulation, expense of solution, and clarity of cause-and-effect relationships. Normally, very simple models are developed in this text. Many simplifying assumptions are made. As a result, the accuracy of the assumptions is sometimes questionable. On the other hand, it is often possible for simple models to yield answers that closely agree with those obtained from a highly sophisticated model closely approximating the "real world."

Of the several approaches suggested in the literature for model validation, one that can be employed with the type of models considered here is to compare the answers obtained from different models of the problem. For example, the solutions obtained using an iconic model, an analog model, a simulation model, a simple mathematical model, and a complex mathematical model could be compared. Obviously, such an approach could prove to be quite expensive. In practice, one might choose two models and compare their solutions. If the answers agree, there is a higher probability that the model is valid. If the answers disagree, one model could still be valid or, for that matter, both could be invalid. In the latter case, a third model might be developed, and its solution compared with the other models. However, one can never be assured that a facility layout and location model is valid in the strictest sense.

The problem of model validation is not easy to resolve. However, this difficulty should not cause the analyst to ignore the problem. Model validation is quite important and should be considered. If the same, or very similar, facility layout and location problem occurs a number of times, the validity of the model can be improved through "hindsight analysis." That is, one can ask the question, "If a different location had been chosen, would costs have been reduced?" If so, make whatever modifications necessary to the model and determine the solution for a subsequent problem. After gaining some experience with the solution, employ "hindsight analysis" again. It is this "evolutionary property" of analytical models that provides strong support for their use, as opposed to a strict reliance on qualitative factors.

1.6 Layout Design Process

Throughout the remainder of the text we address the subject of facility layout and location problems. In particular, we are interested in solving a design problem. In the case of the facility layout problem we seek the "best" layout design; in solving the facility location problem we wish to find the "optimum" arrangement (design) of facilities. Since we shall be faced with a number of design problems, it is important that these problems be approached using the general procedure advocated for the solution of any design problem. Specifically, the following steps are suggested by Krick [6]:

1. Formulation of the problem.
2. Analysis of the problem.
3. Search for alternative solutions.
4. Selection of the solution.
5. Specification of the solution.

In our subsequent discussion each of these steps will be treated briefly. The discussion will concentrate primarily on the problem of designing facility layouts, rather than the related problem of facility location. However, it should be apparent that much of the discussion is equally appropriate to the facility location problem.

1.6.1 *Formulation of the problem*

In some cases the formulation of the problem is obvious. In fact, it may be formulated for you; e. g. , "Go over to the personnel building and determine the best location for a new data-processing machine." In such a case, care should be taken to assure that the proper problem is formulated. In the present case, a change in the design of the paper-work system might eliminate the need for the new data-processing machine!

Rather than having the problem correctly formulated for you, more commonly you will be confronted with a situation, and you must proceed from there. Typically, the situation is the present solution to the problem. This is especially true in re-layout problems. There are rather serious dangers associated with such a formulation. Specifically, it is rather easy to become biased and lose one's objectivity when confronted with an existing layout.

The formulation of the problem is aided by taking a *black-box* approach. As Krick [6] describes the black box, there exists an originating state of affairs (state *A*) and a desired state of affairs (state *B*). Also, a transformation must take place in going from state *A* to state *B*, as shown in Figure 1.1. There is more than one method of accomplishing the transformation from state *A* to state *B*, and there is unequal preferability of these methods (otherwise, no problem exists). The solution to the problem is visualized as a black box of unknown, unspecified contents, having input *A* and output *B*.

The black-box approach facilitates a proper identification of states *A* and *B* during problem formulation. As an illustration, consider the problem

Figure 1.1. Black-box approach.

of locating a lathe in department *X*. The problem could be formulated in terms of states *A* and *B* as follows:

	State A	*State B*
1.	Lathe to be located in department *X*.	Lathe located in department *X*.
2.	Lathe to be located.	Lathe located.
3.	Machine tool to be located.	Machine tool located.
4.	Operations to be performed.	Operations performed.
5.	Customer having need for a product.	Customer's need satisfied.

The first formulation is very narrow. Using it, we would concentrate on determining the best location for the lathe in department *X*. The second formulation allows us to consider location sites outside department *X*. The third formulation suggests the best machine tool might not be the lathe initially specified. The fourth formulation allows us to develop the process best suited for the performance of the set of operations and to design the overall layout. The fifth formulation involves a total systems design. No doubt, there are other formulations having varying degrees of breadth.

Just how broadly should the problem be formulated? In general, we recommend that the problem be formulated as broadly as the economics of the situation, time constraints, and the organizational boundaries permit. In the example just considered, it is not likely that the last formulation given would be justified. The point to be made is that the narrower the formulation the fewer solutions there are to the problem. Thus, by formulating a problem narrowly you might miss a solution that would produce significant cost savings for the total system. In the case of the lathe, it may well be that the product mix has changed sufficiently over time to the point that a complete re-layout is justified. In fact, adding another lathe might be counterproductive for the total system.

1.6.2 *Analysis of the problem*

The *analysis of the problem*, which has been formulated previously, consists of a relatively detailed phrasing of the characteristics of the problem, including restrictions. This phase of the design process involves considerable *fact gathering*. Be careful to separate the real restrictions from the fictitious restrictions. Just because something has been done the same way for the past 20 years does not justify it as a real restriction. The analysis of the problem also involves the identification of the appropriate criteria to be used in evaluating alternative solutions to the problem.

The process of fact gathering brings you face to face with the present solution to the problem. Consequently, it is wise to make every effort to avoid becoming biased in your thinking. Preoccupation with the present solution often results in solutions to the problem that are only slight modifications of the present solution.

If original and creative designs are to be developed, it is essential for the layout designer to concentrate initially on the correct problem formulation and then to analyze the true problem. Although charts and diagrams can be quite useful in layout design, it should be remembered they are aids to, not substitutes for, a sound analysis of the problem.

1.6.3 *Search for alternative solutions*

The *search for alternative solutions* to the layout problem consists of the specification of alternative contents of the black box employed in formulating the problem. Remember, the quality of the final solution can be no better than the quality of the alternative solutions generated during this phase of the design process. Since the final solution to the problem must come from the set of alternatives generated, we suggest that you attempt to maximize the number, quality, and variety of alternative solutions.

In the search for alternative solutions, strive to be creative, and make every effort to divorce your thinking from the present solution. There exist a number of aids to creativeness that can improve your ability to generate more and better alternative designs. Krick [7] has summarized a number of these. We shall draw on his discussion.

1. *Exert the necessary effort.* Set a time limit and force yourself to concentrate on the problem during the allotted time.
2. *Do not get bogged down in details too soon.* Think big initially; the details can be considered later. Of course, you can overdo it. We are reminded of the operations research analyst during World War II who supposedly suggested that the German submarine force could be destroyed by boiling the ocean. When questioned further about the feasibility of the plan, the analyst responded, "I'm supposed to develop solutions, not implement them. I leave the details to someone else."
3. *Make liberal use of the questioning attitude.* The questions what, who, when, where, how, and why should be applied to the problems.

4. *Seek many alternatives.* Establish a goal for the number of ideas to be generated and work to achieve it. Again, it is emphasized that the preferred solution will come only from the set of alternative solutions generated.
5. *Avoid conservatism.* Do not restrict yourself to simple variations of the present solution. Think big.
6. *Avoid premature rejection.* This phase of the design process emphasizes the generation of alternative solutions, not the evaluation of these alternatives. Consequently, do not reject possible solutions, regardless of their apparent feasibility.
7. *Avoid premature satisfaction.* Do not terminate the search for alternative solutions when a seemingly good solution has been generated. Save your evaluation of the solutions until later.
8. *Refer to analogous problems for ideas.* Consult trade magazines, look around you, refer to recent architectural literature, and be aware of the large variety of layout solutions that surround you. Ideas for your layout can come from the layout of communitites, offices, libraries, plants, stores, restaurants, homes, and schools, to list but a few.
9. *Consult others.* Actively seek information and suggestions from urban planners, engineers, workers, supervisors, and others. The more people actively involved in the generation of ideas, the greater the chance of (a) getting new and different solutions, (b) selling the final solution, and (c) stimulating your own thought processes.
10. *Attempt to divorce your thinking from the existing solution.* Employ the black-box approach repeatedly. Many of today's layouts are the product of evolution rather than careful design. As an illustration, consider the following anonymous poem given in [8]:

Path of the Calf

One day through the primeval wood
A calf walked home as good calves should;
But made a trail all but bent askew,
A crooked trail as all calves do.
Since then three hundred years have fled,
And I infer the calf is dead.
But still he left behind his trail,
And thereby hangs my moral tale.
The trail was taken up next day
By a lone dog that passed that way;
And then a wise bell wether sheep
Pursued the trail o'er vale and steep,
And drew the flock behind him, too,
As good bell wethers always do.

And from that day o'er hill and glade
Through those old woods a path was made.

The years passed on in swiftness fleet,
The road became a village street;
And this, before men were aware,
A city's crowded thoroughfare.
And soon the central street was this
Of a renowned metropolis;
And men two centuries and a half
Trod in the footsteps of that calf.
Each day a hundred thousand rout
Followed this calf about
And o'er his crooked journey went
The traffic of a continent,
A hundred thousand men were led
By one calf near three centuries dead.
They followed still his crooked way,
And lost one hundred years a day;
For thus such reverence is lent
To well established precedent.

For men are prone to go it blind
Along the calf-paths of the mind,
And work away from sun to sun
To do what other men have done.

11. *Try the group approach.* This method, popularly known as brainstorming, involves a group of four or five or preferably more persons assembled for the stated purpose of generating layout solutions. No evaluation takes place. Instead, one attempts to maximize the number of solutions generated. The problem should be stated in a broad, unrestricted manner. For example, the actual problem may be that of devising a new and effective way of transporting emergency patients to surgery in a hospital. In this case the initial instructions to the group could be, "generate many ways of moving objects." After a period of brainstorming, this group could be instructed, "now generate many ways of moving emergency patients." By brainstorming a very general formulation first, followed by a more specific formulation, we can improve the benefits of the group approach.
12. *Remain conscious of the limitations of the mind in this process of idea generating.* If you are constantly aware of the tendency to impose fictitious restrictions, to be overly conservative, to accept prematurely and reject, etc., then you have already made an important step toward overcoming the tendencies that stifle creativity.

1.6.4 *Selection of the solution*

Once the set of alternative designs has been generated, the *selection of the solution* is made. This phase of the design process consists of the measurement of the alternative solutions, using the criteria as the basis for comparison. If the criteria are easily quantified, the process of measurement is easily performed. However, in the following discussion we list a number of objectives that are qualitative, rather than quantitative. Realistically, there usually exist multiple criteria that are to be used in evaluating alternative layouts. Some of the criteria probably will be quantifiable, but inevitably some will not be. Thus, the process of evaluating alternative layout designs is not a simple one. In fact, Moore [9] states that this is the most difficult part of the layout process.

A number of the techniques used in evaluating alternative layout designs are given by Muther [10]. Some of these are

1. *List of Pros and Cons.* The simplest way to evaluate alternative layouts is by listing the advantages (pros) and disadvantages (cons) of each.
2. *Ranking.* Select the factors or considerations felt to be important to the layout, list them, and rank the alternatives in numerical order for each factor.
3. *Factor Analysis.* Each factor is assigned a numeric weight and then each alternative is ranked against each factor. The weighted rankings are totaled for each alternative, and the alternative having the best score is chosen.
4. *Cost Comparison.* All costs associated with each alternative are identified as well as cost savings produced, and that alternative which is most economical is chosen.

The listing of pros and cons is probably the easiest way to evaluate alternative layouts. However, it also is probably the least accurate. Its primary purpose is to allow an initial screening of those alternatives that have major deficiencies.

The ranking procedure has the property that all alternatives are compared against the same set of factors. The shortcomings of this method are that some important factors might be overlooked and that a final selection of the preferred design is not easily made. After using the ranking procedure, you still have to combine the rankings in some way so that a choice can be made. The factor-analysis technique makes this process explicit by assigning weights to each factor. Since quantitative and qualitative factors can be included and the preferred solution is made explicit, factor analysis is a very popular method of evaluating alternative layouts.

Both the ranking procedure and the factor-analysis procedure involve a comparison of layout alternatives for each factor. It is important that this comparison be made as objectively as possible. One approach that is used is the paired-comparison technique, where all combinations of two layout alternatives are ranked for each factor. As an illustration, suppose that there are five layout alternatives (A, B, C, D, and E), and the paired-comparison results for a given factor are

$$\begin{array}{ll} A > B, & B < D \\ A > C, & B < E \\ A < D, & C < D \\ A > E, & C < E \\ B < C, & D > E \end{array}$$

where $X > Y$ means X ranks higher than Y, and $X < Y$ means X ranks lower than Y. Combining these comparisons indicates that

$$D > A > E > C > B$$

for the factor considered. If there are n alternatives and m factors, then

$$\frac{mn(n-1)}{2}$$

comparisons are required.

The paired-comparison technique is quite beneficial in testing for inconsistencies. As an example, suppose it is found that someone has provided the rankings

$$A > B, \qquad B > C, \qquad C > A$$

Obviously, there is an inconsistency in the rankings. Such inconsistencies might occur due to imprecise definitions of the factors or to a selection of factors that are too broadly defined.

Some of the most commonly involved factors or considerations are [1]

1. Ease of future expansion.
2. Flexibility of layout.
3. Material-handling effectiveness.
4. Space utilization.
5. Safety and housekeeping.
6. Working conditions.
7. Ease of supervision and control.
8. Appearance, promotional value, public or community relations.

9. Fit with company organization structure.
10. Equipment utilization.
11. Ability to meet capacity or requirements.
12. Investment or capital required.
13. Savings, payout, return, profitability.

A potential weakness of both the ranking procedure and the factor analysis procedure is the possiblity of the *halo effect*. The halo effect refers to the situation in which the analyst ceases to act objectively and lets high rankings on a few factors for a particular layout alternative influence his rankings on a number of other factors. One can detect this by looking for an alternative that clearly dominates all others. If the search process was truly effective, there should not exist a dominant alternative. The design ultimately chosen will normally represent a compromise among the important factors.

A cost comparison of the layout alternatives involves a consideration of investment, operating, and maintenance costs. In making this comparison, one should make explicit the planning horizon (number of years) over which the alternatives are to be compared. An economic analysis can be performed using the time value of money and considering income-tax effects. There are a number of excellent references that can be used as guides in performing this analysis. For example, see [2], [5], and [15].

We have considered four methods of evaluating layouts that are commonly used to assist the layout analyst in arriving at a preferred design. There is one danger in this process of evaluation that should be emphasized. The value system of the analyst might be quite different from that of management. For example, the analyst might choose that alternative which is least costly, when management would prefer that alternative which best fits the present organizational structure of the firm. Thus, it is very important that the analyst determine the goals of management.

1.6.5 *Specification of the solution*

The final phase in the design process involves the *specification of the solution*. After the layout design has been chosen for implementation, a detailed specification of the design is required. Normally, an iconic representation of the layout, along with a written report, is used to specify the solution. Examples of two- and three-dimensional iconic models of a layout are given in Figures 1.2 and 1.3. Depending on the scope of the layout problem, a detailed set of drawings might be used to specify the location of electric, sewage, gas, and water lines. These carefully prepared, detailed, dimensional drawings are often the main vehicle for documenting and communicating the layout solution.

COMPUTER ROOM
TAB ROOM
CUSTOMER ENGINEER AREA
A/C EQUIPMENT ROOM

Figure 1.3. Three-dimensional iconic model.

1.6.6 *Design cycle*

The specification of the solution might be the final phase of the design process, but it seldom represents the end of the analyst's association with the layout problem. There still remain the selling of the solution, installing the design, observing and evaluating the design in use, and detecting the need for a redesign. These functions form the design cycle depicted in Figure 1.4. Implementation consists of *selling* all persons involved in the solution, *documenting* the procedures to be followed, *training* personnel in the performance of new duties, and *monitoring* the operation of the newly designed system.

There have been many excellent layout designs developed that were never implemented. Undoubtedly, there are a number of reasons for this. Layout analysts tend to be problem solvers. As such, many of them bring a great deal of enthusiasm to the analysis until, to them, the problem has been solved. Typically, such analysts turn in their recommendations to management with a statement to the effect that they will be available to answer any questions concerning the proposed layout.

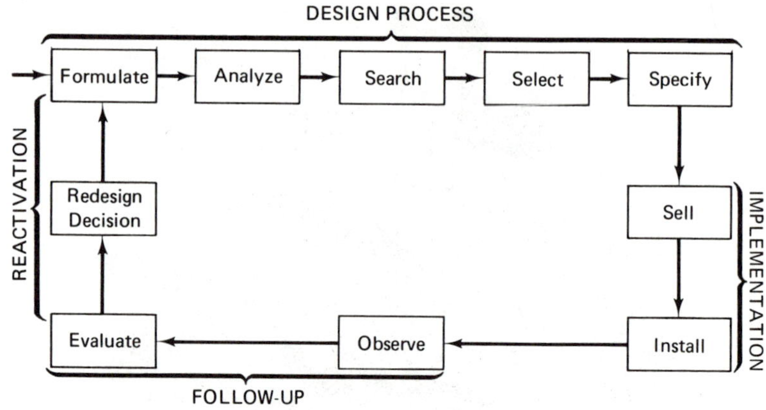

Figure 1.4. Design cycle.

Another reason layout solutions are not implemented is a poor job of selling. Not only must top, middle, and lower management be sold on the solution, but, more critically, the individual employee must be sold. Management may *make* the decisions, but the employee can *break* the decision if he is not sold on it.

A third reason implementation fails is that a good layout analyst is often in high demand. He will have a backlog of requests to be met. As such, management may pressure him to let someone else handle the implementation in order that he will be free to tackle other problems. Such an approach is very shortsighted and will eventually lead to dissatisfaction with the work of the analyst (and a subsequent perusal of the want ads).

One of the major aids in selling a layout is the iconic model of the layout. Admittedly, we have deemphasized the role of the iconic model in developing alternative solutions, but its importance in aiding visualization of the new layout cannot be overemphasized. A two- or three-dimensional representation of the proposed layout can go a long way toward selling the manager on the design. Along with the iconic model, clear plastic overlays are often used to show product flow lines, alternative designs, expansion plans, service lines, etc.

Just as in academic life there is the "publish or perish" philosophy, in industry there should be a "document or die" philosophy. The documentation of the procedures to be followed with the new layout is essential for successful implementation. Recommended flow lines should be specified, the material-handling procedures must be detailed, and any other information pertinent to the new layout design must be documented.

If new procedures are developed to facilitate the layout design, the affected personnel should be properly trained to carry out these procedures.

New rules for moving items in and out of storage, new methods of handling materials, and new flow patterns can result from a change in the layout. Unless the employees affected are able to adjust to these changes, the potential benefits to be derived from the new layout may never be realized.

In addition to seeing that the design is installed, control procedures should be devised to monitor the production system. Management information systems should be designed to indicate when a re-layout study should be conducted. One should make periodic checks for the presence of the problem indicators cited earlier. Also, one should compare the performance of the layout with that which was predicted. If serious discrepancies exist between the two, adjustments should be made in the methods of predicting system performance.

If you compare the steps of the design process with the sequence of steps suggested in a number of the more traditional texts on plant layout, it is apparent that traditional approaches often begin with the analysis of the problem. Consequently, when the traditional approach is followed, one assumes that the correct problem has been formulated. The dangers inherent in this approach should be obvious. We strongly recommend that careful consideration be given to the formulation of the problem before analysis begins.

Recall, the problem should be formulated as broadly as the organizational boundaries and economics permit. One of the major prerequisites to successful layout design continues to be an understanding of how the organization functions and how to work within the organization to bring about desired changes.

In much of our subsequent discussion we shall employ analytical approaches in solving facility layout and location problems. It is emphasized that analytical approaches are used to stimulate the search phase of the design process. Even though normative models are employed, a number of relevent nonquantifiable factors have been ignored. Consequently, in many situations the solutions obtained from the models must undergo separate evaluations by taking into consideration all relevant criteria.

1.7 Classification of Facility Layout and Location Problems

In describing the wide variety of queueing problems, Wagner [17] states, "What is said about birds is also true of waiting line models: their variety and number seem infinite." The same is becoming true for facility layout and location problems. During the past decade over 500 papers have been published or presented at national meetings on the subject of facility layout and location. Based on the variety of problems considered thus far, it is possible to define a number of categories that can be used to classify facility layout and location research.

In a sense, facility layout problems can be considered as a special class of facility location problems. The facility layout problem consists of the determination of the location of, say, offices, as well as a determination of the size and the configuration of the offices. In the subsequent classification we shall treat the facility layout problem as a special type of facility location problem. However, in subsequent chapters we continue to distinguish between facility layout and location problems.

In classifying facility location problems, six major elements are to be considered: new facility characteristics, existing facility locations, new and existing facility interactions, solution-space characteristics, distance measure, and the objective. These elements are depicted in Figures 1.5 through 1.10.

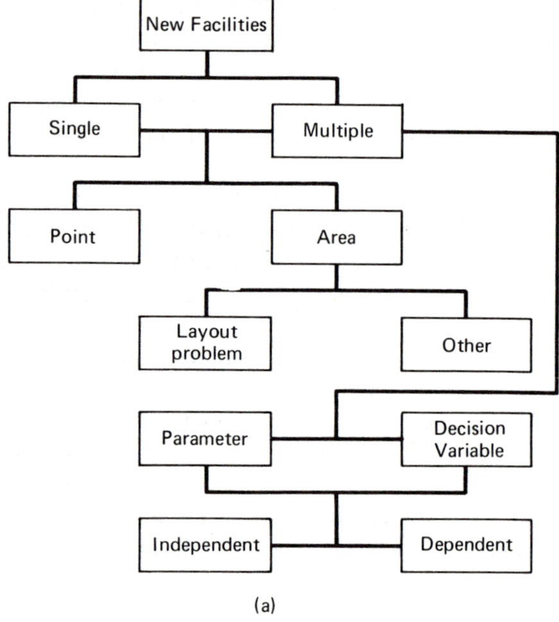

Figure 1.5. Classification of facility layout and location problems—new facility characteristics.

Facility location problems can be classified according to the number of new facilities involved. Furthermore, the new facilities can be considered to occupy either point locations or area locations. In a number of facility location problems, the number of new facilities is a decision variable, rather than being given. Finally, the location of a new facility can be either dependent on or independent of the locations of the remaining new facilities. If area

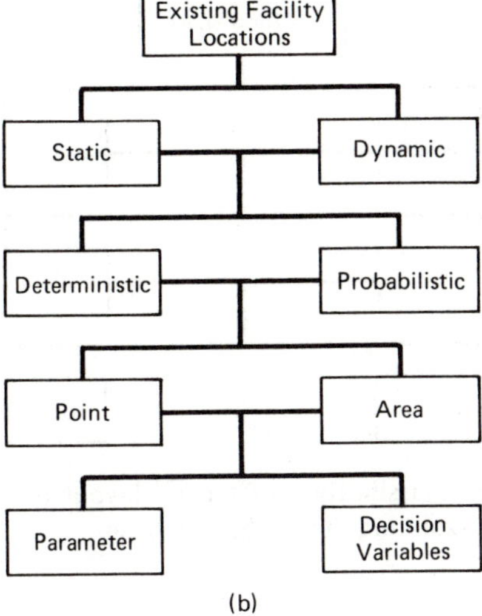

Figure 1.6. Classification of facility layout and location problems—existing facility locations.

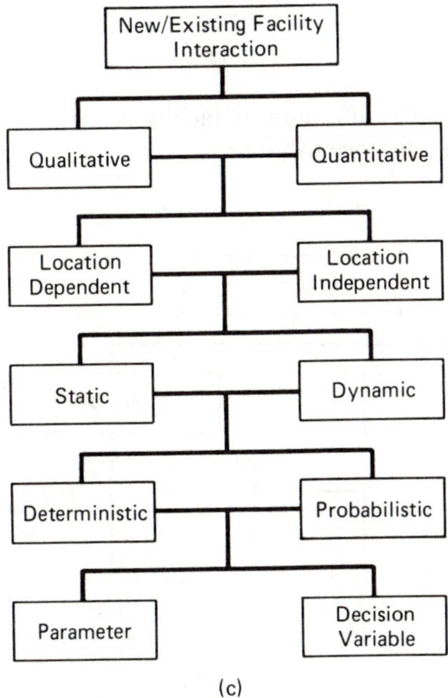

Figure 1.7. Classification of facility layout and location problems—new and existing facility interactions.

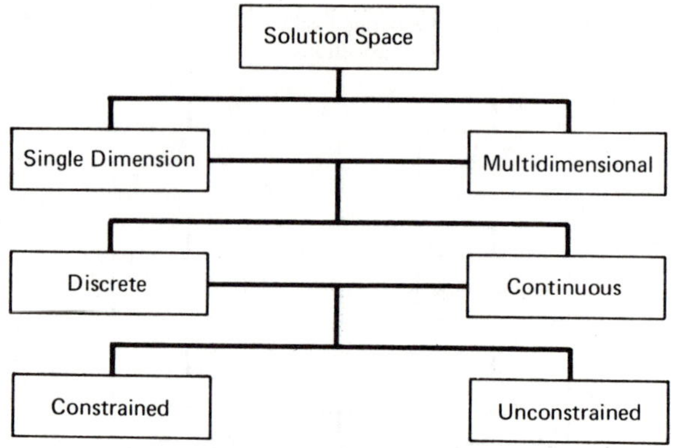

Figure 1.8. Classification of facility layout and location problems—solution–space characteristics.

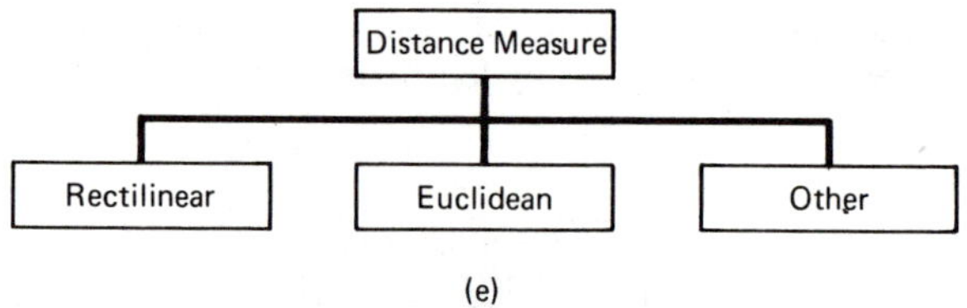

Figure 1.9. Classification of facility layout and location problems—distance measure.

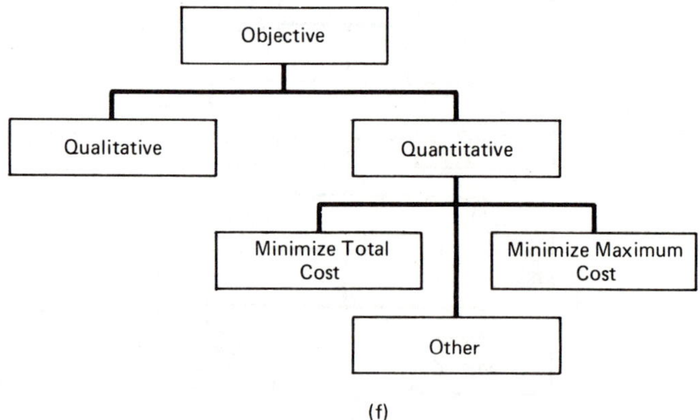

Figure 1.10. Classification of facility layout and location problems—the objective.

locations are to be considered for both new and existing facilities, the facility location problem is often classified as a facility layout problem; in which case, the size and the configuration of the area location are frequently decision variables. In the facility layout problem, the facilities might be plants, offices, warehouses, etc.

The location of an existing facility can be either static or dynamic, as well as deterministic or probabilistic. Additionally, depending on the size of an existing facility, it can be considered to have a point location or an area location. Again, a special class of facility location problems involving area locations is the facility layout problem. When existing facilities are included in the layout problem a re-layout problem can exist; in which case, the configuration and locations of the existing facilities must be determined. A *re-layout* or *relocation* problem arises when the location of an existing facility is a decision variable.

When there exists a quantitative or qualitative relationship between facilities, we consider the facilities to interact. In some cases the degree of interaction is a function of the locations of the facilities. Of course, there are a number of situations in which such interaction is independent of the facility locations. Furthermore, the magnitude of the interaction can be either static or dynamic, either deterministic or probabilistic, and either a parameter or a decision variable.

The fourth category considered in classifying a facility location problem concerns the solution space for the location problem. In some facility location problems the solution space is one dimensional; however, more commonly a two- or three-dimensional solution space exists. Additionally, the solution space can be either discrete or continuous. Typically, the discrete solution space consists of a finite number of possible locations; whereas, a continuous solution space consists of an infinite number of possible locations. In either case, the solution space can be further restricted by one or more constraints. A discrete layout involves discrete modules, such as bays in a warehouse, whereas a continuous layout considers the modules to be points. An example of a finite solution space is a network location problem.

The distance measure involved in the facility location problems provides another basis for classifying facility location problems. In subsequent discussions, rectilinear and Euclidean distances are commonly used. However, there do exist a number of facility location problems in which the distances between facilities cannot be reasonably represented as either rectilinear or Euclidean distances. An illustration is an urban location problem in which streets do not form a rectangular grid.

The final category commonly used to classify facility location problems concerns the objective function employed to evaluate alternative solutions. In the case of facility *layout* problems a number of qualitative objectives are commonly used, whereas quantitative objectives are used to solve facility *location* problems. The quantitative objectives normally encountered in facility location problems are the minimization of some total cost function or the minimization of the maximum cost between pairs of facilities. Of course, there can be other quantitative objectives, but these two occur most frequently in the literature.

Although some facility location problems might not be completely classified using the categories we have suggested, the classification scheme serves to unify a growing body of literature treating the subject of facility layout and location. It might be instructive to refer to Figures 1.5 through 1.10 as we consider in detail in subsequent chapters a number of different facility layout and location problems.

1.8 Overview of the Text

As an introduction to facility layout and location, the traditional subject of plant layout is examined in Chapter 2. The discussion in Chapter 2 is based on the Systematic Layout Planning (SLP) approach developed by Muther [11]. Using the steps of SLP as a base, the subject of plant layout is treated briefly. The discussion in Chapter 2 emphasizes qualitative consideration in plant layout.

In Chapter 3 we treat the subject of computerized layout planning. Even though the development of computer programs to aid in layout design is a relatively recent development, a considerable body of literature has emerged on the subject. This chapter represents the first attempt to summarize and evaluate within a text the growth of computerized layout planning.

Chapters 4 and 5 treat single- and multiple-facility location problems, respectively. In particular, in these two chapters it is assumed that an infinite number of possible location points exists within the plane for the new facilities. In these chapters distances are assumed to be measured either along straight-line paths connecting two points (Euclidean distance) or between two points along an orthogonal (perpendicular) set of axes (rectilinear distance).

In Chapter 6, facility location problems are formulated as assignment and generalized assignment problems. A discrete number of possible locations for the new facilities are considered in Chapter 6. In Chapter 7 the discrete problems of Chapter 6 are formulated as continuous problems, and are presented in a warehouse-layout context.

Chapter 8 presents quadratic assignment problem formulations of facility location problems. Heuristic, as well as exact, procedures for solving quadratic assignment problems are presented. Chapter 9 is a minimax counterpart of Chapters 4 and 5, since single and multiple facility, minimax location problems are described for the cases of rectilinear and Euclidean distance measures. Chapter 10 presents the topics of discrete plant location and set-covering location problems.

The approach that we recommend strongly to facility layout and location problems is to employ analytical approaches where feasible to do so. More specifically, we suggest the use of analytical approaches in developing alternative layout and location designs. Furthermore, we advocate that the steps of the design process be followed in solving facility layout and location problems.

We have already emphasized a number of times that many of the analytical approaches which we describe will yield benchmark designs against which other designs can be compared. This point will be made a number of other times throughout the text, since it is quite important that you adopt this view of analytical design aids. Since a number of important qualitative, as well as quantitative, factors are not normally included in the analytical model, the solution obtained from the model will usually be modified based on a number of considerations not accounted for explicitly in the model.

In a sense, the design obtained from analysis is an *ideal* solution. Our philosophy is that it is better to begin with an idealized solution and modify it based on real-world considerations than attempt to develop the preferred solution directly. Nadler [12] has suggested a similar approach in work place design. He has identified the approach as the "ideals concept." In subsequent discussions we shall repeat our philosophy, since it is important that you place in proper perspective the use of analytical approaches in solving facility layout and location problems.

REFERENCES

1. APPLE, J. M., *Plant Layout and Material Handling*, 2nd ed., The Ronald Press Company, New York, 1963.
2. DEGARMO, E. P., and J. R. CANADA, *Engineering Economy*, 5th ed., The Macmillan Company, New York, 1973.
3. ELMAGHRABY, S. E., *The Design of Production Systems*, Van Nostrand Reinhold Company, New York, 1966.
4. ELMAGHRABY, S. E., "The Role of Modeling in IE Design," *The Journal of Industrial Engineering*, Vol. 19, No. 6, 1968, p. 292.
5. GRANT, E. L., and W. G. IRESON, *Principles of Engineering Economy*, The Ronald Press Company, New York, 1970.
6. KRICK, E. V., *An Introduction to Engineering and Engineering Design*, John Wiley & Sons, Inc., New York, 1965.
7. KRICK, E. V., *Methods Engineering*, John Wiley & Sons, Inc., New York, 1962.

8. LEHRER, R. N., *Work Simplification (Creative Thinking About Work Problems)*, Prentice-Hall, Inc., Englewood Cliffs, N.J., 1957.

9. MOORE, J. M., *Plant Layout and Design*, The Macmillan Company, New York, 1962.

10. MUTHER, R., *Practical Plant Layout*, McGraw-Hill Book Company, New York, 1955.

11. MUTHER, R., *Systematic Layout Planning*, Industrial Education Institute, Boston, Mass., 1961.

12. NADLER, G., *Work Design: A Systems Concept*, Richard D. Irwin, Inc., Homewood, Ill., 1970.

13. REED, R., *Plant Layout: Factors, Principles, and Techniques*, Richard D. Irwin, Inc., Homewood, Ill., 1961.

14. REED, R., *Plant Location, Layout, and Maintenance*, Richard D. Irwin, Inc., Homewood, Ill., 1967.

15. THUESEN, H. G., W. J. FABRYCKY, and G. J. THUESEN, *Engineering Economy*, 4th ed., Prentice-Hall, Inc., Englewood Cliffs, N.J., 1971.

16. VOLLMANN, T. E., and E. S. BUFFA, "The Facilities Layout Problem in Perspective," *Management Science*, Vol. 12, No. 10, June 1966, pp. 450–468.

17. WAGNER, H. M., *Principles of Operations Research*, Prentice-Hall, Inc., Englewood Cliffs, N.J., 1969.

PROBLEMS

1.1. Suppose that you are asked to recommend the location for a hospital which is to be placed in a major city within the state. What criteria would be appropriate in evaluating alternative sites?

1.2. Suppose that you are asked to recommend the number, sizes, and locations for storerooms in an industrial plant. What criteria would be appropriate in evaluating alternative solutions to the problem?

1.3. List six facility location problems and classify them using the scheme provided in Figures 1.5 through 1.10.

1.4. Using the black-box approach, provide alternative formulations of states A and B for Problems 1.1 through 1.3.

1.5. Perform a literature search and obtain the titles, authors, and other bibliographic data for 10 recent facility layout and location papers not included in the list of references given at the end of Chapters 2 through 10.

1.6. Locate four professional journals, other than those listed in Section 1.1, which contain articles dealing with facility layout and location problems.

1.7. Prepare a written critique of the Vollmann–Buffa paper [16].

1.8. Suggest alternative methods of monitoring the operation of a newly designed system to ascertain when a new design is justified.

1.9. Suppose that you are asked to re-layout the library at your university, plant, or town. Formulate the problem using the black-box approach, giving alternative formulations having varying degrees of breadth. List the criteria to be used in evaluating alternative layout designs.

1.10. Prepare a written summary of the "ideals concept" as it relates to work design, and point out how the concept can be used in layout design.

1.11. List a number of common iconic, analog, and symbolic models that are used in your daily activity.

THE PLANT LAYOUT PROBLEM

2.1 Introduction

In this chapter we treat the plant layout problem. More specifically, we are interested in that collection of facility layout and location problems which has received and continues to receive considerable attention from industrial engineers, manufacturing engineers, and production engineers. For the most part, the interest of management scientists and operations researchers in facility layout and location problems has not been concerned primarily with the plant layout problem. Instead, management scientists and operations researchers have tended to be more concerned with plant location and office layout problems, rather than the layout of manufacturing or production plants.

It should be made perfectly clear at the outset that we are using the term "plant layout" in a very literal sense. In particular, we are referring to the design of *layouts* for production *plants*. Consequently, the subsequent discussion will be directed primarily toward industrial applications. If your interest in facility layout and location problems is not oriented toward the layout of production plants, we still believe the subsequent discussion will be beneficial. For example, much of the discussion is appropriate in the design of layouts for hospitals, warehouses, schools, offices, individual

2

work stations, console displays, banks, shopping centers, parking lots, and airports.

There currently exists a large body of literature treating the plant layout problem. For the most part, the current literature represents the accumulation of approaches that have been employed by industrial engineers in solving plant layout problems since the early 1900s. Since analytical approaches have been developed only recently, the traditional solution procedures relied rather heavily on empirical approaches. Now that a number of analytical approaches are available, it is important that we relate the roles of the empirical approach and the analytical approach in plant layout design.

In this chapter we provide a qualitative overview of the entire subject of plant layout. Our objectives are twofold. First, as we emphasized in Chapter 1, qualitative considerations are very important in solving real-world problems. Consequently, we want to recognize those qualitative aspects of facility layout and location that we believe to be worthy of consideration. Second, we wish to place the remainder of this text in context within the overall process of layout design. The layout design process consists of problem formulation, analysis of the problem, search for layout designs, selection of the preferred design, and specification of the layout design to be installed. The analytical approaches described in subsequent chapters are used in the search for solutions to facility layout and location problems. Thus, we see that a number of other steps in the design process must still be performed when analytical approaches are used. Consequently, in this chapter we discuss the overall layout design problem as it relates to the plant layout problem.

In striving to achieve our stated objectives, the chapter is organized as follows: first, we discuss, in very general terms, the overall plant layout problem. Next, we present an approach that is often used in solving plant layout problems. The approach, Systematic Layout Planning, developed by Muther [18], serves as the basis for the subsequent discussion in this chapter. The discussion concludes with a recommendation concerning the position of analytical approaches in layout design.

One major objective of this chapter is to relate the plant layout problem to a number of other problem areas in an industrial setting and to indicate a variety of analytical approaches that can facilitate each of these areas. Consequently, we shall make references to scheduling, inventory control, waiting line, reject allowance, and economic analysis problems, as well as such analytical approaches as simulation, queueing theory, decision theory, and critical-path analysis. Obviously, a detailed discussion of these topics goes well beyond the scope of this text. Therefore, at the appropriate juncture in our discussion we cite appropriate references that treat the particular topic.

Depending on the objectives of the course, as well as your objectives, the subject matter of this chapter might not be emphasized as heavily as subsequent chapters. If such is the case for you, we recommend that you read at least the following sections: 2.2, 2.3, 2.5.3, 2.5.4, 2.6, and 2.10. The material presented on Systematic Layout Planning and from–to charts is referred to a number of times in later discussions.

2.2 Overview of the Plant Layout Problem

Defining the boundaries on a plant layout problem is no easy undertaking. In fact, the boundaries are different from one problem to the next. For example, the layout analyst might be called on to determine the location for a new machine, and then his next assignment might be to develop the layout for a new plant. Consequently, there is a tremendous variety in the types of layout problems one may encounter. This variety is due, in part, to the number of ways plant layout problems develop. For example, a layout problem can arise because of a change in the design of a product, the addition or deletion of a product from the company's product line, a significant increase or decrease in the demand for a product, changes in the design of the process, the replacement of one or more pieces of equipment, the adoption of new safety standards, organizational changes within the company, or a decision to build a new plant. Plant layout problems may also develop because of gradual changes over time that finally manifest themselves in terms of bottlenecks in production, crowded conditions, unexplainable delays and idle time, backtracking, poor housekeeping, excessive temporary storage space, obstacles to materials flow, failure to meet schedules, and a high ratio of material-handling time to production time.

Thus, we see that the plant layout designer might be called on to interact with the product designer, the process designer, and the schedule designer. Furthermore, it is clear that the layout problem can be a very complex systems problem requiring the most sophisticated systems analysis tools in order to develop satisfactory layout solutions.

Plant layout problems can occur in a large number of ways and can have significant effects on the overall effectiveness of the production system. Therefore, it is highly desirable that the optimum plant layout be designed. Unfortunately, the magnitude of the problem is so great that true system optimization is beyond our current capabilities. The approach normally taken in solving the plant layout problem is to try to find a "satisfactory" solution by employing the *component approach*. We define the overall system as a collection of components or subsystems and attempt to obtain "optimum" solutions for the subsystems. Granted, the resulting system will be a suboptimum solution, rather than an optimum solution. However, the solution is believed to be better than that which would otherwise be obtained.

In using the component approach there is the danger of developing a solution for one component that is detrimental to the overall system. To minimize this possibility, it is necessary to coordinate the product, process, schedule, and layout design decisions, as depicted in Figure 2.1.

In searching for a plant layout design that "satisfices," we must agree on the basis for evaluating alternative designs. As mentioned in Chapter 1, one of the criteria commonly used to evaluate alternative layouts is that of material-handling cost. It was also emphasized that for many situations this may not be an appropriate criterion. Typically, one has a number of goals that are important. For example, some of the objectives of the plant layout study may be to

1. Minimize investment in equipment.
2. Minimize overall production time.
3. Utilize existing space most effectively.
4. Provide for employee convenience, safety, and comfort.
5. Maintain flexibility of arrangement and operation.
6. Minimize material-handling cost.
7. Minimize variation in types of material-handling equipment.

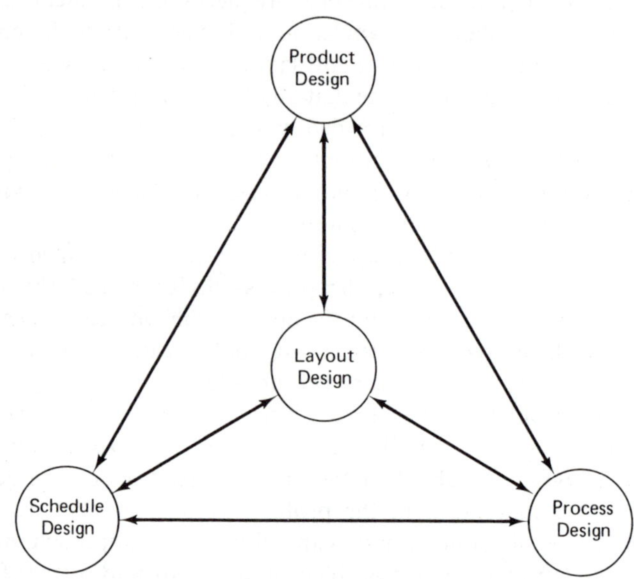

Figure 2.1. Communication links among product, process, schedule, and layout design.

8. Facilitate the manufacturing process.
9. Facilitate the organizational structure.

In addition to there being a variety of objectives that can be used to guide the analyst in solving the plant layout problem, there also may exist a number of constraints on the solution. For example, the Occupational Safety and Health Act (OSHA) has placed restrictions on allowable noise levels. This, along with other standards concerning aisle widths, ventilation, temperature, and lighting, can affect the solution to the layout problem. Building geometry can serve as a restriction on the layout. In the case of an existing building, the layout might be required to fit within the present building. In the case of a new building, the building site can restrict the shape of the building and, consequently, the layout. When there exists a building within which the layout is to be designed, there can exist a large number of restrictions on the solution. For example, the layout solution will be affected by the present location of walls and columns, equipment, footings to support heavy equipment, loading docks, windows, lights, ventilating equipment, storage and office areas, and sewage, water, and power lines. Some of these can be relocated. However, it should be remembered that whenever an existing layout is to undergo a revision the analyst must consider the costs of relocating existing facilities along with the advantages derived from the relocation.

Over the years a number of rules of thumb have been developed as guides to a good plant layout. A number of these have been assembled in the form of checklists. Checklists serve the useful purpose of forcing the analyst to examine a number of important aspects and details pertinent to the design of a successful plant layout. As examples, Apple [1] and Muther [16] provide very useful checklists. However, due to their length we have not included them here.

It is rarely the case that one individual is involved in carrying out all of the analyses involved in solving the plant layout problem. Rather, different individuals at different levels within the organization are involved. Also, it is not uncommon to find that a "layout team" is formed to design the layout. Depending on the organizational structure of the company, it may well occur that the plant layout analyst does not have the freedom to participate in a portion of the decision-making process that establishes the boundaries within which he must operate. Not uncommonly, the product design, process design, and schedule design decisions have been made previously, and the layout analyst must make the best of the situation.

In this chapter we concentrate on the solution of the plant layout problem, assuming certain preliminary decisions have already been made. That is, we assume a number of product, process, and schedule design decisions have previously been made. As shown in Figure 2.1, the functions of product, process, schedule, and layout design should interact. Some authors portray the plant layout function as encompassing a large portion of the other functions. We admit that we are sufficiently biased in our view of the importance of the layout function to lobby for a larger role in the overall decision-making process in production systems design. However, we are also sufficiently cognizant of organizational realities to recognize the facts of life concerning the role of the plant layout analyst. It is unusual for the plant layout analyst to influence significantly product design decisions; he normally has some influence on process design decisions; and he can have a significant influence on schedule design decisions. Furthermore, it is quite common to find that a person who is charged with the responsibility of facilities planning and design is also responsible for carrying out other duties.

2.3 Systematic Layout Planning (SLP)*

An organized approach to layout planning has been developed by Muther [17] and has received considerable publicity due to the success derived

*The material presented in this chapter on SLP is based on a variety of sources made available to us by Richard Muther Associates, Kansas City, Missouri. Their assistance is gratefully acknowledged.

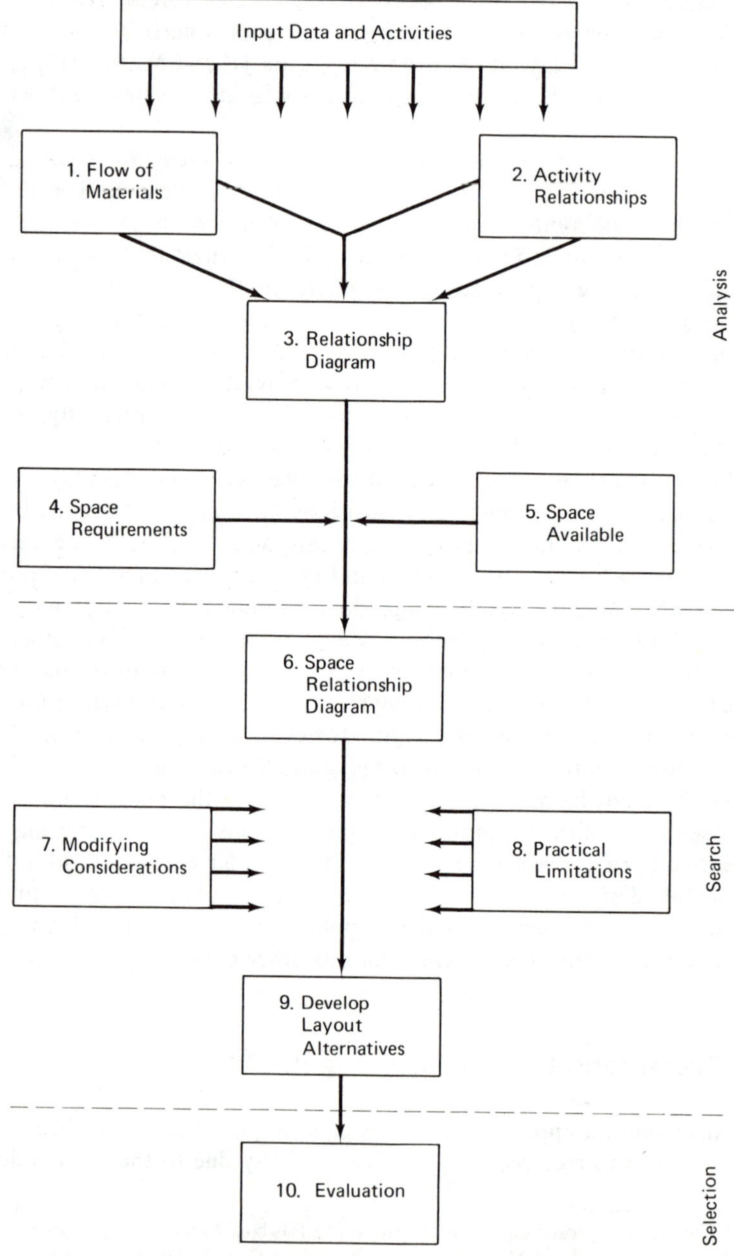

Figure 2.2. Systematic layout planning procedure.

from its application in solving a large variety of layout problems. The approach is referred to as Systematic Layout Planning or simply SLP. SLP has been applied to a variety of problems involving production, transportation, storage, supporting services, and office activities.

The SLP procedure is depicted graphically in Figure 2.2. We see that once the appropriate information is gathered, a flow analysis can be combined with an activity analysis to develop the relationship diagram. Space considerations, when combined with the relationship diagram, lead to the construction of the space-relationship diagram. Based on the space-relationship diagram, modifying considerations, and practical limitations, a number of alternative layouts are designed and evaluated. In comparison with the steps in the design process, we see that SLP begins after the problem is formulated. The first five steps of SLP involve the analysis of the problem. Steps 6 through 9, including the generation of alternative layouts, constitute the search phase of the design process. The selection phase of the design process coincides with step 10 of SLP.

Our subsequent discussion will be based on the steps of SLP depicted in Figure 2.2. In particular, in Section 2.4 we shall examine the data-gathering step of SLP. In Section 2.5 we discuss flow analysis and activity analysis. The development of the relationship diagram is treated in Section 2.6. A consideration of space requirements and space availability takes place in Section 2.7. In Section 2.8 we consider the development of the layout. The discussion in Section 2.8 includes steps 6 through 9 of SLP. In Section 2.9 we consider the selection, specification, implementation, and follow-up steps in the design cycle. Finally, in Section 2.10 we take a look at the remainder of the book as it relates to SLP.

2.4 Information Gathering

For the plant layout analyst to perform effectively, he must obtain certain information pertaining to the product, process, and schedule. The data requirements may not coincide with data availabilities in some cases. However, we shall assume such data exist; otherwise, the necessary management information system to obtain these data must be developed.

Data regarding product design decisions can significantly affect the layout. However, the major effect of product design decisions is felt by the process designer. Whether a part is made of aluminum or plastic will influence processing decisions, which ultimately affect the layout. Thus, product design decisions can indirectly affect the layout. Product design decisions can also have a direct effect on the layout. The design of the product affects the sequence of assembly operations, and this sequence can influence

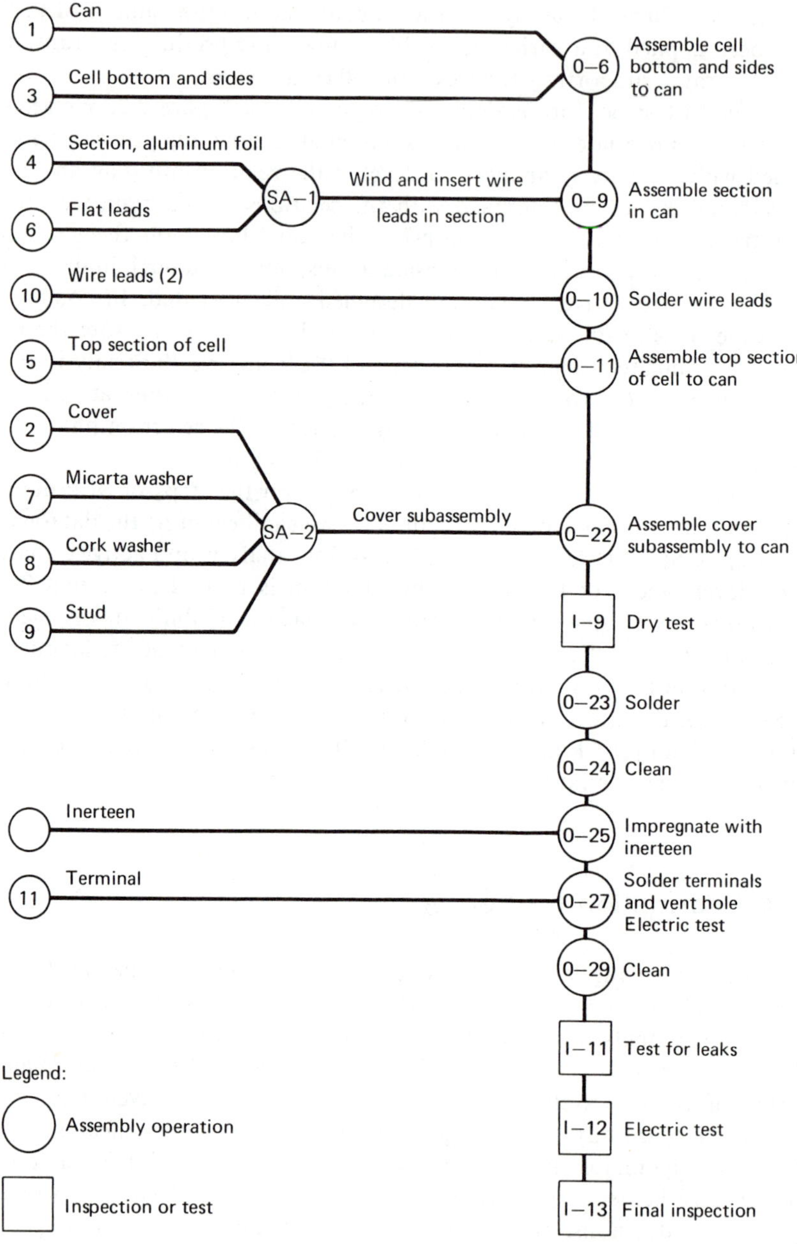

Figure 2.3. Assembly chart for capacitor. (After E.S. Buffa, *Models for Production and Operations Management*, John Wiley & Sons, Inc., New York; by permission of the publisher.)

the layout. Therefore, it is important to have available data concerning the design of the product. Basic product design data can be obtained from production drawings, assembly charts, parts lists, bills of materials, and prototypes of the product. As an illustration, an *assembly chart* is given in Figure 2.3. Notice, depending on the type of product involved, that the layout could be governed by the flow lines from an assembly chart.

Process design decisions determine whether a part will be purchased or produced, how the production of a part will be achieved, what equipment will be used, and how long it will take to perform each operation. This information is typically summarized on a *route sheet* (Figure 2.4) or *operation sheet*. A separate route sheet is normally required for each part. Therefore, the *operation process chart* (Figure 2.5) is often used to supplement the route sheet. The operation process chart is an analog model of the operations and inspections involved in producing a product. It depicts the operations with a circle and inspections with a square. Flow lines connect the various operations and inspections. In some cases the operation process chart is the primary basis for the layout.

Schedule design decisions provide the answers to the questions, how much to produce? and when to produce? The market forecast is converted into production demand, and decisions are made concerning the production rate. Associated with these decisions is the determination of the number of machines of each type required to meet the production volume. Depending on the product mix (number of different products) for the firm and the required production rate, the decision may be made to produce each product on a continuous basis or an intermittent basis. The plant layout analyst is often involved in this decision. The degree of his involvement is dependent on the organizational policies of the firm. In any case, the layout will be affected by the decisions concerning the number of machines and the production schedule. Basic schedule design data can normally be obtained from the master schedule. Production schedules are also sometimes given using an analog model called the Gantt chart (Figure 2.6).

Many of the decisions to be made in designing the production schedule are dependent on layout decisions. Thus, it is not uncommon to find the layout analyst working closely with the schedule analyst.

Throughout the information-gathering phase you are again reminded to avoid a preoccupation with the present solution. Most information sources available portray yesterday's and, possibly, today's data. We are interested in tomorrow's data, since we are developing a solution to be used tomorrow. Therefore, some of the information available must be modified. Remember, the product, process, and schedule data available from the sources we have listed are probably representative of the present solution, not the preferred solution.

If you have taken previously a quantitative course in production manage-

Part name ___Shower-head face___ Drawing No. ___DSH–6–1___

Part model ___SHF–4___ Part No. ___PSH–12___ Date effective ___12/10___

No. per model ______ Material ___Steel___ Replaces issue of ______

Economic lot size ______ Unit weight ______ Sheet ______ of ______

Oper. No.	Dept. or prod. cen.	Description of operation	Machine	Tools and jigs	U.T. (min)		Oper. spec.
					Set up	Prod.	
SHF–4–1	Press	Forge and inspect	Nat. Max-Press #12		0.340	0.334	
SHF–4–2	Press	Pierce six holes	Bliss 74 1/2 #16		0.460	0.075	
SHF–4–3	Drill	Rough ream and chamfer	L & G Drill Press #19	1/8 in. T.S. reamer	0.900	0.334	
SHF–4–4	Drill	Drill three 13/64 in. holes	Avey Drill Press #21	13/64 in. T.S. drill	0.870	0.152	
SHF–4–5	Lathe	Turn stem and face	W & S #3	5/16 in. R.H. side facing and 5/16 in. R.H. round nose.	0.985	0.552	
SHF–4–6	Lathe	From outside diameter and face back	W & S #4	5/16 in. R.H side facing and 5/16 in. R. H. round nose	0.985	0.648	
SHF–4–7	Press	Stamp identification	Bliss 20B #18		0.300	0.097	

Figure 2.4. Route sheet.

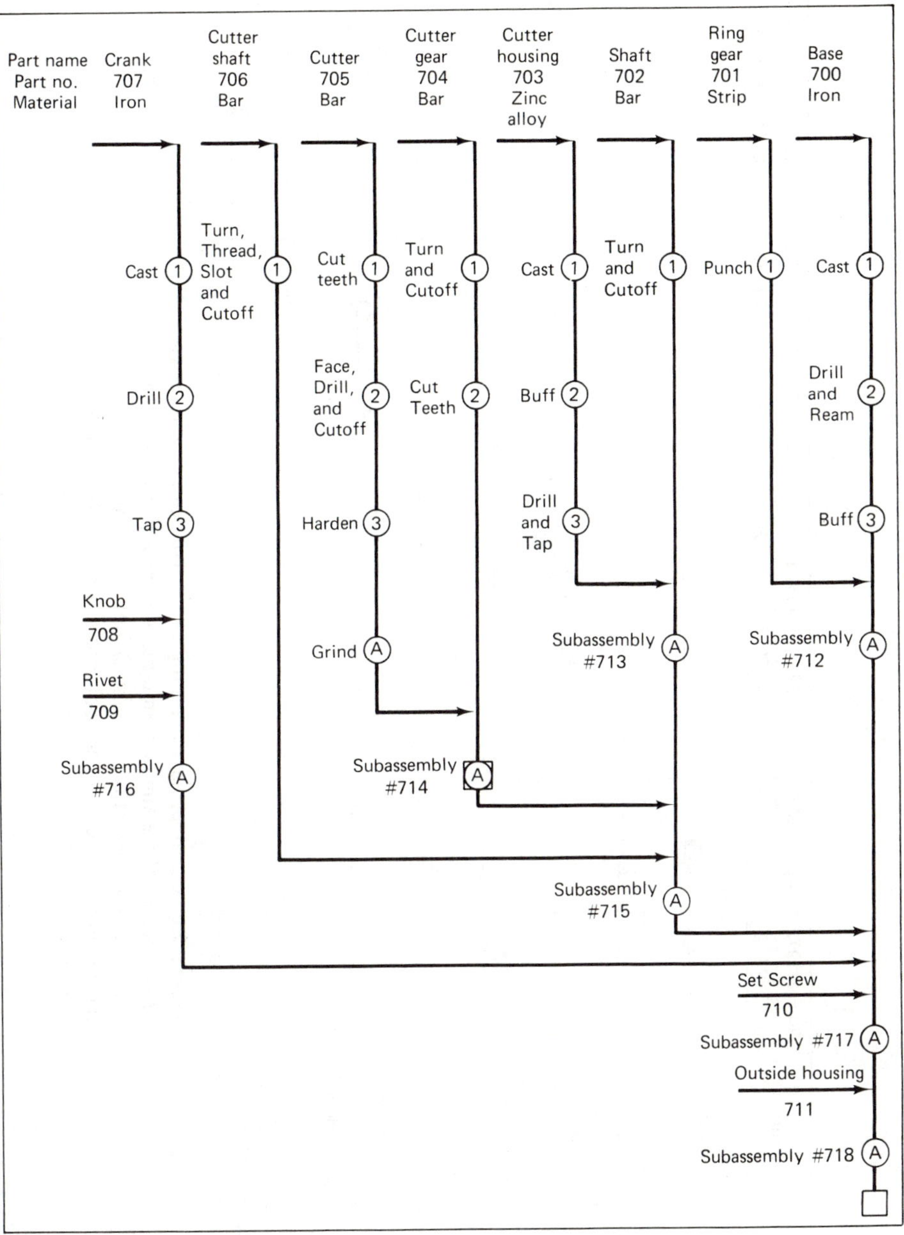

Figure 2.5. Operation process chart of the manufacturing process required to produce a pencil sharpener. (After E. V. Krick, *Methods Engineering*, John Wiley & Sons, Inc., New York; by permission of the publisher.)

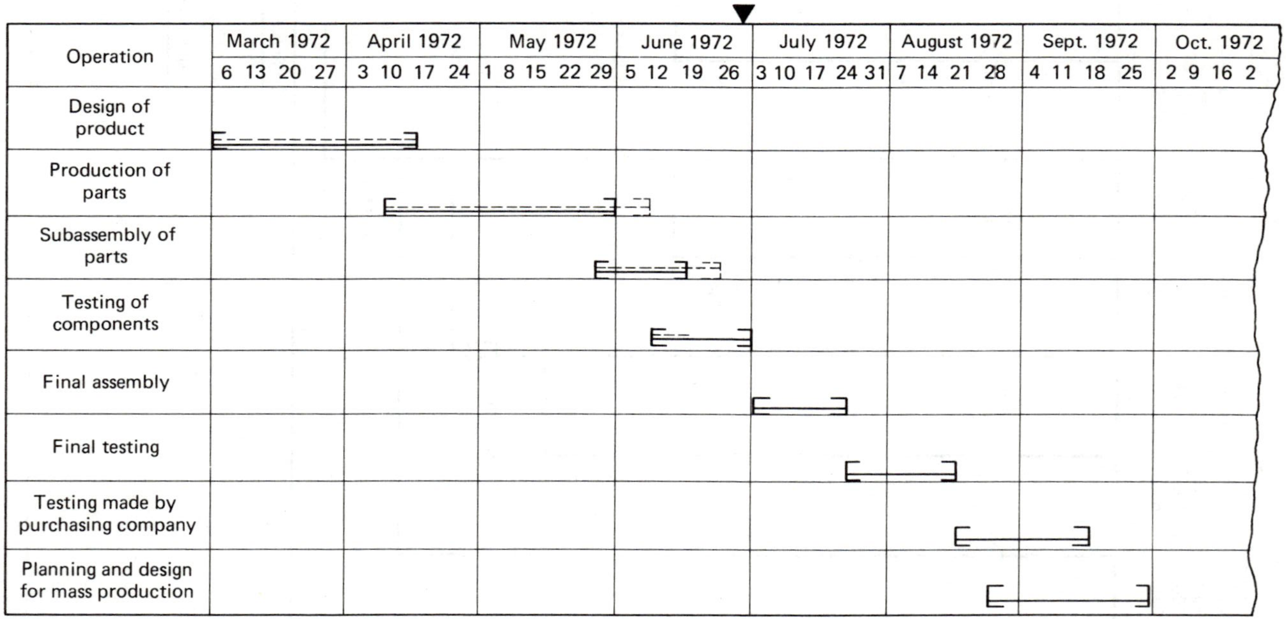

Operation	March 1972	April 1972	May 1972	June 1972	July 1972	August 1972	Sept. 1972	Oct. 1972
	6 13 20 27	3 10 17 24	1 8 15 22 29	5 12 19 26	3 10 17 24 31	7 14 21 28	4 11 18 25	2 9 16 2
Design of product								
Production of parts								
Subassembly of parts								
Testing of components								
Final assembly								
Final testing								
Testing made by purchasing company								
Planning and design for mass production								

Figure 2.6. Gantt project planning chart. (Indicates current week of operation, the estimated amount of time a particular operation will take, and the actual amount of time that particular operation has taken.) We see that the project is 1 week behind schedule.

ment or production planning and control, you are aware that analytical approaches are available to the product, process, and schedule designer in making many of the decisions which affect the plant layout. We have not treated these approaches, since our concern is with the design of the layout. Regardless of how product, process, and schedule design decisions are made, the layout designer should be aware of the decisions. We have listed some sources for this information.

2.5 Flow Analysis and Activity Analysis

In this section we describe flow-analysis techniques and then describe activity-analysis techniques. Flow analysis concentrates on some quantitative measure of movement between departments or activities, whereas activity analysis is primarily concerned with the nonquantitative factors that influence the location of departments or activities.

Once certain basic data have been obtained concerning the process, product, and schedule, the plant layout analyst is in a position to analyze the flow of materials, equipment, and personnel. Since the layout is designed to facilitate the flow of the product, from raw material to the finished product, we are primarily concerned with the flow of materials.

Some of the factors that affect the flow pattern are given by Apple [1] as

1. External transportation facilities.
2. Number of parts in product.
3. Number of operations on each part.
4. Sequence of operations on each part.
5. Number of subassemblies.
6. Number of units to be produced.
7. Necessary flow between work areas.
8. Amount and shape of space available.
9. Influence of processes.
10. Types of flow patterns.
11. Product versus process type of layout.
12. Location of service areas.
13. Production department locations.
14. Special requirements of departments.
15. Material storage.
16. Desired flexibility.
17. The building.

2.5.1 *Types of flow patterns*

Most of the factors listed require little additional discussion. However, we shall examine two of these factors in greater detail. First, consider the various types of flow patterns. We can classify flow patterns as being either *horizontal* or *vertical*. There are at least five basic types of horizontal flow patterns, as shown in Figure 2.7. A number of other flow patterns can be developed by combining these basic flow patterns. (In fact, we can visualize A, B, C, D, E, F, G, H, J, K, M, N, P, Q, R, T, V, W, X, Y, and Z flow patterns. No doubt, other alphabetic systems would suggest even more flow patterns, e.g. , π, λ, ϕ, and γ.)

Straight-line flow is the simplest form of flow. However, when employed in a plant, separate receiving and shipping crews are normally required. The L-shaped flow pattern is usually adopted when either straight-line flow cannot be accommodated in an existing facility or construction costs do not permit straight-line flow. The U-shaped flow pattern is very popular, since it is simple to administer and facilitates a combination of receiving and shipping activities. Circular flow is applicable when it is desired to terminate the flow very near the point where the flow originated. The serpentine flow pattern is used when the production line is so long that zigzagging on the production floor is required.

Vertical flow patterns exist both in single-story and multistory buildings.

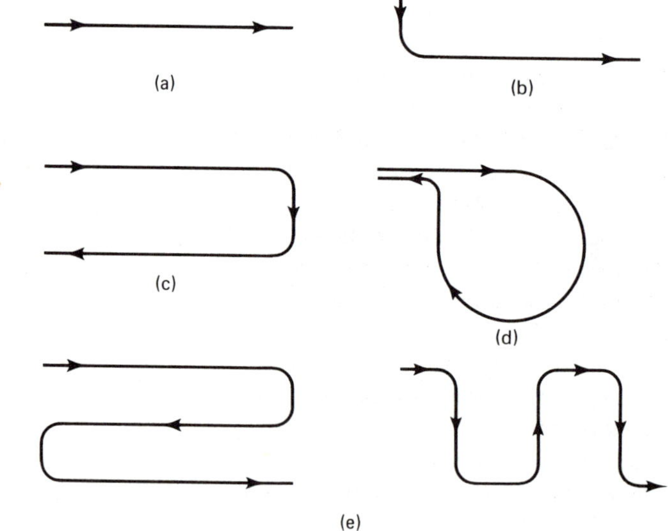

Figure 2.7. Basic horizontal flow pattern: (a) straight, or I flow; (b) L flow; (c) U flow; (d) circular, or O flow; (e) serpentine, or S flow.

Utilization of overhead space with conveyorized material flow has focused greater attention on the design of vertical flow patterns. We shall examine vertical flow patterns as they commonly exist in multistory structures. It should be recognized that similar patterns also occur in single-story buildings. Figure 2.8 presents six different vertical flow patterns in a three-story building.

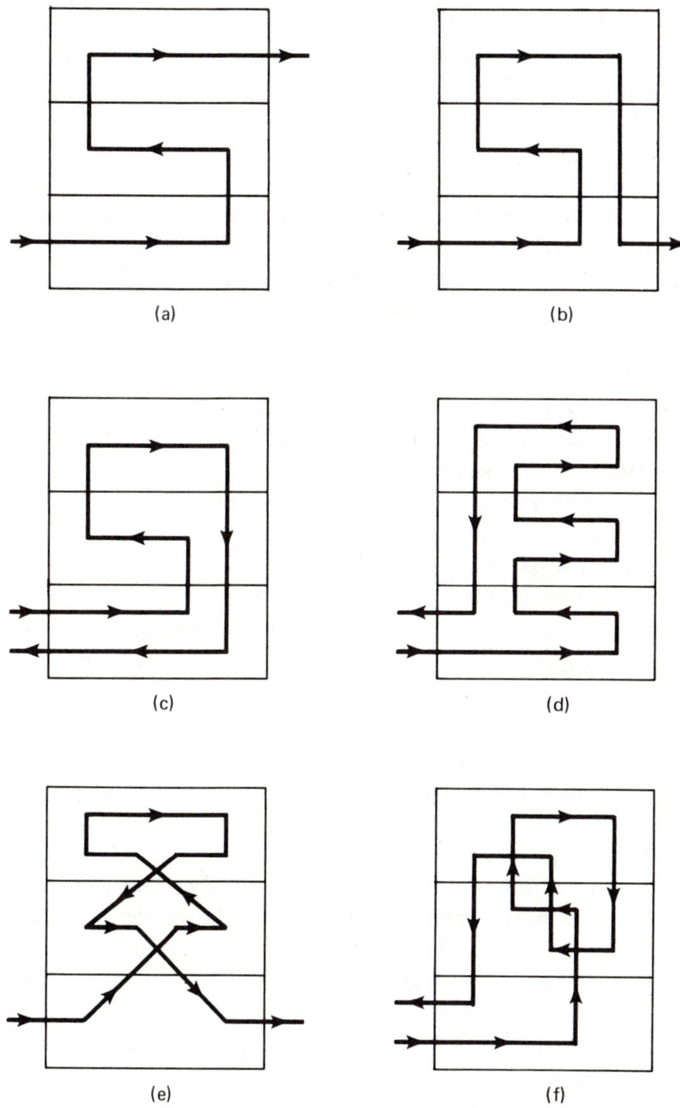

Figure 2.8. Vertical flow pattern.

Vertical flow pattern (a) is often used when there is flow between buildings and there exists an elevated connection between the buildings. Pattern (b) is used when ground-level ingress (entry) and egress (exit) are required. In (c), ground-level ingress and egress occur on the same side of the building. In (a), (b), and (c), decentralized elevation is present, since travel between floors can occur on either side of the building. Centralized elevation exists in (d), since travel between floors occurs on the same side of the building. Inclined flow occurs in (e); some bucket and belt conveyors and escalators result in inclined flow. Backtracking occurs in (f) due to the return to the top floor.

2.5.2 *Product versus process layout*

The second factor affecting the flow pattern that we wish to examine is the choice of a product or a process type of layout. The product layout is also referred to as the production-line layout and results when machines and auxiliary services are located according to the processing sequence for the product. Product layout results when a high production volume exists for the product, such that continuous production is justified. In a strict product layout, machines are not shared by different products. Therefore, the production volume must be sufficient to achieve satisfactory utilization of the equipment. A process layout results when similar machines and services are located together. Therefore, in a process layout all drills are located in one area of the plant, and all turning machines are located in another area of the plant. Process layouts normally result when the production volume is not sufficient to justify a product layout. Typically, job shops employ process layouts due to the variety of products manufactured and their low production volumes. Some of the major advantages and limitations of product and process layouts are given in Table 2.1.

A third type of layout is the static product layout, or layout by fixed position, in which the physical characteristics of the product dictate that machines and men are brought to the product. The shipbuilding industry commonly employs a static product layout. Since the static product layout is not justified except in unusual situations, we shall not consider it further.

As an illustration of a product layout and a process layout, consider a job shop that manufactures three products. The processing sequences for each product are given in Table 2.2. A product layout for the job shop is given in Figure 2.9(a), with a process layout given in Figure 2.9(b). Assuming rectilinear travel to the points denoted by crosses in each department and equal annual demand for each product gives a total distance traveled equal to 564 feet for the product layout and 724 feet for the process layout. The utilization of personnel and equipment will be the same in both layouts for all but the lathe department and the inspection department. If the same

Table 2.1. ADVANTAGES AND LIMITATIONS OF PRODUCT AND PROCESS LAYOUTS

Product Layout

Advantages

1. Since the layout corresponds to the sequence of operations, smooth and logical flow lines result.
2. Since the work from one process is fed directly into the next, small in-process inventories result.
3. Total production time per unit is short.
4. Since the machines are located so as to minimize distances between consecutive operations, material handling is reduced.
5. Little skill is usually required by operators at the production line; hence, training is simple, short, and inexpensive.
6. Simple production planning and control systems are possible.
7. Less space is occupied by work in transit and for temporary storage.

Limitations

1. A breakdown of one machine may lead to a complete stoppage of the line that follows that machine.
2. Since the layout is determined by the product, a change in product design may require major alterations in the layout.
3. The "pace" of production is determined by the slowest machine.
4. Supervision is general, rather than specialized.
5. Comparatively high investment is required, as identical machines (a few not fully utilized) are sometimes distributed along the line.

Process Layout

Advantages

1. Better utilization of machines can result; consequently, fewer machines are required.
2. A high degree of flexibility exists relative to equipment or manpower allocation for specific tasks.
3. Comparatively low investment in machines is required.
4. The diversity of tasks offers a more interesting and satisfying occupation for the operator.
5. Specialized supervision is possible.

Limitations

1. Since longer flow lines usually result, material handling is more expensive.
2. Production, planning, and control systems are more involved.
3. Total production time is usually longer.
4. Comparatively large amounts of in-process inventory result.
5. Space and capital are tied up by work in process.
6. Because of the diversity of the jobs in specialized departments, higher grades of skill are required.

number of lathes and inspectors is required for both layouts, there will be no difference in utilization. However, suppose that product A requires 4.3 lathes, product B requires 1.8 lathes, and product C requires 2.4 lathes. Using the product layout will result in 5 lathes being available for A and 5

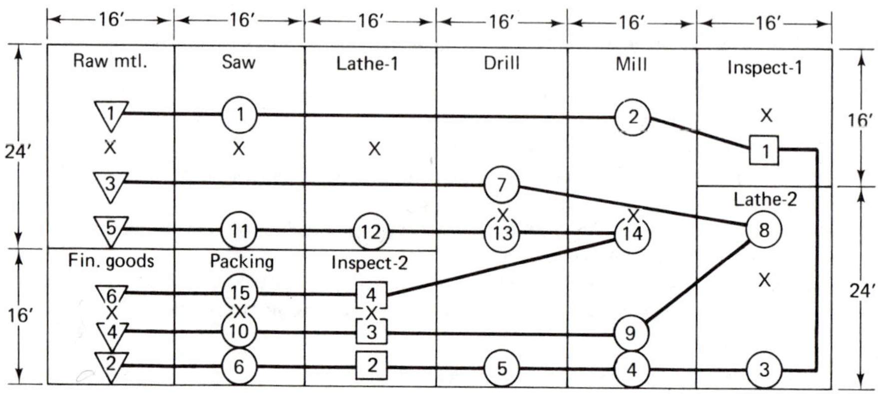

(a) Product Layout

Operation

Inspection

Storage

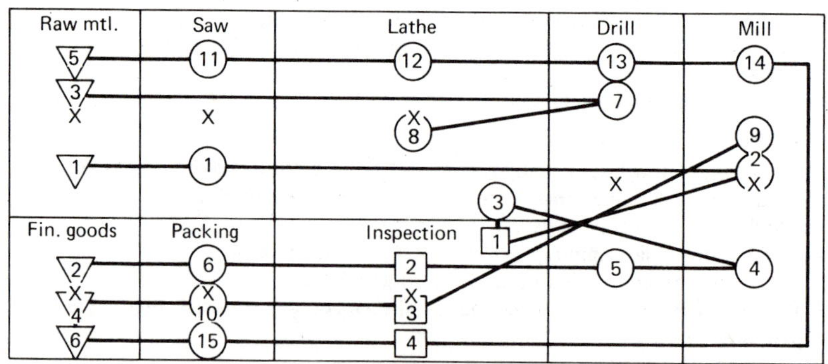

(b) Process Layout

Figure 2.9. Product (a) and process (b) layouts for the example problem.

Table 2.2. EXAMPLE PROCESSING SEQUENCES FOR THREE PRODUCTS

Product	*Processing Sequence*
A	Saw, mill, inspect, turn, mill, drill, inspect, package
B	Drill, turn, mill, inspect, package
C	Saw, turn, drill, mill, inspect, package

lathes being available for *B* and *C*. If a process layout is used, only 9 lathes are required for *A*, *B*, and *C*. Therefore, higher utilization of the equipment would result with the process layout.

The decision to use a product layout usually results when the required production quantity is very high, the product is easily transported, and the processing sequence is similar for all products. On the other hand, if a large variety of products is to be made in relatively small quantities using significantly different processing sequences, then a process layout is selected. Now, how do you decide which of these conditions exists? In fact, does either exist? There do not exist clear-cut decision rules that indicate if a product or a process layout should be used.

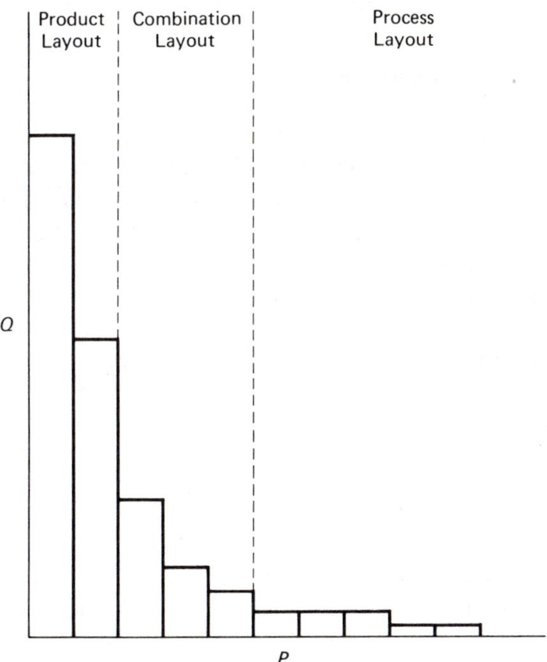

Figure 2.10. *P–Q* chart indicating that different layouts are justified.

As you might expect, many layouts tend to be a combination product–process layout. A combination is sought that achieves a compromise in the advantages and limitations of each of the layout types. However, rather than immediately choosing one layout for all products, you should also consider the possibility of using different types of layouts for different products.

One approach that is taken to assist the layout planner in his choice of the type of layout to use is "*P*–*Q* analysis." Muther [18] recommends the construction of a chart relating production quantity *Q* to the number of products *P*. The chart is given in the form of an ordered frequency histogram. Using some appropriate measure of *Q* (unit loads produced per year, number of units produced per year, pounds produced per year, etc.), we arrange the products in descending order of *Q* to form a curve similar to that given in Figure 2.10. A product layout is considered for those products having a high proportion of the total production quantity. The products having a low proportion of *Q* are normally considered for a process layout. The remaining products are candidates for a combination layout.

We suggest that your search for alternative layouts include at least one product layout, one process layout, and one combination layout. Even though a product layout might not be justified on the basis of today's production quantities, future production volumes might warrant such a layout.

2.5.3 *Flow-analysis approaches*

The preceding discussion has provided motivation for the importance of flow analysis in layout design. It still remains to consider some techniques that are commonly used in analyzing the flow of materials, men, and machines. The most popular method of analyzing flow is to use charts and diagrams. These descriptive, analog models are used to assist the layout analyst's visualization of the required flow. The assembly chart and the operations process chart have already been described. Additional charts and diagrams that have been found to be useful in flow analysis are

1. Flow process chart.
2. Multiproduct process chart.
3. Flow diagram.
4. From–to chart.

The flow process chart is an analog model that substitutes circles for operations, squares for inspections, arrows for transportations, triangles for storages, and the capital letter D for delays. Vertical flow lines connect these symbols in the sequence in which they are performed. A sample flow process chart is given in Figure 2.11.

The multiproduct process chart is used to conveniently combine the

Job Assemble Slab — wooden pencil

Follow the ☐ Product ☐ Main ☐ Material ☐ Form

Chart begins Slabs in storeroom

Chart ends Assembled and clamped

Charted by P.O.E. Date 9/29

No. ___ Page ___ of ___

Summary

600 Assemblies	Present		Proposed		Differance	
	No.	Time	No.	Time	No.	Time
○ Operations	7	304.8	7	70.0	–	234.8
⇨ Transportations	10	4.2	4	0.5	6	3.7
□ Inspections	–		1	–	+1	–
D Delays	–		2		+2	
▽ Storages	3	v	1	v	–2	
Totals	20	309	15	70.5	–5	238.5
Distance travelled	417 ft		80 ft		337 ft	

Details of Present / Proposed Method	Operation Transport Inspection Delay Storage	Distance (ft)	Quantity	Est. time (min.)	Notes
1. Stored in storeroom	○ ⇨ □ D ▽				
2. To slotter-groover by hand truck	○ ⇨ □ D ▽	25	1,200	0.25	Finished stock thinner one box contains 1,200 four-stock slabs (2,400)
3. Slot cut in bottom and four grooves in top	○ ⇨ □ D ▽		1,200	30.00	One pass thru tandem set machines
4. To lead-laying machine (one-half lot — see 9)	○ ⇨ □ D ▽	25	600	0.13	Hand truck
5. Wait for lead layer	○ ⇨ □ D ▽		600	v	Stock delay between lots all four-grove run before starting next size
6. Loaded in machine magazine	○ ⇨ □ D ▽		600	–	Loaded during machine operation
7. Lead layed in slab	○ ⇨ □ D ▽		600	20.00	Push-bar mach. pushes slabs from bottom of mag. under lead hopper.
8. Inspected for full leads. Moved to topper (see 12)	○ ⇨ □ D ▽				Inspected by machine tender on steel bench slide on way to topper.
	~~○ ⇨ □ D ▽~~				During machine time
9. To glue topper (one-half lot — seee 4)	○ ⇨ □ D ▽	30	600	0.15	Hand truck
10. Wait for glue topper	○ ⇨ □ D ▽		600	v	Refer to 5
11. Loaded in glue machine magazine	○ ⇨ □ D ▽		600	2.40	Glue topper loads 25 slabs at time into mag. = 24 loads @ 10 min/load
12. Glued	○ ⇨ □ D ▽		600	–	Push-bar mach. pushes slab over glue wheel into topping position
13. Topped and turned	○ ⇨ □ D ▽		600	11.60	Topper places glued slab on leaded slab and turns on edge
14. Assembled slabs Clamped by topper	○ ⇨ □ D ▽		600	6.00	Topper clamps unit of 25 assem. slabs = 24 units (topper paced by layer)
	○ ⇨ □ D ▽				
	○ ⇨ □ D ▽				

Figure 2.11. Flow process chart of a proposed method of assembling pencil slabs. (After G. C. Close, *Work Improvement*, John Wiley & Sons, Inc., New York; by permission of the publisher.)

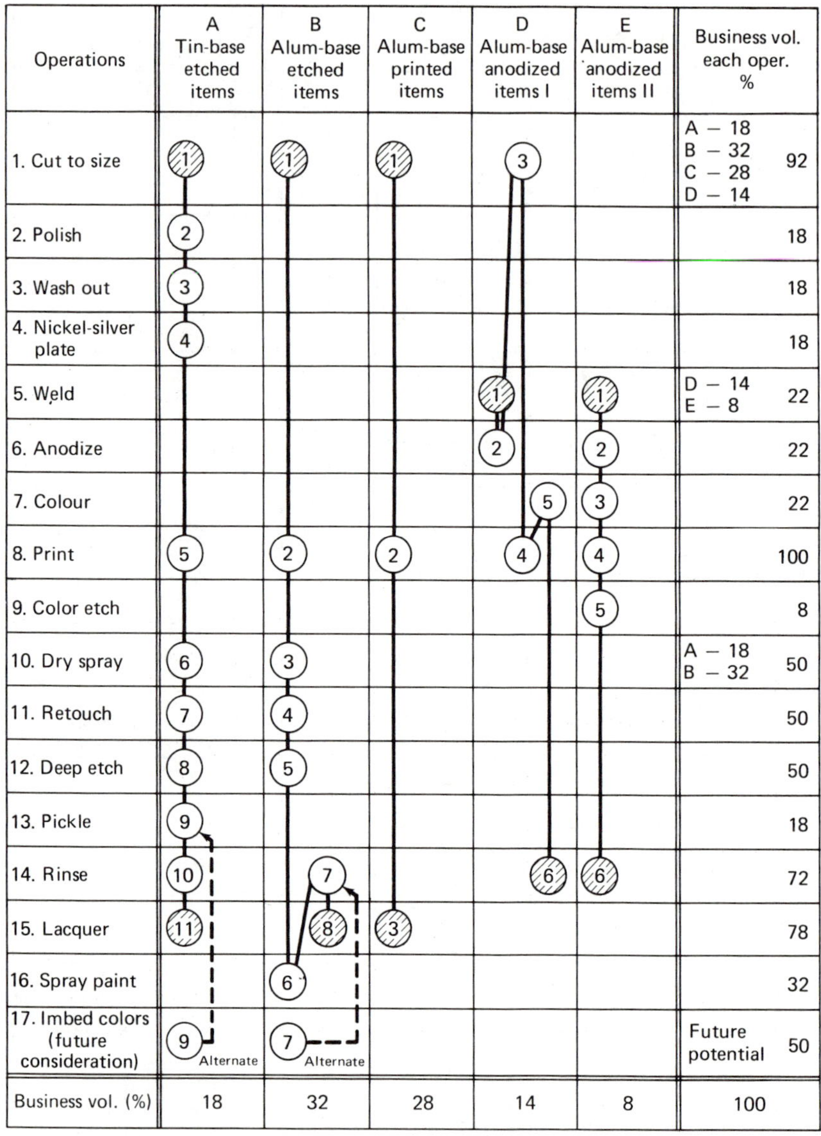

Operations	A Tin-base etched items	B Alum-base etched items	C Alum-base printed items	D Alum-base anodized items I	E Alum-base anodized items II	Business vol. each oper. %
1. Cut to size	1	1	1	3		A – 18, B – 32, C – 28, D – 14; 92
2. Polish	2					18
3. Wash out	3					18
4. Nickel-silver plate	4					18
5. Weld				1	1	D – 14, E – 8; 22
6. Anodize				2	2	22
7. Colour				5	3	22
8. Print	5	2	2	4	4	100
9. Color etch					5	8
10. Dry spray	6	3				A – 18, B – 32; 50
11. Retouch	7	4				50
12. Deep etch	8	5				50
13. Pickle	9					18
14. Rinse	10	7		6	6	72
15. Lacquer	11	8	3			78
16. Spray paint		6				32
17. Imbed colors (future consideration)	9 Alternate	7 Alternate				Future potential 50
Business vol. (%)	18	32	28	14	8	100

Figure 2.12. Multiproduct process chart for five products. (After Richard Murther, *Practical Plant Layout*, McGraw-Hill, Inc., New York; by permission of the publisher.)

operation process charts for more than one product. A sample multiproduct process chart is given in Figure 2.12.

A flow diagram consists of the flow lines superimposed on the floor plan of the area under study. An example of a flow diagram is given in Fig-

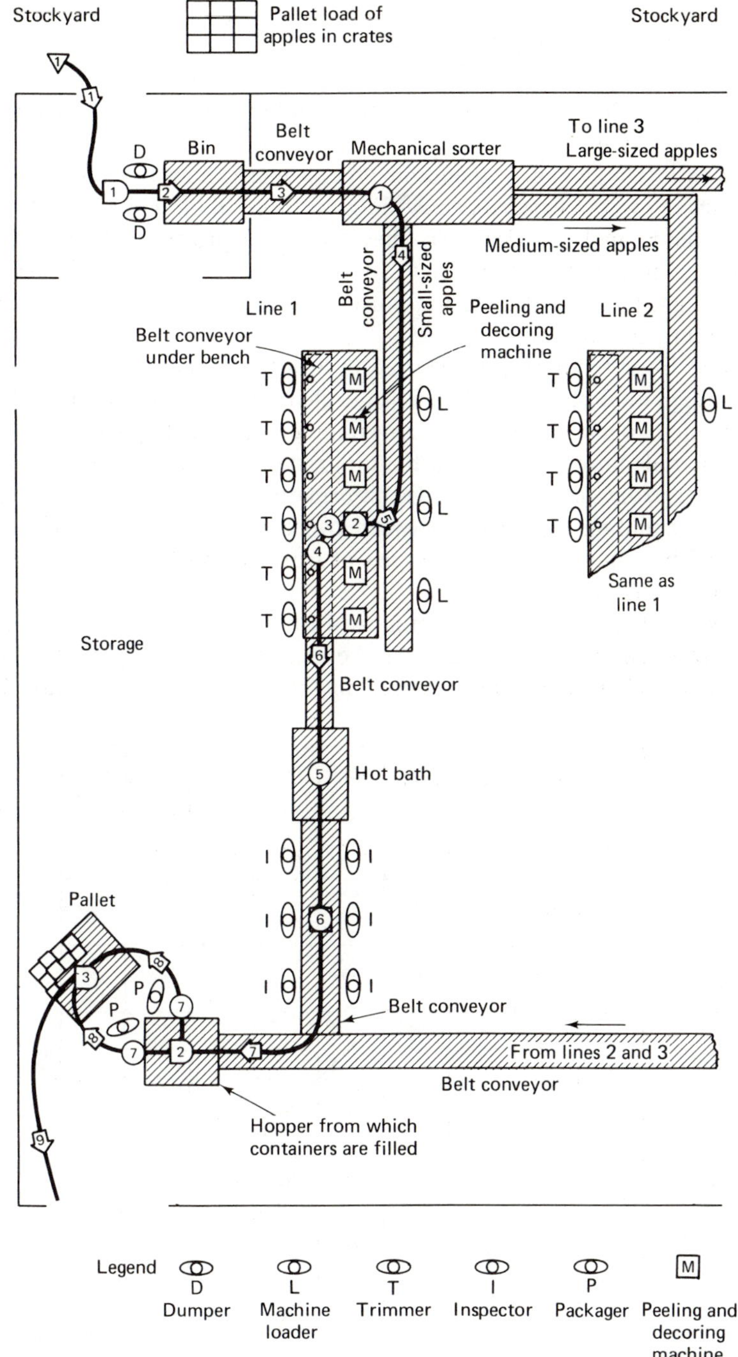

Figure 2.13. Flow diagram for process of preparing apple sections for freezing. (After E. V. Krick, *Methods Engineering*, John Wiley & Sons, Inc., New York; by permission of the publisher.)

ure 2.13. The flow diagram is quite beneficial in evaluating the efficiency of an existing layout.

The assembly chart, operation process chart, flow process chart, multiproduct process chart, and flow diagram were initially developed as aids for methods analysis. However, they have also proved to be beneficial in plant layout as aids in flow analysis. The travel chart, on the other hand, was designed specifically for the purpose of assisting the layout analyst in developing layout designs.

The from–to chart, also referred to as a travel chart and a cross chart, is an adaptation of the familiar mileage chart appearing on most road maps. The from–to chart normally contains numbers representing some measure of the materials flow between two buildings, departments, or machines. Typically, the from–to chart provides information concerning the number of material-handling trips made between two centers of activity and the total material-handling distance. Examples of from–to charts are given in Figure 2.14(a) through (c).

Normally, the from–to chart is used to analyze the flow in process layouts. The item movement that occurs over some specified period of time is totaled for all products and entered in the from–to chart. Figure 2.14(a)

From \ To	R.M.	Saw	Lathe	Drii;	Mill	Ins.	Pkg.	F.G.
R.M.		16	40	72	88	60	36	20
Saw	16		24	56	72	44	20	36
Lathe	40	24		32	48	20	44	60
Drill	72	56	32		16	36	60	76
Mill	88	72	48	16		52	76	92
Ins.	60	44	20	36	52		24	40
Pkg.	36	20	44	60	76	24		16
F.G.	20	36	60	76	92	40	16	

(a)

Figure 2.14(a). From–to chart showing distances between centers of departments as given in Figure 2.9(b).

gives the distances between departments for the process layout given in Figure 2.9(b). Assuming material-handling volumes for the three products, as given in Table 2.3, Figure 2.14(b) gives the number of pallet loads transported between departments. Combining the data in Figures 2.14(a) and (b), we obtain the material-handling distance traveled per day for the proposed layout as shown in Figure 2.14(c).

Table 2.3. FLOW RATES FOR THE PRODUCTS CONSIDERED IN TABLE 2.2

Product	*Flow Rate* (*pallet loads/day*)
A	8
B	3
C	5

From–to charts are descriptive models. As such, the construction of a from–to chart does not result directly in the solution of a layout problem. Rather, the from–to chart is a convenient means of reducing a large volume of data into a workable form. By inspecting the data displayed in the from–to chart, the layout analyst can identify the departments having large volumes

From \ To	R.M.	Saw	Lathe	Drill	Mill	Ins.	Pkg.	F.G.
R.M.		13		3				
Saw			5		8			
Lathe				5	11			
Drill			3		5	8		
Mill				8		16		
Ins.			8				16	
Pkg.								16
F.G.								

(b)

Figure 2.14(b). From–to chart showing number of material-handling trips per day between departments.

To / From	R.M.	Saw	Lathe	Drill	Mill	Ins.	Pkg.	F.G.	Total
R.M.		208		216					424
Saw			120		576				696
Lathe				160	528				688
Drill			96		80	288			464
Mill				128		832			960
Ins.			160				384		544
Pkg.								256	256
F.G.									0
Total	0	208	376	504	1,184	1,120	384	256	4,032

(c)

Figure 2.14(c). From–to chart showing distance traveled per day using the process layout as given in Fig. 2.9(b).

of item movement and can develop a layout design in which these departments are located near one another.

As mentioned earlier, from–to charts are used in conjunction with process layouts. They have been found to be useful in

1. Selling the layout.
2. Analyzing material movement.
3. Developing departmental block plans.
4. Developing detailed layout arrangements.
5. Evaluating layout alternatives.
6. Demonstrating the dependency of one area upon another.
7. Improving the use of floor space.
8. Showing the interrelationship of product lines.

The from–to chart also is used in Chapter 3 to provide necessary input for a computerized layout program. (Notice that the from–to chart can also be portrayed as a to–from chart, depending on where you are from.)

The method of analyzing material flow is dependent on the production volume for the product. Referring to the *P–Q* chart, the high-volume products are candidates for a product layout. Therefore, the assembly chart, operation process chart, and flow process chart are used to analyze the flow of materials for the high-volume items. When there are several high- and medium-volume products to be produced, the multiproduct process chart

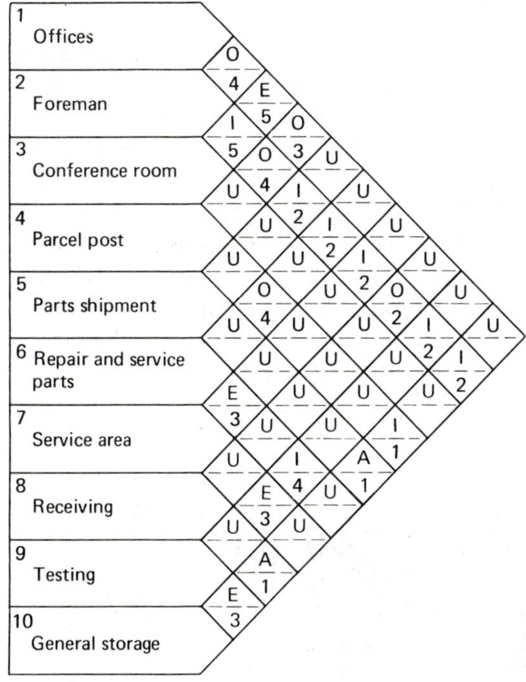

Code	Reason
1	Flow of materials
2	Ease of supervision
3	Common personnel
4	Contact necessary
5	Convenience
6	
7	
8	
9	
10	

Rating	Definition
A	Absolutely necessary
E	Especially Important
I	Important
O	Ordinary closeness OK
U	Unimportant
X	Undesirable

(a)

Figure 2.15(a). Activity relationship chart.

is used to examine the interrelationships between products. The from–to chart is used in conjunction with process layout analysis. Consequently, it is used to analyze the material flow for the low-volume products. The flow diagram is used in both product and process layout, as well as their variations.

2.5.4 *Activity relationship analysis*

Flow analysis tends to relate various activities on some quantitative basis, and, as mentioned, the relationship is normally expressed as some function of material-handling cost. We emphasized in Chapter 1, as well as in an earlier discussion in this chapter, that a number of factors other than material-handling cost might be of primary concern in layout design. The activity relationship chart, or REL chart, was designed to facilitate a consideration of qualitative factors. As developed by Muther [18], the REL chart replaces the numbers in a from–to chart by a qualitative *closeness* rating. A typical REL chart is shown in Figure 2.15(a). Notice that all pairwise combinations of relationships are evaluated, and a closeness rating (A, E, I, O, U, or X) is assigned to each combination. When evaluating activity relationships for N activities there are $N(N-1)/2$ such evaluations. Along

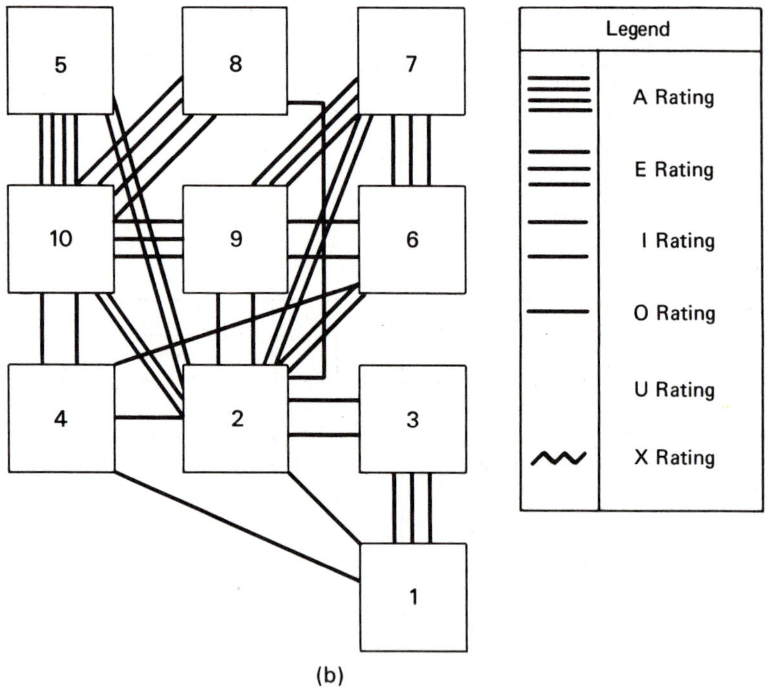

(b)

Figure 2.15(b). Activity relationship diagram.

with each closeness rating, other than a U rating, is provided a numeric code giving the reason(s) for the particular closeness rating.

2.6 The Relationship Diagram

If activity relationships alone are considered, we go directly to the construction of the activity relationship (REL) diagram. If the flow of materials is the dominant factor, we develop a flow diagram such as given in Figure 2.9. If both the flow of materials and the relationship of activities are to be considered, they must be combined and a REL diagram developed. A sample REL diagram is given in Figure 2.15(b) for the REL chart given in Figure 2.15(a).

In the REL diagram each activity is represented by an equal-sized square. The squares are connected by a number of lines corresponding to the closeness rating. The squares are shifted around until the proper relationship between activities is obtained.

2.7 Space Requirements and Availability

Once consideration has been given to the flow of materials and the relationship of activities, and the appropriate relationship diagram has been constructed, we are in the position to evaluate the space requirements for the layout. Ideally, we would like to develop the layout and then construct the building around the layout. However, from a practical viewpoint, we often find that our solution is constrained by the amount and configuration of available space. The constraint can be in the form of an existing building, a limitation on the size of the building site, or the availability of capital for new construction. For this reason, we must consider not only space requirements, but space availability as well.

2.7.1 *Production rate determination*

One of the major determinants of the amount of space required is the desired production rate. We have assumed the desired production rate has been previously determined by the schedule designer. Recall, we previously pointed out that schedule design decisions are closely related to layout decisions. The production rate was used in our *P–Q* analysis to guide us in our choice of a product layout or a process layout. We use the production rate to develop estimates of space requirements.

How is the production rate determined? In brief, a marketing forecast is translated into required production quantities. As an illustration, consider

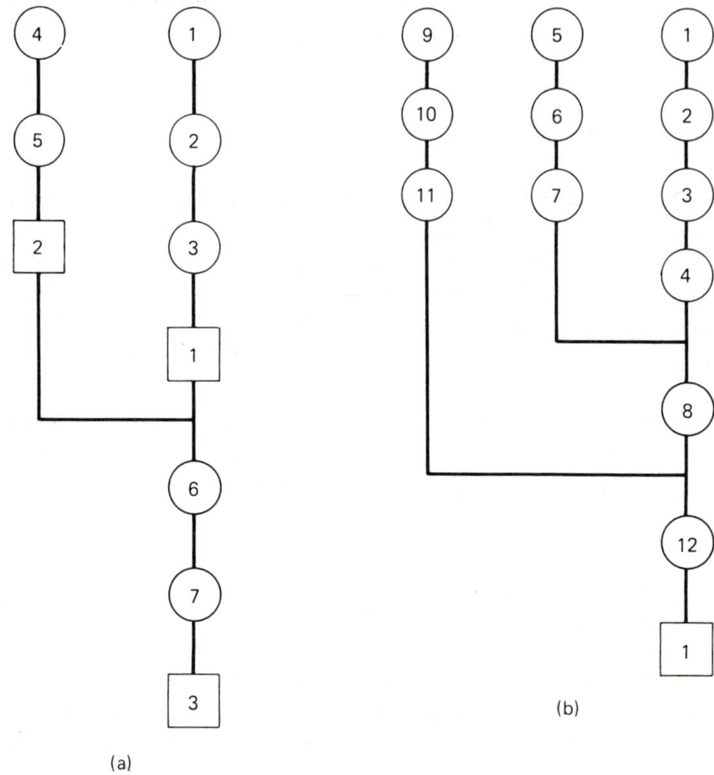

Figure 2.16. Operation process charts of manufacturing processes.

an example involving seven operations and three inspections, as shown in Figure 2.16(a). We assume that bad-quality parts are rejected only at the inspections, and that the fraction defective equals 3, 2, and 4%, respectively, at the three inspection stations.

Suppose that there are 2,000 operating hours in a year, and the annual demand for the product is forecast to be 180,000 units. Production planning and marketing agree that 200,000 good units should be produced to protect against forecasting errors. Consequently, 100 good units are to be produced per hour in order to maintain a balanced production line.

Since 100 good units must leave the third inspection station and, on the average, 4% of the parts entering the station are bad, then $100/0.96 \doteq 104.2$ parts should enter the third inspection station. Consequently, operations 6 and 7 must produce 104.2 parts per hour. The second inspection station rejects 2% of the incoming product. Therefore, $104.2/0.98 \doteq 106.3$ parts must be produced per hour by operations 4 and 5. Operations 1, 2, and 3 must produce $104.2/0.97 \doteq 107.4$ parts per hour.

In a process layout a given machine can be used to process a variety of different products. Typically, a given job is produced for a period of time, and then a new job is produced. The scheduling problem in batch production is a complex problem. If we are producing to inventory when equipment capacity far exceeds the demand level, then optimum batch production quantities might be computed using an appropriate inventory-control model [9]. There still remains the problem of scheduling the batches on each machine. Since this is not a layout problem, the scheduling problem is not pursued further in this book. However, depending on the organization, you might not be functioning strictly as a layout analyst and might also be involved in scheduling decisions. If so, we refer you to [3], [5], and [8].

Process layouts are also used in job shops where "one-shot" jobs are received and processed. Rather than producing to inventory, the order is processed and shipped to the customer. In such a situation we might be tempted to determine the production quantity in much the same way as for the product layout. As an illustration, suppose that an order is received to produce 50 units of a production item having the processing sequence given in Figure 2.16(b). Suppose that, on the average, 10% of the units inspected are rejected. Consequently, we might schedule $50/0.90 \doteq 55.56$, or 56 units to be produced. What is wrong with this approach?

In the case of the product layout, a large number of items are being produced over a long period of time. An additional 20,000 units were added to the annual production rate as a safety factor against forecasting errors. Consequently, it was reasonable to use "expected values" in the calculations. In the job shop, we are producing the batch only once. If 56 units are produced there might be only 40 good units available. In this case another setup must be made and the remainder produced. But, how many should be scheduled? In fact, the production lead time might be so large that no additional units could be produced in time to meet the contracted delivery date. Suppose it is found that there are 55 good units in the 56 produced. What is done with the excess?

The point to be made is that the production quantity should be based on the economics of the situation. One must balance the costs of producing too few good units against the costs of producing too many good units. This problem is referred to as the reject-allowance problem. For a discussion of the problem, as well as a simplified formulation of the problem, see [22].

2.7.2 *Equipment requirements*

Given the desired production rate at each processing stage, one can determine the number of machines required. To carry out this calculation, we need to know the operation efficiencies for the equipment and the standard production times for each operation. To make the process explicit, let

P_{ij} = desired production rate for product i on machine j, measured in pieces per production period
T_{ij} = production time for product i on machine j, measured in hours per piece
C_{ij} = number of hours in the production period available for the production of product i on machine j
M_j = number of machines of type j required per production period
n = number of products

Therefore, M_j can be expressed as

$$M_j = \sum_{i=1}^{n} \frac{P_{ij}T_{ij}}{C_{ij}} \tag{2.1}$$

As an illustration of the use of Equation (2.1), consider the data displayed in Table 2.4. The same type of machine is used in the processing of six different products or parts. The standard number of pieces to be produced per hour, $1/T_{ij}$, is given for each product, along with the desired production rate, P_{ij}. The number of hours in a month is 150. Therefore, for this example M_j equals 2.643. In this case, we will probably use three machines for the production of the six products. If an increase in demand is anticipated or if protection is to be provided against loss of production due to machine breakdown, four machines might be purchased.

All the calculations were based on deterministic values for P_{ij} and T_{ij}. Realistically, there will be some variation in the number of units to be produced and the processing times. If such variation is believed to be significant, one might develop the probability distribution for M_j and determine the optimum number of machines to purchase, based on the costs involved [15]. However, such determinations seldom are made by the layout analyst.

Table 2.4. SAMPLE DATA FOR MACHINE REQUIREMENT CALCULATION

Product Number i	*Required Production Rate* P_{ij}	*Standard Production Rate* $1/T_{ij}$	*Hours/Month* C_{ij}	$P_{ij}T_{ij}/C_{ij}$
2501	6,000	120	150	0.333
2502	9,000	150	150	0.400
3104	15,000	100	150	1.000
3206	2,000	100	150	0.133
3617	8,000	120	150	0.444
3618	4,000	80	150	0.333
				2.643

2.7.3 *Employee requirements*

We have previously considered ways of determining the desired production rate and the number of machines required to satisfy this rate. In this section we discuss the determination of the number of employees required.

In the case of manual assembly operations, we can determine the number of employees required in the same way we calculated machine requirements. That is,

$$A_j = \sum_{i=1}^{n} \frac{P_{ij}T_{ij}}{C_{ij}}$$

where A_j = number of operators required for assembly operation j

P_{ij} = desired production rate for product i and assembly operation j, pieces per day

T_{ij} = standard time to perform operation j on product i, minutes per piece

C_{ij} = number of minutes available per day for assembly operation j on product i

n = number of products

The number of machine operators required is dependent on the number of machines tended by one or more operators. In many cases this number is determined by the existing labor contract and the requirements of the job. However, if highly automatic equipment is used, there is the possibility of one operator tending a number of machines. Again, depending on the particular organization, the determination of the number of machines supervised by one operator will be the function of the scheduling analyst or the methods and standards analyst, not the layout analyst. If such is not the case, then the following discussion might be beneficial to you in this determination.

A determination of the number of machines to be supervised by one operator can take two approaches. First, you can assume all time values are deterministic. As an alternative, you can treat the activity times as random variables, and perform a probabilistic analysis.

One deterministic approach used to determine the assignment of operators to machines is to employ the man–machine chart. The man–machine chart is a descriptive, analog model showing the man–machine relationships graphically against a time scale. An example of the use of the man–machine chart is given in Figure 2.17. When the man–machine chart is used, there still remains the choice of the criterion to be used in determining the number of operators to assign.

The man–machine chart is especially useful in analyzing the man–machine relationships when nonidentical machines are being supervised by

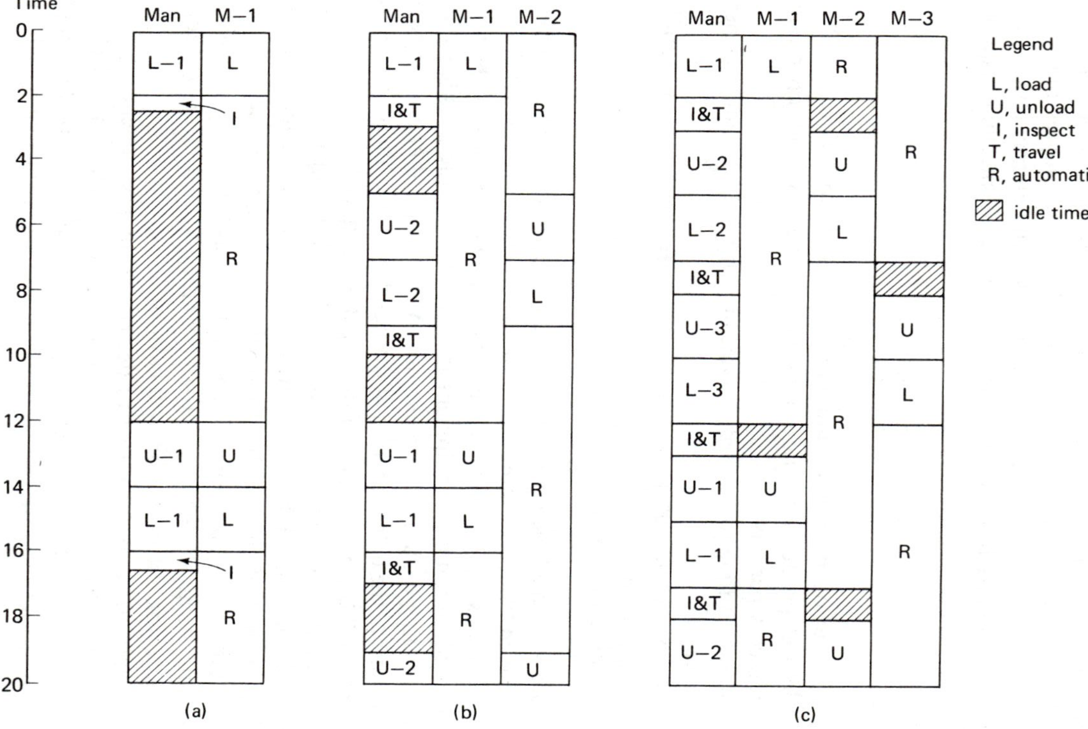

Figure 2.17. Man–machine chart analysis of multimachine assignment.

one operator. Also, the man–machine chart has been used in selling all concerned on the feasibility of a particular assignment.

As an alternative to the man–machine chart when identical machines are involved, we present a prescriptive, symbolic model that can be used to determine the number of machines to assign an operator [7]. Let

a = concurrent activity time, e. g., loading, unloading
b = independent operator activity time, e. g., walking, inspecting, packaging
t = independent machine activity time, e. g., automatic run time
n' = number of machines to assign an operator for neither machine nor operator idle time
m = number of machines assigned an operator
T_c = repeating cycle time
I_o = idle operator time during a repeating cycle
I_m = idle time per machine during a repeating cycle
$TC(m)$ = cost per unit produced, based on an assignment of m machines per operator
C_1 = cost per operator-hour
C_2 = cost per machine-hour

Notice that it takes $a + b$ time units for an operator to perform the work content required on a single machine during one complete production cycle. Also, it takes the machine $a + t$ time units to complete a production cycle. Consequently, the ideal assignment, n', would be

$$n' = \frac{a + t}{a + b} \tag{2.2}$$

For the example illustrated in Figure 2.17, a equals 4 minutes, t equals 10 minutes, and b equals 1 minute if we include both inspection and travel between machines. Consequently, n' equals $\frac{14}{5}$, or 2.8 machines.

Since a fractional number of machines cannot be assigned to an operator, consider the effects of assigning m machines to an operator, where m is integer valued and $m < n'$. In this case, the operator will be idle and the machine will be kept busy. Consequently, the repeating cycle will equal $a + t$. On the other hand, if $m > n'$, then the operator will be kept busy, and it will require $m(a + b)$ time units to complete a cycle. Therefore,

$$T_c = \begin{cases} a + t, & m \leq n' \\ m(a + b), & m > n' \end{cases} \tag{2.3}$$

Furthermore, it follows that

$$I_o = \begin{cases} (a + t) - m(a + b), & m \leq n' \\ 0, & m > n' \end{cases} \tag{2.4}$$

and

$$I_m = \begin{cases} m(a+b) - (a+t), & m > n' \\ 0, & m \leq n' \end{cases} \tag{2.5}$$

For our example problem, if m equals two machines, the repeating cycle is 14 minutes, and the operator is idle 4 minutes during a repeating cycle. If m equals three machines, the repeating cycle is 15 minutes, and each machine is idle 1 minute during a repeating cycle.

If we wish to determine the cost per unit produced by an m machine assignment, notice that the cost of such an assignment per unit time equals $C_1 + mC_2$. Furthermore, during a repeating cycle m units are produced. Consequently,

$$TC(m) = (C_1 + mC_2)\frac{T_c}{m} \tag{2.6}$$

Substituting (2.3) in (2.6), we see that

$$TC(m) = \begin{cases} \dfrac{(C_1 + mC_2)(a+t)}{m} & m \leq n' \\ (C_1 + mC_2)(a+b), & m > n' \end{cases} \tag{2.7}$$

From (2.7), we see that if $TC(m)$ is to be minimized, then m should be made as large as possible when $m \leq n'$, and m should be made as small as possible when $m > n'$. Consequently, if n' is integer valued, then n' machines should be assigned to minimize (2.7). However, if n' is not integer valued, such that $n < n' < n + 1$, where n is the integer portion of n', then either n or $n + 1$ machines should be assigned, depending on whether $TC(n) \leq TC(n + 1)$, or vice versa.

Since we must determine which is more economic, an n machine assignment or an $n + 1$ machine assignment, form the ratio of $TC(n)$ and $TC(n + 1)$ such that

$$\begin{aligned} \Phi &= \frac{TC(n)}{TC(n+1)} \\ &= \frac{(C_1 + nC_2)(a+t)}{[C_1 + (n+1)C_2]n(a+b)} \end{aligned} \tag{2.8}$$

Letting $\epsilon = C_1/C_2$ and substituting (2.2) in (2.8) gives

$$\Phi = \frac{\epsilon + n}{\epsilon + n + 1}\,\frac{n'}{n} \tag{2.9}$$

Consequently, if $\Phi < 1$, assign n machines; $\Phi > 1$, assign $n + 1$ machines; and if $\Phi = 1$, assign either n or $n + 1$ machines.

For the example given in Figure 2.22, suppose that C_1 equals \$3 per hour and C_2 equals \$10 per hour for each machine. From (2.2), n' equals 2.8. Therefore, n equals 2.0 and $n + 1$ equals 3.0. From (2.9), Φ equals 6.44/6.60. Since $\Phi < 1$, we would assign two machines per operator.

The deterministic machine-assignment model is considered further in the problems. The prescriptive, symbolic model is appropriate when identical machines are to be assigned, times are deterministic, and cost per unit produced is to be minimized. If we wish to consider nonidentical machine assignment and/or two or more operators jointly operating two or more machines, then the man–machine chart, a deterministic, descriptive, analog model, can be used.

If we wish to perform a probabilistic analysis, we might use appropriate queueing models or Monte Carlo simulation to determine the number of machines to assign an operator. In fact, when random variation is present, we might find that it is more economical to establish a pool of operators who service a group of machines, rather than have each operator assigned to specific machines.

Queueing theory and simulation are two subjects that can be extremely valuable to the layout analyst. Not only can they be used in determining the number of operators required to tend the production equipment, but they are also useful in balancing assembly lines, designing conveyor systems, and determining maintenance crew sizes, receiving and shipping crew sizes, toolroom attendant requirements, material-handling crew sizes, material-handling equipment pool sizes, receiving and shipping dock sizes, and rest room facility requirements, to list a few applications that quickly come to mind.

Both queueing theory and simulation are sufficiently important to the layout analyst to justify a separate and detailed treatment. Unfortunately, to undertake a detailed discussion of these subjects here goes beyond the intended scope of this text. Consequently, we refer you to [6], [20], and [23].

2.7.4 *Space determination*

Once we know how many machines will be tended by a single operator, we can determine space requirements for equipment. Also, the space requirements for the assembly department can be determined, since we know the number of assembly operators required.

Some commonly used methods of determining space requirements are introduced and discussed below.

1. *Production-Center Method.* The production center consists of a single machine plus all the associated equipment and space required for its operation. Work space (front, rear, left side, right side), additional maintenance space, and storage space are added to the space requirements for the machine.

All equipment and storage locations are arranged in the production center, and the floor space is determined. The space requirement is multiplied by the number of similar pieces of equipment to determine the total space requirement for, say, Monarch engine lathes.

2. *Converting Method.* Using this method, the present space requirements are converted to those required for the proposed layout. If this method is used, be very careful about the assumptions. Remember that total space required is not a linear function of the production quantity. Therefore, just because production doubles, twice as much space is not necessarily needed. Furthermore, you may not be utilizing the present space most efficiently. This method is commonly used to determine space requirements for supporting service and storage areas, whereas the production-center method is used to determine the space requirements for manufacturing areas.

3. *Roughed-Out Layout Method.* Templates or models are placed on the layout to obtain an estimate of the general configuration and space requirements.

4. *Space-Standards Method.* In certain cases industry standards can be used to determine space requirements. Additionally, standards may be established based on past successful applications. The use of such standards without an understanding of their underlying assumptions is dangerous. Such standards adopted by others should be closely scrutinized and compared with the present layout.

5. *Ratio Trend and Projection Method.* This method is limited to general space requirements. It is probably the least accurate of the methods presented. To use this method, one establishes a ratio of square feet to some other factor that can be measured and predicted for the proposed layout. Examples are square feet per direct labor hour, square feet per unit produced, and square feet per supervisor. In making the space determination, space must be included for

1. Raw material storage.
2. In-process inventory storage.
3. Finished goods storage.
4. Aisles, cross aisles, and main aisles.
5. Receiving and shipping.
6. Material-handling equipment storage.
7. Toolrooms and tool cribs.
8. Maintenance.
9. Packaging.
10. Supervision.

11. Quality control and inspection.
12. Health and medical facilities.
13. Food service.
14. Lavatories, wash rooms, etc.
15. Offices.
16. Employee and visitor parking.
17. Receiving and shipping parking.
18. Other storage.

As an illustration of the calculation of space requirements, consider the determination of production space requirements given in Table 2.5. To

Table 2.5. PRODUCTION SPACE REQUIREMENTS

Process	*Equipment*	*No.*	*Machine Center Dimensions per Machine (ft) Depth Width*	*Machine Center Area per Machine (ft²)*	*Total Process Area (ft²)*
Saw	Armstrong hack saw	3	10 × 19	190	570
Mill	K & T plain mill	5	$13\frac{1}{2} \times 10\frac{1}{2}$	142	710
	Vertical mill	7	$11 \times 10\frac{1}{4}$	113	791
	Hand mill	4	$7\frac{1}{4} \times 9\frac{3}{4}$	71	284
Drill	2 Spindle Avey	2	$8\frac{1}{4} \times 6\frac{1}{2}$	54	108
	1 Spindle Delta	2	$7\frac{1}{2} \times 4\frac{1}{2}$	34	68
	6 Spindle Delta	1	$7\frac{3}{4} \times 10\frac{1}{2}$	82	82
Turn	Gisholt	1	$9\frac{1}{4} \times 17\frac{3}{4}$	164	164
	Monarch	2	$14 \times 6\frac{1}{4}$	88	176
	Hardinge	1	$9\frac{1}{2} \times 5$	48	48
	W & S turret	1	$8\frac{1}{2} \times 20\frac{1}{4}$	173	173
	B & S automatic	1	$7\frac{1}{2} \times 15\frac{1}{2}$	116	116
Form	Gas furnance	1	8 × 7	56	56
	Arbor press X				
Paint	Dip tank	2	7 × 12	84	168
	Spray booth	1	9 × 11	99	99
Clean	Tumble	1	7 × 6	42	42
Assemble	Bench	1	8 × 7	56	56
	Bench	1	8 × 7	56	56
	Avey drill	2	$8\frac{1}{4} \times 6\frac{1}{2}$	54	108
Packaging	Bench	1	8 × 7	56	56
			Total square feet required		3,931
			40% aisle space		1,572
			Production space required		5,503

Table 2.6. NONPRODUCTION ACTIVITY SPACE REQUIREMENTS

Activity	*Area* (ft^2)
Storage	
Warehouse	180
Other	180
Office	
Main office	500
Hallway	120
Rest rooms	100
Locker rooms	
Men	84
Women	60
Foreman	
Desk	24
Maintenance	
Desk	20
Parts	80
Tool crib	50
Receiving and shipping	50
Total space required	1,448

the 5,503 square feet of floor space for production, we must add the 1,448 square feet shown in Table 2.6 to give an estimate of 6,951 square feet of floor space required in total.

2.8 Designing the Layout

Having analyzed the flow of materials and the relationship of activities and determined the space requirements, alternative layouts can be designed. In terms of the phases in the design process, we have completed the analysis phase and are entering the search phase. A number of alternative layouts are to be developed, based on the analysis of flow, activity relationships, and space requirements. (At this point it is suggested you review the earlier discussion in Chapter 1 concerning creativity.)

In brief, the overall layout is designed by first combining space considerations with the REL diagram. Following the SLP approach, the combination of space considerations with the REL diagram is accomplished by developing the space-relationship diagram. Recalling the REL diagram in Figure 2.15, the corresponding space-relationship diagram is given in Figure 2.18(a). Next, the layout design is modified based on practical limitations and other considerations. At this point an overall layout design has been generated in the form of a block plan. The block plan is a scaled diagrammatic represen-

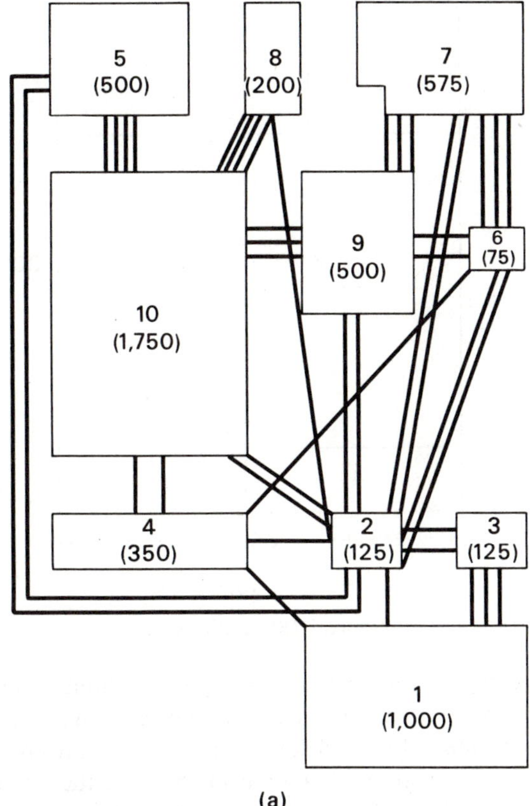

Figure 2.18(a). Space relationship diagram.

tation of the building, and normally shows the locations of internal partitions and columns. Detailed locations of machinery, equipment, and facilities are not usually included in the block plan. An example of a block plan based on Figure 2.18(a) is given in Figure 2.18(b). Once a number of block plans have been generated, they must be converted to detailed layouts.

A detailed layout analysis consists of basically the same steps involved in developing the overall design. Now, instead of analyzing the flow and/or activity relationships between departments, we are concerned with the relationships between work centers.

In developing both the overall layout and the detailed layout we must remember to consider the possibility of future expansion and the possibility of future change. It is extremely important that the layout be flexible enough to accommodate, say, changes in product design, process design, and schedule design. At the beginning of this chapter we enumerated a number of causes for layout problems. If we design a flexible layout initially, we can postpone the need for a major redesign until a much later point in time.

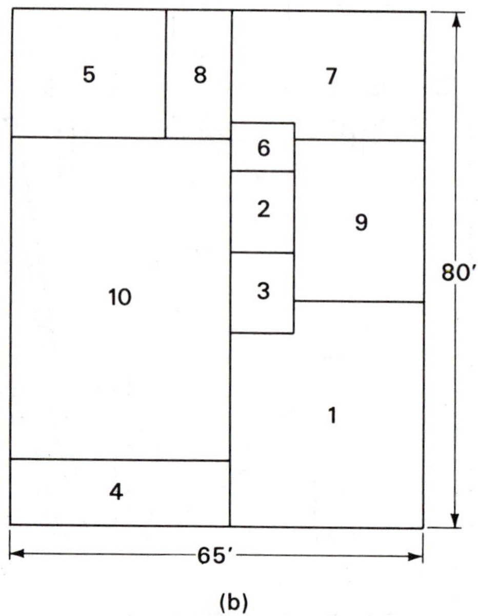

Figure 2.18(b). Block plan.

A detailed layout should not be designed without giving consideration to material-handling requirements. The choice of handling methods and equipment is an integral part of layout design. It is extremely important to incorporate effective material-handling methods in the layout.

The design of the material-handling system follows basically the same sequence of steps as outlined for the design process. Many of the tools that we have employed in the analysis of the layout problem are commonly used in analyzing material-handling problems. However, the search phase of the design process requires a high degree of familiarity with the types, capabilities, and limitations of material-handling equipment.

In designing the material-handling system, we recommend you follow the steps of the design process. Furthermore, we suggest you consult references [2], [4], and [10]. As you can see, the design of material-handling systems is a subject unto itself beyond the scope of this book and worthy of separate discussion.

We shall make the assumption that the material-handling system has been designed. In fact, we assume this process has taken place in parallel with the development of the layout design. Now, we have several layout designs, including their associated material-handling system. But, how are they presented? In what form do the designs exist?

Basically, there are three methods of visually representing layouts:

1. Drawings or sketches.

2. Two-dimensional iconic models.
3. Three-dimensional iconic models.

Drawings and sketches have the advantage of being easy to make. This, of course, is not necessarily the case if we are referring to engineering or architectural draftings. However, in layout design here is no longer much need for detailed drafting, due to the availability of cross-section paper. A major disadvantage of drawings is their lack of flexibility. If a change is made in the design, the entire layout might need to be redrawn. One approach taken to guard against this possibility is to make detail drawings of individual departments and then assemble the drawings in the same way you assemble puzzles.

Two-dimensional iconic models, commonly called templates, currently are the most popular method of presenting layout designs. Templates are scaled representations of physical objects in a layout. They commonly are prepared from cardboard, paper, sheet metal, plastic, and wood. A wide variety of templates are available commercially, along with adhesive tapes

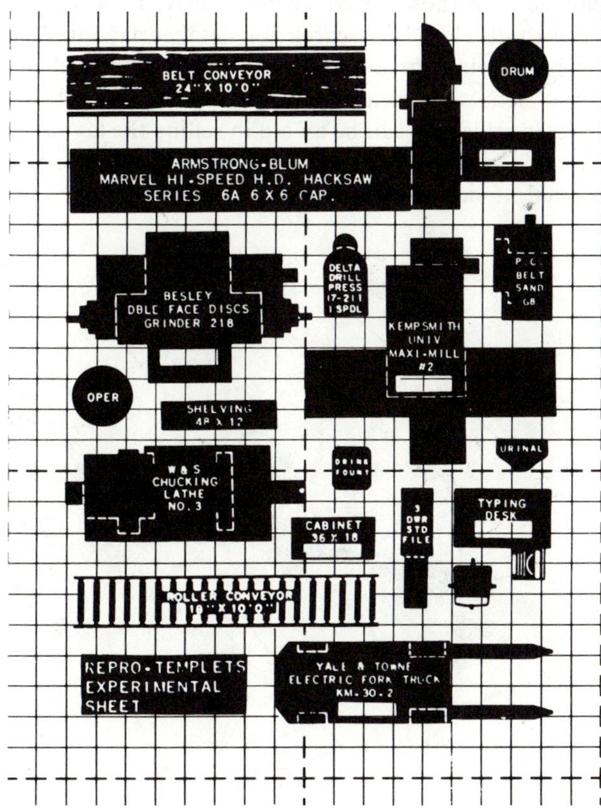

Figure 2.19. Sample templates.

that can be used to denote walls, aisles, columns, pallets, tables, benches, etc. Some sample templates are shown in Figure 2.19.

When templates are used, normally they are placed on a plastic sheet having grid lines. Transfer lettering kits and die-cut symbols can supply the finishing touch to the layout. Granted, the use of templates, tapes, grid sheets, transfer lettering, and die-cut symbols can become expensive. However, the sales advantage realized by developing a layout with a professional appearance can more than offset the added expense over, say, drawings. One disadvantage of two-dimensional templates is the inability to visualize the vertical requirements of the layout.

Two-dimensional templates that are available commercially provide an accurate and detailed layout. Actual floor space relationships are clearly indicated. Reproductions are easily made. Overlays and underlays are commonly used, along with color-coded tapes. Normally, changes in the layout can be accommodated quickly.

As shown by Figure 2.20, three-dimensional scale models of the layout are more realistic than those described previously. However, they are also more expensive to develop. The search for detailed layout alternatives can be facilitated by moving three-dimensional models around in the overall layout. Overhead details and clearances are easily visualized. Unquestionably, the three-dimensional layout can be an aid in selling the layout design. However, in some cases visualization of flow lines can be hindered by the use of three-dimensional models. Because of this, it is not uncommon to see only

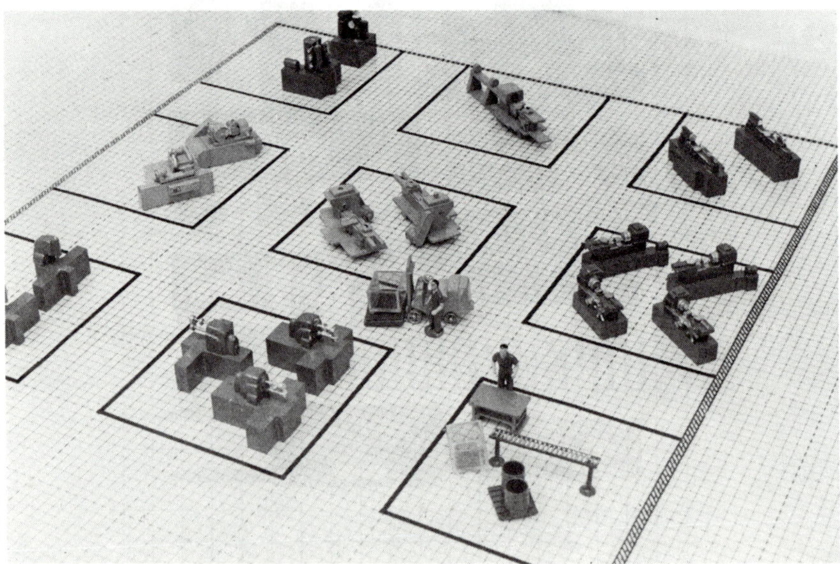

Figure 2.20. Three-dimensional model.

the machines, equipment, and personnel represented by three-dimensional models and have walls and columns represented by tapes.

The ultimate decision as to the method of presenting the layout might not be yours to make. Precedent might dictate the approach you will use. However, if you are free to choose, we suggest that you order some catalogs from the suppliers of layout materials and evaluate the economics involved. The ultimate choice should be influenced by the scope of the problem. If you are designing a new plant, three-dimensional models are probably justified. If you are determining the layout for, say, a toolroom, two-dimensional models should suffice.

It is easy to lose your perspective when dealing with the iconic models. It is our opinion that too much emphasis has been placed on the use of iconic models as aids in developing layouts. We believe their major contribution lies in the selling of the design. However, they also can be quite valuable in the evaluation of the practicality of alternative layout designs.

2.9 Selection, Specification, Implementation, and Follow-Up

Now that alternative layout designs have been prepared, it still remains to select that design from among the alternatives which best meets your objectives. The design selected must be specified, all concerned must be sold, and the layout must be installed, observed, and periodically evaluated in an operating environment. We shall treat these steps briefly, realizing they were considered in Chapter 1 in a discussion of the total design cycle.

The selection of the "best" layout normally means the selection of the design that results in the most favorable compromise among competing objectives. Among these objectives was the minimization of cost. Thus far, little has been said about the cost of the design. If costs are to be considered, we are interested in estimating the cost of installing the new design, as well as the long-run operating cost resulting from the new layout. If there is an existing layout design, then we shall be interested in comparing the performance of the new design with that of the old design. We might find that the cost of installing the new design offsets any of the benefits it provides. In making this evaluation, remember that the new design is intended for the future. Performance costs for the existing system probably reflect present and past operating levels. Therefore, it will be necessary to forecast future costs for both the new and the old designs.

If costs are an important consideration in the evaluation of alternative layouts, it is necessary that the relevant costs be measured. This is no easy undertaking for several reasons. First, we are interested in incremental costs, rather than standard costs. The latter include overhead items, which many times will be unaffected by the layout design. Second, we are interested in future costs, rather than present or past costs. Furthermore, in the case of a new layout we have no prior experience on which we can base our estimates of future costs.

Many times costs are not the major consideration in evaluating layout designs. Typically, a number of alternative layouts will have approximately the same costs, and other considerations are used in choosing the preferred design. We enumerated a number of these factors in Chapter 1 and in Section 2.2, too!

Although we previously discussed ways of performing the evaluation step in the design process, one additional factor should be recognized. That is, the layout design which is chosen must be sold. Consequently, it is prudent at this point to consider the amount of resistance to change that will accompany each design. No matter how good the design might be, if certain individuals are opposed to it, once it is implemented they can make the design look so bad that a new layout study will be required. Therefore, you should strive to assess and reduce such resistance and anticipate the amount of resistance that will be encountered for each alternative.

In the final analysis, the ability to sell the layout design is strongly influenced by the ability to cope with resistance to change. Some specific causes of resistance to change on the part of persons having a veto power over the proposed layout design are given by Krick [12] as

1. Inertia.
2. Uncertainty.
3. Failure to see the need for the proposed change.
4. Failure to understand the proposal.
5. Fear of obsolescence.
6. Loss of job content.
7. Personality conflict with the analyst.
8. Resentment of outside help, or interference.
9. Resentment of criticism.
10. Lack of participation in the formulation of the proposed change.
11. Tactless approach on the part of the layout analyst.
12. Lack of confidence in the analyst.
13. Inopportune timing.

Some methods of minimizing resistance to change suggested by Krick [12] are

1. Convincingly explain the need for the change.
2. Thoroughly explain the need for the change.

3. Facilitate participation or at least the feeling of participation in formulation of the proposed method.
4. Use a tactful approach in introducing your proposal.
5. Watch your timing.
6. In the case of major changes, if possible introduce the change in stages.
7. Capitalize on the features that provide the most personal benefit to the person(s) you are trying to sell.
8. If possible, by appropriate questioning maneuver a prospective rejector into "thinking" of the (your) idea himself.
9. Show a personal interest in the welfare of the person directly affected by the change.
10. Whenever possible have changes announced and introduced by the immediate supervisor of those affected.

Assuming a layout design has been accepted by the appropriate persons within the organization, it must be installed. When installing the layout it is important to remember that a considerable amount of planning must precede the actual location and installation of equipment. Once the plans are made, all activities should be scheduled. Since the installation of the layout can involve a number of activities, a project scheduling model, such as the critical-path method (CPM), can be quite useful. We shall not describe CPM, but, instead, refer you to [13].

Once the layout has been installed, you should follow up to see that the layout was installed as designed. If modifications were made during installation, either they should be accepted or arrangements should be made to correct the discrepancies. Periodic checks should be made to see that the layout is performing satisfactorily. Also, you should be on the lookout for the problem indicators we pointed out earlier so that a redesign of the layout is begun when justified.

2.10 Where to from Here?

Please give some serious thought to the following questions:

1. Would it surprise you to learn that some people actually believe that all, or certainly most, layout problems are solved by moving templates around until a design is found that looks good?
2. Would it surprise you to learn that a very large number of layout problems are solved in this way?
3. Do you believe the solution obtained from the template-shuffling approach can be improved upon by applying a systematic approach?

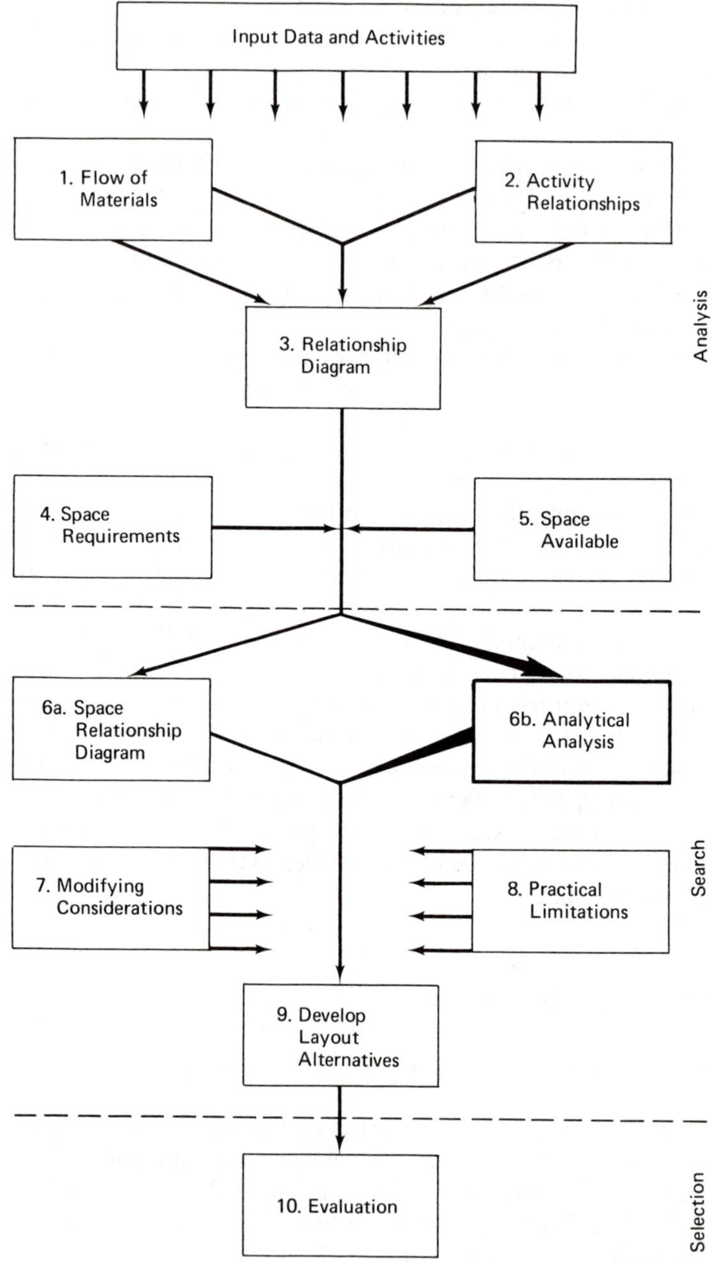

Figure 2.21. Modified SLP procedure incorporating analytic analysis.

4. Do you believe that analytical approaches can assist the layout designer in developing even better layout designs than are obtained when using the traditional approach?

The answers to the first two questions probably provide some measure of your familiarity with the real world of plant layout. However, it is your answers to the last two questions that concern us most. If your answer to the third question is No!, then you certainly wasted your time in reading this chapter. In fact, if your answer to the fourth question is also No!, then, unless you are willing to revise your belief, you will probably waste your time by reading further.

Now, for those of you who are still with us, here is where *we* are going from here. The remainder of this book is devoted to the presentation of analytic approaches that can be used to assist the designer in developing both overall and detailed layouts. We assume in our subsequent discussions that the preliminary steps of problem formulation and analysis have been performed. It is our opinion that traditional approaches have placed too much emphasis on intuition alone. Of course, depending on the scope of the problem, commonsense solutions are sometimes preferred to a more costly analysis. However, when very complex problems are encountered, analytic approaches can serve as very helpful aids to design. Furthermore, it has been our experience that an individual's intuition can be wrong. Also, what might pass for a *commonsense* solution may, in reality, be a *nonsense* solution when subjected to the test of analysis.

We view the use of analytical models in layout planning to be an activity that parallels the development of the space-relationship diagram. Just as modifying considerations and practical limitations influence the development of layout alternatives in traditional SLP, they are also necessary steps when using analytical approaches. Consequently, we recommend that the steps shown in Figure 2.21 be followed in *analyzing* the problem, *searching* for alternative layout designs, and *selecting* the preferred design. Furthermore, it is recommended that all the steps in the design cycle be followed.

REFERENCES

1. APPLE, J. M., *Material Handling Systems Design*, The Ronald Press Company, New York, 1973.
2. APPLE, J. M., *Plant Layout and Material Handling*, 2nd ed., The Ronald Press Company, New York, 1963.
3. BIEGEL, J. E., *Production Control: A Quantitative Approach*, 2nd ed., Prentice-Hall, Inc., Englewood Cliffs, N. J., 1971.

4. Bolz, H. A., and G. E. Hageman (editors), *Materials Handling Handbook*, The Ronald Press Company, New York, 1958.

5. Conway, R. W., W. L. Maxwell, and L. W. Miller, *Theory of Scheduling*, Addison-Wesley Publishing Company, Inc., Reading, Mass., 1967.

6. Cooper, R. B., *Introduction to Queueing Theory*, The Macmillan Company, New York, 1972.

7. Eilon, S., *Elements of Production Planning and Control*, The Macmillan Company, New York, 1962.

8. Elmaghraby, S. E., *The Design of Production Systems*, Van Nostrand Reinhold Company, New York, 1966.

9. Hadley, G., and T. M. Whitin, *Analysis of Inventory Systems*, Prentice-Hall, Inc., Englewood Cliffs, N.J., 1967.

10. Immer, J. R., *Material Handling*, McGraw-Hill Book Company, New York, 1953.

11. Ireson, W. G., *Factory Planning and Plant Layout*, Prentice-Hall, Inc., Englewood Cliffs, N.J., 1952.

12. Krick, E. V., *Methods Engineering*, John Wiley & Sons, Inc., New York, 1962.

13. Moder, J. J., and C. R. Phillips, *Project Management with CPM and PERT*, Van Nostrand Reinhold Company, New York, 1964.

14. Moore, J. M., *Plant Layout and Design*, The Macmillan Company, New York, 1962.

15. Morris, W. T., *Analysis of Management Decisions*, Richard D. Irwin, Inc., Homewood, Ill., 1964.

16. Muther, R., *Practical Plant Layout*, McGraw-Hill Book Company, New York, 1955.

17. Muther, R., *Systematic Layout Planning*, Industrial Education Institute, Boston, Mass., 1961.

18. Reed, R., *Plant Layout: Factors, Principles, and Techniques*, Richard D. Irwin, Inc., Homewood, Ill., 1961.

19. Reed, R., *Plant Location, Layout and Maintenance*, Richard D. Irwin, Inc., Homewood, Ill., 1967.

20. Schmidt, J. W., and R. E. Taylor, *Simulation and Analysis of Industrial Systems*, Richard D. Irwin, Homewood, Ill., 1970.

21. Wagner, H. M., *Principles of Operations Research*, Prentice-Hall, Inc., Englewood Cliffs, N.J., 1969.

22. White, J. A., "On Absorbing Markov Chains and Optimum Batch Production Quantities," *AIIE Transactions*, Vol. 2, No. 1, March 1970, pp. 82–88.

23. White, J. A., J. W. Schmidt, and G. K. Bennet, *Analysis of Queueing Systems*, Academic Press, New York, forthcoming.

PROBLEMS*

2.1. Briefly describe each of the following, giving their application to layout planning.

(a) Assembly chart
(b) Operation process chart
(c) Route sheet
(d) Flow process chart
(e) From–to chart
(f) Multiproduct process chart
(g) Activity relationship chart
(h) Flow diagram

2.2. Discuss the advantages and disadvantages of using standard times, rather than actual times, in making calculations in layout planning.

2.3. Discuss the use of analytical approaches in designing plant layouts. Contrast the use of analytical approaches with traditional approaches. Relate both the traditional approach and the analytical approach to the design process. Compare the use of iconic models when traditional approaches and analytical approaches are used.

2.4. Prepare assembly charts for the chair on which you are sitting and the lamp on your desk. (If you are standing and do not have either a desk or a lamp on your desk then you are exempt from this question.)

2.5. Prepare an assembly chart for one of the following products:

(a) Charcoal grill
(b) Bicycle
(c) Toy wagon

2.6. Compare the primary layout objectives for the following situations:

(a) Grocery store
(b) Drugstore
(c) Doctor's office
(d) Bank
(e) Sheriff's office
(f) Department store
(g) Warehouse
(h) Parking lot
(i) Post office
(j) Elementary school
(k) Movie theater
(l) Laundromat
(m) Beauty shop
(n) Automotive repair shop
(o) Restaurant
(p) Carry-out sandwich shop

2.7. Develop alternative layout designs for one or more of the situations listed in Problem 2.6.

2.8. Using the black-box approach, identify states *A* and *B* for a layout design problem encountered in one of the following situations:

(a) Golf course
(b) University campus planning
(c) Library
(d) Grocery store
(e) Bank
(f) Carry-out sandwich shop
(g) Warehouse
(h) Grocery store
(i) Meat-processing plant
(j) Printing shop

---

*Additional problems requiring SLP solutions are given at the end of Chapter 3.

2.9. An assembly chart for a simple pipe valve is given in Figure P2.9(a). Processing information for the pipe valve is given in Figure P2.9(b). The following scheduling information is available:

(1) All designing and planning for the production of the pipe valves has been completed.

(2) Patterns for the various cast parts can be obtained in 4 working days.

(3) Two days are required for the production of castings. Castings cannot be produced in the foundry until after receipt of the patterns.

(4) Various sizes of bar stock can be obtained from a local metal supply house 3 days after orders for stock have been placed.

(5) Nuts, screws, packing, etc., can be obtained in 3 days.

(6) The design engineers have specified that the body, bushing, and handle be made of cast bronze, and that the cap and stem of the valve be machined from standard brass bar stock.

(7) The bill of material indicates that the fiber packing and the nut are purchased parts.

(8) The machine shop estimates the following fabrication times for processing batches of 500 parts:

Body—3 days
Bushing—3 days
Stem—1 day
Cap—1 day
Handle—2 days

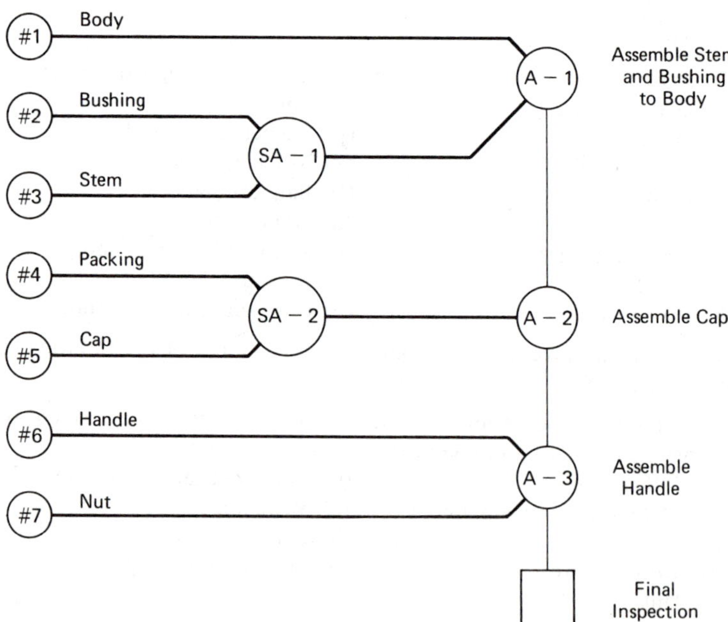

Figure P2.9(a)

Parts List

Parts	Part No.	Material
Body	001	Cast bronze
Bushing	002	Cast bronze
Stem	003	3/8 in. bar stock
Fiber packing	004	(Purchased part)
Cap	005	3/4 in. hex bar stock
Handle	006	Cast bronze
Nut	007	(Purchased part)

Assembly Operations

No.	Operation	Machine	Standard time (min/part)
1	Final assembly	Bench	2.00
2	Clean	Solvent tank	0.60
3	Inspect (pressure test)	Water test stand	1.20
4	Pack in boxes	Bench	0.12

Fabrication Operations

Part No.	Operation No.			Standard time (min/part)
001	001–1	Cast	Bench mold	1.50
	001–2	Clean	Tumble barrel	0.40
	001–3	Machine-thread and face three surfaces	Turret lathe	2.40
002	002–1	Cast	Bench mold	0.75
	002–2	Clean	Tumble barrel	0.20
	002–3	Machine all i.d. and o.d.	Turret lathe	1.00
003	003–1	Machine all surfaces and cut off	Automatic screw machine	0.30
005	005–1	Machine all surfaces	Automatic screw machine	0.12
006	006–1	Cast	Bench mold	1.20
	006–2	Clean	Tumble barrel	0.30
	006–3	Machine two surfaces	Turret lathe	0.75
	006–4	Broach square hole	Broach	0.75

Note: 100% inspection is required for all fabricated parts prior to delivery to final assembly.

Figure P2.9(b). (Based on a problem given in A. L. Roberts, *Production Management Workbook,* John Wiley & Sons, Inc., New York, 1962, p, 22.)

Construct an operation process chart, route sheets, and a Gantt chart for the production of the pipe valve. Develop a summary of equipment requirements for the production of 500 pipe valves per hour.

2.10. A highway construction firm is working on a cut-and-fill stretch of roadway for which dump trucks and power shovels are the primary equipment being used. Time studies reveal the following time values (minutes)

Time to load a truck	10
Travel time to dumping point	10
Dumping time	2
Return time	9

(a) What is the minimum number of trucks that will prevent idle time on the part of the power shovel?

(b) Suppose that it costs $20 per hour to operate the power shovel and $12 per hour to operate each truck. How many trucks should be assigned to the job to minimize the cost per truck load hauled?

(c) Construct a man–machine chart for the most economical arrangement between dump trucks and the power shovel.

2.11. Consider a toaster that toasts one side of each of two pieces of bread at the same time. It takes two hands to insert or remove each slice. To turn the slice over, it is necessary to push the toaster door all the way down and allow the spring to bring it back. Thus, both slices can be turned at the same time, but only one slice can be inserted or removed at one time. The time required to toast one side of a slice of bread is 0.50 minute. The time required to turn a slice over is 0.02 minute. It takes 0.05 minute to remove a toasted slice and place it on a plate. The time required to secure a piece of bread and place it in the toaster is 0.05 minute. Determine the minimum amount of time required to toast three slices of bread on both sides. Begin with three untoasted slices of bread on a plate and end with all three slices of bread toasted and placed on a plate. Illustrate your solution with a man–machine chart.

2.12. The Malt Shop specializes in milk shakes and malts. It takes 1 minute to place the required ingredients in the mixing cup and 0.5 minute to pour the completed shake or malt into a paper cup, place a lid on the cup, and place the cup in the freezer. Mixing time equals 3 minutes. How many mixers can be tended by one person without exceeding the 3-minute mixing time? Construct a man–machine chart for the assignment.

2.13. Ye Olde Fish and Chips Shop uses deep-fry baskets for cooking fish and chips. The time required to load and unload a basket is 3 minutes. Cooking time equals 8 minutes. The time required to separate a basket load into individual orders equals 2 minutes. Determine the number of basket loads that can be processed by one attendant without exceeding the cooking time. Construct a man–machine chart for the assignment.

2.14. Presently, one operator is tending five identical machines. Each machine is used to produce similar products. With five machines it has been observed that during a cycle a total of 20 machine-minutes is consumed in machines waiting for service to begin. In the past, it was observed that the operator was idle for 10 minutes each cycle when assigned to tend three machines. If machining time is 25 minutes and independent operator time is 1 minute per machine cycle, what is the value for concurrent activity? If C_1 equals $4 per hour and C_2 equals $6 per hour, what is the economic assignment? What is the value of the minimum cost per unit produced?

2.15. The Moonshiner's Mixing Company has several large mixing vats that must be tended periodically by an operator. The following man–machine chart,

Time	Operator	Mixing vat
0	Unload batch	Unloading
2	Read *Playboy*	Idle
4		
6		
8		
10	Load batch	Loading
12	Sample last batch	Run
14	Read *Playboy*	
16		
18		
20		
22		
24	Go for cup of coffee and return	
26		
28		Idle
30		

Figure P2.15

Figure P2.15, describes the present relationship between the operator and one mixing vat. Time values are given in minutes.

(a) What is the maximum number of vats that could be operated and have no idle time for the vats? Show your solution with a man–machine chart.

(b) What should be the average number of batches to be produced per 8 hours of repeating cycle operation? Assume there are 12 vats to be tended using the service combination obtained in part (a).

(c) How many idle minutes should occur per mixing vat if the man were assigned one more vat than that obtained in part (a)?

(d) If two vats are assigned per operator, how many idle man-minutes should occur per cycle? How many idle minutes should occur per vat per cycle?

(e) For five vats per operator, what should be the repeating cycle length?

2.16. An operator is currently operating two machines. Because of an increase in sales the company wishes to have the operator run the maximum number of machines without having machine idle time. The union opposes such a move, arguing that the operator should have idle time during a cycle. After considerable controversy, the company and union agree the operator will handle the maximum number of machines that will allow the operator to be idle at least 3 minutes per cycle. A time study is made, and the following standard times are determined for a cycle under the existing assignment of two machines:

Activity	Standard Time (min)
Unload, load, and start M-1	2.00
Inspect parts and package	0.80
Walk to M-2	0.20
Wait for M-2 to stop	3.50
Unload, load, and start M-2	2.00
Inspect part and package	0.80
Walk to M-1	0.20
Wait for M-1 to stop	3.50

(a) Based on the time-study data and the agreement between the company and the union, how many machines should be assigned the operator?

(b) What is the added cost per unit produced due to the union demand? Assume the operator costs $5 per hour and each machine costs $15 per hour.

2.17. With a man–machine chart show how one man can handle during a repeating cycle two machines of type A and one machine of type B using the following data.

Machine A		Machine B	
Activity	Time (min)	Activity	Time (min)
Load	1	Load	$1\frac{1}{2}$
Inspect	$\frac{1}{2}$	Inspect	0
Travel	$\frac{1}{2}$	Travel	$\frac{1}{2}$
Machining	7	Machining	$6\frac{1}{2}$
Unload	1	Unload	1

2.18. The Wonderful Widgit Manufacturing Company has an operator who currently operates four machines: A, B, C, and D. A time study has been made of the present assignment with the following results:

Operator's Activities	Standard Time (min)
Load M-A and start	2.00
Inspect part from M-A	0.20
Package part	0.20
Travel to M-B	0.10
Unload M-B	1.50
Load M-B and start	2.00
Inspect part from M-B	0.20
Package part	0.20
Travel to M-C	0.10
Unload M-C	1.50
Load M-C and start	2.00
Inspect part from M-C	0.20
Package part	0.20

Operator's Activities	Standard Time (min)
Travel to M-D	0.10
Unload M-D	1.50
Load M-D and start	2.00
Inspect part from M-D	0.20
Package part	0.20
Travel to M-A	0.10
Unload M-A	1.50

It has been observed that total machine idle time per cycle is 10 minutes. An attempt to secure cost data was not as successful as hoped. However, the following has been determined:

$$\$2 \leq \text{cost/hour for operator} \leq \$12$$

$$\$4 \leq \text{cost/hour for machine} \leq \$30$$

(a) What is the most economic number of machines based on total cost per unit produced?

(b) What would be the savings (cost) over the present assignment if the operator were assigned two machines with $C_1 = \$5$ per hour and $C_2 = \$20$ per hour?

2.19. The Lotta-Nuthin Construction Company is building a shopping center near White Pine, Tennessee. Concrete is being poured at the job site using wheelbarrows. The company wishes to have a mixing truck with a supply of concrete at the site at all times. Furthermore, they want to have sufficient laborers available with wheelbarrows to eliminate the possibility of the truck standing idle once the unloading begins. The following data are available:

Truck	Time (min)
Travel from supply to job site	30
Travel from job site to supply	30
Load truck at supply	10
Unload truck at job site	20

Laborer	Time (min)
Load wheelbarrow	1
Unload wheelbarrow	$\frac{1}{2}$
Travel from truck to pour	3
Travel from pour to truck	1

(a) Determine the minimum number of trucks required and the minimum number of laborers with wheelbarrows to meet the company objectives.

(b) How many minutes will a truck have to wait at the job site before the unloading begins?

(c) How many times, on the average, will a given laborer obtain concrete from any given truck?

2.20. (a) Determine the economic number of machines to assign an operator based on the following:

Activity	Time (min)
Load	4
Unload	3
Run (automatic)	46
Inspect and package	1.5
Travel	0.5

$$C_1 = \$10/\text{h}$$
$$C_2 = \$12/\text{h}$$

(b) What value of C_1/C_2 is the break-even value for assigning n or $n + 1$ machines?

(c) What are the respective idle minutes per cycle for each partner (man and machine) based on an assignment of one machine? eight machines?

2.21. The Grande Sombrero Company is considering the manufacture of freshman beanies. Filipe Methodo has been given the task of determining whether an operator should service one or two machines. He has determined that the work time for a single operator is

Load machine	0.15 min
Start machine	0.10 min
Unload machine	0.15 min

Filipe has also determined that machines may be placed in such a fashion that it will require 0.05 minute to walk between adjacent presses. The baking time required of each press is 1.50 minutes. The following costs are known:

Machine (in operation)	\$ .45/h
Machine (idle)	.30/h
Operator	1.80/h

(a) Construct man–machine charts for the operation with one and two machines. (Show a representative cycle on each chart.)

(b) Specify the number of machines each operator should service for minimum cost per piece. Give the minimum cost. Show all calculations.

2.22. The North Phiggins Manufacturing Company manufactures Phig-neutrons, which are specialized sheet-metal fixtures used for baking cookies to ensure that the baked cookies are the proper size and shape. The fixtures are supplied to the National Bakery Company for use in their kitchens. The process of manufacturing the fixtures is largely machine controlled, with a total time for stamping, punching, and rolling the sheet metal into a finished product of 1.40 minutes. The entire operation is done on individual fully automatic machines, requiring an operator only for loading, starting, and unloading the machines. The times required for the operator activities are

Load machine	0.20 min
Start machine	0.05 min
Unload machine	0.15 min

In addition, if an operator is to service more than one machine, 0.10 minute walking time is required to go from one machine to the next. The known costs of the operation are

Machine (in operation)	=	\$3.00/h
Machine (idle)	=	1.80/h
Man	=	2.40/h

(A machine is idle during the time an operator is loading, unloading, or starting it.)

(a) Determine the most economical number of machines an operator can service (based on minimum cost per piece). Specify this minimum cost showing all calculations.

(b) Construct a man–machine chart for the most economical method of servicing the machines.

(c) Find the labor cost for which the cost per piece using n or $n + 1$ machines is the same.

2.23. The Acme Chemical Company has a drag-line conveyor system that operates between two manufacturing buildings. Large hoppers are attached to the conveyor to transport chemical mixes from building A to building B for further batch processing. Empty hoppers are returned to building A on the same loop conveyor. Empty hoppers are cleaned automatically on the return trip to A. Activities and times for this process are

Activities	Average Time (min)
Unload empty hoppers from conveyor at A, fill and place on conveyor	40
Travel from A to B	10
Unload full hopper at B, empty, and place on conveyor	50
Travel from B to A (including cleaning)	15

To ensure that chemical mixes are always available at building B, what is the minimum number of hoppers required? If the chemical mix cannot stay in the hopper longer than 20 minutes from the time loading is completed and unloading begins, what is the maximum number of hoppers that could be used? Demonstrate graphically that your solution is feasible.

2.24. It takes 2 minutes to load and 2 minutes to unload a machine. Inspection, packaging, and travel between machines equal a total of 1 minute. The machine runs automatically 8 minutes. Operators cost \$4 per hour, each. Each machine costs \$6 per hour.

(a) What is the maximum number of machines one man can handle and not have machine idle time during a repeating cycle?

(b) What is the assignment that will minimize cost per unit produced?

(c) What is the cost per unit produced, based on a four-machine assignment?

(d) If the unload and load time had not been known, for what range of values for part (a) would the economic assignment be 2? 3?

(e) If operator cost is normally distributed with a mean of 4 and a variance of 2, and machine cost is normally distributed with a mean of 6 and a variance of 2, what is the probability that the economic assignment equals 2?

(f) If the company has seven identical machines, how should they be distributed among operators to minimize cost per unit produced?

2.25. An operator is currently operating three identical machines. The operator is utilized 78% of the repeating cycle. Concurrent activity equals 8 minutes and independent operator activity equals 5 minutes. Each operator costs $5 per hour and each machine costs $6 per hour.

(a) What is the cost per unit produced based on the minimum cost assignment?

(b) For what range of values for concurrent activity will the economic assignment equal three machines?

2.26. The KLM Job Shop has requested that a new layout be designed for their operation in Wilmot, Arkansas. There are 12 departments involved. The de-

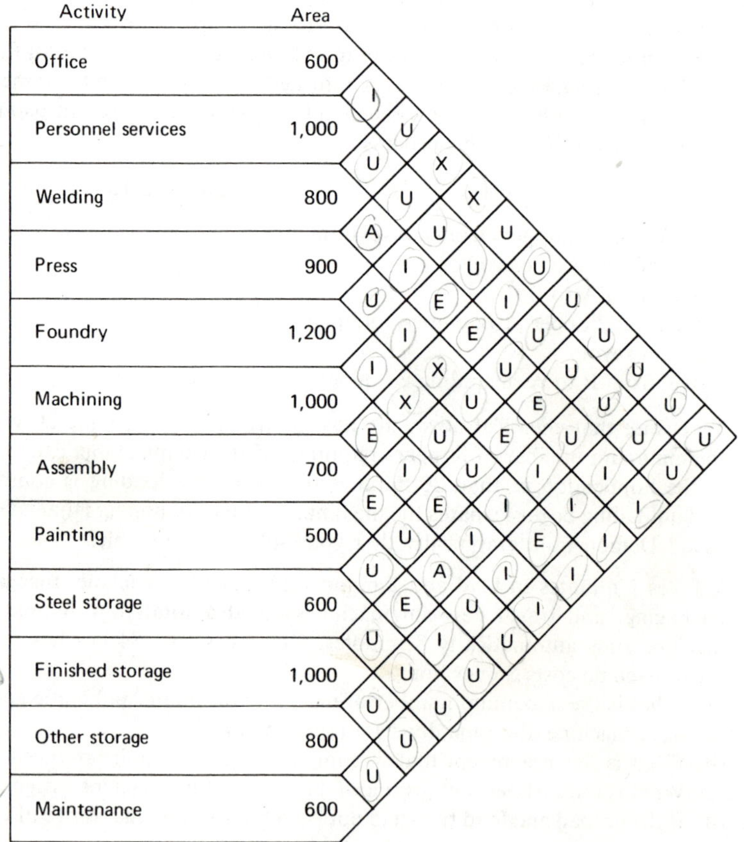

Activity
Area
Office
600
Personnel services
1,000
Welding
800
Press
900
Foundry
1,200
Machining
1,000
Assembly
700
Painting
500
Steel storage
600
Finished storage
1,000
Other storage
800
Maintenance
600

Figure P2.26

partment areas (in square feet) and activity relationships for the job shop are summarized in Figure P2.26. Design a block layout using the SLP approach.

2.27 The Multiple Products Company manufactures multiple products. Six departments are involved in the processing required for the products. A summary of the processing sequences required for the 10 major products and the monthly production volumes for the products is given in Figure P2.27, along with the departmental area.

Product	Processing sequence	Mo. prodn.
1	A B C D E F	800
2	A B C B E D C F	1,000
3	A B E F	600
4	A B C E B C F	2,000
5	A C E F	1,500
6	A B C D E F	400
7	A B D E C B F	2,000
8	A B C B D B E B F	2,500
9	A B C D F	800
10	A B D E F	1,000

Dept.	Area (ft^2)
A	1,000
B	1,200
C	800
D	1,500
E	2,500
F	1,500

Figure P2.27

(a) Develop the from–to chart giving number of units per month moving between combination of departments.

(b) Develop a layout design using SLP.

(c) Compute the total distance traveled per month based on the design obtained in part (b). Assume rectilinear travel between departmental centroids and assume moves between departments are made on the basis of production lots of 100 units per move.

2.28. The Edsel Auto Parts Warehouse in Bluefield, West Virginia, has requested that a new layout be designed for their main warehouse located in metropolitan Bluefield. The warehouse has 10 major activity "centers." The current building has the dimensions of 150 by 225 feet. Other pertinent data are summarized in Figure P2.28. Using SLP, design a layout to be contained in the current building.

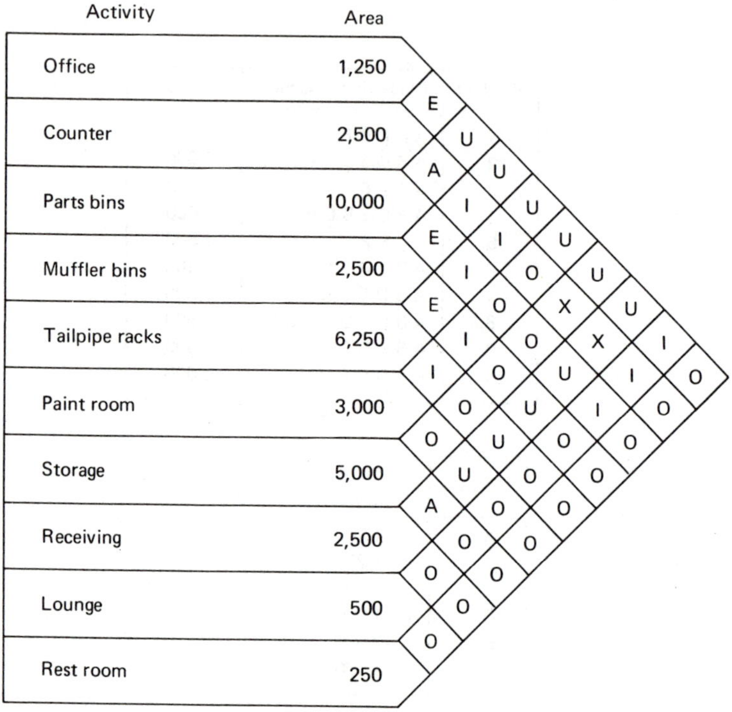

Activity	Area	Office	Counter	Parts bins	Muffler bins	Tailpipe racks	Paint room	Storage	Receiving	Lounge	Rest room
Office	1,250		E	U	U	U	U	U	U	I	O
Counter	2,500			A	I	I	O	X	X	I	O
Parts bins	10,000				E	I	O	O	U	I	O
Muffler bins	2,500					E	I	O	U	O	O
Tailpipe racks	6,250						I	O	U	O	O
Paint room	3,000							O	U	O	O
Storage	5,000								A	O	O
Receiving	2,500									O	O
Lounge	500										O
Rest room	250										

Figure P2.28

2.29. The Original Architectural Designs (TOAD) Company specializes in designing houses to meet a client's desires. A client is asked to specify the number, type, and size of rooms required and to assign closeness ratings to pairwise combinations of the rooms. Mr. and Mrs. Snob have supplied the data shown in Figure P2.29. Using SLP, obtain a relationship diagram, space-relationship diagram, and a block plan based on the REL chart. Hallways, closets, and baths will be added after meeting with the client a second time.

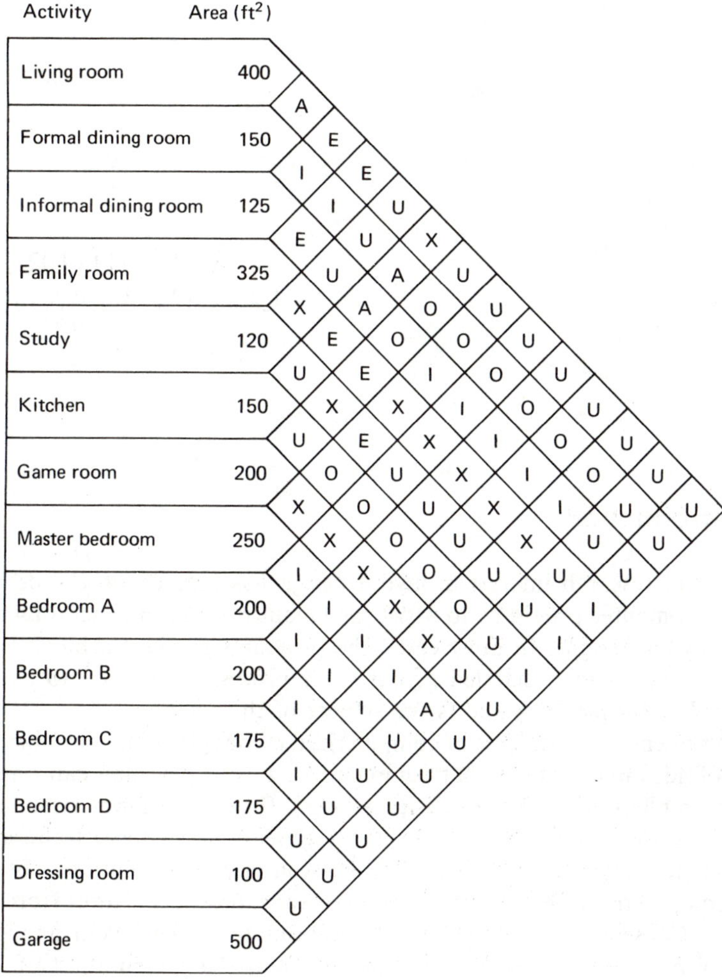

Figure P2.29

COMPUTERIZED LAYOUT PLANNING

3.1 Introduction

Recent research interest in layout design has focused on the development of computer programs to assist the layout planner in generating alternative layout designs. In most cases, the programs produce a block layout design, rather than a detailed layout. This process is referred to as *computerized layout planning* and is the subject of this chapter.

In presenting the subject of computerized layout planning, our objectives are twofold. First, it is felt that computerized layout planning can improve the search phase of the layout design process. By using computerized layout algorithms, the layout analyst can quickly generate a number of alternative solutions to a layout problem. As will be seen later, the solutions obtained from computerized algorithms often represent radical departures from convention. Consequently, computerized solutions force the layout analyst to consider new and provocative designs. Second, the presentation of computerized models provides a logical transition from the traditional plant layout approach to the analytical approaches in subsequent chapters.

If you are familiar with the use of computers, you no doubt recognize one of the hazards in describing computerized layout algorithms in a text. By the time you have obtained a copy of this text, any layout programs

3

we describe will probably have undergone at least one major revision. Furthermore, the number of such programs will probably have doubled or tripled. Consequently, some of the material we provide will be obsolete before it is ever published.

Computerized layout algorithms can be classified according to the way the final layout is generated. Specifically, some *construct* the layout by building up a solution "from scratch." *Construction* algorithms consist of the successive selection and placement of activities (departments) until a layout design is achieved. The second type of algorithm is of the *improvement* type. In this case a complete existing layout is required initially, and locations of activities (departments) are interchanged so as to *improve* the layout design. Admittedly, this classification scheme does not cover all computerized algorithms. For example, RUGR [11] is based upon the mathematics of graph theory. However, for the most part, a computerized layout algorithm can be classified as being either a construction or an improvement algorithm.

We shall describe two construction algorithms: CORELAP (Computerized Relationship Layout Planning) and ALDEP (Automated Layout Design Program). There are other construction algorithms available, such as

PLANET [1], RMA Comp I [17], LSP [28], and LAYOPT [13]. However, CORELAP and ALDEP are believed to be representative of the construction algorithms. Furthermore, copies of these programs are relatively easy to obtain. The only improvement algorithm we will describe is CRAFT (Computerized Relative Allocation of Facilities Technique). Although there are other improvement algorithms (see Chapter 8), in general none have been shown to be superior to CRAFT in *layout* design.

Both ALDEP and CORELAP are concerned with the construction of a layout based on the closeness ratings given by the REL chart described in Chapter 2. CRAFT is concerned with the minimization of a linear function of the movement between departments. Typically, CRAFT employs an improvement procedure to obtain a layout design based on the objective of minimizing material-handling cost. One of the inputs required for CRAFT is the from–to chart. Consequently, in terms of the modified SLP procedure outlined in Figure 2.21, CRAFT is used when the flow of materials is dominant; ALDEP and CORELAP are used when activity relationships are a major consideration and when constantly changing conditions prohibit the collection of precise numerical data.

3.2 Measurement Scales

As you might recall from the discussion in Chapter 2, the REL chart is normally based on qualitative considerations. We attempt to measure these qualitative aspects of activity relationships by making pairwise evaluations of the importance for two activities to be located close together in the layout. Closeness ratings of A, E, I, O, U, and X are assigned, where

A = *absolutely essential* for two activities to be located close together
E = *essential* for two activities to be located close together
I = *important* for two activities to be located close together
O = *ordinary* closeness of two activities is preferred
U = *unimportant* for two activities to be located close together
X = *undesirable* for two activities to be located close together

Basically there are two approaches that can be taken in assigning numerical values to closeness preferences. First, we can assign numerical values to the equivalence classes given by the A, E, I, O, U, and X ratings. In this case, we determine the numerical value of an A rating, the numerical value of an E rating, etc. The second approach consists of the assignment of a value to the closeness preference for each pairwise combination of activities being considered. Obviously, in a number of applications the first approach is much simpler, due to a fewer number of values to be assigned. However, it is unlikely that, say, an A rating between activities U and V and an A rating between activities X and Y really means each is equally preferred. Rather, there probably exists an internal ranking within a given closeness rating classification. If this is true, the second approach should be used.

Due to the definitions of the closeness ratings, it is clear that the ratings have the ranking

$$\mathrm{A} > \mathrm{E} > \mathrm{I} > \mathrm{O} > \mathrm{U} > \mathrm{X}$$

in terms of importance for two activities to be located close together. Since a ranking or ordinal scale is employed in measuring closeness, it is very important that the limitations associated with such a measurement scale be understood.

Three properties of numbers that are important for measurement are *identity*, *rank order*, and *additivity*. The properties are made explicit by the following nine axioms, which are based on those of Campbell [5], Siegal [22], and Stevens [23], and are presented by Hall [8].

1. Either $A = B$ or $A \neq B$.	identity
2. If $A = B$, then $B = A$.	
3. If $A = B$ and $B = C$, then $A = C$.	
4. If $A > B$, then $B \not> A$.	rank order
5. If $A > B$ and $B > C$, then $A > C$.	
6. If $A = P$ and $B > 0$, then $A + B > P$.	additivity
7. $A + B = B + A$.	
8. If $A = P$ and $B = Q$, then $A + B = P + Q$.	
9. $(A + B) + C = A + (B + C)$	

The axioms are used to distinguish four levels of measurement: *nominal*, *ordinal*, *interval*, and *ratio* scales. It is important that we understand the differences in these scales, since the higher the level of scale, the more statistical and mathematical operations can be performed using the numbers obtained from measurement.

3.2.1 *Nominal scales*

The logical basis for nominal scales is found in axioms 1, 2, and 3. Consequently, we see that the identity property is obtained when a nominal scale is constructed. As an illustration of numbers obtained from measurement by a nominal scale, consider the numbers assigned to football and basketball players. The numerals are simply a convenient means of distinguishing different players.

Nominal scales are basically qualitative. In our subsequent discussion we shall employ a nominal scale in identifying the various departments to be placed in the layout.

3.2.2 *Ordinal scales*

An ordinal scale is simply a ranking. Ordinal data have both identity and rank-order properties. As mentioned previously, an ordinal scale is employed in measuring preferences for departments to be located close together. Consequently, if the same closeness rating is assigned to two different combinations of departments, they are equally preferred.

In simple ordering scales each item must rank higher or lower than every other. However, in many practical cases ties are permitted, due to limitations on our ability to discriminate. The restriction of closeness ratings to one of six classes is based on the premise that we cannot be expected to discriminate accurately beyond six classifications.

The ordering axioms permit the calculation of frequencies, modes, medians, percentiles, and rank-order coefficients of correlation. Items on ordinal scales are not necessarily spaced equally along the scale. Consequently, arithmetical and all other statistical operations beyond those cited are not allowed.

In the case of closeness ratings, we have stated that

$$\mathrm{A} > \mathrm{E} > \mathrm{I} > \mathrm{O} > \mathrm{U} > \mathrm{X}$$

If we assign the set of numbers 6, 5, 4, 3, 2, 1, the order is preserved. Similarly, assigning the numbers 64, 16, 4, 1, 0, −1,024 also preserves the order.

The calculation of sums and averages that are to be used in comparing alternative layout designs has little meaning with ordinal data. As an example, suppose that we arbitrarily assign numerical values to the closeness ratings:

$$\begin{array}{lll} \mathrm{A} = 6, & \mathrm{I} = 4, & \mathrm{U} = 2 \\ \mathrm{E} = 5, & \mathrm{O} = 3, & \mathrm{X} = 1 \end{array}$$

It is not valid to state that having one A and one U is better than having one I and one O.

3.2.3 *Interval scales*

An interval scale has all the properties of an ordinal scale, plus the properties of having an arbitrary zero point and a constant unit of measurement. Temperature scales are examples of interval scales.

Interval scales do not achieve the important properties of additivity as defined by axioms 6 through 9. Consequently, none of the basic arithmetical operations can be applied. Interval-scale transformations must be linear and monotonic. Therefore, any interval scale and its transform, $y = ax + b$, $a \neq 0$, have the same descriptive accuracy.

In the case of measuring "closeness" preferences on an interval scale, we must be able to standardize a unit of preference. Whether this is possible remains an open and debated question.

3.2.4 *Ratio scales*

A ratio scale has an absolute zero point and a constant unit of measurement, as well as identity and rank-order properties. Consequently, all nine axioms apply in the case of a ratio scale. Weight, distance, and cost are measured using a ratio scale.

All the arithmetic and statistical operations are valid for ratio scales. The only transformation that leaves a ratio scale unchanged is $y = ax$, $a \neq 0$.

3.3 Preference Measurement

The problem of measuring preferences is depicted by Hall [8] as being a twofold problem. First, there exists the problem of devising a scale that will have the mathematical properties needed for the work it is to do. Second, we must show that the scale can be put into practice. We need to know not only how to measure individual preferences, but also how to aggregate the preferences of a number of individuals into a single measurement. According to Hall [8], neither problem has been satisfactorily solved. However, more progress has been made on measuring individual preferences than on aggregating them. We shall consider briefly one approach proposed for measuring individual preferences. Our discussion will continue to be based on the measurement of closeness preferences.

The classical work *Theory of Games and Economic Behavior* by Von Neumann and Morgenstern [26] established the foundation for much of the work currently being performed in the area of decision theory. Von Neumann and Morgenstern proposed an interval scale for measuring individual utilities or preferences. Their system was based on the following assumptions: an individual can order all outcomes based on his preferences and he can express preferences for combinations of chance outcomes.

To illustrate the Von Neumann–Morgenstern approach, consider the closeness ratings A, E, I, O, U, and X. We shall ask the individual (whose preferences are being measured) to answer a series of questions. To begin, we assume that the following combinations of departments have been assigned the indicated closeness ratings by the individual.

Departments	*Rating*
1, 2	A
1, 3	E
1, 4	I
2, 3	O
2, 4	U
3, 4	X

The following situation is posed. A layout is to be designed, but only two departments can be located adjacently in the layout due to physical restrictions imposed by a number of other departments, which are fixed in location. The layout can either be designed randomly or according to the specifications of the individual. If the layout is designed randomly, there is a probability of p that departments 1 and 2 will be located adjacently and a probability of $1 - p$ that departments 3 and 4 will be located adjacently. As an alternative, the layout can be designed so that two specified departments are located adjacently. The individual is asked to supply the value of p such that he would be indifferent between the randomly generated layout and the specified layout. He must supply a value of p for each combination of departments shown.

Notice that a rational decision maker would assign a value of 1 to p in order to be indifferent between the randomly generated layout and a layout which specifies that departments 1 and 2 will be located adjacently. Furthermore, a value of 0 would be assigned to p in order to be indifferent between the randomly generated layout and the layout that has departments 3 and 4 located adjacently. The values assigned to p based on the remaining combination of departments are dependent on the values of the individual's closeness preferences. As an illustration, suppose that the individual is asked to assign a value to p such that he is indifferent between the randomly generated layout and a specified layout having departments 1 and 3 located adjacently. The individual responds that he will be indifferent if p equals 0.95. When faced with the choice of the randomly generated layout and a specified layout having departments 1 and 4 located adjacently, he assigns a value of 0.90 to p. Continuing the process, the individual assigns values of 0.75 and 0.50 to the combinations of departments 2 and 3 and departments 2 and 4. Consequently, the values of the individual's closeness preferences are 1.00 (A), 0.95 (E), 0.90 (I), 0.75 (O), 0.50 (U), and 0.00 (X).

Notice that we have established an arbitrary zero point and a constant unit of measurement by assigning a value of 1.00 to an A rating and a value of 0.00 to an X rating. Consequently, we have obtained an interval scale. Since an interval scale is obtained, the values can be transformed accordingly; e. g. , letting $y = 5 + 20x$ be the transformation produces the values 25, 24, 23, 20, 15, and 5.

There are a number of limitations to the Von Neumann–Morgenstern approach that should be noted. First, a hypothetical situation is posed to the individual. His responses must be consistent with his assignment of closeness ratings in an actual layout design situation. Second, he must be able to think in terms of probabilities and mathematical expectations. Third, the hypothetical situation is posed in terms of closeness meaning adjacent locations. Such a definition of closeness might not agree with that used by the individual when he assigned closeness ratings.

The hypothetical situation posed is analogous to the "reference contract" or "standard gamble" employed to obtain an individual's utility function in decision theory. For further discussion of the approach in a decision-theory context, see Morris [16].

Another procedure developed for measuring individual preferences is the "approximate measure of value" technique due to Churchman and Ackoff [6]. Although some claim the procedure results in an interval scale, Hall [8] contends that the scale produced is "an ordinal scale with certain constraints placed on the distances between items." He states that "the constraints are not sufficient to guarantee an interval scale."

The point to be made by all of this concern over measurement scales is that one should not (1) arbitrarily limit the number of preference classes to six or (2) arbitrarily assign numerical values to the preference classes if an interval or ratio measurement scale is required. Furthermore, at present the best we can do is achieve an interval scale in measuring closeness preferences. There do not exist procedures that guarantee a ratio scale. Recall that arithmetic operations are only allowed with ratio scales. You might keep this in mind as we discuss ALDEP and CORELAP.

3.4 ALDEP[1]

ALDEP was developed within IBM and was originally presented by Seehof and Evans [19]. ALDEP is primarily a construction program. However, due to the evaluation process employed in accepting or rejecting a given layout, it can also be considered to be an improvement program. It constructs layouts without the need for an existing layout like construction

[1]For a copy of the ALDEP program, contact the IBM Corporation, Program Information Department, 40 Saw Mill River Road, Hawthorne, New York 10532, U.S.A. Ask for Program Order No. 360D-15.0.004.

layout programs, but it also compares solutions which are produced in a manner similar to that used in improvement programs.

Although variations of ALDEP have been reported, we shall describe the random-selection version. In brief, the random-selection version of ALDEP develops a layout design by randomly selecting a department and placing it in the layout. Next, the REL chart is scanned, and a department having a high closeness rating (e.g., A or E) is placed in the layout. This process is continued until either all departments are placed or no departments available for placement have a high closeness rating with departments already placed. In the latter situation, a department is randomly selected from among those departments available for placement, and it is placed in the layout. The selection process continues until all departments are placed in the layout. The score for the layout is determined by totaling for adjacent departments the numerical values assigned to the closeness ratings. The entire process is repeated a specified number of times.

The ALDEP program we are familiar with has fixed numerical values assigned to the closeness ratings. The values assigned are

$$\begin{aligned} A &= 4^3 = 64, \qquad & O &= 4^0 = 1 \\ E &= 4^2 = 16, \qquad & U &= 0 \\ I &= 4^1 = 4, \qquad & X &= -4^5 = -1{,}024 \end{aligned}$$

ALDEP has the capability of handling up to 63 departments or activities, and can generate multistory layouts up to three floors. Furthermore it is possible to place restrictions on the solution such that the layout is designed around such areas as aisles, elevator shafts, stairwells, lobbies, and existing departments.

The input requirements for ALDEP include

1. Length, width, and area requirements for each floor.
2. Scale of layout printout.
3. Number of departments in the layout.
4. Number of layouts to be generated.
5. Minimum allowable score for an acceptable layout.
6. Minimum department preference.
7. REL chart for the departments.
8. Location and size of restricted area for each floor.

Since the length, width, and area requirements for each floor must be specified, it is apparent that the building outline must be known. In the case of an existing facility within which the design must be placed, the building outline would coincide with that of the present facility. If the design is not constrained by an existing facility, some preliminary work is required in establishing the desired building outline.

The scale of the layout is partially determined by the maximum dimensions of the layout printout and the areas of the individual departments. ALDEP can handle layouts up to 30 by 50 in size. Suppose that we have a building outline that is 30 by 60 feet. The scale of the layout would have to be at least 1.2 feet per measurement unit. In this case we would probably choose a scale of 1.5 feet per measurement unit and specify a building outline of size 20 by 40. Of course, we also could have specified a scale of, say, 2.0 feet per measurement unit and obtained building dimensions of 15 by 30. Suppose that we also have department areas of 100, 200, 500, and 1,000 square feet. In this case, a scale of 2.0 feet per measurement unit would yield integer department areas of 25, 50, 125, and 250 square measurement units.

ALDEP scores the layouts it generates and compares the score against some minimum allowable score for the layout to be acceptable. Normally, on the first run the minimum allowable score is set equal to zero. The ALDEP program is designed to generate a specified number of layouts. Only those layouts having a score of at least zero are given in the printout. For the second run, the maximum layout score achieved on the first run is used as the minimum allowable score on the second run. This process is continued until no layouts are found that have scores at least as great as the minimum allowable score. (Since an X rating has a numerical value of −1,024 and the minimum allowable score cannot be negative, it is possible for there to be no acceptable layouts generated. This will occur when there are a large number of X ratings in the REL chart.)

Once a department is placed in the layout, all remaining departments are candidates for the next placement in the layout. If one of these departments has the desired *minimum department preference* (e.g., an E rating) with a department already located in the building, then it is placed in the layout; otherwise, a department is selected randomly. Typically, the minimum department preference is given as either an A or an E rating.

The REL chart is entered using the alphabetic rating system. The chart is entered as a triangular matrix, as shown in Figure 3.1 for a 10-department

111	S									
112	U	S								
113	U	U	S							
114	I	O	I	S						
115	U	O	U	I	S					
116	U	U	U	I	E	S				
117	U	U	U	O	U	U	S			
118	U	U	U	I	I	E	U	S		
119	U	I	A	I	U	U	A	E	S	
120	E	O	U	O	U	U	U	U	U	S

Figure 3.1. Relationship chart for example problem, entered as triangular matrix.

example. (A three-digit department number is required.) The letter S is entered on the main diagonal to represent the fact that it is the same department.

For every layout developed, ALDEP scores the layout based on the adjacency of departments. According to Seehof and Evans [19, p. 693], "The layout score is the summation of the preference value for adjacent departments. For each module (square) of the building, the preference value of the eight surrounding modules is added to the layout score. Then the preference value is set to zero so that it is included only once in the layout score." Since department adjacency is a very important factor in tabulating the total score, we find that ALDEP-designed layouts can have very irregularly shaped departments. Such configurations tend to be impractical from an operational viewpoint, as well as expensive to construct if walls are to be used to separate departments.

ALDEP is designed to avoid extreme zigzagging borders by using a vertical scan method of placing departments. Basically, the layout area is filled by using vertical strips having a specified width and a length equal to the depth of the layout. The method can be visualized by considering a roll of tape with a specified width. A length of tape is cut from the roll. The area of the tape corresponds to the floor space area of the department. The tape is placed on the layout. If the length of the tape exceeds the depth of the layout, the tape is cut and the remainder is placed alongside the piece of tape that was placed previously. The scanning pattern used is basically as shown in Figure 3.2. Despite this attempt to avoid unusually shaped borders, they can still occur, as we shall see later.

Two disadvantages of the ALDEP program are the method of scoring and the fixed numerical values assigned to the closeness ratings. The major attraction of ALDEP is the method of selecting entering departments. So long as its limitations are understood by the layout analyst, we feel that ALDEP can be of assistance in generating a large number and variety of layout solutions during the search phase of the design process.

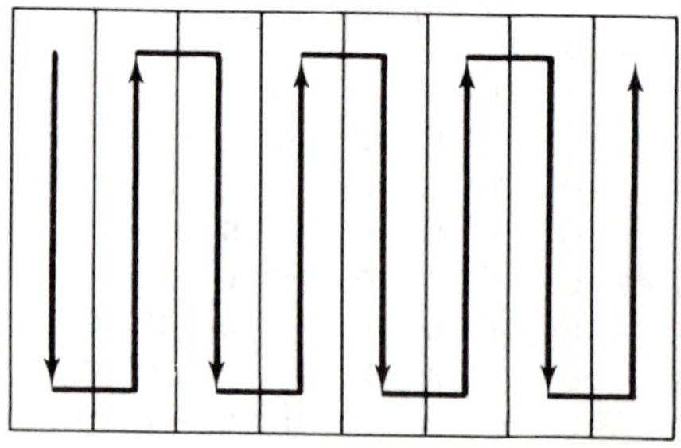

Figure 3.2. Vertical scanning pattern used by ALDEP.

To illustrate the use of ALDEP, consider the example problem involving 10 departments. The associated REL chart is given in Figure 3.1. The floor area required for each department is given in Table 3.1. A single-story building will have a total area of 5,120 square feet and will be twice as long as it is wide. The layout scale will be 10 square feet of floor area per unit square of output. Based on this layout scale, the number of unit squares required per department is also given in Table 3.1. It is desired to reserve approximately 220 square feet of floor space for expansion.

Table 3.1. DEPARTMENTAL FLOOR SPACE REQUIREMENTS FOR EXAMPLE PROBLEM

Department	*Required Area*	*No. Squares*
111	120.000	12
112	340.000	34
113	410.000	41
114	130.000	13
115	60.000	6
116	570.000	57
117	170.000	17
118	450.000	45
119	1,400.000	140
120	1,250.000	125
Departments available for random placement = 10		

We stipulate a minimum department preference of an A rating, and initially stipulate a minimum allowable score of zero. Twenty layouts are generated. Of these, all 20 are acceptable, i. e., have a score of at least zero. Figure 3.3 is a sample layout that is generated. It has a score of 414.

The score of 414 is obtained by noting the following combinations of adjacent departments: (10-3), (10-4), (10-2), (10-1), (1-9), (1-2), (2-9), (2-3), (2-4), (4-3), (3-9), (9-6), (9-8), (9-7), (7-8), (8-5), (8-6), and (6-5). From the REL chart in Figure 3.1, the closeness ratings associated with the combinations of adjacent departments are U, O, O, E, U, U, I, U, O, I, A, U, E, A, U, I, E, and E, respectively. Summing the numerical values of the closeness ratings gives a score of 207. Since the ALDEP program treats $(X - Y)$ and $(Y - X)$ as different adjacent combinations, 207 is multiplied by 2 to obtain the score of 414.

The maximum score obtained on the first run is 414. For the second run a minimum allowable score of 414 is stipulated, a new random number seed is supplied, and 20 additional layouts are generated. In the second run, only one acceptable layout is generated. The layout is shown in Figure 3.4 and has a total score of 420. For the third run a minimum allowable score of 420 is stipulated, a new random number seed is supplied, and 20 more layouts

TRIAL LAYOUT 11A SCORE = 414

TOP FLOOR

0	0	0	0	0	0	0	0	0	0	0	0	0	0	0	0	0	0
0	10	10	1	1	1	1	9	9	9	9	8	8	8	8	0	0	0
0	10	10	10	1	1	1	9	9	9	9	7	7	8	8	0	0	0
0	10	10	10	10	1	1	9	9	9	9	7	7	8	8	0	0	0
0	10	10	10	10	1	1	9	9	9	9	7	7	8	8	0	0	0
0	10	10	10	10	1	2	9	9	9	9	7	7	8	8	0	0	0
0	10	10	10	10	2	2	9	9	9	9	7	7	8	8	0	0	0
0	10	10	10	10	2	2	9	9	9	9	7	7	8	8	0	0	0
0	10	10	10	10	2	2	9	9	9	9	7	7	8	8	0	0	0
0	10	10	10	10	2	2	9	9	9	9	7	7	8	8	0	0	0
0	10	10	10	10	2	2	9	9	9	9	9	7	8	8	0	0	0
0	10	10	10	10	2	2	9	9	9	9	9	9	8	8	0	0	0
0	10	10	10	10	2	2	9	9	9	9	9	9	8	8	5	5	0
0	10	10	10	10	2	2	9	9	9	9	9	9	8	8	5	5	0
0	10	10	10	10	2	2	9	9	9	9	9	9	8	8	5	5	0
0	10	10	10	10	2	2	9	9	9	9	9	9	8	8	6	6	0
0	10	10	10	10	2	2	3	9	9	9	9	9	8	8	6	6	0
0	10	10	10	10	2	2	3	3	9	9	9	9	8	8	6	6	0
0	10	10	10	10	2	2	3	3	9	9	9	9	8	8	6	6	0
0	10	10	10	10	2	2	3	3	9	9	9	9	8	8	6	6	0
0	10	10	10	10	2	2	3	3	9	9	9	9	8	8	6	6	0
0	10	10	10	10	2	2	3	3	9	9	9	9	8	8	6	6	0
0	10	10	10	10	4	2	3	3	9	9	9	9	6	8	6	6	0
0	10	10	10	10	4	4	3	3	9	9	9	9	6	6	6	6	0
0	10	10	10	10	4	4	3	3	9	9	9	9	6	6	6	6	0
0	10	10	10	10	4	4	3	3	9	9	9	9	6	6	6	6	0
0	10	10	10	10	4	4	3	3	9	9	9	9	6	6	6	6	0
0	10	10	10	10	4	4	3	3	9	9	9	9	6	6	6	6	0
0	10	10	10	10	4	4	3	3	9	9	9	9	6	6	6	6	0
0	10	10	10	10	3	3	3	3	9	9	9	9	6	6	6	6	0
0	10	10	10	10	3	3	3	3	9	9	9	9	6	6	6	6	0
0	10	10	10	10	3	3	3	3	9	9	9	9	6	6	6	6	0
0	10	10	10	10	3	3	3	3	9	9	9	9	6	6	6	6	0
0	0	0	0	0	0	0	0	0	0	0	0	0	0	0	0	0	0

GROUND FLOOR

Figure 3.3. Sample ALDEP layout for example problem.

are generated. In the third run, two layouts having scores of 420 and 422 are generated. The layout having a score of 422 is shown in Figure 3.5. On a subsequent run no layouts were obtained having a score of 422 or more. Consequently, the search for layouts was terminated.

It is strongly emphasized that Figure 3.5 might be better or worse than the remaining 21 acceptable designs generated. The evaluation of these designs should be performed separately. Due to the questions we have raised previously concerning measurement scales and ALDEP's method of scoring, we do not believe the total score given by ALDEP should be used as the only basis for evaluating layouts. Instead, we recommend that you examine

TRIAL LAYOUT 16B SCORE = 420

TOP FLOOR

0	0	0	0	0	0	0	0	0	0	0	0	0	0	0	0	0	0
0	3	3	9	9	9	9	8	8	6	6	10	10	10	10	0	0	0
0	3	3	9	9	9	9	8	8	6	6	10	10	10	10	0	0	0
0	3	3	9	9	9	9	8	8	6	6	10	10	10	10	0	0	0
0	3	3	9	9	9	9	8	8	6	6	10	10	10	10	0	0	0
0	3	3	9	9	9	9	8	8	6	6	10	10	10	10	0	0	0
0	3	3	9	9	9	9	8	8	6	6	10	10	10	10	0	0	0
0	3	3	9	9	9	9	8	8	6	6	10	10	10	10	0	0	0
0	3	3	9	9	9	9	8	8	6	6	10	10	10	10	0	0	0
0	3	3	9	9	9	9	8	8	6	6	10	10	10	10	0	0	0
0	3	3	9	9	9	9	8	8	6	6	10	10	10	10	0	0	0
0	3	3	9	9	9	9	8	8	6	6	10	10	10	10	0	0	0
0	3	3	9	9	9	9	8	8	6	6	10	10	10	10	2	2	0
0	3	3	9	9	9	9	8	8	6	6	10	10	10	10	2	2	0
0	3	3	9	9	9	9	8	8	6	6	10	10	10	10	2	2	0
0	3	3	9	9	9	9	8	8	6	6	10	10	10	10	2	2	0
0	3	3	9	9	9	9	8	8	6	6	10	10	10	10	2	2	0
0	3	3	9	9	9	9	8	8	6	6	10	10	10	10	2	2	0
0	3	3	9	9	9	9	8	8	6	6	10	10	10	10	2	2	0
0	3	3	9	9	9	9	8	8	6	6	10	10	10	10	2	2	0
0	3	3	9	9	9	9	8	8	6	6	10	10	10	10	2	2	0
0	3	9	9	9	9	9	8	8	6	6	10	10	10	10	2	2	0
0	9	9	9	9	9	9	8	8	6	6	10	10	10	10	2	2	0
0	9	9	9	9	9	9	8	4	6	6	10	10	10	10	2	2	0
0	9	9	9	9	9	9	4	4	6	6	10	10	10	10	2	2	0
0	9	9	9	9	9	9	4	4	6	6	10	10	10	10	2	2	0
0	9	9	9	9	9	9	4	4	6	6	10	10	10	10	2	2	0
0	9	9	9	9	9	7	4	4	6	6	10	1	10	10	2	2	0
0	9	9	9	9	7	7	4	4	6	6	1	1	10	10	2	2	0
0	9	9	9	9	7	7	4	4	6	5	1	1	10	10	10	10	0
0	9	9	9	9	7	7	7	7	5	5	1	1	10	10	10	10	0
0	9	9	9	9	7	7	7	7	5	5	1	1	10	10	10	10	0
0	9	9	9	9	7	7	7	7	5	1	1	1	10	10	10	10	0
0	0	0	0	0	0	0	0	0	0	0	0	0	0	0	0	0	0

GROUND FLOOR

Figure 3.4. Sample ALDEP layout for example problem.

several of the layouts generated and employ one of the evaluation procedures described in Chapter 2.

3.5 CORELAP

CORELAP was originally presented by Lee and Moore [12], and was subsequently improved by Sepponen [20, 21]. A time-shared version of CORELAP, called Interactive CORELAP, has been developed by Moore [14]. We shall describe CORELAP 8 and Interactive CORELAP. Due to

TRIAL LAYOUT

15C SCORE = 422

0	0	0	0	0	0	0	0	0	0	0	0	0	0	0	0	0	0
0	7	7	9	9	9	9	6	6	6	6	10	10	10	10	0	0	0
0	7	7	9	9	9	9	6	6	6	6	10	10	10	10	0	0	0
0	7	7	9	9	9	9	6	6	6	6	10	10	10	10	0	0	0
0	7	7	9	9	9	9	6	6	6	6	10	10	10	10	0	0	0
0	7	7	9	9	9	9	6	6	6	6	10	10	10	10	0	0	0
0	7	7	9	9	9	9	6	6	6	6	10	10	10	10	0	0	0
0	7	7	9	9	9	9	6	8	6	6	10	10	10	10	0	0	0
0	7	7	9	9	9	9	8	8	6	6	10	10	10	10	0	0	0
0	7	9	9	9	9	9	8	8	6	6	10	10	10	10	0	0	0
0	9	9	9	9	9	9	8	8	6	6	1	10	10	10	0	0	0
0	9	9	9	9	9	9	8	8	6	6	1	1	10	10	0	0	0
0	9	9	9	9	9	9	8	8	6	6	1	1	10	10	10	10	0
0	9	9	9	9	9	9	8	8	6	6	1	1	10	10	10	10	0
0	9	9	9	9	9	9	8	8	6	6	1	1	10	10	10	10	0
0	9	9	9	9	9	3	8	8	6	6	1	1	10	10	10	10	0
0	9	9	9	9	3	3	8	8	6	6	2	1	10	10	10	10	0
0	9	9	9	9	3	3	8	8	6	6	2	2	10	10	10	10	0
0	9	9	9	9	3	3	8	8	6	6	2	2	10	10	10	10	0
0	9	9	9	9	3	3	8	8	6	6	2	2	10	10	10	10	0
0	9	9	9	9	3	3	8	8	6	6	2	2	10	10	10	10	0
0	9	9	9	9	3	3	8	8	6	6	2	2	10	10	10	10	0
0	9	9	9	9	3	3	8	8	6	6	2	2	10	10	10	10	0
0	9	9	9	9	3	3	8	8	5	5	2	2	10	10	10	10	0
0	9	9	9	9	3	3	8	8	5	5	2	2	10	10	10	10	0
0	9	9	9	9	3	3	8	8	5	5	2	2	10	10	10	10	0
0	9	9	9	9	3	3	8	8	4	4	2	2	10	10	10	10	0
0	9	9	9	9	3	3	8	8	4	4	2	2	10	10	10	10	0
0	9	9	9	9	3	3	8	8	4	4	2	2	10	10	10	10	0
0	9	9	9	9	3	3	8	8	4	4	2	2	10	10	10	10	0
0	9	9	9	9	3	3	3	3	4	4	2	2	10	10	10	10	0
0	9	9	9	9	3	3	3	3	4	4	2	2	10	10	10	10	0
0	9	9	9	9	3	3	3	3	4	2	2	2	10	10	10	10	0
0	0	0	0	0	0	0	0	0	0	0	0	0	0	0	0	0	0

GROUND FLOOR

Figure 3.5. Sample ALDEP layout for example problem.

the degree of interest that exists concerning CORELAP, it is quite likely that further improvements will be made in both CORELAP and Interactive CORELAP before this discussion is published.

3.5.1 *CORELAP 8*[2]

CORELAP 8 is a construction program. Like ALDEP, CORELAP employs the REL chart in constructing layouts. Up to 70 departments can be handled, and the scale of the layout is limited by the maximum dimension allowable for the final layout, 40 by 40.

A building outline is not required for CORELAP. Additionally, it is possible to place a constraint on the length-to-width ratio of the final layout. Another feature of CORELAP is the ability to preassign some departments in the layout. However, this can be done only along the periphery of the layout.

We shall briefly describe the process used by CORELAP to construct layouts. The brevity of our discussion is motivated by the fact that a detailed discussion of CORELAP is provided in the "CORELAP User's Manual" [20]. Also, you are probably more interested in what CORELAP does than in how and why it does it. Furthermore, there is that obsolescence issue we cited previously.

The minimum input requirements for CORELAP include

1. Relationship chart for the departments.
2. Number of departments.
3. Area of each department.
4. Weights for REL chart entries.

Optional input parameters for CORELAP include

1. Scale of output printout.
2. Building length to width ratio.
3. Punch or CALCOMP plotting of the final layout.
4. Department preassignment.

CORELAP constructs layouts by locating rectangular-shaped departments when the departmental area and layout scale permit a rectangular representation of the departmental area. The REL chart provides the basis for the order in which departments enter the layout. The placement of the department within the layout is based on both the REL chart and the numerical weighted rating assigned to the closeness ratings.

---

[2]Hereafter, we shall refer to CORELAP 8 as CORELAP. For a copy of the program contact Engineering Management Associates, Room 590, United Realty Building, 360 Huntington Avenue, Boston, Mass. 02115, U.S.A.

In CORELAP the following numerical values are assigned to the closeness ratings:

$$A = 6, \quad E = 5, \quad I = 4, \quad O = 3, \quad U = 2, \quad X = 1$$

These values are used to calculate a total closeness rating for each department. If we let $V(r_{ij})$ be the numerical value assigned to the closeness rating for departments i and j, and there are m departments, then the total closeness rating (TCR) for department i is defined as

$$TCR_i = \sum_{j=1}^{m} V(r_{ij})$$

where $V(r_{ii})$ equals zero.

Assuming no preassigned departments, CORELAP chooses that department having the greatest TCR as the first department to be placed in the layout. For the sake of discussion, suppose that department 1 is the first department placed in the layout. Next, the REL chart is scanned to see if any department has an A rating with department 1. If so, it is placed in the layout next to department 1. If no A ratings exist, a check is made for E ratings with department 1. If no E ratings are found, a check is made for I ratings, followed by O ratings. If ties develop, the department having the largest TCR is chosen. If no department can be found that has at least an O rating with department 1, the unassigned department with the greatest TCR is chosen.

At this point two departments have been placed in the layout. Suppose that the second department placed in the layout is department 2. The REL chart is scanned to see if there exists an unassigned department with an A rating with department 1. If one exists, it is placed in the layout next to department 1. If one does not exist, we look for an unassigned department having an A rating with department 2. If one does not exist, we look for an unassigned department with an E rating with department 1. If one is found, it is placed in the layout; otherwise, we try to place in the layout a department having an E rating with department 2. This process continues until either a department is placed in the layout or no department is found that has a U rating with either department 1 or 2. In the latter case, the unassigned department having the greatest TCR is placed in the layout.

Now, we have placed three departments in the layout. The same search process is continually used to select entering departments until finally all departments are placed in the layout. As with ALDEP, departments are placed in the layout one at a time. However, whereas CORELAP uses the TCR to select a department for placement in the layout, ALDEP makes a random selection.

ALDEP and CORELAP also differ in the way departments are located in the layout. ALDEP uses a vertical scan routine and places the departments in the layout in a manner analogous to the placement of strips of adhesive tape. CORELAP evaluates a number of possible locations for the rectangular-shaped department, as well as a number of different rectangular shapes for the department. The evaluation phase employs a *placing rating* and a *boundary length.*

The *placing rating* is the sum of the *weighted ratings* between the new department to be placed in the layout and its neighbors in the layout. A neighbor is defined to be any department already in the layout that has a common border with the new department. The weighted rating values are specified by the user of the program. *Boundary length* refers to the length of the boundaries common to the new department and all departments already in the layout.

As an illustration of the calculation of the placing rating and the boundary length, consider Figure 3.6. Departments 1, 2, and 3 are already located in the layout, and department 4 has been selected for placement in the layout. Assume weighted rating values have been assigned as follows: A = 600, E = 200, I = 50, O = 10, U = 0, X = −200. Furthermore, suppose that the closeness ratings between department 4 and departments 1, 2, and 3 are

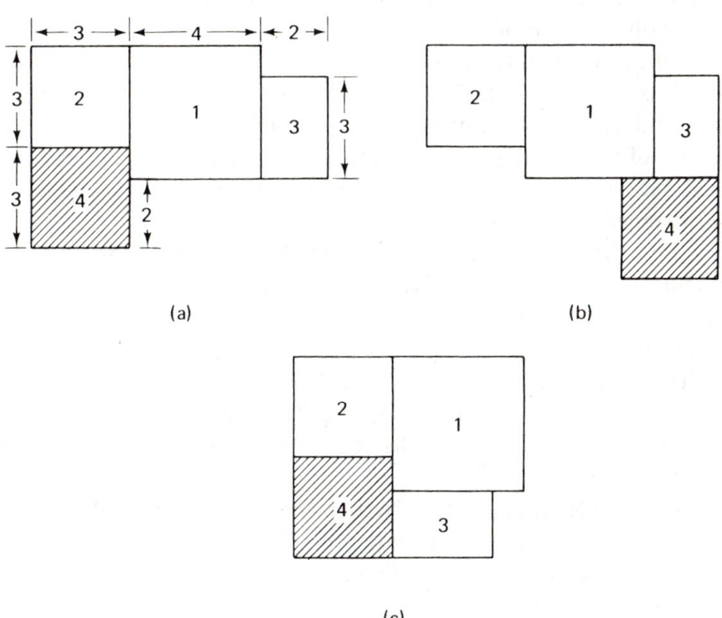

Figure 3.6. CORELAP's placement method.

A, E, and I, respectively. Using the placement given in Figure 3.6(a), the placing rating is 800. The placement given in Figure 3.6(b) has a placing rating of 650. Were it not for the fact that department 3 is already located in the layout, we could obtain a placing rating of 850 with the placement given in Figure 3.6(c). The boundary lengths are 4, 3, and 6 for placements (a), (b), and (c), respectively, in Figure 3.6.

The values of the weighted ratings are assigned by the user. Notice that these values are treated as though they were obtained from a ratio measurement scale. As pointed out earlier, we have no guarantee that the weighted ratings have the properties of ratio scales. Furthermore, we do not have the flexibility of assigning the numerical values to the closeness ratings used in calculating the total closeness rating (*TCR*). Since the value of *TCR* influences the order of entry into the layout for the departments, and arithmetic operations are used in calculating *TCR*, it would be very desirable to have the flexibility of measuring the closeness ratings on a ratio scale.

CORELAP produces a single final layout using a deterministic approach. Consequently, running CORELAP a second time with the same data will produce the same final layout. To obtain a number of different layouts for a subsequent, independent evaluation, it is necessary to change either the entries in the REL chart, the weighted rating values, the values of the departmental areas, the layout scale, the value of the length-to-width ratio, or a combination of these.

As an illustration of the use of CORELAP, consider the following problem, which is given in [14]. There are 10 departments to be located with the areas and closeness ratings given in Figure 3.7. A layout scale of 10 feet per side of the square is chosen to represent the smallest element of area. Weighted rating values of 243 (A), 81 (E), 27 (I), 9 (O), 1 (U), and −729 (X) are assigned.

Based on the data for the problem, *TCR* values are computed and the departments ordered accordingly. Likewise, the integer number of unit squares to be used to represent each department is calculated, as well as the widths and lengths of the rectangles that will be used in locating each department. These values are given in Figure 3.8. Also given in Figure 3.8 is a rearranged REL chart. The columns of the REL chart are ordered according to the *TCR* values. The rows of the REL chart are arranged according to department number.

Based on the REL chart, the departments enter in the order 19, 13, 17, 18, 16, 15, 14, 12, 11, and 20 to give the final layout shown in Figure 3.9(a). If we are constrained by an existing facility and want to ensure that the building length is no greater than, say, twice the width, the CORELAP solution will be as given in Figure 3.9(b). If we change the layout scale to 15 feet per side of a square, the resulting layout is given in Figure 3.9(c) and can be

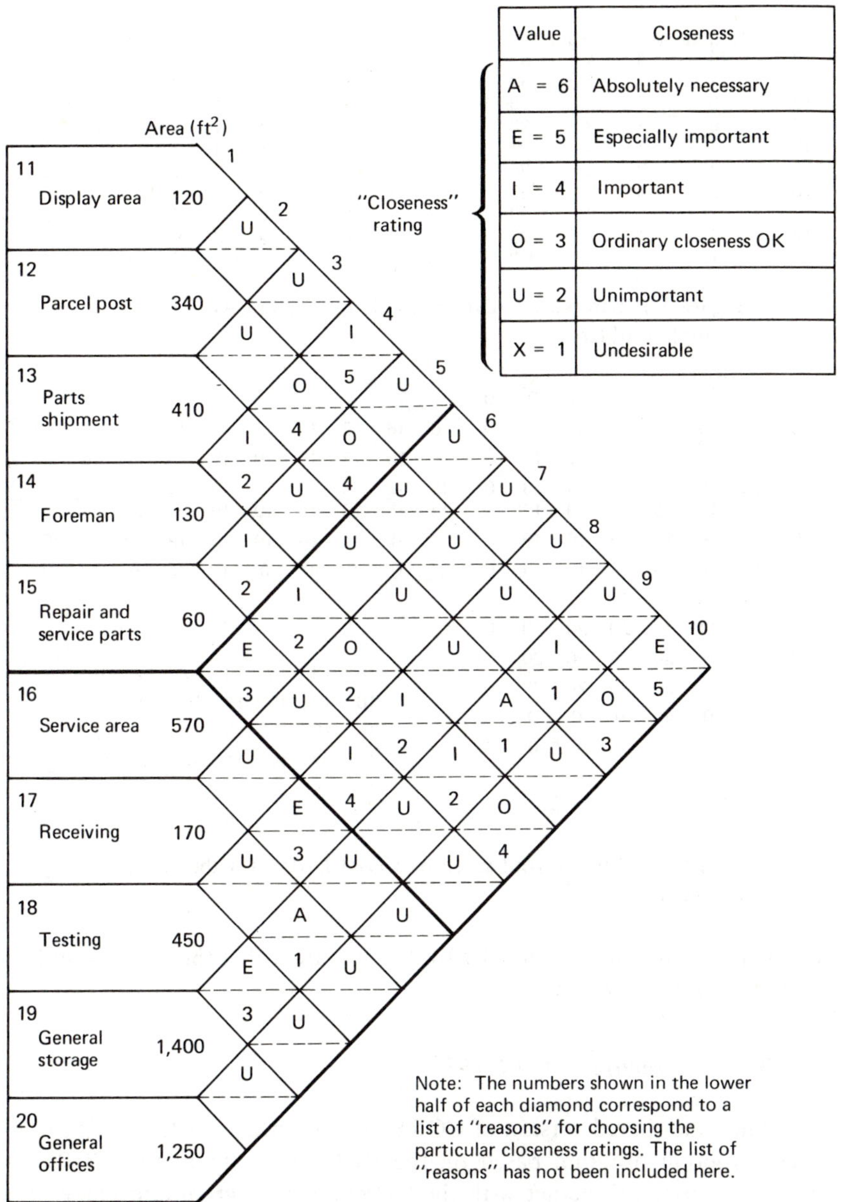

Figure 3.7. Relationship chart for CORELAP example problem. (After *Industrial Engineering*, Aug. 1971. Copyright American Institute of Industrial Engineers, Inc., 25 Technology Park/Atlanta, Norcross, Ga. 30071.)

	No. units	Width	Length	TCR
19	14	4	4	33
14	1	1	1	33
18	5	2	3	28
16	6	2	3	26
15	1	1	1	26
13	4	2	2	24
20	13	4	4	23
12	3	1	3	23
17	2	1	2	23
11	1	1	1	23

No.	19	14	18	16	15	13	20	12	17	11
11	2	4	2	2	2	2	5	2	2	0
12	4	3	2	2	3	2	3	0	2	2
13	6	4	2	2	2	0	2	2	2	2
14	4	0	4	4	4	4	3	3	3	4
15	2	4	4	5	0	2	2	3	2	2
16	2	4	5	0	5	2	2	2	2	2
17	6	3	2	2	2	2	2	2	0	2
18	5	4	0	5	4	2	2	2	2	2
19	0	4	5	2	2	6	2	4	6	2
20	2	3	2	2	2	2	0	3	2	5

Figure 3.8. Sample calculations made by CORELAP 8 for the example problem.

(a)

0	0	0	0	0	0	0	0
0	0	18	13	13	0	0	0
0	18	18	13	13	17	17	0
0	18	18	19	19	19	19	0
0	16	16	19	19	19	19	0
0	16	16	19	19	19	19	0
0	16	16	19	19	19	19	0
0	15	14	12	12	12	0	0
0	0	11	20	20	20	20	0
0	0	0	20	20	20	20	0
0	0	0	20	20	20	20	0
0	0	0	20	0	0	0	0
0	0	0	0	0	0	0	0

(b)

16	16	13	13	17	17	12	12	12	0	0 0
16	16	13	13	19	19	19	19	20	20	20
16	16	18	18	19	19	19	19	20	20	20
14	15	18	18	19	19	19	19	20	20	20
11	0	0	18	19	19	19	19	20	20	20
0	0	0	0	0	0	0	0	20	0	0

(c)

16	18	13	13	17	20	20
16	18	19	19	19	20	20
16	15	19	19	19	20	20
14	12	12	11	0	0	0

Figure 3.9. Final layout given by CORELAP 8 for the example problem.

compared with that in Figure 3.9(b). It is apparent that the layout solution is sensitive to the scale used.

3.5.2 *Interactive CORELAP*[3]

The time-shared version of CORELAP, Interactive CORELAP, is also a construction program based on the REL chart. Interactive CORELAP allows the user to interact with the CORELAP program and make modifications to the solution. The interaction can take place during intermediate stages of the layout development.

---

[3]Information on Interactive CORELAP is available from Dr. James Moore, Room 302, Whittemore Hall, Virginia Tech, Blacksburg, Va. 24061, U.S.A.

Interactive CORELAP employs the original version of CORELAP, rather than CORELAP 8, in developing the layout. Since the original version did not attempt to place departments in rectangular configurations, the resulting department configurations can be very irregular. Also, the original version of CORELAP did not use the weighted rating values to assist in the selection of the location for the departments. Consequently, unusual building outlines might be generated. However, with the interaction mode the user can rearrange the department and building configurations to obtain more realistic solutions.

Interactive CORELAP has a number of other significant differences from the batch-processing version, CORELAP 8. According to Moore [14], Interactive CORELAP

1. Handles both new plant and existing plant problems.
2. Permits fixed department locations at positions other than the periphery of the layout.
3. Scores any layout alternative.
4. Simplifies modifying adjustments.

Since Interactive CORELAP is a construction program, no initial layout is required. Consequently, the solution obtained is not constrained by an existing building configuration. However, when an existing structure is present, the user can interact with the computer and "massage" the solution until it fits in the existing structure.

Interactive CORELAP fixes department locations by inserting the department in the layout during the execution of the program. Once a department is added into the layout, the computer program will not cause it to be moved. Only the user can move the department.

A feature of Interactive CORELAP is its ability to score any layout at any time during the program execution. The scoring takes place as follows: the shortest rectilinear distances between the borders of all pairs of departments are multiplied by the numerical values of the closeness ratings between the departments, and a total score is computed. As an illustration of the scoring technique, consider the layout given in Figure 3.10(a), with its associated REL chart given in Figure 3.10(b). The shortest rectilinear distances between the borders of all pairs of departments are given in Figure 3.10(c). Interactive CORELAP assigns numerical values of 6, 5, 4, 3, 2, and 1 to A, E, I, O, U, and X ratings, respectively. Consequently, a calculation establishes that the total score for the layout shown is 95.

Three aspects of the scoring technique should be noted:

1. Shortest rectilinear distances to departmental borders are used as a measure of closeness.

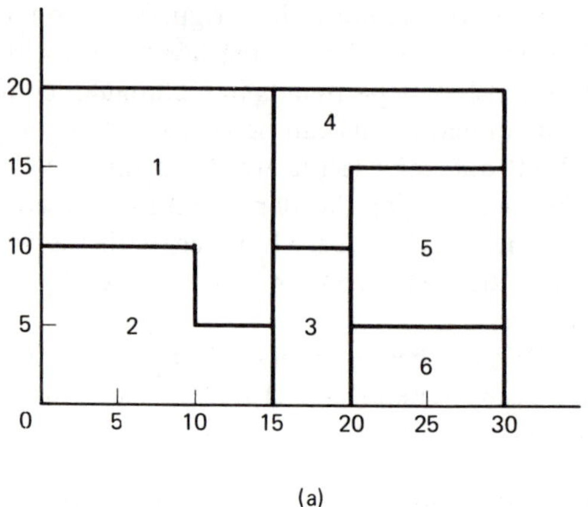

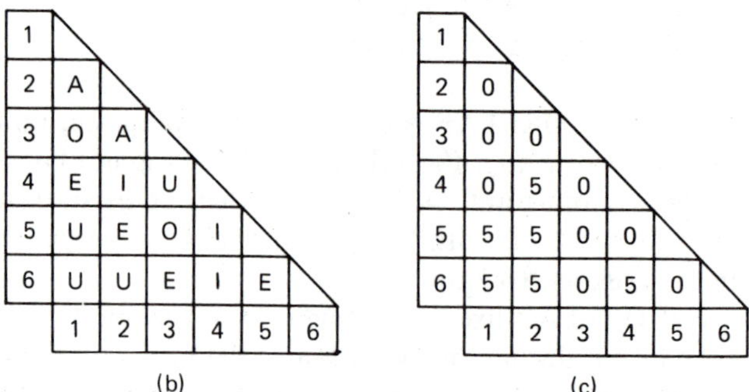

(b)

	1	2	3	4	5	6
1						
2	A					
3	O	A				
4	E	I	U			
5	U	E	O	I		
6	U	U	E	I	E	

(c)

	1	2	3	4	5	6
1						
2	0					
3	0	0				
4	0	5	0			
5	5	5	0	0		
6	5	5	0	5	0	

Figure 3.10. (a) Sample layout for CORELAP scoring with (b) relationship chart and (c) shortest rectilinear distances between department borders.

2. Arbitrary numerical values are assigned to ordinal closeness ratings, and arithmetic operations are performed on a distance-related function.
3. It is assumed that "closeness" is a linear function of shortest rectilinear distance.

In a number of practical situations closeness can probably be operationally defined as the shortest rectilinear distance between department borders. However in some cases such a definition might be inappropriate. If this definition is accepted and understood by the person(s) making the closeness ratings, we are agreeable to its use. Consequently, in eliciting closeness ratings from managers the shortest rectilinear distance definition of closeness should be made very clear if Interactive CORELAP's score is to be used.

We considered previously the problem of ordinal measurement scales being treated as either interval or ratio scales. Here, we are performing operations only allowed with ratio scales. Consequently, the total score given by Interactive CORELAP should not be the only yardstick used in evaluating layouts.

The scoring system used by Interactive CORELAP is based on the assumption that the overall effectiveness of a layout is a linear function of shortest rectilinear distances. Consequently, a layout that has departments 1 and 2 located 30 feet apart and departments 1 and 3 located 20 feet apart is considered to be equivalent to a layout in which departments 1 and 2 are adjacent and departments 1 and 3 are located 60 feet apart if departments 1 and 2 have an I rating and departments 1 and 3 have an A rating. In a number of applications the linear assumption might not be valid.

Despite our objections to the scoring technique used, we believe that Interactive CORELAP provides a very useful and worthwhile approach in the development of layouts. However, we do suggest that you use your own evaluation scheme and not rely completely on the "total score" for the layout.

Since the user is involved in the generation of layouts, steps 7, 8, and 9 of the modified SLP procedure, given in Figure 2.21, are performed, to some extent, simultaneously with step 6b. The final layout obtained from Interactive CORELAP has already undergone revisions based on modifying considerations and practical limitations.

Input requirements for Interactive CORELAP include the

1. REL chart.
2. Number and area of departments.
3. Layout scale.

A number of other input options are also available. However, most of these are entered by the user during the interaction mode of operation. Examples of optional input are fixed departments, existing building configuration, modifying considerations, and practical limitations.

As an illustration of the use of Interactive CORELAP, consider the example problem previously solved using CORELAP 8. Based on the *TCR* and departmental area, department 19 (general storage) enters the layout first. The next four departments to enter (in order of entry) are department 13 (parts shipment), department 17 (receiving), department 18 (testing), and department 16 (service area). Departments 13 and 17 had A ratings with department 19. Department 18 had an E rating with department 19. Department 16 had an E rating with department 18. However, as department 16 is placed in the layout, the computer heuristic used to locate departments encounters some difficulties in placing the department and prints out the layout for the first five departments. The reason for the difficulty in placing department 16 is that no high closeness rating exists between itself and more than one department already placed.

As shown in Figure 3.11, the computer prints out the question, "Do you wish a score for this layout?" If the answer is "No"," a 0 is typed by the user; otherwise, a 1 is typed. The user indicates a score is desired, and the total score of 13 is computed. The next question asked is, "Do you wish to make changes?" On examining the layout the user decides he wants to rearrange elements from departments 16, 17, and 18 to obtain better departmental configurations. Therefore, the user types a 1. Since changes are to be made, the question is asked. "Which change option do you desire, 1, 2, or 3?" An option 1 change is a pairwise exchange of two squares and is indicated using the following format:

$$U, V, W, X, Y, Z$$

where

$U =$ row number
$V =$ column number
$W =$ number to be entered at position (U, V)
$X =$ row number
$Y =$ column number
$Z =$ number to be entered at position (X, Y).

An option 2 change is a row change and is entered using the following format:

$$Q, R, S, T$$

where

$Q =$ row number
$R =$ starting column number
$S =$ ending column number
$T =$ number to be entered in row Q, columns R through S, inclusive.

Column changes are made using option 3. The format used is

$$K, L, M, N$$

where

$K =$ column number
$L =$ starting row number
$M =$ ending row number
$N =$ number to be entered in column K, row L through M, inclusive.

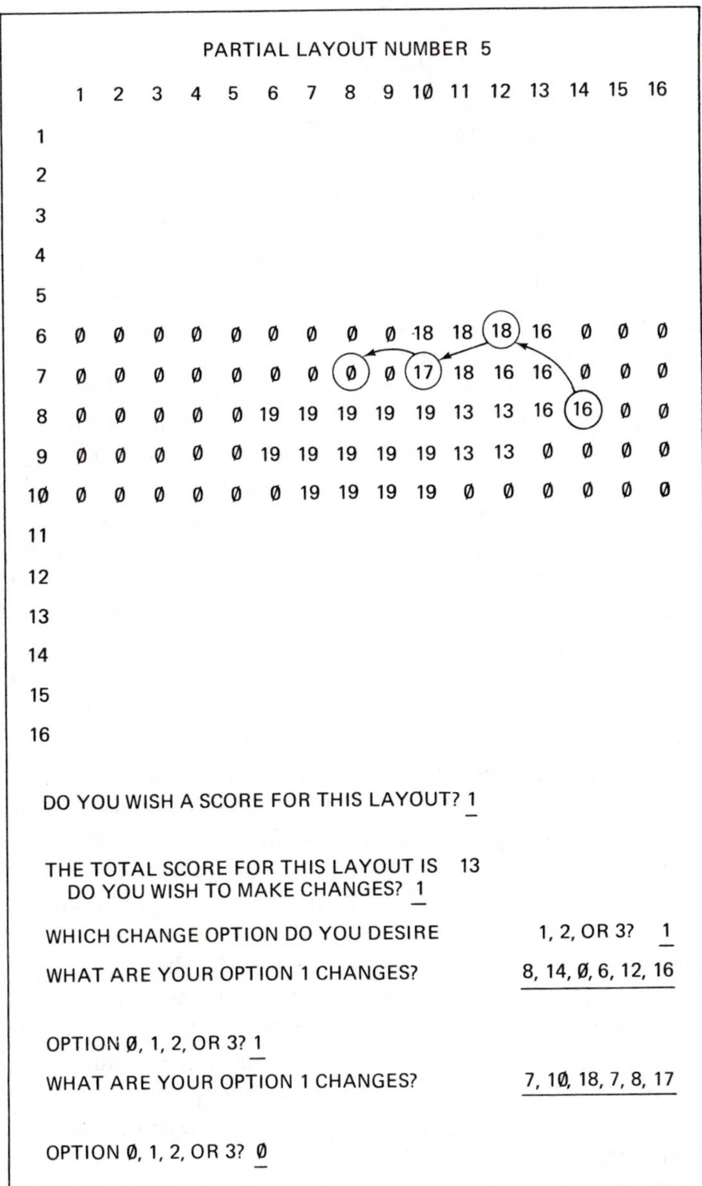

PARTIAL LAYOUT NUMBER 5

	1	2	3	4	5	6	7	8	9	1Ø	11	12	13	14	15	16
1																
2																
3																
4																
5																
6	Ø	Ø	Ø	Ø	Ø	Ø	Ø	Ø	Ø	18	18	(18)	16	Ø	Ø	Ø
7	Ø	Ø	Ø	Ø	Ø	Ø	Ø	(Ø)	Ø	(17)	18	16	16	Ø	Ø	Ø
8	Ø	Ø	Ø	Ø	Ø	19	19	19	19	19	13	13	16	(16)	Ø	Ø
9	Ø	Ø	Ø	Ø	Ø	19	19	19	19	19	13	13	Ø	Ø	Ø	Ø
1Ø	Ø	Ø	Ø	Ø	Ø	Ø	19	19	19	19	Ø	Ø	Ø	Ø	Ø	Ø
11																
12																
13																
14																
15																
16																

DO YOU WISH A SCORE FOR THIS LAYOUT? 1

THE TOTAL SCORE FOR THIS LAYOUT IS 13
DO YOU WISH TO MAKE CHANGES? 1

WHICH CHANGE OPTION DO YOU DESIRE 1, 2, OR 3? 1

WHAT ARE YOUR OPTION 1 CHANGES? 8, 14, Ø, 6, 12, 16

OPTION Ø, 1, 2, OR 3? 1

WHAT ARE YOUR OPTION 1 CHANGES? 7, 1Ø, 18, 7, 8, 17

OPTION Ø, 1, 2, OR 3? Ø

Figure 3.11. Computer printout presents an arrangement for the first five departments to be entered. (After *Industrial Engineering*, Aug. 1971. Copyright American Institute of Industrial Engineers, Inc., 25 Technology Park/Atlanta, Norcross, Ga. 30071.)

Using option 1 changes, elements from departments 16, 17, and 18 are exchanged. Since no further changes are desired, a 0 is typed and the computer continues constructing the layout.

Department 15 (repair and service) is entered because of an E rating with department 16. Next, department 14 (foreman's office) is entered on the basis of the I ratings with department 15. At this point another partial layout is printed out, as shown in Figure 3.12. A score of 80 is obtained. On examining Figure 3.12 it is seen that the foreman's office is located on the periphery of the layout. Since the *TCR* value for the foreman's office is high, it is decided to move it to a central position in the layout. Consequently, it is decided to make the changes indicated on the layout in Figure 3.12.

The computer adds department 12 (parcel post) and department 11 (display area) to the layout. However, since department 11 only has a high closeness rating with one department in the layout, department 14, the computer asks for help by printing out the partial layout shown in Figure 3.13.

A score of 150 is obtained. The user realizes all departments have been located but one, department 20 (general office), and there is an E rating between departments 11 and 20. If department 11 is left in its present location, department 20 can be located near it. Therefore, the user decides to leave department 11 alone. However, he notices that the present building size would be better accommodated by moving two elements of department 19 to the positions shown. These changes are made, and the computer completes the construction process by placing department 20 in the layout.

The layout obtained is given in Figure 3.14 with its associated score of 188. On inspection of the layout it is noted that department 13 (parts shipment) is not located on the periphery of the building as it should be for access to external transportation facilities. Also, the location of department 15 produces an irregular building shape. Consequently, elements from departments 13, 15, 16, and 20 are moved, based on modifying considerations and practical limitations.

Since no more departments are to be added, when the computer asks "Which partial layout should be printed next"?, we type number 99. Partial layouts are numbered according to the number of departments placed in the layout. Therefore, we can type any number greater than the number of departments and the completed layout will be printed out. (Interactive CORELAP can handle 45 departments.) The final layout is given in Figure 3.15 with a final score of 165.

As we mentioned earlier, we do not feel that the total score of 165 is too meaningful. Rather, the important contribution of Interactive CORELAP is its interaction mode. The layout analyst can work with the computer to develop a large number of layout alternatives in a rather short amount of time. The effectiveness of the search phase of the design process can be improved significantly using Interactive CORELAP.

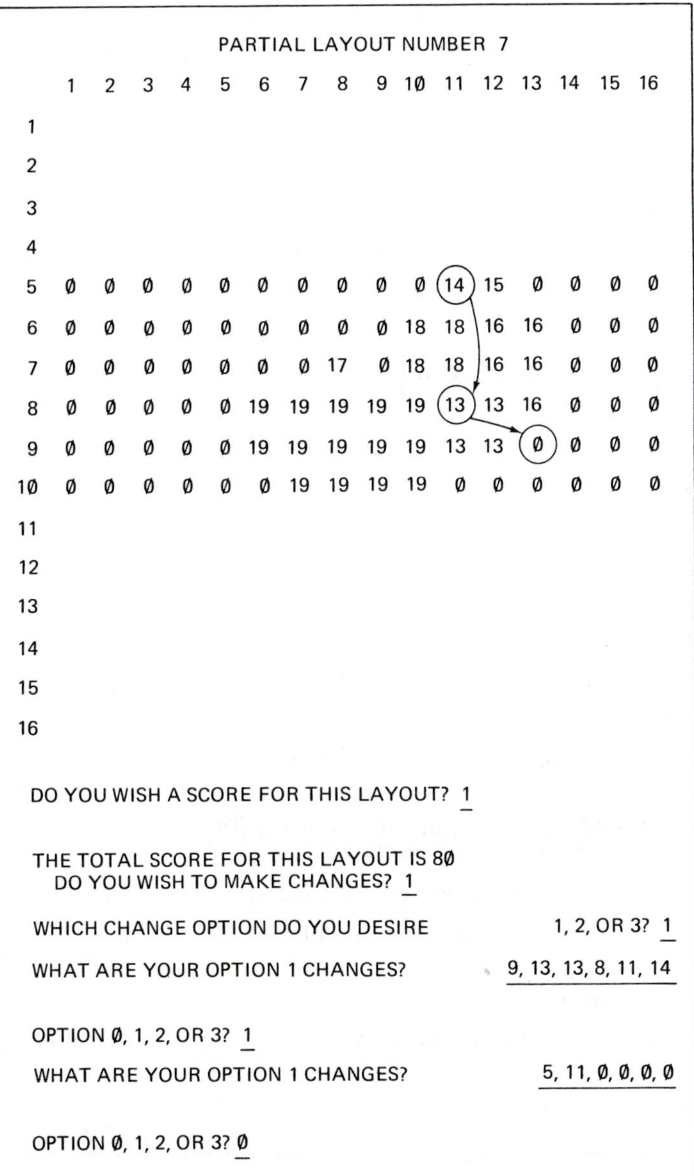

PARTIAL LAYOUT NUMBER 7

	1	2	3	4	5	6	7	8	9	1Ø	11	12	13	14	15	16
1																
2																
3																
4																
5	Ø	Ø	Ø	Ø	Ø	Ø	Ø	Ø	Ø	Ø	14	15	Ø	Ø	Ø	Ø
6	Ø	Ø	Ø	Ø	Ø	Ø	Ø	Ø	Ø	18	18	16	16	Ø	Ø	Ø
7	Ø	Ø	Ø	Ø	Ø	Ø	Ø	17	Ø	18	18	16	16	Ø	Ø	Ø
8	Ø	Ø	Ø	Ø	Ø	19	19	19	19	19	13	13	16	Ø	Ø	Ø
9	Ø	Ø	Ø	Ø	Ø	19	19	19	19	19	13	13	Ø	Ø	Ø	Ø
1Ø	Ø	Ø	Ø	Ø	Ø	Ø	19	19	19	19	Ø	Ø	Ø	Ø	Ø	Ø
11																
12																
13																
14																
15																
16																

DO YOU WISH A SCORE FOR THIS LAYOUT? 1

THE TOTAL SCORE FOR THIS LAYOUT IS 8Ø
DO YOU WISH TO MAKE CHANGES? 1

WHICH CHANGE OPTION DO YOU DESIRE 1, 2, OR 3? 1

WHAT ARE YOUR OPTION 1 CHANGES? 9, 13, 13, 8, 11, 14

OPTION Ø, 1, 2, OR 3? 1

WHAT ARE YOUR OPTION 1 CHANGES? 5, 11, Ø, Ø, Ø, Ø

OPTION Ø, 1, 2, OR 3? Ø

Figure 3.12. After some departments are moved and others entered by the heuristic, the computer prints layout. (After *Industrial Engineering*, Aug. 1971. Copyright American Institute of Industrial Engineers, Inc., 25 Technology Park/Atlanta, Norcross, Ga. 30071.)

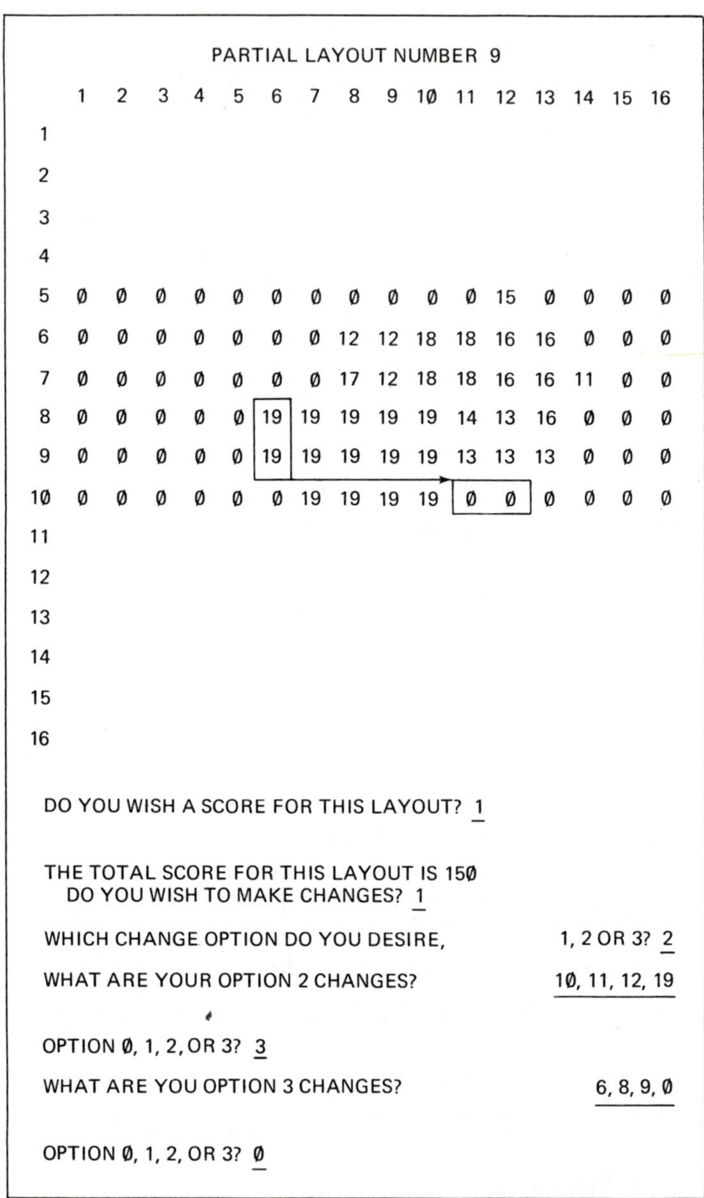

PARTIAL LAYOUT NUMBER 9

	1	2	3	4	5	6	7	8	9	1Ø	11	12	13	14	15	16
1																
2																
3																
4																
5	Ø	Ø	Ø	Ø	Ø	Ø	Ø	Ø	Ø	Ø	Ø	15	Ø	Ø	Ø	Ø
6	Ø	Ø	Ø	Ø	Ø	Ø	Ø	12	12	18	18	16	16	Ø	Ø	Ø
7	Ø	Ø	Ø	Ø	Ø	Ø	Ø	17	12	18	18	16	16	11	Ø	Ø
8	Ø	Ø	Ø	Ø	Ø	19	19	19	19	19	14	13	16	Ø	Ø	Ø
9	Ø	Ø	Ø	Ø	Ø	19	19	19	19	19	13	13	13	Ø	Ø	Ø
1Ø	Ø	Ø	Ø	Ø	Ø	Ø	19	19	19	19	Ø	Ø	Ø	Ø	Ø	Ø
11																
12																
13																
14																
15																
16																

DO YOU WISH A SCORE FOR THIS LAYOUT? 1

THE TOTAL SCORE FOR THIS LAYOUT IS 15Ø
DO YOU WISH TO MAKE CHANGES? 1

WHICH CHANGE OPTION DO YOU DESIRE, 1, 2 OR 3? 2

WHAT ARE YOUR OPTION 2 CHANGES? 1Ø, 11, 12, 19

OPTION Ø, 1, 2, OR 3? 3

WHAT ARE YOU OPTION 3 CHANGES? 6, 8, 9, Ø

OPTION Ø, 1, 2, OR 3? Ø

Figure 3.13. Layout printout before moving two units of Department 19. (After *Industrial Engineering*, Aug. 1971. Copyright American Institute of Industrial Engineers, Inc., 25 Technology Park/Atlanta, Norcross, Ga. 30071.)

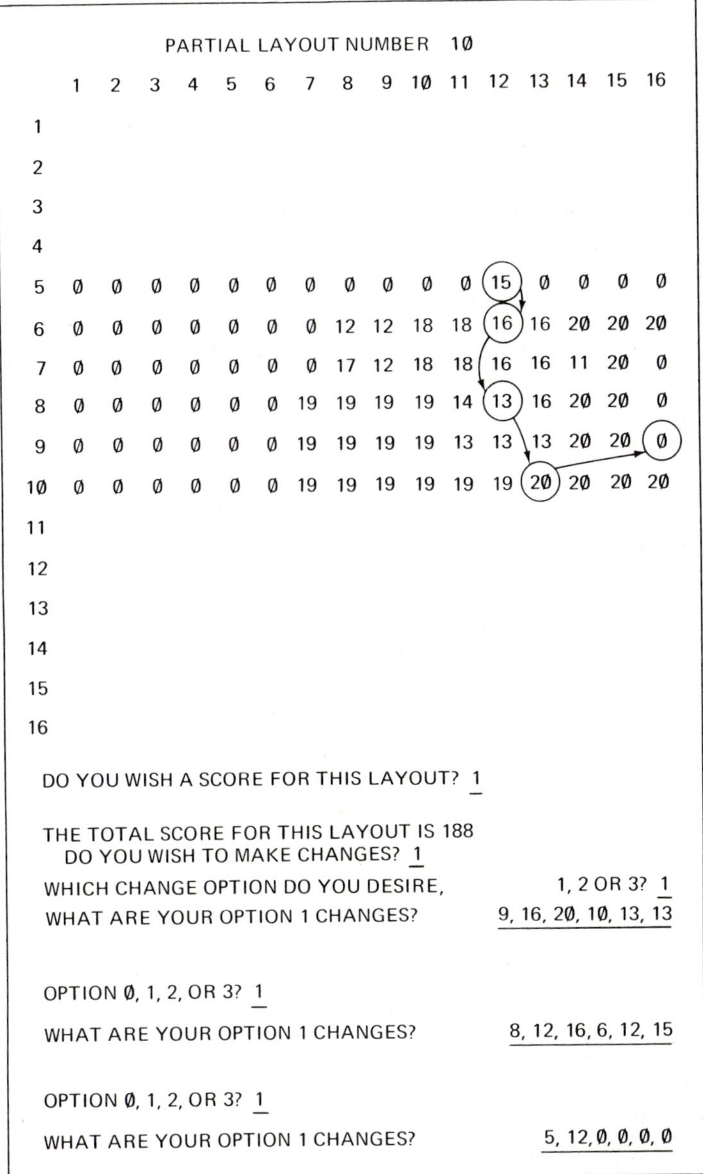

PARTIAL LAYOUT NUMBER 1Ø

	1	2	3	4	5	6	7	8	9	1Ø	11	12	13	14	15	16
1																
2																
3																
4																
5	Ø	Ø	Ø	Ø	Ø	Ø	Ø	Ø	Ø	Ø	Ø	15	Ø	Ø	Ø	Ø
6	Ø	Ø	Ø	Ø	Ø	Ø	Ø	12	12	18	18	16	16	2Ø	2Ø	2Ø
7	Ø	Ø	Ø	Ø	Ø	Ø	Ø	17	12	18	18	16	16	11	2Ø	Ø
8	Ø	Ø	Ø	Ø	Ø	Ø	19	19	19	19	14	13	16	2Ø	2Ø	Ø
9	Ø	Ø	Ø	Ø	Ø	Ø	19	19	19	19	13	13	13	2Ø	2Ø	Ø
1Ø	Ø	Ø	Ø	Ø	Ø	Ø	19	19	19	19	19	19	2Ø	2Ø	2Ø	2Ø
11																
12																
13																
14																
15																
16																

DO YOU WISH A SCORE FOR THIS LAYOUT? 1

THE TOTAL SCORE FOR THIS LAYOUT IS 188
DO YOU WISH TO MAKE CHANGES? 1
WHICH CHANGE OPTION DO YOU DESIRE, 1, 2 OR 3? 1
WHAT ARE YOUR OPTION 1 CHANGES? 9, 16, 2Ø, 1Ø, 13, 13

OPTION Ø, 1, 2, OR 3? 1
WHAT ARE YOUR OPTION 1 CHANGES? 8, 12, 16, 6, 12, 15

OPTION Ø, 1, 2, OR 3? 1
WHAT ARE YOUR OPTION 1 CHANGES? 5, 12, Ø, Ø, Ø, Ø

Figure 3.14. The last department, Department 20, is entered in the layout. (After *Industrial Engineering*, Aug. 1971. Copyright American Institute of Industrial Engineers, Inc., 25 Technology Park/Atlanta, Norcross, Ga. 30071.)

```
DO YOU WISH A SCORE FOR THIS LAYOUT? 1

THE TOTAL SCORE FOR THIS LAYOUT IS 165
  DO YOU WISH TO MAKE CHANGES? Ø

WHICH PARTIAL LAYOUT SHOULD BE PRINTED NEXT? 99

                                  CORELAP BLOCK PLAN LAYOUT

     1   2   3   4   5   6   7   8   9  1Ø  11  12  13  14  15  16
 1
 2
 3
 4
 5
 6   Ø   Ø   Ø   Ø   Ø   Ø   Ø  12  12  18  18  15  16  2Ø  2Ø  2Ø
 7   Ø   Ø   Ø   Ø   Ø   Ø   Ø  17  12  18  18  16  16  11  2Ø   Ø
 8   Ø   Ø   Ø   Ø   Ø   Ø  19  19  19  19  14  16  16  2Ø  2Ø   Ø
 9   Ø   Ø   Ø   Ø   Ø   Ø  19  19  19  19  13  13  13  2Ø  2Ø  2Ø
1Ø   Ø   Ø   Ø   Ø   Ø   Ø  19  19  19  19  19  19  13  2Ø  2Ø  2Ø
11
12
13
14
15
16

PROGRAM STOP AT Ø

USED 11.34 UNITS
GOODBYE
ØØ16.43 CRU     ØØØØ.74 TCH     ØØ11.57 KC

OFF AT 112Ø EDT 1Ø/13/7Ø
```

Figure 3.15. The final layout is obtained after manipulation to conform with the shape of the existing building. (After *Industrial Engineering*, Aug. 1971. Copyright American Institute of Industrial Engineers, Inc., 25 Technology Park/Atlanta, Norcross, Ga. 30071.)

3.6 CRAFT[4]

CRAFT was originally presented by Armour and Buffa [2] and subsequently tested, refined, and applied by Buffa, Armour, and Vollmann [4]. Of the three computerized layout models we are considering in this chapter, a considerably larger body of literature treats the CRAFT layout program than either ALDEP or CORELAP. No doubt to some extent this is due to its computational efficiency in providing heuristic solutions to quadratic assignment problems (as discussed in Chapter 8), the availability of materials describing the algorithm, its capability for considering multiple material-handling systems, the fact that CRAFT was developed before either ALDEP or CORELAP, and CRAFT's criterion for evaluating layouts.

The criterion employed in CRAFT is the minimization of the cost of item movement, where this cost is expressed as a linear function of distance traveled. This particular criterion coincides with that employed in subsequent chapters of this book. Furthermore, the criterion is one that is quite commonly used when the flow of materials is a significant factor to be considered in layout design. Whereas ALDEP and CORELAP were developed to facilitate the design of layouts based on activity relationship, CRAFT facilitates the design of layouts based on flow considerations. Due to this major distinction, CRAFT is sometimes referred to as a quantitative layout program, and ALDEP and CORELAP are referred to as qualitative layout programs.

Although CRAFT was originally developed for use in designing layouts when material-handling costs were a major consideration, by broadening the interpretation of materials flow CRAFT also can be an aid in designing layouts for nonmanufacturing activities [24]. In our subsequent discussion of CRAFT we shall continue to refer to the minimization of material-handling cost. However, you should not infer from this that CRAFT can only be applied when the minimization of material-handling cost is the criterion. In fact, a form of the heuristic imbedded in the CRAFT algorithm has been used to develop office layouts [25].

CRAFT is an improvement program. As such, it seeks an optimum design by making improvements in the layout in a sequential fashion. CRAFT first evaluates a given layout and then considers what the effect will be if department locations are interchanged. If improvements can be made by making parwise exchanges, the exchange producing the greatest improvement is made. The process continues until no improvement can be made by

---

[4]For a copy of the CRAFT program, contact the IBM Share Library System. Ask for Program Order No. SDA 3391.

pairwise exchanges. Only departments with common borders or of the same area are considered for exchanges of locations.

At this point it is instructive to consider the input information required for CRAFT. Specifically, the input includes

1. Initial spatial array (layout).
2. Flow data.
3. Cost data.
4. Number and location of fixed departments.

CRAFT requires that an initial layout be specified. If available, an existing layout is normally given as the initial layout. If no existing layout is present, one must be specified. Therefore, some preliminary analysis might be required to specify an initial layout. Since the building outline given in the initial layout will not change, the configuration of the final layout can be significantly affected by the configuration of the initial layout.

Associated with the specification of an initial layout is the specification of the number of departments and their respective floor areas. CRAFT can handle up to 40 different departments, and the layout scale must be sufficient to meet the maximum allowable dimensions of 30 by 30, plus the requirement that no department can use more than 75 squares in the layout.

In addition to a spatial array, two from–to charts are required. The first from–to chart provides flow data, and the second from–to chart contains cost data. The flow data are expressed in terms of the number of trips per time period that are made between combinations of departments. The cost data reflect the material-handling cost required to move one unit of distance between combinations of departments. Since material handling can be performed in a variety of ways, the cost elements need not be the same. As an example, material handling might be performed by fork truck between departments A and F and by hand truck between departments A and B. Since the cost data are expressed as cost per unit distance, it is necessary that the distance unit be the same as the layout scale. Thus, if the layout scale is such that an individual square within the layout has an area of 100 square feet, then the cost elements in the from–to chart must be the cost per 10 feet traveled.

If conveyors are used, cost is considered to be proportional to conveyor length and is not a linear function of flow. Consequently, if conveyors connect, say, department A and department C, the flow between these departments is set equal to unity, and the cost element for material handling between these departments will be the cost per unit time per unit length of conveyor. Again, the unit length of conveyor must agree with the layout scale used.

CRAFT has the flexibility of allowing certain departments to have fixed locations in the layout. Also, CRAFT can be used to evaluate individual layouts without searching for improved designs.

To illustrate the approach used by CRAFT to develop layouts, consider the initial layout and the from–to chart for the flow data given in Figure 3.16. For simplicity, we assume all cost elements are equal to unity in the from–to chart for the cost data.

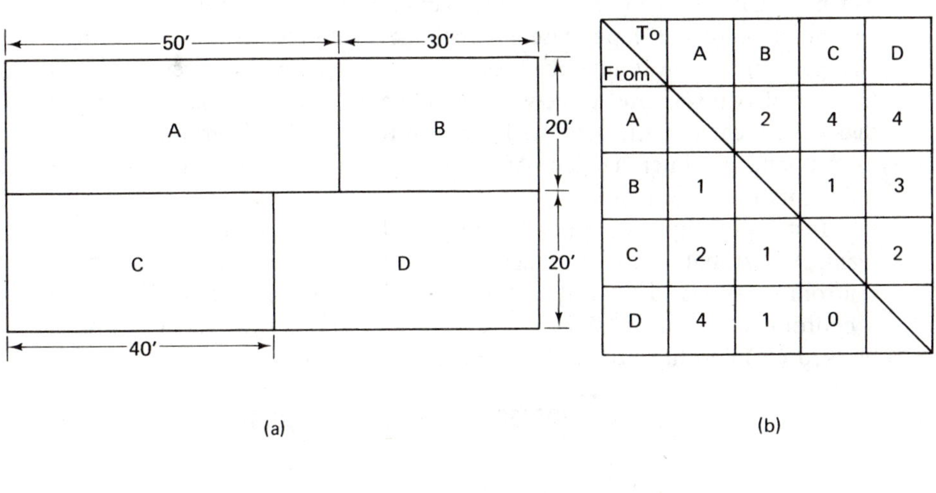

From \ To	A	B	C	D
A		2	4	4
B	1		1	3
C	2	1		2
D	4	1	0	

(a)

(b)

From \ To	A	B	C	D
A		40	25	55
B	40		65	25
C	25	65		40
D	55	25	40	

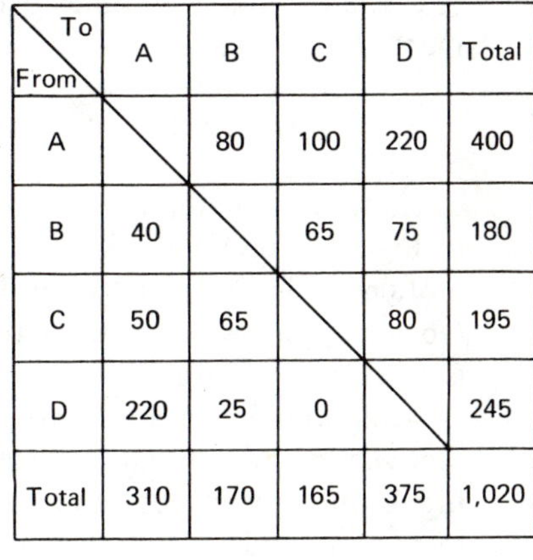

From \ To	A	B	C	D	Total
A		80	100	220	400
B	40		65	75	180
C	50	65		80	195
D	220	25	0		245
Total	310	170	165	375	1,020

(c)

(d)

Figure 3.16. (a) Initial layout, (b) flow data, (c) distance data, and (d) total cost for CRAFT example problem.

CRAFT computes the total cost for the initial layout by developing a from–to chart of distances between departments. The distances are computed as the rectilinear distances between department centroids. For the initial layout the distance chart is given in Figure 3.16(c). The total distance traveled per unit time equals 1,020 feet for the initial layout, as shown in Figure 3.16(d).

Next, CRAFT considers exchanges of locations for those departments which either are the same area or have a common border. The layout analyst can have CRAFT consider (1) only pairwise interchanges, (2) only three-way interchanges, (3) pairwise interchanges followed by three-way interchanges, (4) three-way interchanges followed by pairwise interchanges, and (5) best of two-way and three-way interchanges. The interchange that produces the greatest anticipated reduction in total cost is determined and the departments are interchanged. As an illustration, suppose that we consider only pairwise interchanges.

Instead of physically interchanging the locations of all pairs of departments, we estimate the cost reduction that will be achieved by interchanging centroid locations. To facilitate this determination, we place the layout on a coordinate system as shown in Figure 3.17. The centroid locations are marked ($\times$) and have the following coordinate values:

$$(x_A, y_A) = (25, 30), \qquad (x_C, y_C) = (20, 10)$$
$$(x_B, y_B) = (65, 30), \qquad (x_D, y_D) = (60, 10)$$

The rectilinear distance between, say, the current centroids for departments A and B is

$$|x_A - x_B| + |y_A - y_B| = |25 - 65| + |30 - 30| = 40$$

as shown in Figure 3.17(b). Notice that Figure 3.17(b) is a rearrangement of Figure 3.16(c), with A and B interchanged. The total cost is estimated to be 1,060 if departments A and B are exchanged, as shown in Figure 3.17(c). Obviously, we will not want to interchange these departments, due to an anticipated increase of 40 in total cost.

Suppose that A and C are interchanged. If we develop the resulting distance chart and compute the total cost, we obtain a cost of 955. Interchanging A and D produces a total cost of 1,095. Notice that B and C are not the same area, and do not have a common border. Thus, an interchange of B and C is not considered. Interchanging B and D results in a total cost of 945. Interchanging C and D produces a total cost of 1,040. Consequently, the best interchange is B and D.

On interchanging B and D we obtain the layout given in Figure 3.18(a). Since B and D are not the same size, we notice that D is no longer rectangu-

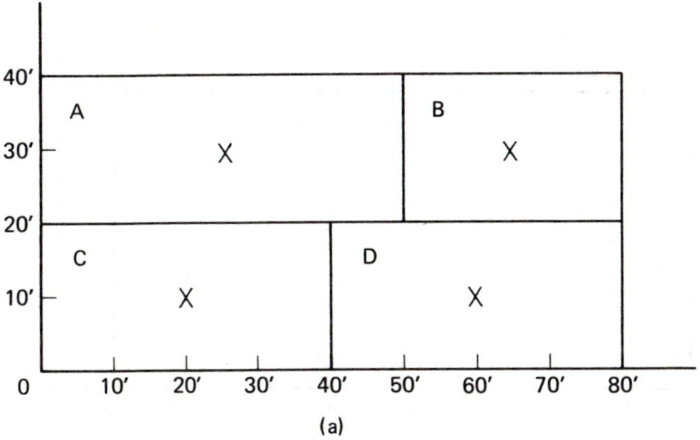

(a)

To / From	A	B	C	D
A		40	65	25
B	40		25	55
C	65	25		40
D	25	55	40	

(b)

To / From	A	B	C	D	Total
A		80	260	100	440
B	40		25	165	230
C	130	25		80	235
D	100	55	0		155
Total	270	160	285	345	1,060

(c)

Figure 3.17. Initial layout with department centroids shown. Distance chart and total cost chart if departments A and B are interchanged in the initial layout.

larly shaped and that the department centroids were not exchanged exactly. Rather, a calculation shows that

$$(x_A, y_A) = (25, 30), \qquad (x_C, y_C) = (20, 10)$$
$$(x_B, y_B) = (55, 10), \qquad (x_D, y_D) = (67.5, 25)$$

Since the centroid locations have changed, our estimate of total cost might be incorrect. Therefore, we develop a distance chart for the new layout, as given in Figure 3.18(b). As before, these distances are the rectilinear distances between department centroids. On multiplying the flow values in

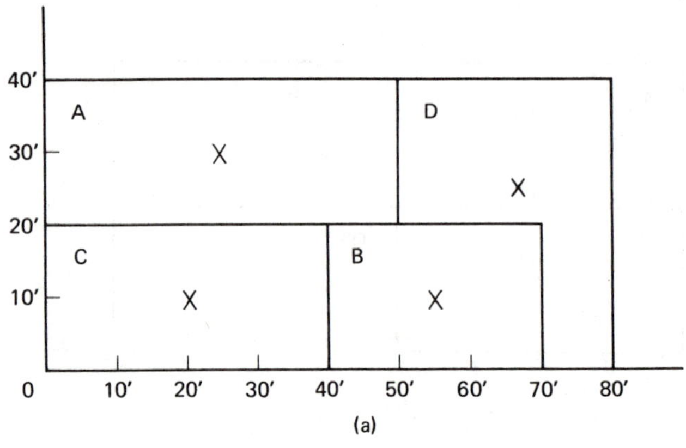

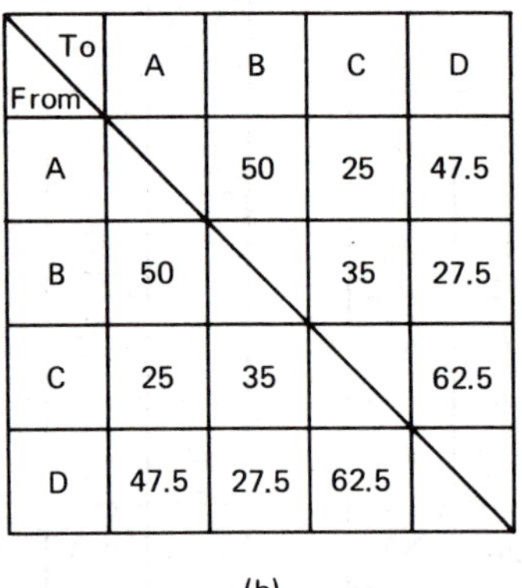

From \ To	A	B	C	D
A		50	25	47.5
B	50		35	27.5
C	25	35		62.5
D	47.5	27.5	62.5	

(b)

Figure 3.18. First improved layout and its associated distance chart.

Figure 3.16 and the distance values in Figure 3.18, we obtain a total cost of 985, rather than the estimated cost of 945.

CRAFT applies the pairwise interchange to the improved layout. In doing so we find that interchanging A and B gives the greatest estimated cost reduction with an *estimated* total cost of 945. Figure 3.19(a) gives the second improved layout having the following centroid locations:

$$(x_A, y_A) = (49, 18), \qquad (x_C, y_C) = (20, 10)$$
$$(x_B, y_B) = (15, 30), \qquad (x_D, y_D) = (67.5, 25)$$

A computation of the actual cost yields a value of 969, rather than the estimated value of 945.

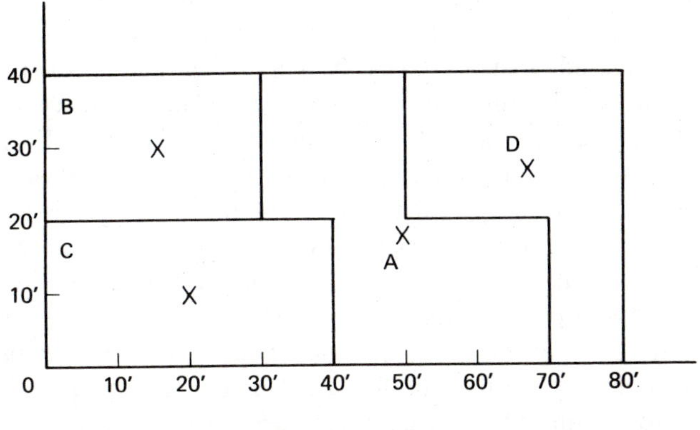

(a) Second Improved Layout

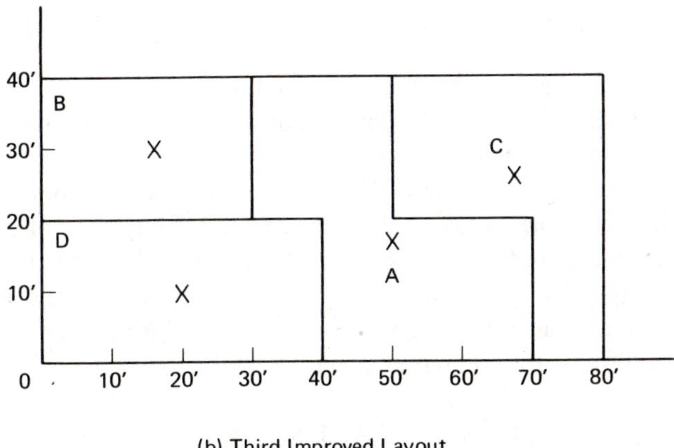

(b) Third Improved Layout

Figure 3.19. Layout design obtained from CRAFT.

Applying the pairwise interchange technique to the second improved layout indicates C and D should be interchanged. The estimated cost is 927 for an exchange of C and D. Since C and D have the same area, their centroids will be exchanged exactly, and the true total cost will be 927. Making the indicated exchange of department locations yields the layout shown in Figure 3.19(b). On making the pairwise interchange calculations for the third improved layout, we find that no pairwise interchange will produce an *estimated* total cost less than 927. Consequently, our search terminates with the layout given in Figure 3.19(b).

The use of the CRAFT program does not guarantee that the least-cost layout will be found, since all possible interchanges (four or more at a time) are not considered. Instead, CRAFT is a suboptimal, heuristic procedure which produces a layout that cannot be easily improved upon. (The term "heuristic" is used to mean an appealing rule of thumb.) Furthermore, it should be noted that the CRAFT solution obtained is path dependent. By path dependent, we mean that the final solution obtained is dependent on the starting solution. For this reason it is common practice to specify a number of different initial layouts and choose the best final solution generated.

Several assumptions made in the CRAFT algorithm have been criticized in the literature. For example, the use of centroid locations in measuring distances might not be realistic for some practical applications. In fact, it is possible to obtain some unusual layout designs because of this assumption. As an illustration, consider the layout given in Figure 3.20(a). The centroid locations are

$$(x_A, y_A) = (40, 7), \qquad (x_C, y_C) = (40, 27.50)$$
$$(x_B, y_B) = (40, 35.833), \qquad (x_D, y_D) = (40, 16.5625)$$

On multiplying the distance values by the flow values in Figure 3.16, a total cost of 383.6235 is obtained. This is considerably better than the cost of 927 obtained previously. In fact, solutions can be obtained having costs equal to zero. But, how practical are such solutions? The point to be made is this—it is possible to take advantage of the assumptions of the program. The centroid assumption becomes less valid as the shape of the department becomes "less square." As long as you are aware of this and interpret the CRAFT solution as a design aid or benchmark solution against which more realistic solutions are compared, then CRAFT can serve a very useful purpose.

Further assumptions of CRAFT are that material-handling costs are significant, known, and linear in distance traveled. How realistic are these assumptions? In some cases, they are not valid at all. Does this mean the CRAFT solutions are useless in such cases? Of course not! As long as you remember that we recommend the use of the computerized layout planning programs in carrying out the *search* phase of the design process, *not* the *search and selection* phases, CRAFT can be quite useful in designing layouts.

As an illustration of the use of the CRAFT computer program, consider an example involving 12 departments. A process layout is to be developed such that material-handling cost is minimized. There exists a facility within which the layout is to be installed. The floor space requirements for the departments are given in Table 3.2, with approximately 576 square feet of the floor space restricted. A scale of 16 square feet per square yields the number of squares given in Table 3.2. A computer printout of the initial layout

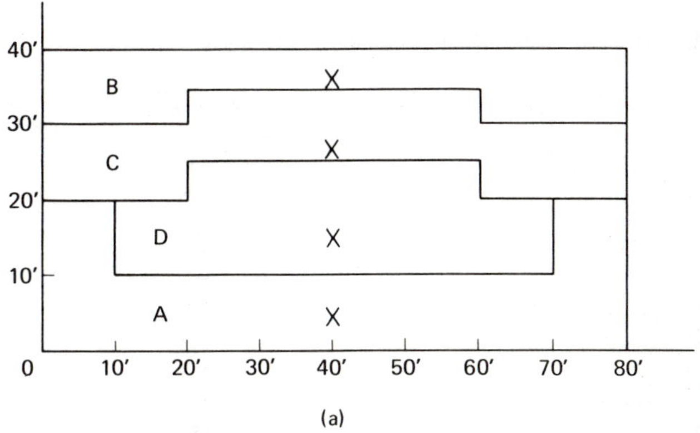

From \ To	A	B	C	D
A		26.833	18.50	9.5625
B	26.833		8.333	19.2708
C	18.500	8.333		10.9375
D	9.5625	19.2708	10.9375	

(b)

Figure 3.20. Sample layout for CRAFT evaluation with its associated distance chart.

is given in Figure 3.21, with the restricted area designated as department M. Flow data and cost data for the example problem are given in Figure 3.22 (a) and (b).

Notice the initial layout is evaluated by the CRAFT program and found to have a total cost of 110,696.88. Using a pairwise interchange option, it

Table 3.2. DATA FOR THE CRAFT EXAMPLE PROBLEM

Department	*Area* (ft^2)	*No. Squares*
A. Receiving	600	38
B. Machining	425	27
C. Grinding	200	13
D. Welding	250	16
E. X ray	210	13
F. Inspection	175	11
G. Sanding	125	8
H. Facing	275	17
I. Painting	285	18
J. Cleaning	150	9
K. Labeling	75	5
L. Storage and shipping	715	45
M. Restricted area	576	36

LOCATION PATTERN

	1	2	3	4	5	6	7	8	9	10	11	12	13	14	15	16
1	A	A	A	A	A	A	A	C	C	C	I	I	I	I	I	I
2	A					A	A	C		C	I					I
3	A					A	C	C	C	C	I	I	I	I	I	I
4	A					A	C	C	C	H	H	H	H	J	J	J
5	A					A	G	G	G	H			H	J		J
6	A	A	A	A	A	A	G	G	G	H			H	J	J	J
7	B	B	B	B	B	B	G	G	H	H	H	H	H	K	K	K
8	B				B	B	F	F	L	L	L	L	L	L	K	K
9	B				B	F	F	F	L					L	L	L
10	B				B	F		F	L							L
11	B	B	B	B	B	F	F	F	L							L
12	D	D	D	D	E	L	L	L	L	L	L	L	L	L	L	L
13	D		D	E	E	E	L	L	L	L	M	M	M	M	M	M
14	D		D	E		E	M	M	M	M	M					M
15	D		D	E		E	M									M
16	D	D	D	E	E	E	M	M	M	M	M	M	M	M	M	M

TOTAL COST 110696.88 EST. COST REDUCTION 0.0 MOVEA MOVEB MOVEC INTERATION 0

Figure 3.21. Initial layout for CRAFT solution.

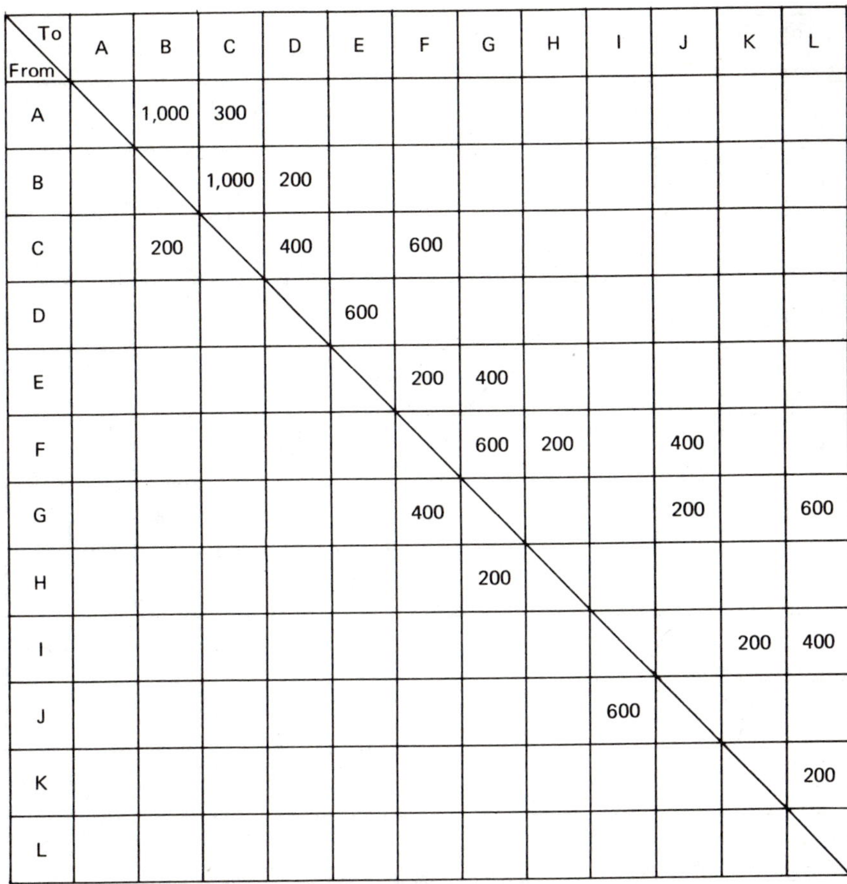

From \ To	A	B	C	D	E	F	G	H	I	J	K	L
A		1,000	300									
B			1,000	200								
C		200		400		600						
D					600							
E						200	400					
F							600	200		400		
G						400				200		600
H							200					
I											200	400
J									600			
K												200
L												

(a)

Figure 3.22(a). Flow data (given in trips per month).

is found that interchanging departments H and L produces the greatest estimated cost reduction, with a value of 10,573.84. Making this exchange results in the layout shown in Figure 3.23. The CRAFT program evaluates the actual total cost for the layout obtained on iteration 1 and obtains a value of 106,874.06. Thus, instead of the estimated reduction of 10,573.84, there actually resulted a reduction of 3,822.82.

The layout obtained on the second iteration is given in Figure 3.24 and resulted from the interchange of departments C and L. There was an estimated cost reduction of 11,054.95, but an actual reduction of 13,436.25, to give a total cost of 93,437.81.

To / From	A	B	C	D	E	F	G	H	I	J	K	L
A		3.00	1.50									
B			1.23	0.98								
C		0.98		1.23		1.10						
D					2.10							
E						1.80	1.50					
F							1.00	1.00		1.20		
G						1.00				1.00		2.40
H							1.00					
I											1.30	2.10
J									1.70			
K												2.00
L												

(b)

Figure 3.22(b). Cost data (given in units of $0.001 per 4 ft of travel) for the example problem.

On the third iteration departments C and F are interchanged to give the layout in Figure 3.25, which has a total cost of 88,743.81. Instead of a cost reduction of 5,457.52 as estimated, interchanging departments C and F produced a cost reduction of 4,694.00.

The final CRAFT layout is obtained on the fourth iteration when departments I and J are interchanged to give a total cost of 87,963.81. The final layout is given in Figure 3.26. The last iteration produced a cost reduction of 780.00, rather than the estimated reduction of 1,410.00.

Notice the irregular configurations for some departments in the final layout, e. g., department L. Obviously, some manual adjustments are re-

	1	2	3	4	5	6	7	8	9	10	11	12	13	14	15	16
1	A	A	A	A	A	A	A	C	C	C	I	I	I	I	I	I
2	A					A	A	C		C	I					I
3	A					A	C	C	C	C	I	I	I	I	I	I
4	A					A	C	C	C	L	L	L	L	J	J	J
5	A					A	G	G	G	L			L	J		J
6	A	A	A	A	A	A	G	G	G	L			L	J	J	J
7	B	B	B	B	B	B	G	G	L	L			L	K	K	K
8	B				B	B	F	F	L			L	L	L	K	K
9	B				B	F	F	F	L			L	H	H	H	H
10	B				B	F		F	L			L	H			H
11	B	B	B	B	B	F	F	F	L		L	L	H			H
12	D	D	D	D	E	L	L	L	L	L	L	H	H	H	H	H
13	D		D	E	E	E	L	L	L	L	M	M	M	M	M	M
14	D		D	E		E	M	M	M	M	M					M
15	D		D	E		E	M									M
16	D	D	D	E	E	E	M	M	M	M	M	M	M	M	M	M

TOTAL COST 106874.06 EST. COST REDUCTION 10573.85

MOVEA H MOVEB L MOVEC INTERATION 1

Figure 3.23. First improved layout.

LOCATION PATTERN

	1	2	3	4	5	6	7	8	9	10	11	12	13	14	15	16
1	A	A	A	A	A	A	A	L	L	L	I	I	I	I	I	I
2	A					A	A	L		L	I					I
3	A					A	L	L		L	I	I	I	I	I	I
4	A					A	L	L	L	L	L	L	L	J	J	J
5	A					A	G	G	G	L			L	J		J
6	A	A	A	A	A	A	G	G	G	L			L	J	J	J
7	B	B	B	B	B	B	G	G	L	L			L	K	K	K
8	B				B	B	F	F	L			L	L	L	K	K
9	B				B	F	F	F	L			L	H	H	H	H
10	B				B	F		F	L	L	L	L	H			H
11	B	B	B	B	B	F	F	F	L	C	C	C	H			H
12	D	D	D	D	E	C	C	C	C	C	C	H	H	H	H	H
13	D		D	E	E	E	C	C	C	C	M	M	M	M	M	M
14	D		D	E		E	M	M	M	M	M					M
15	D		D	E		E	M									M
16	D	D	D	E	E	E	M	M	M	M	M	M	M	M	M	M

TOTAL COST 93437.81 EST. COST REDUCTION 11054.95
MOVEA C MOVEB L MOVEC ITERATION 2

Figure 3.24. Second improved layout.

LOCATION PATTERN

	1	2	3	4	5	6	7	8	9	10	11	12	13	14	15	16
1	A	A	A	A	A	A	A	L	L	L	I	I	I	I	I	I
2	A					A	A	L		L	I					I
3	A					A	L	L		L	I	I	I	I	I	I
4	A					A	L	L	L	L	L	L	L	J	J	J
5	A					A	G	G	G	L			L	J		J
6	A	A	A	A	A	A	G	G	G	L			L	J	J	J
7	B	B	B	B	B	B	G	G	L	L			L	K	K	K
8	B				B	B	C	C	L			L	L	L	K	K
9	B				B	C	C	C	L			L	H	H	H	H
10	B				B	C		C	L	L	L	L	H			H
11	B	B	B	B	B	C	C	C	L	F	F	F	H			H
12	D	D	D	D	E	C	C	F	F	F	F	H	H	H	H	H
13	D		D	E	E	E	F	F	F	F	M	M	M	M	M	M
14	D		D	E		E	M	M	M	M	M					M
15	D		D	E		E	M									M
16	D	D	D	E	E	E	M	M	M	M	M	M	M	M	M	M

TOTAL COST 88743.81 EST. COST REDUCTION 5457.52
MOVEA F MOVEB C MOVEC INTERATION 3

Figure 3.25. Third improved layout.

LOCATION PATTERN

	1	2	3	4	5	6	7	8	9	10	11	12	13	14	15	16
1	A	A	A	A	A	A	A	L	L	L	J	J	J	I	I	I
2	A					A	A	L		L	J		J	I		I
3	A					A	L	L		L	J	J	J	I		I
4	A					A	L	L	L	L	L	L	L	I		I
5	A					A	G	G	G	L			L	I		I
6	A	A	A	A	A	A	G	G	G	L			L	I	I	I
7	B	B	B	B	B	B	G	G	L	L			L	K	K	K
8	B				B	B	C	C	L			L	L	L	K	K
9	B				B	C	C	C	L			L	H	H	H	H
10	B				B	C		C	L	L	L	L	H			H
11	B	B	B	B	B	C	C	C	L	F	F	F	H			H
12	D	D	D	D	E	C	C	F	F	F	F	H	H	H	H	H
13	D		D	E	E	E	F	F	F	F	M	M	M	M	M	M
14	D		D	E		E	M	M	M	M	M					M
15	D		D	E		E	M									M
16	D	D	D	E	E	E	M	M	M	M	M	M	M	M	M	M

TOTAL COST 87963.81 EST. COST REDUCTION 1410.00
MOVEA J MOVEB I MOVEC INTERATION 4

Figure 3.26. Fourth (and final) improved layout.

quired to obtain a feasible layout. Furthermore, since a number of qualitative factors have been ignored in developing the final layout, the "minimum-cost" layout obtained from CRAFT is not necessarily a "better" layout than those obtained on preceding iterations. Remember, we have recommended that the computerized layout programs be used to stimulate the search process. It still remains to adjust the layouts obtained based on practical limitations and modifying considerations. Likewise, a separate evaluation from that performed within the computer program is strongly recommended.

3.7 Summary

The use of computers to generate layout designs is a relatively new concept. For example, CRAFT was proposed as recently as 1963, followed by ALDEP and CORELAP in 1967. Over a 5-year period there have been at least two major revisions of CORELAP, including the development of a time-shared version, Interactive CORELAP, and an advanced version utilizing graphical input and output devices with a PDP 15/30 computer along with two cathode-ray tubes such that the user can interact graphically with the CORELAP heuristic [14]. Also, in the past decade a number of other computerized layout approaches have been developed [1, 11, 13, 17, 27, 28]. The Center for Environmental Research [7] has compiled a catalog of over 80 computer programs of interest to layout analysts and architects.

There exists considerable interest in computerized layout planning. Furthermore, there is every reason to believe that this interest will continue in the future. As an illustration, consider the following forecast given by Anderson:[5]

> Imagine, if you will, the office of a plant layout engineer in the year 2000. Television screens flood one wall with 3-dimensional images of maintenance-free, totally automated assembly lines and conveyor systems. Miniaturized, desk-top computers monitor material handling traffic flow conditions and accurately predict future bottlenecks.
>
> Sensitivity testing in relation to changes in product mix and product volume and their effects on the layout efficiency is accomplished in a matter of minutes. Proposals for building layout arrangements, as well as relayout arrangements, are instantaneously coded into a preprogrammed project selection package; in a matter of seconds a feasible, accurate solution is obtained. Specific department relocation proposals can be analyzed for any given combinations, and outputs range from travel footage saved by material handling equipment to an associated cost savings. Finally, in our futuristic environment, the layout engineer can periodically evaluate an existing layout arrangement and arrive at

---

[5] Anderson, D. M., "New Plant Layout Information System," *Industrial Engineering*, Vol. 5, No. 4, (April 1973), pp. 32–37.

> an optimal pattern that will eliminate traffic congestion and increase plant efficiency—again with the help of the computer aided system. In short, the layout engineering team of the year 2000 will be comprised of one individual, armed with computerized technology, who can derive more meaningful and less costly solutions, more accurately and by far more quickly than his counterpart of the early 1970's.

At present, it appears that the major contribution of computerized approaches is in the *search phase* of the layout design process. Even though some of the computer programs include an *evaluation* of the layouts generated, we do not recommend that they be taken to be absolute. Both ALDEP and CORELAP employ arithmetic operations on data obtained from an ordinal measurement scale. CRAFT employs an objective function that is linear in the rectilinear distance traveled between department centroids. Obviously, ALDEP, CORELAP, and CRAFT will be better suited for use in some situations than others. All three programs should be viewed as design aids. The solutions obtained also can serve as benchmarks against which manually developed layouts are compared.

ALDEP and CORELAP are used when a layout is to be developed based on the REL chart. CRAFT is used when material-handling movement is a primary concern in layout design. Thus, we see that the development of computerized layout methods can be dichotomized into those which are concerned with flow of materials and those which are concerned with activity relationships. Consequently, we are brought face to face with the classical choice between qualitative and quantitative approaches in layout design. As we discussed in Chapter 2, there are a number of layout objectives that can only be expressed qualitatively. In such cases, we recommend the use of the REL chart as a design aid in developing layouts when qualitative considerations dominate. However, since there are also several layout objectives that can be quantitatively expressed, we recommend that analytical approaches be used as aids in designing the layout. In the remaining chapters of this book, we concentrate on those facility layout and location problems that are to be solved on the basis of quantitative objectives.

Quite obviously, we have not provided an exhaustive treatment of computerized layout planning. To the contrary, our discussion has been rather abbreviated. We only described three computerized layout procedures and, in some cases, our descriptions did not include a number of the optional features of the computer programs. We feel that our approach is quite justified, due to the obsolescence rate described previously. Consequently, if you desire a detailed understanding of how these computerized layout approaches work, we strongly recommend you obtain copies of the programs and acquire first-hand experience in their use.

The discussion of computerized layout planning should serve a useful purpose even if you are unable to use the computer programs. That is, the construction approaches of ALDEP and CORELAP and the improvement approach of CRAFT can be used when layouts are to be developed manually. The scoring techniques used by ALDEP and CORELAP can be modified to fit your own situation and manual operations employed in developing the layout.

Obviously, a number of plant layout and location problems do not justify computer analysis using, say, ALDEP, CORELAP, and/or CRAFT. As an example, if a single new facility is to be located in an existing layout, computerized layout programs probably would neither be justified nor appropriate. Likewise, if several new facilities are to be located in an existing layout, ALDEP, CORELAP, and CRAFT might not be appropriate solution techniques.

ALDEP, CORELAP, and CRAFT are used when there are, say, n new facilities (departments) of given areas to be located relative to each other. Since a number of layout problems involve the location of, say, n new machines relative to each other and m existing machines, alternative approaches are to be used in solving the location problem. Our subsequent discussion concerns the latter problem. Specifically, we shall employ analytic approaches in solving a number of facility location problems.

In closing, we feel that computerized layout planning can significantly improve the search phase of the layout design process. Since the solutions obtained from computerized algorithms can deviate significantly from traditional design approaches, the analyst is forced to consider new and provocative designs. Consequently, by using the computerized layout programs the analyst might obtain a significantly better design than would have been obtained otherwise.

Admittedly, we have pointed out the weaknesses in each of the computerized algorithms. This was done for two reasons. First, we did not want you to feel that the final layout given by the computer was necessarily the one best design. There exist too many people who believe that anything obtained from a computer is sacred and unquestionably correct. We wanted to be sure you did not join that group. Second, only by pointing out the weaknesses of the present methods can improvements be made. Who knows, perhaps this discussion will stimulate you to write your own computerized layout algorithm. On that optimistic note we close the discussion on computerized layout planning.

REFERENCES

1. Apple, J. M., and M. P. Deisenroth, "A Computerized Plant Layout Analysis and Evaluation Technique (PLANET)," *Proceedings*, American Institute of Industrial Engineers, 23rd Annual Conference and Convention, Anaheim, Calif., 1972.
2. Armour, G. C., and E. S. Buffa, "A Heuristic Algorithm and Simulation Approach to Relative Location of Facilities," *Management Science*, Vol. 9, No. 1, 1963, pp. 294–309.
3. Buffa, E. S., *Production-Inventory Systems: Planning and Control*, Richard D. Irwin, Inc., Homewood, Ill., 1968.
4. Buffa, E. S., G. C. Armour, and T. E. Vollmann, "Allocating Facilities with CRAFT," *Harvard Business Review*, Vol. 42, No. 2, 1964, pp. 136–159.
5. Campbell, N. R., *Measurement and Calculation*, Longman Group Ltd., Essex, England, 1928.
6. Churchman, C. W., and R. L. Ackoff, "An Approximate Measure of Value," *Operations Research*, Vol. 2, No. 2, 1954, pp. 172–181.
7. Denholm, D. H., and G. H. Brooks, "A Comparison of Three Computer Assisted Plant Layout Techniques," *Proceedings*, American Institute of Industrial Engineers, 21st Annual Conference and Convention, Cleveland, 1970.
8. Hall, A. D., *A Methodology for Systems Engineering*, Van Nostrand Reinhold Company, New York, 1962.
9. Kaiman, Lee, *Computer Architecture Programs* (3 vols., loose leaf), Center for Environmental Research, 955 Park Square Building, Boston, 1970.
10. Kaiman, Lee, "Computer Programs for Architects and Layout Planners," *Proceedings*, American Institute of Industrial Engineers, 22nd Annual Conference and Convention, Boston, 1971.
11. Krejeirik, M., "RUGR Algorithm," Technical Paper, Computer Aided Plant Layout and Design Seminar, Helsinki, March 1969 (reprinted in *Computer Aided Design*, Autumn 1969).
12. Lee, R. C., and J. M. Moore, "CORELAP-Computerized Relationship Layout Planning," *Journal of Industrial Engineering*, Vol. 18, No. 3, 1967, pp. 195–200.
13. Matto, R., "LAYOPT-General Purpose Layout Optimizing Program," unpublished working paper, Technical University, Helsinki, 1969.
14. Moore, J. M., "Computer Program Evaluates Plant Layout Alternatives," *Industrial Engineering*, Vol. 3, No. 8, 1971, pp. 19–25.
15. Moore, J. M., "Computer Aided Facilities Design: An International Survey," 2nd International Conference on Production Research (Copenhagen), Taylar & Francis Ltd., London, 1973.
16. Morris, W. T., *Analysis of Management Decisions*, Richard D. Irwin, Inc., Homewood, Ill., 1964.

17. MUTHER, R., and K. MCPHERSON, "Four Approaches to Computerized Layout Planning," *Industrial Engineering*, Vol. 2, 1970, pp. 39–42.

18. ROBERTS, S. D., "Computerized Facilities Design—An Evaluation," Department of Industrial and Systems Engineering, The University of Florida, Gainesville, Fl., TR30, Project THEMIS, ARO-D Contract No. DAH/CO4/68C/0002, Nov. 1969.

19. SEEHOF, J. M., and W. O. EVANS, "Automated Layout Design Program," *The Journal of Industrial Engineering*, Vol. 18, No. 12, 1967, pp. 690–695.

20. SEPPONEN, R., "CORELAP 7 User's Manual," unpublished working paper, Technical University, Helsinki, 1969.

21. SEPPONEN, R., "CORELAP 8 User's Manual," Department of Industrial Engineering, Northeastern University, Boston, 1969.

22. SIEGAL, S., *Nonparametric Statistics for the Behavioral Sciences*, McGraw-Hill Book Company, New York, 1956.

23. STEVENS, S. S., "Mathematics, Measurement, and Psychophysics," Chap. 1, *Handbook of Experimental Psychology*, John Wiley & Sons, Inc., New York, 1951.

24. VOLLMANN, T. E., and E. S. BUFFA, "The Facilities Layout Problem in Perspective," *Management Science*, Vol. 12, No. 10, 1966, pp. 450–468.

25. VOLLMANN, T. E., C. E. NUGENT, and R. L. ZARTLER, "A Computerized Model for Office Layout," *The Journal of Industrial Engineering*, Vol. 18, No. 7, 1968, pp. 321–327.

26. VON NEUMANN, J., and O. MORGENSTERN, *Theory of Games and Economic Behavior*, Princeton University Press, Princeton, N.J., 1953.

27. WHITEHEAD, B., and M. Z. ELDARS, "The Planning of Single Story Layouts," *Building Science*, Vol. 1, 1965, pp. 127–139.

28. ZOLLER, K., and K. ADENDORFF, "Layout Planning by Computer Simulation," *AIIE Transactions*, Vol. 4, No. 2, 1972, pp. 116–125.

PROBLEMS

3.1. Given the initial layout, flow matrix, and cost matrix shown in Figure P3.1, use the CRAFT pairwise exchange technique to obtain the final CRAFT layout.

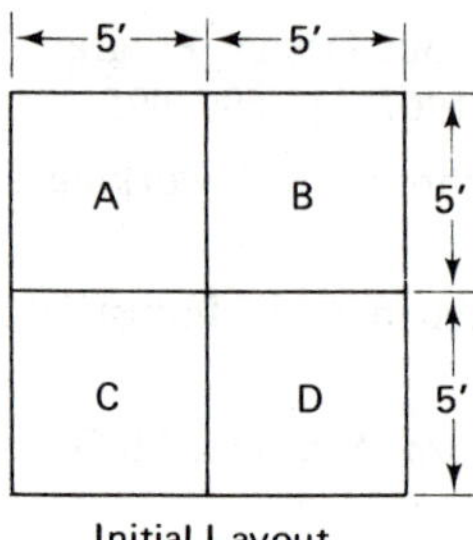

Initial Layout

From \ To	A	B	C	D
A		1	1	3
B	2		2	1
C	0	2		2
D	1	0	0	

Flow Matrix

From \ To	A	B	C	D
A		1	1	2
B	1		2	1
C	1	2		1
D	2	1	1	

Cost Matrix

Figure P3.1

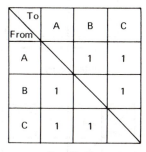

3.2. Given the initial layout, flow matrix, and cost matrix shown in Figure P3.2, use the CRAFT pairwise exchange technique to obtain the optimum layout.

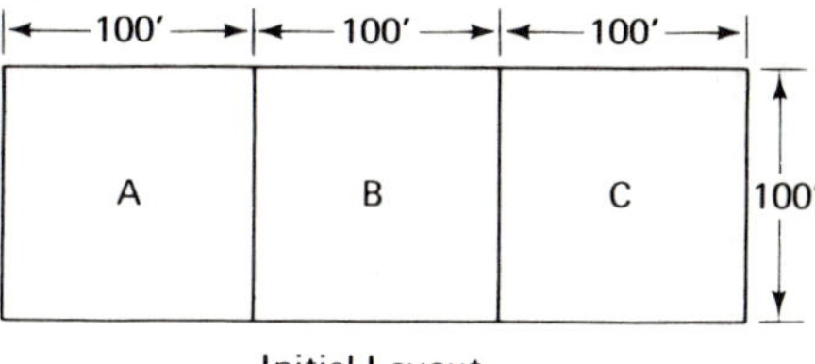

Initial Layout

From \ To	A	B	C
A		1	3
B	1		3
C	4	2	

Flow Matrix

From \ To	A	B	C
A		1	1
B	1		1
C	1	1	

Cost Matrix

Figure P3.2

3.3. Given the initial layout, flow matrix, and cost matrix shown in Figure P3.3, use the CRAFT pairwise exchange technique to obtain the final CRAFT layout. If there are multiple optima, specify each.

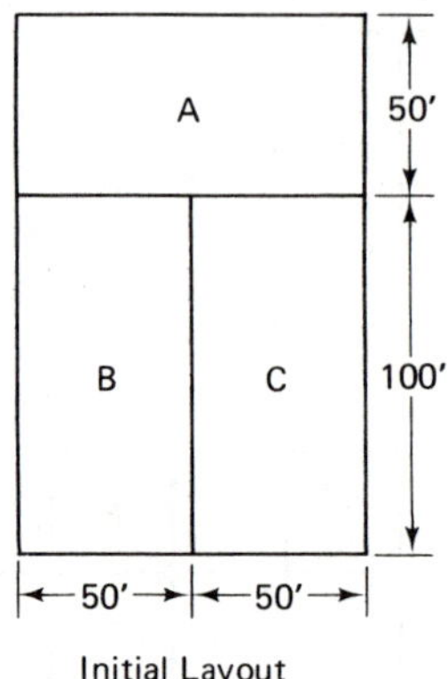

Initial Layout

To / From	A	B	C
A		4	1
B	2		2
C	2	1	

Flow Matrix

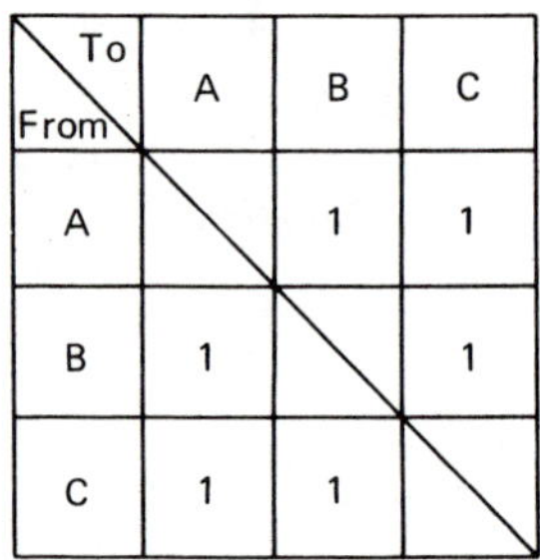

To / From	A	B	C
A		1	1
B	1		1
C	1	1	

Cost Matrix

Figure P3.3

3.4. A manufacturing concern has four departments located in two buildings, as shown in Figure P3.4. The firm is contemplating a rearrangement of departments. They wish to obtain a final CRAFT layout using the pairwise interchange technique. The initial layout and other pertinent data are given below:

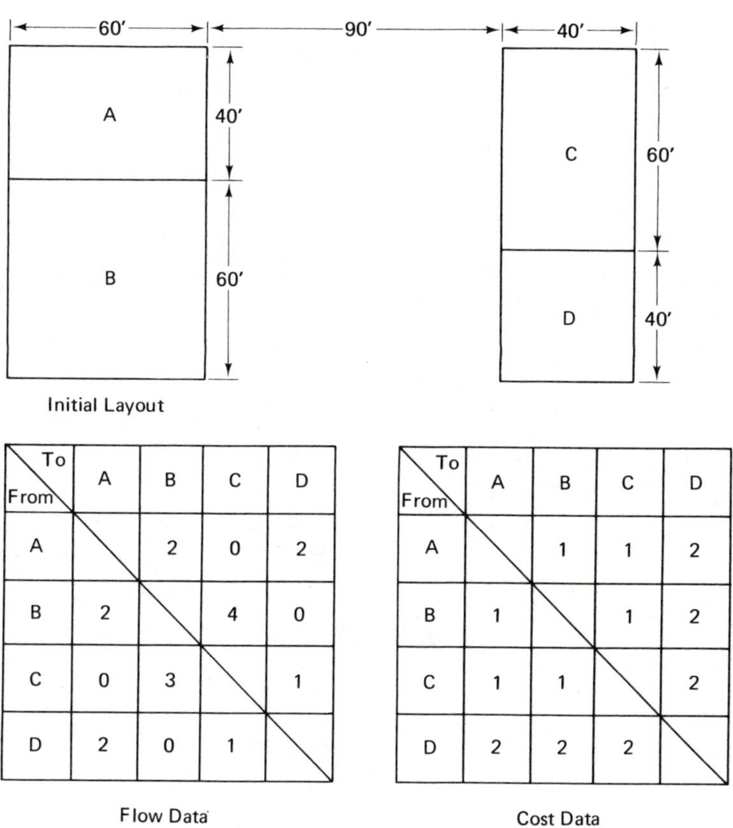

From \ To	A	B	C	D
A		2	0	2
B	2		4	0
C	0	3		1
D	2	0	1	

Flow Data

From \ To	A	B	C	D
A		1	1	2
B	1		1	2
C	1	1		2
D	2	2	2	

Cost Data

Figure P3.4

3.5. Follow the steps of the CRAFT algorithm and develop a final CRAFT layout using the pairwise interchange technique and the data shown in Figure P3.5.

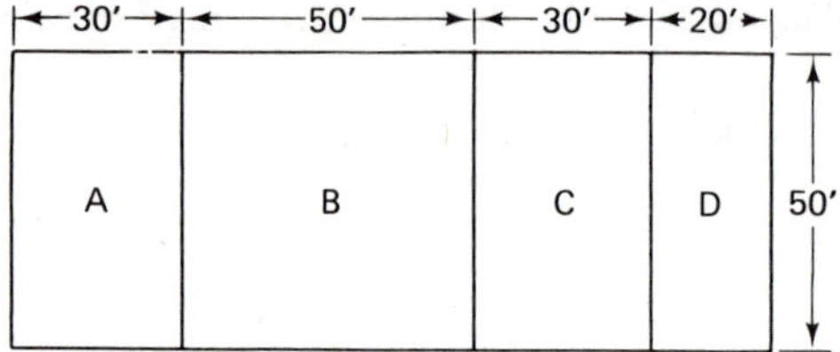

Initial Layout

From \ To	A	B	C	D
A		3	4	0
B	2		0	4
C	2	0		0
D	0	2	0	

Cost Data

From \ To	A	B	C	D
A		1	2	1
B	1		1	2
C	2	1		1
D	1	2	1	

Flow Data

Figure P3.5

3.6. Based on the data given in Figure P3.6, use the CRAFT pairwise interchange technique to obtain the final CRAFT layout. Departments C and E are to be fixed in location.

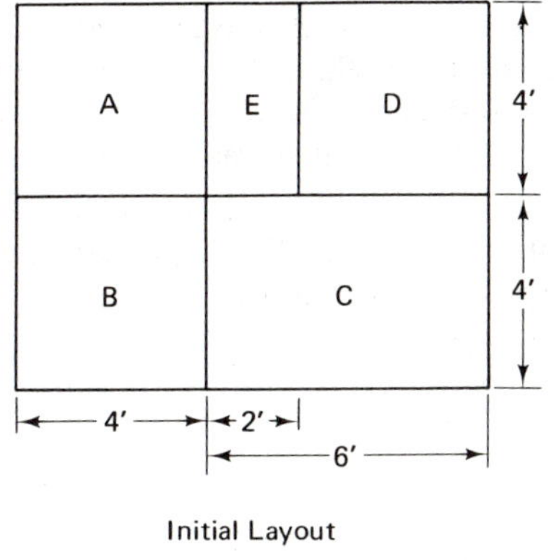

Initial Layout

From \ To	A	B	C	D	E
A		1	2	2	0
B	0		2	0	0
C	2	0		0	0
D	3	0	1		0
E	0	0	0	0	

Flow Data

From \ To	A	B	C	D	E
A		1	1	1	1
B	1		1	1	1
C	1	1		1	1
D	1	1	1		1
E	1	1	1	1	

Cost Data

Figure P3.6

3.7. Solve the ALDEP example problem given in Section 3.4 by using (a) the CORELAP program and (b) the CRAFT program by assigning numerical values to the closeness ratings. Compare the solutions obtained.

3.8. Solve the CORELAP example problem given in Section 3.5.1 by using (a) the ALDEP program and (b) the CRAFT program by assigning numerical values to the closeness ratings. Compare the solutions obtained.

3.9. Consider the CRAFT example problem associated with Figure 3.17. Design a layout that would have a zero cost using the CRAFT objective function. Why is such a solution not likely to occur when the CRAFT program is used?

3.10. Manually adjust the layout given in Figure 3.26 such that departments F and L take on more realistic shapes, and use the modified layout as the initial layout for a CRAFT solution. Compare the new "final layout" with that given in Figure 3.26. How much, if any, cost reduction was obtained?

3.11. The Flimsy Furniture Company consists of 10 departments having areas on the REL chart given below. The initial layout, Figure P3.11(a), and the flow data, and the cost data, Figure P3.11(b), are given below.

(a) Obtain a recommended layout using SLP.
(b) Solve part (a) using ALDEP.
(c) Solve part (a) using CORELAP.

H–Shipping 20′ x 40′
E–Fabric Storage 60′ x 16′
I–Offices 20′ x 40′
G–Sewing 16′ x 40′
D–Upholstery 40′ x 28′
J–General Storage 40′ x 12′
F–Fabric cutting 24′ x 40′
C–Framing 32′ x 40′
B–Receiving 32′ x 16′
A–Wood cutting 32′ x 40′

Current Layout

Figure P3.11(a)

(d) Solve part (a) using CRAFT.

(e) Modify the results obtained, perform a subjective evaluation, and recommend the preferred layout.

	A	B	C	D	E	F	G	H	I	J
A		2	1						2	
B	6				6	2				2
C	1	2		4					2	
D	1	1	4			4	2	3	2	2
E						5	3		2	
F				4	5		4		2	
G				2	2	4			2	
H					3				2	
I	1	1	1	1	1	1	1	1		1
J				2			1	1	2	

Flow Data

	A	B	C	D	E	F	G	H	I	J
A		2	2						1	
B	3				2	2				2
C	1	1		5					1	
D	1	1	2			2	2	3	1	1
E						4	3		1	
F				1	3		2		1	
G				4	2	2			1	
H					2				1	
I	1	1	1	1	1	1	1	1		1
J				1			1	2	1	

Cost Data

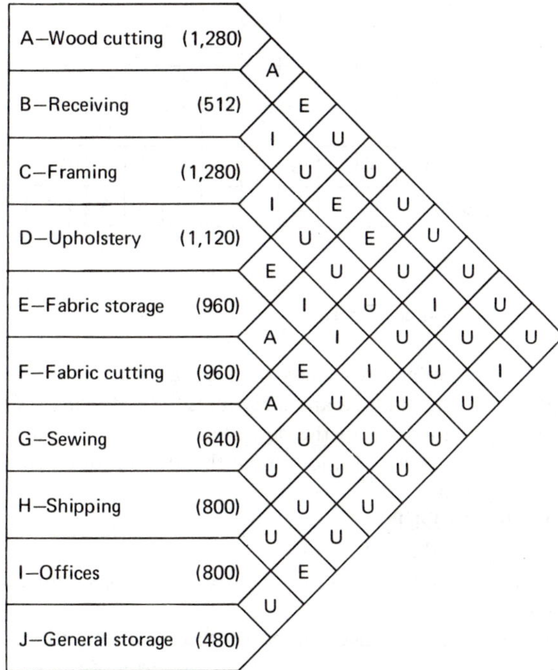

Relationship Chart

Figure P3.11(b)

3.12. The Weird Widget Works is planning on locating their new production plant in the suburbs of Prices Fork, Virginia. Eleven departments are to be located in the production plant. Activity relationships and departmental areas for the new plant are summarized in Figure P3.12.

(a) Develop a layout using the SLP approach.
(b) Use the ALDEP heuristic to design the layout.
(c) Use the CORELAP heuristic to design the layout.
(d) Assign numerical values to the closeness ratings, and use the CRAFT heuristic to design the layout.
(e) What assumptions are inherent in the CRAFT solution?

Dept. No.	Department	Area (ft^2)										
11	Lobby	600										
12	General manager's office	400	A									
13	Bookkeeping	350	E	E								
14	Supervisor's office	150	U	E	U							
15	Snack bar	350	O	U	O	O						
16	Maintenance	150	U	U	U	E	O					
17	Receiving and shipping	200	X	U	O	O	U	U				
18	Storage	800	U	U	O	O	U	O	A			
19	First aid	200	U	U	O	O	U	I	O	O		
20	Rest rooms	350	U	U	O	O	I	O	O	O	O	
21	Production	5,000	U	O	U	A	I	E	A	A	E	I

Relationship Chart

Figure P3.12

3.13. Carry-On Baggage, Inc. manufactures carry-on luggage for airline travel. A new production plant is being planned and a layout is to be designed. Fourteen activities are to be provided for in the new plant. The relationships for the activities are given in Figure P3.13, along with the area requirements.[6]

(a) Design a layout using SLP.
(b) Design a layout using ALDEP.
(c) Design a layout using CORELAP.
(d) Assign appropriate numerical values to the closeness ratings, and design a layout using CRAFT.
(e) Modify the layout designs obtained (based on practical limitations and modifying considerations) and recommend your preferred design.

[6]Based on a problem entitled "Novelty Luggage" developed by Richard Muther, Richard Muther and Associates, Kansas City, Mo.

Activity	Area (ft^2)													
Cutting	1,500													
Art area	500	U												
Dark room	500	U	E											
Silk screen	2,000	E	O	E										
Inspection	1,500	U	U	U	U									
Subassembly	1,000	I	U	U	U	U								
Final assembly	6,000	U	U	U	E	E	E							
Receiving and shipping	2,500	U	U	X	O	I	U	U						
Material storage	1,500	A	U	U	U	I	I	U	A					
Finished stores	2,500	U	U	U	U	E	U	U	A	I				
Maintenance	500	U	U	U	U	U	O	I	U	U	U			
Office	2,500	I	O	O	I	O	U	O	I	O	O	U		
Rest rooms	500	U	U	U	O	U	O	O	O	U	U	U	O	
Lunch room	1,500	U	U	U	O	U	O	O	O	U	U	U	O	O

Figure P3.13

3.14. The town council of the Burg of Black has decided to construct a new municipal building to replace the old building, which was recently condemned. A study has been made and data were collected. The number of daily interdepartmental personnel contacts, departmental area requirements, closeness ratings between departments, and relative ratings of the importance of each department have been obtained. The importance rating is a composite evaluation based on the number of people in the department and the average hourly wage of the departmental personnel.

(a) Using the REL chart data, Figure P3.14(a), design a layout for the new municipal building using (1) ALDEP and (2) CORELAP.

(b) Using the interdepartmental personnel flow data, the departmental areas, and the departmental ratings, Figures P3.14(a) and (b), design a layout for the new municipal building using CRAFT.

Dept.	Area (ft²)	Ratings	M	L	K	J	I	H	G	F	E	D	C	B	A
A	1,500	12	U	I	U	U	U	U	U	U	O	O	I	A	
B	650	5	U	U	U	U	U	U	U	U	U	U	I		
C	3,800	5	I	U	U	U	U	U	U	U	U	A			
D	300	11	U	U	U	U	U	U	U	U	U				
E	900	8	U	U	U	U	U	U	U	E					
F	900	8	U	U	U	U	I	U	E						
G	900	8	U	U	I	E	U	E							
H	900	8	U	U	I	U	E								
I	900	10	U	U	I	A									
J	2,300	13	U	U	I										
K	900	8	U	U											
L	300	9	A												
M	750	6													

Relationship Chart

Figure P3.14(a)

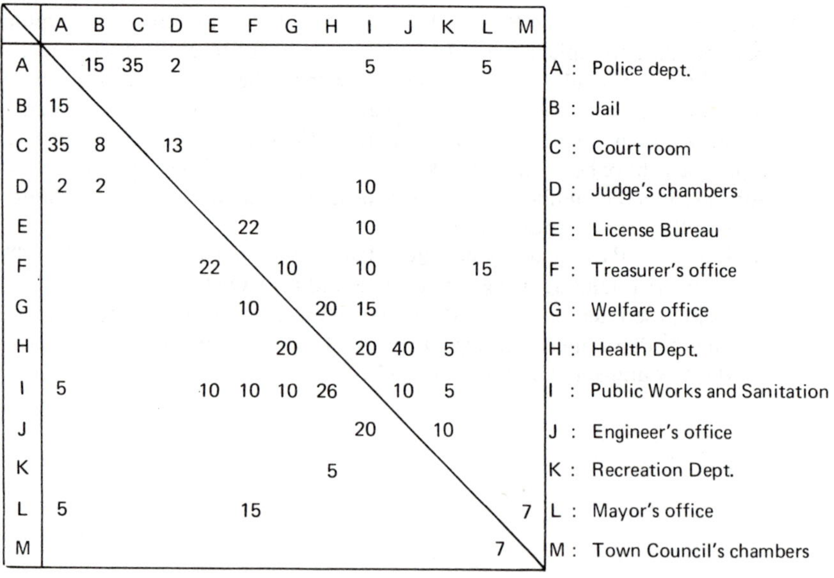

	A	B	C	D	E	F	G	H	I	J	K	L	M
A		15	35	2					5			5	
B	15												
C	35	8		13									
D	2	2							10				
E						22			10				
F					22		10		10			15	
G						10		20	15				
H							20		20	40	5		
I	5				10	10	10	26		10	5		
J									20		10		
K								5					
L	5					15							7
M												7	

A : Police dept.
B : Jail
C : Court room
D : Judge's chambers
E : License Bureau
F : Treasurer's office
G : Welfare office
H : Health Dept.
I : Public Works and Sanitation
J : Engineer's office
K : Recreation Dept.
L : Mayor's office
M : Town Council's chambers

Interdepartmental Personnel Flow Data

Figure P3.14(b)

3.15. A layout is to be designed for the engineering department of the G. W. Quarles, Corp. A study of the face-to-face contacts per month between the members of the 10 groups within the department yielded the data shown below in Figure P3.15(a). Also shown are the area requirements and average wage rates for the groups. The current layout is also given, Figure 3.15(c). Obtain the

Area Requirements and Average Wage Rate Data

No.	Group	Area (ft^2)	Average Wage ($hr)
01	Filing	400	3.00
02	Supervision	600	10.50
03	Blueprint	750	3.50
04	Product support	750	4.00
05	Structural design	1,500	8.50
06	Electrical design	750	8.50
07	Hydraulic design	1,250	8.50
08	Mechanical design	1,000	8.50
09	Systems design	750	8.50
10	Safety design	750	8.50
11	Production liaison	1,250	4.00
12	Detailing and checking	1,250	6.50
13	Secretarial pool	1,000	3.50

Figure P3.15(a)

From \ To	01	02	03	04	05	06	07	08	09	10	11	12	13
01	–	15	0	0	0	5	5	10	15	10	10	0	15
02	40	–	25	40	100	90	80	70	40	50	160	85	60
03	0	20	–	0	0	0	0	0	0	0	0	0	0
04	10	60	0	–	0	0	20	30	210	100	280	0	10
05	50	120	600	0	–	40	0	60	0	50	0	340	50
06	5	90	350	0	40	–	0	20	60	40	160	270	60
07	10	80	400	40	20	0	–	30	0	10	140	320	45
08	10	70	350	30	10	10	60	–	10	20	150	310	40
09	15	110	215	60	10	15	10	20	–	60	210	180	110
10	30	85	75	85	120	115	105	135	60	–	180	150	105
11	30	160	0	450	160	350	380	300	400	100	–	0	680
12	0	300	700	40	550	510	400	410	250	370	0	–	20
13	20	90	0	10	50	60	50	30	410	320	680	30	–

Number of Face-to-Face Contacts per Month

Figure P3.15(b)

final CRAFT layout using (a) two-way exchanges, (b) three-way exchanges, (c) best of two-way and three-way exchanges.[7]

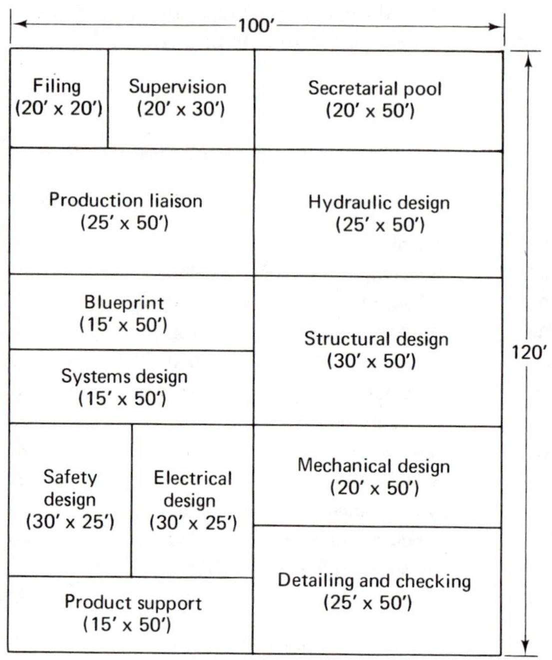

Initial Layout

Figure P3.15(c)

[7]Based on a problem given by E. S. Buffa, *Modern Production Management*, 3rd Edition, John Wiley & Sons, Inc., New York, 1969.

3.16. Manufactured Products, Inc., has requested a layout study of their White Pine, Tennessee, plant. The plant consists of 17 departments. Fork lift trucks are used for all material handling. A summary of the number of truck loads delivered per day between combinations of departments is given in Figure P3.16. Obtain the final CRAFT layout using pairwise exchanges.

No.	Department	Area (ft^2)	01	02	03	04	05	06	07	08	09	10	11	12	13	14	15	16	17
1	Receiving	500		100															
2	Raw material storage	1,000			25	22	30	62		15							18		
3	Shearing	200											20	5					
4	Sawing	200						10	5				5						
5	Automatic screw machine	4,000						5	10		7	5					3		
6	Turret lathe	2,000							9		5	3	4	2			31		
7	Engine lathe	500						1			3		10	8			20		
8	Punch press	1,000											15	10	5	7	8		
9	Hobbing	400							2			2	5		8				
10	Broaching	300						2	3					2	3				
11	Milling	3,000						2						30	25	7	1		
12	Drilling	1,200						24	15			2	5		5	2	5		
13	Heat treating	500						2								15	30		
14	Plating	300							1								32		
15	Assembly	700																175	
16	Finished goods storage	1,000																	175
17	Shipping	500																	

Figure P3.16

3.17. The Agee Mechanical Works includes 10 departments with areas shown on the REL chart given in Figure P3.17.[8]

(a) Develop a layout for the plant using SLP.
(b) Solve part (a) using the ALDEP heuristic.
(c) Solve part (a) using the CORELAP heuristic.
(d) Solve part (a) using the CRAFT heuristic by assigning numerical values to the closeness ratings and by treating the converted REL chart as flow data.
(e) Consider the designs obtained in parts (a) through (d) and make appropriate modifications to obtain a preferred layout.

Dept. No.	Department	Area (ft^2)									
301	Lobby	600									
302	Plant manager's office	400	A								
303	Accounting	330	E	E							
304	Supervisor's office	180	U	O	U						
305	Snack bar	340	O	U	O	O					
306	Maintenance	140	U	U	U	E	O				
307	Receiving and storage	690	X	U	O	O	U	U			
308	First aid	210	U	U	O	O	U	O	O		
309	Rest rooms	350	U	U	O	O	I	O	O	O	
310	Production	4,800	U	O	U	A	I	E	A	E	I

Relationship Chart

Figure P3.17

3.18. The information systems center of a metropolitan bank consists of 9 departmental areas. The current layout, flow matrix, and cost matrix for the information center are shown below in Figures P3.18(a) and (b). Flow data for the information center are based on the interaction of the various departments in terms of the number of trips made per day between the departments. The cost data reflect the average salaries of those employed in the various departments and the relative importance of their roles in the operation of the center. (The cost data were approved by C. B. Quarles, Vice President of Operations for the bank.)

(a) Use the CRAFT program, with two-way exchanges, to obtain the final CRAFT layout.
(b) Solve part (a) using three-way interchanges.
(c) Solve part (a) assuming Department I is to be fixed in location.

[8]Based on a problem entitled "ABC Mechanical Works," developed by Richard Muther, Richard Muther and Associates, Kansas City, Mo.

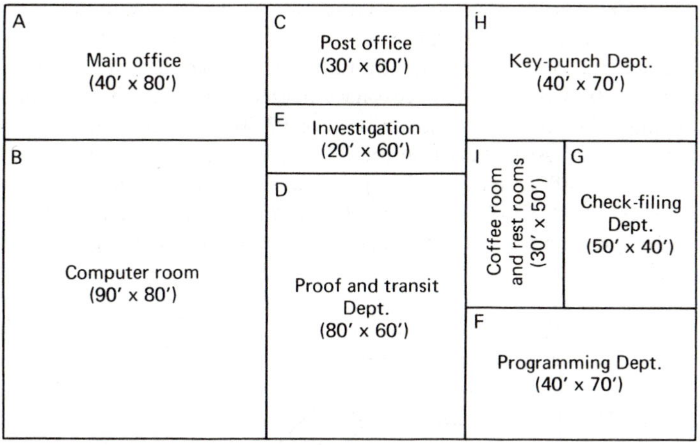

A Main office (40′ x 80′)

C Post office (30′ x 60′)

H Key-punch Dept. (40′ x 70′)

E Investigation (20′ x 60′)

B Computer room (90′ x 80′)

D Proof and transit Dept. (80′ x 60′)

I Coffee room and rest rooms (30′ x 50′)

G Check-filing Dept. (50′ x 40′)

F Programming Dept. (40′ x 70′)

Current Layout

Figure P3.18(a)

	A	B	C	D	E	F	G	H	I
A		15	1	3	2	4	3	2	2
B	10		2	5	3	15	20	30	9
C	1	4		40	6	1	14	5	6
D	4	8	25		10	5	15	7	10
E	6	5	2	20		1	20	2	5
F	7	40	1	3	1		20	10	0
G	2	8	4	25	15	4		0	15
H	5	30	6	3	2	30	5		15
I	0	0	0	0	0	0	0	0	

Flow Data

	A	B	C	D	E	F	G	H	I
A		5	5	5	5	5	5	5	5
B	3		3	3	3	3	3	3	3
C	1	1		1	1	1	1	1	1
D	2	2	2		2	2	2	2	2
E	2	2	2	2		2	2	2	2
F	4	4	4	4	4		4	4	4
G	2	2	2	2	2	2		2	2
H	3	3	3	3	3	3	3		3
I	0	0	0	0	0	0	0	0	

Cost Data

Figure P3.18(b)

3.19. Use the CRAFT program to obtain an improvement over the process layout given in Figure 2.9.

3.20. The JAW Corporation owns a large plot of land in the mountains of Louisiana and plans to design a planned community. The community will offer housing and other facilities to suit all types of people and families. Also, the community will offer the necessary emergency, recreational, and commercial facilities. The REL chart given, Figure P3.20, was developed by a group of city planners. Also shown on the REL chart are the estimated land areas (in acres) to be allocated to each type of facility. Compare the designs obtained from ALDEP, CORELAP, and CRAFT. Do you feel that computerized layout programs can be beneficial in community planning? If so, how? If not, why not?

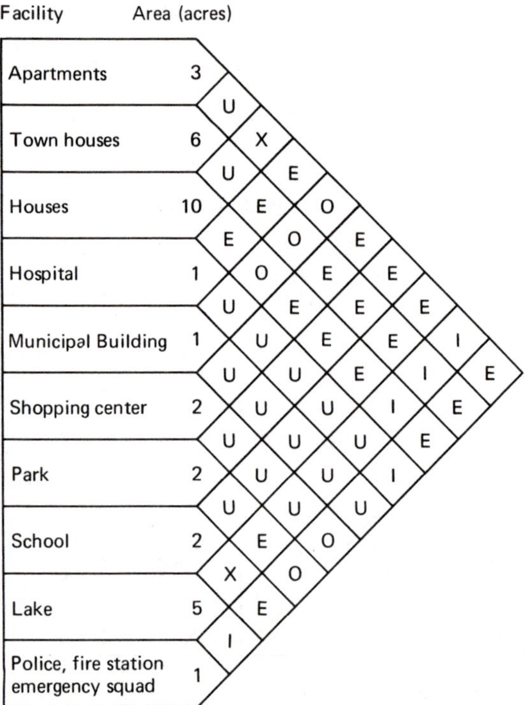

Facility
Area (acres)
Apartments 3
Town houses 6
Houses 10
Hospital 1
Municipal Building 1
Shopping center 2
Park 2
School 2
Lake 5
Police, fire station emergency squad 1
U
X
U E
E O
E O E
O E E
U E E E
U E E I
U U E I E
U U I E
U U U E
U U I
U U U
E O
X O
E
I

Figure P3.20

3.21. The V.I.P. University Library is located in a single-story building and is logically divided into 16 activities. The area requirements for the activities, as well as the REL chart, are given in Figure P3.21.

(a) Using SLP, develop a layout for the library.
(b) Using ALDEP, develop a layout for the library.
(c) Using CORELAP, develop a layout for the library.
(d) Using CRAFT, develop a layout for the library.
(e) Compare the results obtained in parts (a) through (d).

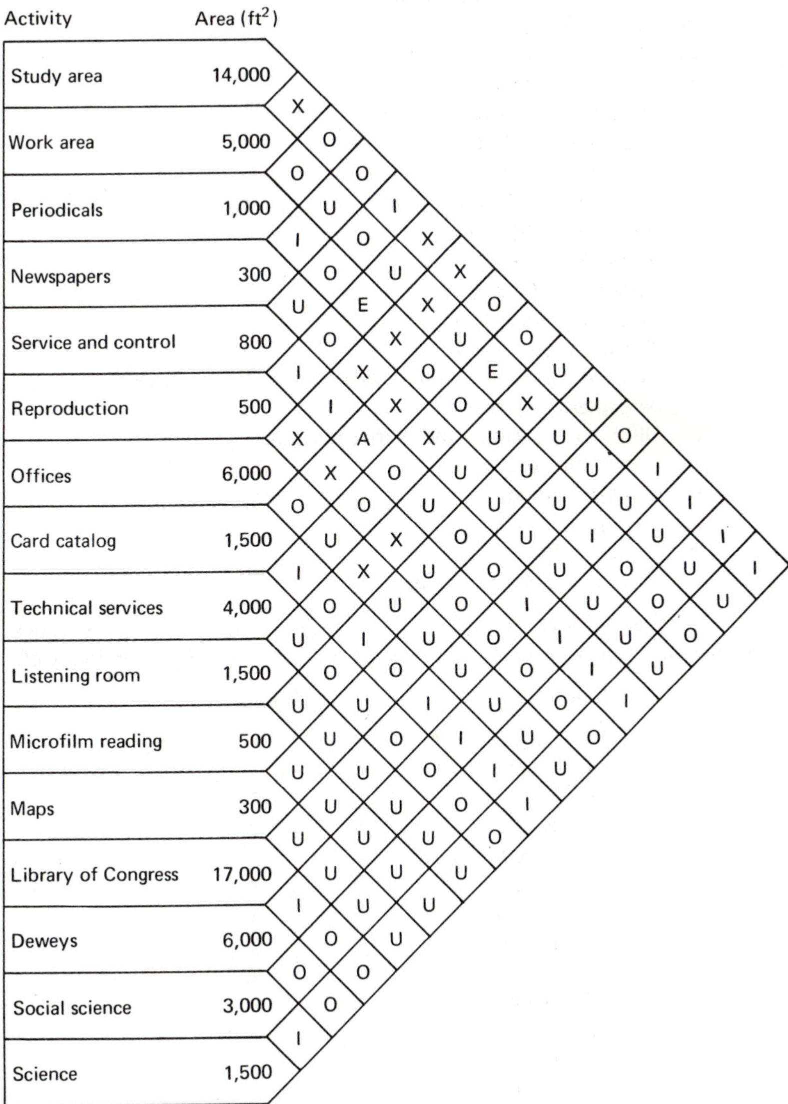

Figure P3.21

3.22. Obtain an ALDEP and CORELAP solution to Problem 2.26.

3.23. Obtain a CRAFT solution to Problem 2.27.

3.24. Obtain an ALDEP and CORELAP solution to Problem 2.28.

3.25. Obtain an ALDEP, CORELAP, and CRAFT solution to Problem 2.29.

3.26. A department in a university is to get new offices and associated facilities. A prior analysis has determined that the following activities should each be placed in a different room, or be treated separately for some reason:

A. Full professor 1 and chairman
B. Full professor 2
C. Full professor 3
D. Full professor 4
E. Associate professor 1
F. Associate professor 2
G. Associate professor 3
H. Associate professor 4
I. Assistant professor 1
J. Assistant professor 2
K. Assistant professor 3
L. Teaching associate 1
M. Teaching associate 2
N. Main departmental office, with equipment for three secretaries, and two small couches
P. Office supply room and Xerox machine
Q. Conference room
R. Student lounge
S. Men's room
T. Ladies' room
U. Four entrances and exits

Annual salaries (hypothetical) of faculty and teaching associates are shown below, as well as the time, in days per week, actually spent in the offices:

Salary ($)	Time in Days
A. 25,000	4
B. 20,000	3
C. 20,000	2
D. 16,000	2.5
E. 17,000	3
F. 16,000	3
G. 16,000	4
H. 16,000	3.5
I. 15,000	4
J. 15,000	4
K. 15,000	3.5
L. 11,000	2
M. 11,000	2

Faculty member A requires in his office one desk and chair, three file cabinets, one table to accommodate eight people, one working table, 20 linear feet of book shelves, and one storage locker. Other faculty require one desk and chair, two file cabinets, one working table, two chairs for visitors, 20 linear feet of book shelves, and one storage locker. Faculty member B needs one extra working table, as well as a couch. It is your job to estimate the space requirements of each activity.

Two hundred students are enrolled in the department; 100 of the students take an average of three courses per term in the department, 50 take an average of four, and 50 take an average of five. It is your job to estimate the time students spend in the lounge, as well as each student's "value" per hour. Faculty members G and I are designated student advisors and, on the average, each advises five students per day.

The conference room should have tables and seating space for 25 people. Hallways should be 8 feet wide. Two of the entrances and exits handle 80% of the traffic (divided equally between the two); the remaining 20% is divided equally between the other two entrances and exits.

Relationship Chart

	A	B	C	D	E	F	G	H	I	J	K	L	M	N	P	Q	R	S	T
A		I	I	O	O	O	O	O	O	O	O	O	O	A	E	A	O	O	U
B			I	O	I	O	E	O	O	O	E	O	O	O	I	O	O	O	U
C				O	O	I	O	I	O	O	X	O	O	O	I	O	O	O	U
D					I	X	E	O	O	O	O	O	O	I	I	O	I	O	U
E						I	I	O	O	O	O	O	O	O	I	O	O	O	U
F							I	I	O	E	X	O	O	O	I	O	O	O	U
G								I	E	O	O	O	O	O	I	O	E	O	U
H									O	O	O	O	O	O	I	O	I	O	U
I										E	I	O	O	O	I	O	E	O	U
J											I	O	O	O	I	O	O	O	U
K												O	O	O	E	O	O	O	O
L													E	O	E	O	O	O	U
M														O	I	O	O	O	U
N															E	U	U	O	I
P																U	U	U	U
Q																	U	O	U
R																		O	U
S																			A

Figure P3.26(a)

	A	B	C	D	E	F	G	H	I	J	K	L	M	N	P	Q	R	S	T
A		2	3	.5	.25	.1	.1	.05	.6	.1	2	.1	.1	10	1	3	1	4	0
B			2	1	.1	.1	3	.6	.05	1	3	1	1	2	.5	.3	1	4	0
C				1	1	3	.1	.2	.05	.1	.05	.3	.2	3	3	.2	1	3	0
D					2	.04	3	.7	.05	.9	.9	.1	.2	4	2	.3	3	6	0
E						2	2	.7	.6	.4	.8	.1	.1	3	2	.2	2	4	0
F							2	2	.1	3	.02	.1	.2	2	3	.2	1	4	0
G								2	4	.4	.3	.2	.2	3	4	.4	6	3	0
H									.3	.4	.1	.2	.1	2	2	.2	5	4	0
I										4	2	.2	.2	2	1	.5	6	3	0
J											2	1	1	1	2	.2	3	3	0
K												1	.5	2	6	.2	1	0	6
L													3	2	7	.3	1	2	0
M														1	1	.2	2	2	0
N															20	2	7	0	10
P																.1	0	0	6
Q																	0	0	0
R																		20	10
S																			0

Figure P3.26(b)

There are a number of other activities the department carries on, of course, but these take place elsewhere, and need not concern you.

An REL chart for the activities is given below, see Figure P3.26(a), activity U has been intentionally omitted; it is your problem to determine how to treat activity U. Also given below is a from–to chart, Figure P3.26(b) giving the estimated number of round trips per day between activities; the data on faculty are only for those days when the faculty member is in the office. Again, activity U has been intentionally omitted.

(a) Develop a layout using the SLP approach.
(b) Develop a layout using the ALDEP program.
(c) Develop a layout using the CORELAP program.
(d) Develop a layout using the CRAFT program.
(e) Evaluate the layouts obtained, pointing out the advantages and disadvantages of each.
(f) Obtain a recommended layout design.

3.27. Use ALDEP, CORELAP, and CRAFT to design alternative layouts for a problem of your choice. Typical sources of layout problems at a university are as follows: computing center, infirmary, library, accounting, cafeteria, book store, post office, warehouse, and academic buildings.

3.28. Apply the computerized layout programs in designing a campus plan (layout) for the university.

3.29. Apply the computerized layout programs in designing layouts for local establishments, such as the following: bank, doctor's office, dentist's office, grocery store, library, municipal building, lumberyard, automotive service department, fire station, school, and church.

3.30. If you were asked to write a computer program to aid in the design of layouts, what features of ALDEP, CORELAP, and CRAFT would you incorporate in your program? What new features would you add to your program?

3.31. In presenting the subject matter of this chapter to a group of practicing engineers, one engineer commented, "We do not have access to a computer, so we can't use computerized layout programs." Another commented, "You have to make too many modifications to the output when you use computerized layout programs. You are better off doing the whole thing manually." How would you respond to these comments? (*Hint:* You flunk the course if you agree with them!)

3.32. Rather than employ the CRAFT objective of minimizing the sum of the material handling costs between departments, conside the objective of minimizing the maximum cost between the departments. Manually, obtain a "minimax" layout solution for the example problem depicted in Figure 3.16.

SINGLE-FACILITY LOCATION PROBLEMS

4.1 Introduction

In Chapter 3 we described the use of computerized layout programs as design aids in developing alternative layout designs. In that discussion, we pointed out that not all layout problems justify the use of ALDEP, CORELAP, or CRAFT. As an example, a number of layout and location problems involve the location of a single new facility in an existing layout or the design of a layout for a single department.

In our previous discussion we have emphasized the fact that qualitative, as well as quantitative, factors should be considered in solving facility layout and location problems. However, we also have noted that an explicit consideration of both qualitative and quantitative factors in developing the "optimum" solution is a difficult undertaking. Qualitative factors are not easily measured on an interval or ratio scale. Consequently, it is difficult to combine qualitative and quantitative factors in a quantitative manner such that a satisfactory evaluation of alternative solutions is obtained.

Typically, we find that a number of layout and location problems are solved using a quantitative objective. Subsequently, the analytical solution is modified based on qualitative considerations. It is our belief that this process is normally quite satisfactory, and superior to an approach that

4

neglects quantitative considerations. The analytic solution serves as a benchmark against which other solutions can be compared. Consequently, henceforth we concentrate on developing analytic solutions to facility layout and location problems with the view that such solutions are to be interpreted as design aids and might undergo subsequent revisions based on qualitative factors. We view the role of analytical analysis to be a stimulant to the search phase of the layout design process described in Chapter 1.

In this chapter we consider the problem of determining the location of a single new facility with respect to a number of existing facilities. The location we seek is that which minimizes an appropriately defined total cost function. As discussed in Chapter 1, we consider total cost to be proportional to distance.

A number of interesting one-facility location problems exist and are amenable to the analysis presented in this chapter. Some typical examples of one-facility location problems are the location of a

1. New lathe in a manufacturing job shop.
2. Tool crib in a manufacturing facility.

3. New warehouse relative to production facilities and customers.
4. Hospital, fire station, police station, or library in a metropolitan area.
5. New classroom building on a college campus.
6. New airfield to be used to provide supplies for a number of military bases.
7. New pump in a chemical operation.
8. Component in an electrical network.
9. Wrecker along a section of highway in anticipation of traffic accidents.
10. New component on a control panel.
11. Loading dock for a warehouse.
12. New appliance in a kitchen.
13. Water fountain in an office building.
14. New power generating plant.
15. Copying machine in a library.

These examples suggest a reason for the interdisciplinary interest in facility location and design discussed in Chapter 1.

A general formulation of the problem considered in this chapter may be given as follows: m existing facilities are located at known distinct points $\mathbf{P}_1, \ldots, \mathbf{P}_m$; a new facility is to be located at a point $\mathbf{X}$; costs of a transportation nature are incurred that are directly proportional to an appropriately determined distance between the new facility and existing facility i. Letting $d(\mathbf{X}, \mathbf{P}_i)$ represent the distance traveled per trip between points $\mathbf{X}$ and $\mathbf{P}_i$, and letting w_i represent the product of cost per unit distance traveled and number of trips made per year between the new facility and existing facility i, the total annual cost due to travel between the new facility and all existing facilities is given by

$$f(\mathbf{X}) = \sum_{i=1}^{m} w_i d(\mathbf{X}, \mathbf{P}_i) \tag{4.1}$$

where the w_i terms are sometimes referred to as "weights." The one-facility location problem is to determine the location of the new facility, say $\mathbf{X}^*$, that minimizes $f(\mathbf{X})$, the total annual transportation cost.

Dimensionally, $f(\mathbf{X})$ is expressed as dollars per year, w_i is expressed as (dollars per distance)(trips per year), and $d(\mathbf{X}, \mathbf{P}_i)$ has the dimensions distance per trip. Thus, if $d(\mathbf{X}, P_i)$ equals 5 miles per trip and w_i equals the product of \$0.10 per mile and 700 trips per year, then $w_i d(\mathbf{X}, \mathbf{P}_i)$ equals \$350 per year.

In many applications the cost per unit distance is a constant. Thus, the minimization problem often reduces to a determination of the location that minimizes distance.

As a very simple example of a one-facility location problem, the new facility might be a factory supplying warehouses, which are the existing facilities, and transportation costs are incurred that are proportional to the distances between the factory and the warehouses; the term w_i might be the total transportation cost per unit distance between the factory and warehouse i for some given time period. As another example, and one that will receive a good deal of attention subsequently, the new facility might be a new machine to be located in a plant layout, while the existing facilities might be existing machines in the layout; as products travel back and forth between the new and existing machines, costs are incurred that are directly proportional to the distances involved. Alternatively, the new facility might be a point in an electrical network to which a number of wires, running from known points in the network (the existing facilities), are to be connected; the location of the point that will minimize the total cost of wire is to be determined.

Obviously, the single-facility location model is a more valid representation of the physical situation in some contexts than in others. Even in contexts where a least-cost new-facility location does not solve the actual physical problem of interest, there are several points to be made in its favor. First, it forces anyone attempting to use the model to quantify the problem he is studying; in so doing he often gains considerable insight into the problem. Second, a least-cost new-facility location provides a benchmark solution in the sense that the solution is as good as can be done, disregarding other factors; a consideration of other factors may then suggest an obvious modification of the least-cost solution that will be more realistic.

The question remains of what is meant by an "appropriately determined distance." The distance that immediately suggests itself is the *straight-line*, or *Euclidean*, distance. If the coordinates for the new facility are x and y and for the existing facility i are a_i and b_i, so that $\mathbf{X} = (x, y)$ and $\mathbf{P}_i = (a_i, b_i)$, the Euclidean distance between $\mathbf{X}$ and $\mathbf{P}_i$ is defined as

$$d(\mathbf{X}, \mathbf{P}_i) = [(x - a_i)^2 + (y - b_i)^2]^{1/2} \tag{4.2}$$

Euclidean distance applies for some network location problems as well as some instances involving conveyors and air travel. Some electrical wiring problems and pipeline design problems are also examples of Euclidean distance problems.

In most machine location problems, travel occurs along a set of aisles arranged in a rectangular pattern parallel to the walls of the building. In such a situation, the appropriate distance is variously referred to as the *rectilinear*, *rectangular*, metropolitan, or Manhattan distance. We choose the former and define the rectilinear distance between X and P_i as

$$d(\mathbf{X}, \mathbf{P}_i) = |x - a_i| + |y - b_i| \tag{4.3}$$

Rectilinear distance is appropriate in some urban location analyses where travel occurs along an orthogonal set of streets. Additionally, a number of offices employ a rectilinear set of aisles and hallways to facilitate the travel of personnel.

In some facility location problems, cost is not a simple linear function of distance. As an example, the cost associated with the response of a fire truck to a fire is expected to be nonlinear with distance. Depending on the location problem, $f(\mathbf{X})$ can take on a number of different formulations. One nonlinear form of $f(\mathbf{X})$ treated in this chapter is the *gravity* problem. Suppose that cost is proportional to the square of the Euclidean distance between $\mathbf{X}$ and $\mathbf{P}_i$. Thus, (4.1) becomes

$$f(\mathbf{X}) = \sum_{i=1}^{m} w_i[(x - a_i)^2 + (y - b_i)^2] \tag{4.4}$$

Location problems having the formulation given by (4.4) are referred to as *gravity* problems. The reason for this particular name becomes clear when conditions for an optimum location are provided in a subsequent discussion.

4.2 Rectilinear-Distance Location Problems

Interestingly, the rectilinear-distance location problem combines the property of being a very appropriate distance measure for a large number of location problems and the property of being very simple to treat analytically. While rectilinear distance is analytically simpler to work with than Euclidean distance, its properties are found by some to be nonintuitive, as may soon be seen.

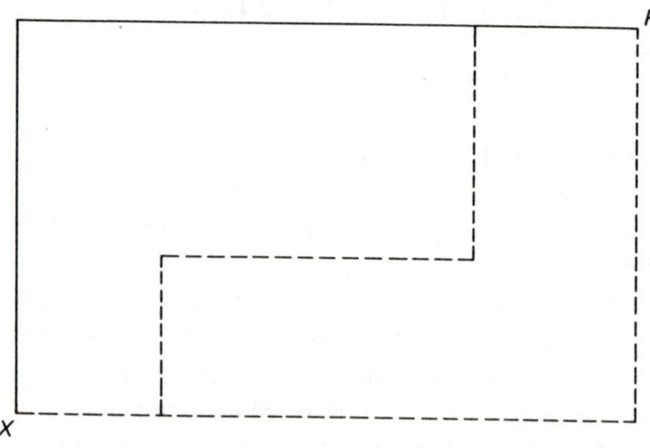

Figure 4.1. Different rectilinear paths between X and P_i having the same rectilinear distances.

Figure 4.1 illustrates several different paths between **X** and $\mathbf{P}_i$ for which the rectilinear distance is the same. The number of such paths is, of course, infinite. Such is not the case with Euclidean distance; there is a unique path.

The rectilinear-distance location problem can be stated mathematically as

$$\underset{x,y}{\text{minimize}}\, f(x, y) = \sum_{i=1}^{m} w_i(|x - a_i| + |y - b_i|) \tag{4.5}$$

From (4.5) it is seen that the problem can be equivalently stated as

$$\underset{x,y}{\text{minimize}}\, f(x, y) = \min_x \sum_{i=1}^{m} w_i |x - a_i| + \min_y \sum_{i=1}^{m} w_i |y - b_i|$$

where each quantity on the right-hand side can be treated as separate optimization problems:

$$\underset{x}{\text{minimize}}\, f_1(x) = \sum_{i=1}^{m} w_i |x - a_i| \tag{4.6}$$

and

$$\underset{y}{\text{minimize}}\, f_2(y) = \sum_{i=1}^{m} w_i |y - b_i| \tag{4.7}$$

Some properties of an optimum solution to the rectilinear-distance location problem (which we shall prove in Section 4.3) are

1. The x coordinate of the new facility will be the same as the x coordinate of some existing facility. Similarly, the y coordinate of the new facility will coincide with the y coordinate of some existing facility. Of course, it is not necessary that both coordinates be for the same existing facility.
2. The optimum x coordinate (y coordinate) location for the new facility is a *median location.* A median location is defined to be a location such that no more than one half the item movement is to the left of (below) the new facility location and no more than one half the item movement is to the right of (above) the new facility.

Perhaps a more appropriate term for the optimum location would be a *half-sum location*, rather than a median location, since the term "median" implies that one half the existing facilities lie to the left (below) and to the right (above) of the median point. Clearly, this is not the intention of the median-location rule. We use the term median location since it is commonly employed in the literature treating the rectilinear location problem.

As an example of a rectilinear-distance location problem, consider the location of a new data-processing machine in a university computer center. Suppose that there are four existing machines which have a material-handling relationship with the new machine. Let the existing facilities have locations of (4, 2), (8, 5), (11, 8), and (13, 2). Furthermore, suppose that the cost per foot traveled is the same between the new machine and each existing machine, and the number of trips per hour made between the new machine and the existing machines are 1/6, 1/3, 1/3, and 1/6, respectively.

From (4.6) and (4.7) it is apparent that the optimum x coordinate value and optimum y coordinate value are independently determined. It is instructive to consider the number of material-handling trips between the new facility and each existing facility over a 6-hour period. Thus, we make one trip to the existing facility located at (4, 2), two trips to the location (8, 5), two trips to the location (11, 8), and one trip to the location (13, 2). Considering the x coordinates, we weight each according to the number of trips to give the sequence 4, 8, 8, 11, 11, 13. The median location is any point in the closed interval [8, 11]. You might want to compare several points in this interval against the definition of a median location to satisfy yourself that this is the optimum x coordinate location. Considering the y coordinate, we have the sequence 2, 5, 5, 8, 8, 2. Arranging these in increasing order, we have 2, 2, 5, 5, 8, 8. Thus, the median location is any point in the closed interval [5, 5], or the optimum y coordinate location is $y = 5$.

For location problems involving a large number of existing facilities and large values of the weights, there is a more efficient way of determining the median location. We divide the accumulated weights by 2 and determine the location such that no more than one half the weights are to the right of the location and no more than one half are to the left. An example will serve to illustrate the procedure. A rent-a-car company has five offices located in a large city. Customers desiring a car can pick up and deliver the car to any of the five offices. The company wishes to locate a maintenance facility in the city to service the cars. The coordinate locations of the five offices are (0, 0), (3, 16), (18, 2), (8, 18), and (20, 2). The numbers of cars transported per day between the new maintenance facility and the offices equal 5, 22, 41, 60, and 34, respectively. What location for the maintenance facility will minimize the distance cars are transported per day?

The optimum x coordinate location will be determined first. The existing facilities are ordered according to their x coordinate values. Next the weights are accumulated. For this example, the total number of cars trans-

Table 4.1. x COORDINATE SOLUTION

Existing Facility	*x-Coordinate Value*	*Weight*	*Cumulative Weight*
1	0	5	5
2	3	22	27 < 81
4	8	60	87 > 81
3	18	41	128
5	20	34	162

$x^* = 8$

ported per day equals 162. Thus, the median location corresponds to a cumulative weight of 162 ÷ 2, or 81. As Table 4.1 shows, the corresponding x coordinate value for the new facility is 8, since the cumulative weight first exceeds 81 at x equal to 8.

A similar procedure is employed to obtain the optimum y coordinate value for the new facility. As shown below, the optimum y coordinate value is 16. The new maintenance facility should be located at the point (8, 16) to minimize the distance cars are transported.

An obvious question to be asked at this point is, "What if the point (8, 16) is not available as a location site?" The point may, for example, be

Table 4.2. y COORDINATE SOLUTION

Existing Facility	*y-Coordinate Value*	*Weight*	*Cumulative Weight*
1	0	5	5
3, 5	2	41 + 34	80 < 81
2	16	22	102 > 81
4	18	60	162
		$y^* = 16$	

inaccessible or may coincide with the location of another structure, a river, or a municipal park. Since such an occurrence is likely for a number of facility location problems, the question deserves additional consideration. One approach that may be employed is to construct contour lines of the cost function. A contour line is a line of constant cost in the plane (since we are considering planar locations). Thus, locating the new facility at any point on a given contour line results in the same total cost.

Contour lines provide considerable insight into the shape of the surface of the total cost function, as well as a useful means of evaluating alternative locations for the new facility. Contour lines (also called iso-cost lines or level curves) indicate at a glance the cost penalty associated with the choice of a nonoptimum location. This is important in practice, since all factors important in the location decision have not been included in the objective function. The final decision, in which subjective judgment enters, can be simplified through the construction of contour lines.

A procedure for constructing contour lines for the rectilinear distance problem is given next. Figure 4.2 should be consulted in going through the procedure.

1. Plot the points $(a_1, b_1), \ldots, (a_m, b_m)$, and draw perpendicular lines (parallel to the x and y axes) through each point.

2. Consider the vertical lines to be numbered $1, 2, \ldots, p$ from left to right, and the horizontal lines to be numbered $1, 2, \ldots, q$ from bottom to top.

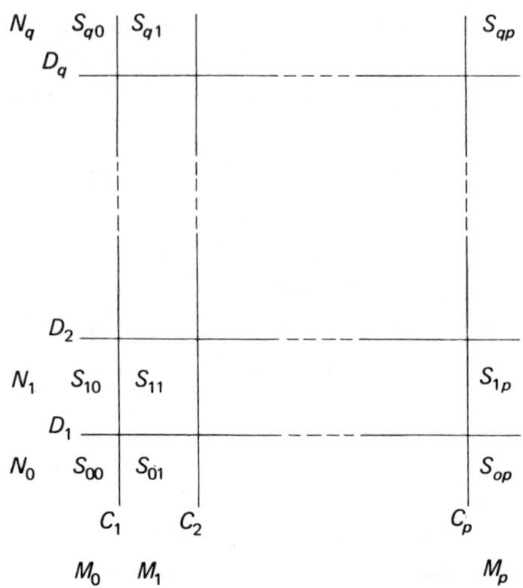

Figure 4.2. Procedure for constructing contour lines.

3. Call the x intercept of the jth vertical line c_j, and the y intercept of the ith horizontal line d_i. Denote the region delimited by vertical lines j and $j + 1$ and horizontal lines i and $i + 1$ by $[i, j]$. (In order that all regions are numbered, *imagine* there is a vertical line numbered 0 to the left of vertical line 1, a vertical line numbered $p + 1$ to the right of vertical line p, a horizontal line numbered 0 below horizontal line 1, and a horizontal line $q + 1$ above horizontal line q.)

4. Let C_j and D_i be the sums of the weights associated with vertical line j and horizontal line i, respectively. [For example, if the points (5, 3) and (5, 10) have the associated weights of 6 and 8, respectively, and the second vertical line has an x intercept of 5, then $C_2 = 14$.] It is convenient to place the numbers C_j at the bottom of the corresponding vertical lines and the numbers D_i at the left of the corresponding horizontal lines.

5. Compute the numbers

$$
\begin{aligned}
M_0 &= -\sum_{j=1}^{p} C_j = -\sum_{i=1}^{m} w_i & N_0 &= -\sum_{i=1}^{q} D_i = -\sum_{i=1}^{m} w_i \\
M_1 &= M_0 + 2C_1 & N_1 &= N_0 + 2D_1 \\
M_2 &= M_1 + 2C_2 & N_2 &= N_1 + 2D_2 \\
&\vdots & &\vdots \\
M_p &= M_{p-1} + 2C_p = \sum_{i=1}^{m} w_i & N_q &= N_{q-1} + 2D_q = \sum_{i=1}^{m} w_i
\end{aligned}
$$

and place them as indicated in Figure 4.2.

6. The slope S_{ij} of any contour line passing through region $[i, j]$ is computed as follows:

$$S_{ij} = -\frac{M_j}{N_i}$$

When N_i is zero, the contour line is vertical in region $[i, j]$.

7. To find a point (x^*, y^*) that minimizes the total cost expression (4.5), there are four cases to consider:

(a) $M_{t-1} < 0, M_t > 0, N_{s-1} < 0, N_s > 0$; take $x^* = c_t, y^* = d_s$.
(b) $M_{t-1} < 0, M_t = 0, N_{s-1} < 0, N_s > 0$; take x^* to be any point between c_t and c_{t+1}, $y^* = d_s$.
(c) $M_{t-1} < 0, M_t > 0, N_{s-1} < 0, N_s = 0$; take $x^* = c_t$; take y^* to be any point between d_s and d_{s+1}.
(d) $M_{t-1} < 0, M_t = 0, N_{s-1} < 0, N_s = 0$; take x^* to be any point between c_t and c_{t+1}, and y^* to be any point between d_s and d_{s+1}.

8. Given the above information, a contour line can be constructed, beginning at any point except a minimum point. As a check, the line should always "end" at the same point where it "begins."

As an example to illustrate the procedure, consider the problem previously treated in which the existing facilities have locations of (4, 2), (8, 5), (11, 8), and (13, 2) with associated weights of 1/6, 1/3, 1/3, and 1/6, respectively. Figure 4.3 shows the slopes in the different regions, while several contour lines for this problem are plotted in Figure 4.4. Any point (x^*, y^*) where x^* is between $c_2 = 8$ and $c_3 = 11$, and $y^* = d_2 = 5$ is a minimum cost location.

The justification for all the steps required in constructing the contour lines, with the exception of step 8, is straightforward, and will now be given. After going through step 4, the expression (4.5) becomes

$$f(x, y) = \sum_{j=1}^{p} C_j|x - c_j| + \sum_{i=1}^{q} D_i|y - d_i| \tag{4.8}$$

where $c_1 < c_2 < \ldots < c_p$ and $d_1 < d_2 < \ldots . < d_q$.

Suppose (x, y) is any point in region $[s, t]$, so that $c_t \leq x \leq c_{t+1}$ and $d_s \leq y \leq d_{s+1}$; then

$$\begin{aligned} f(x, y) &= \sum_{j=1}^{t} C_j(x - c_j) + \sum_{j=t+1}^{p} C_j(c_j - x) \\ &\quad + \sum_{i=1}^{s} D_i(y - d_i) + \sum_{i=s+1}^{q} D_i(d_i - y) \\ &= M_t x + N_s y + C_{st} \end{aligned}$$

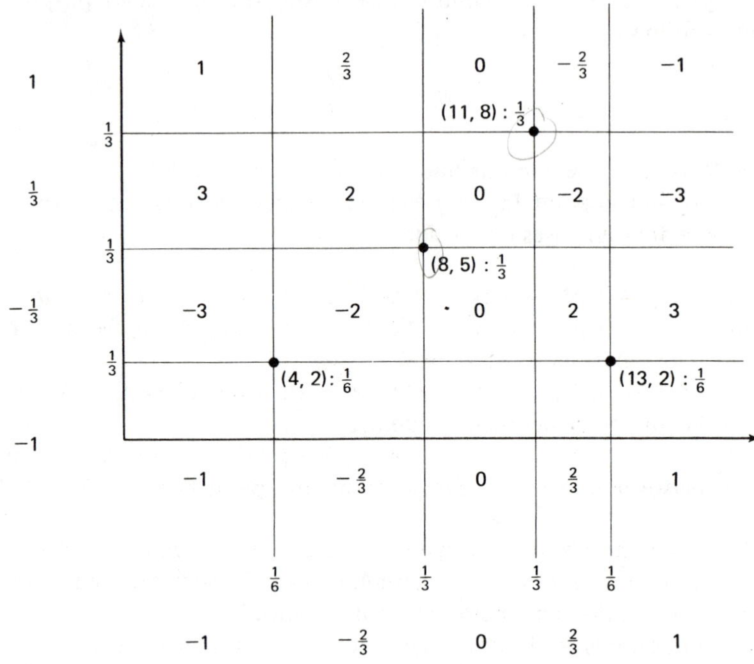

Figure 4.3. Data required for the construction of contour lines.

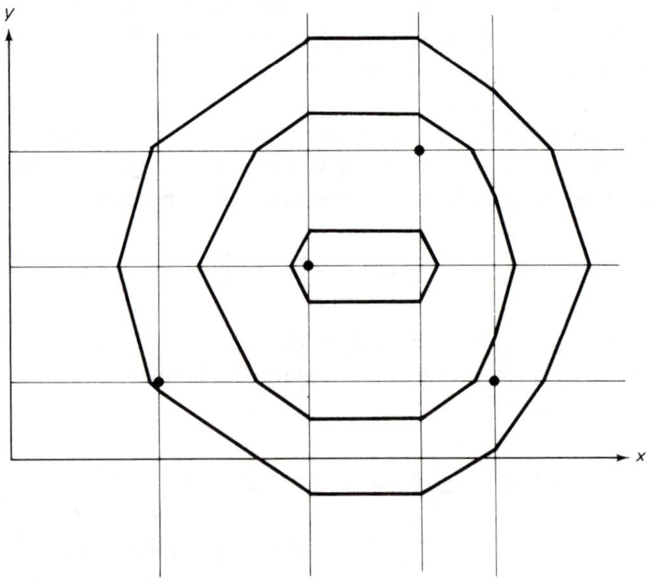

Figure 4.4. Sample contour lines for the example problem.

where

$$M_t = \sum_{j=1}^{t} C_j - \sum_{j=t+1}^{p} C_j \tag{4.9}$$

$$N_s = \sum_{i=1}^{s} D_i - \sum_{i=s+1}^{q} D_i \tag{4.10}$$

and C_{st} consists of the remaining constant terms. It is left as an exercise to show that the definitions of M_t and N_s given by (4.9) and (4.10), respectively, are consistent with those of step 5 of the procedure. Since a contour line is being plotted, $f(x, y)$ takes on a constant value, say k, so that

$$k = M_t x + N_s y + C_{st}$$

Solving the latter equation for y gives

$$y = -\frac{M_t}{N_s}x + \frac{k - C_{st}}{N_s}$$

so that the slope in region $[s, t]$ is given by

$$S_{st} = -\frac{M_t}{N_s}$$

which agrees with the expression in step 6 of the procedure on replacing s and t by i and j, respectively.

A rigorous justification of step 7 of the procedure takes a bit more work, although the ideas involved are quite simple. As a motivation for step 7, consider again the example of the previous section; the expression (4.8) may be rewritten

$$f(x, y) = f_1(x) + f_2(y)$$

where $f_1(x) = \frac{1}{6}|x - 4| + \frac{1}{3}|x - 8| + \frac{1}{3}|x - 11| + \frac{1}{6}|x - 13|$ (4.11)

and $f_2(y) = \frac{1}{3}|y - 2| + \frac{1}{3}|y - 5| + \frac{1}{3}|y - 8|$ (4.12)

Graphs of $f_1(x)$ and $f_2(y)$ are shown in Figure 4.5.

An observation of Figure 4.5(b) shows that $f_2(y)$ is continuous, piecewise linear, that the slope increases strictly from one interval to the next, and that the function is minimized at the point $y^* = 5$, where the slope changes from negative to nonnegative (positive in this case). An observation of Figure 4.5(a) shows the $f_1(x)$ has the same general features as $f_2(y)$, except that the function is minimized for any point in the interval between 8 and 11; it is still the case, however, that a point which minimizes $f_1(x)$ is one where the slope changes from negative to nonnegative.

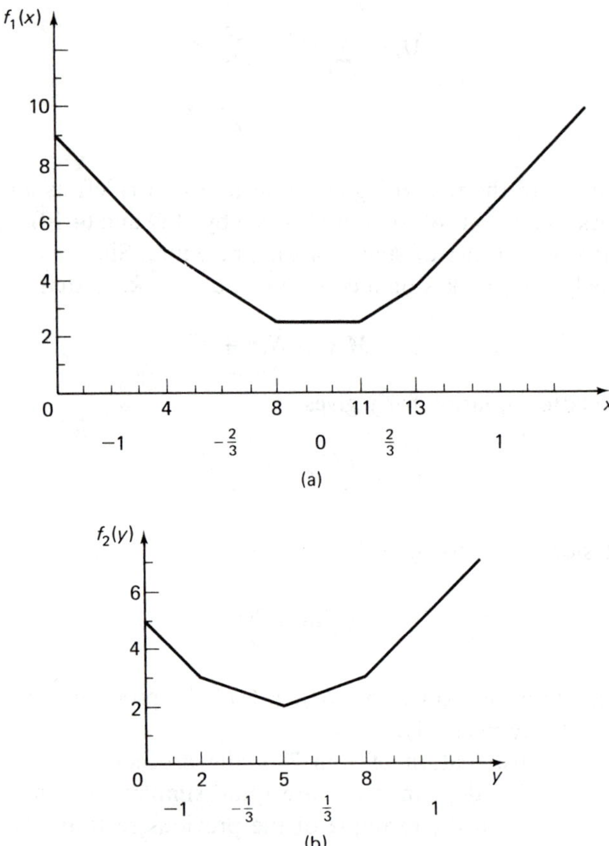

Figure 4.5. Plots of $f_1(x)$ and $f_2(y)$ for the example problem.

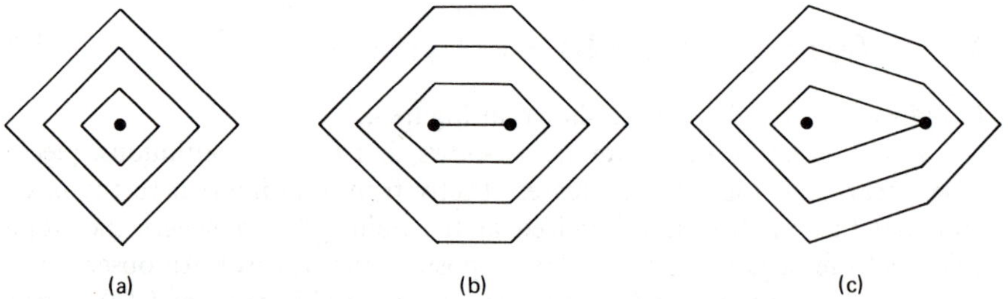

Figure 4.6. Contour lines for three simple rectilinear location problems.

Returning to the rent-a-car example, plotting contour lines for the total cost function will provide an indication of the areas within the city that should be investigated first for potential location sites. The final decision will incorporate a number of factors, one of which is property cost.

Contour lines are given in Figure 4.6 for a number of simple rectilinear example problems. In Figure 4.6(a) there is a single existing facility. In Figure 4.6(b) there are two existing facilities with equal item movement between the new facility and each existing facility. In Figure 4.6(c) there are two existing facilities with there being twice as much item movement between the new facility and existing facility 1 as between the new facility and existing facility 2.

4.3 Duality Relationships*

A rigorous proof will be developed in this section to justify step 7 of the procedure for constructing contour lines and determining the optimum location for the new facility. Notice, first, that (4.8) may be written as

$$f(x, y) = f_1(x) + f_2(y) \tag{4.13}$$

where

$$f_1(x) = \sum_{j=1}^{p} C_j |x - c_j| \tag{4.14}$$

and

$$f_2(y) = \sum_{i=1}^{q} D_i |y - d_i| \tag{4.15}$$

so that

$$\min_{x,y} f(x, y) = \min_{x} f_1(x) + \min_{y} f_2(y)$$

Furthermore, $f_1(x)$ and $f_2(y)$ as defined by (4.14) and (4.15), respectively, have the same form. Thus, any procedure that applies to minimizing (4.14) will also apply to minimizing (4.15).

The optimization problem $\min_x f_1(x)$ can be represented equivalently as a linear-programming problem by defining two new variables r_j and s_j such that

$$r_j = \begin{cases} x - c_j, & x \geq c_j \\ 0, & x < c_j \end{cases}$$

and

$$s_j = \begin{cases} 0, & x \geq c_j \\ c_j - x, & x < c_j \end{cases}$$

*This section requires a background in duality relationships of linear programming. It can be omitted by the reader not interested in the development of the median conditions.

On introducing the variables r_j and s_j, the optimization problem becomes

$$\begin{aligned} &\text{minimize} \quad f(x, r, s) = \sum_{j=1}^{p} C_j(r_j + s_j) \\ &\text{subject to} \\ &\quad x - r_j + s_j = c_j, \quad j = 1, \ldots, p \\ &\quad r_j \geq 0, s_j \geq 0, \quad j = 1, \ldots, p \end{aligned} \tag{4.16}$$

By introducing the constraint $x - r_j + s_j = c_j$, we are assured that for a given value of j both r_j and s_j will not take on positive values in the optimum solution to the linear-programming problem. This is due to the fact that some basic feasible solution will be a minimum feasible solution. If the constraints are represented in matrix form, the columns of the matrix corresponding to r_j and s_j are linearly dependent. Therefore, any solution involving positive values for r_j and s_j will not be a basic solution, establishing the claim that r_j and s_j will not both take on positive values in the optimum solution to the linear-programming problem.

Suppose that r_j is positive valued. Since s_j must be zero, r_j equals the amount by which x exceeds c_j. Suppose that s_j is positive valued. By a similar argument, s_j equals the amount by which c_j exceeds x. The equivalence of the two optimization problems should now be apparent. (The preceding technique is employed a number of times throughout this text. Thus, the reader is advised to develop confidence in its use. The technique is described in greater detail in Section 5.2.)

To minimize (4.14), we formulate the *dual problem* to (4.16) as follows:

$$\begin{aligned} &\text{maximize} \quad \sum_{j=1}^{p} c_j z_j \\ &\text{subject to} \\ &\quad \sum_{j=1}^{p} z_j = 0 \\ &\quad |z_j| \leq C_j, \quad j = 1, \ldots, p \end{aligned} \tag{4.17}$$

It is convenient to introduce several definitions for the dual problem. A *feasible solution* to the dual problem will be any numbers $z_1, \ldots, z_p$ that satisfy the constraints of the dual problem. A *maximum feasible solution* will be a feasible solution that maximizes the objective function.

Several properties relating $f_1(x)$ and the dual problem will now be stated and proved.

Property 1: For any feasible solution $z_1, \ldots, z_p$ of the dual problem, and any x,

$$\sum_{j=1}^{p} c_j z_j \leq f_1(x)$$

Proof: Let $z_1, \ldots, z_p$ satisfy the constraints of the dual problem; then, for any x,

$$\sum_{j=1}^{p} c_j z_j = \sum_{j=1}^{p} c_j z_j - x(0) = \sum_{j=1}^{p} c_j z_j - \sum_{j=1}^{p} x z_j$$
$$= \sum_{j=1}^{p} z_j(c_j - x)$$

But
$$z_j(c_j - x) \leq |z_j(c_j - x)| = |z_j||x - c_j|$$

and
$$|z_j||x - c_j| \leq C_j|x - c_j|$$

so it follows that

$$\sum_{j=1}^{p} c_j z_j \leq \sum_{j=1}^{p} C_j|x - c_j| = f_1(x)$$

An immediate consequence of Property 1 is

Property 2: Suppose that $z_1^*, \ldots, z_p^*$ is a feasible solution to the dual problem and x^* is such that

$$\sum_{j=1}^{p} c_j z_j^* = f_1(x^*)$$

Then x^* minimizes $f_1(x)$, and $z_1^*, \ldots, z_p^*$ is a maximum feasible solution to the dual problem.

Property 2 follows from Property 1 because the value of any feasible solution to the dual problem is a lower bound on $f_1(x)$, while, for any x, $f_1(x)$ is an upper bound on the value of any feasible solution to the dual problem. If the lower and upper bounds are equal, then the value of $f_1(x)$ is equal to the lower bound, and so is as small as possible, while the value of the objective function of the dual problem is equal to the upper bound, and so is as large as possible.

In the property to follow, the convention is adopted that sums extending from 1 to 0 or $p + 1$ to p are defined to be zero.

Property 3: Suppose that $x^* = c_t$, where the index t is such that

$$\sum_{j=1}^{t-1} C_j - \sum_{j=t}^{p} C_j < 0 \tag{4.18}$$

and
$$\sum_{j=1}^{t} C_j - \sum_{j=t+1}^{p} C_j \geq 0 \tag{4.19}$$

and define $z_1^*, \ldots, z_p^*$ as follows:

$$z_j^* = -C_j, \quad j = 1, \ldots, t - 1 \tag{4.20}$$
$$= C_j, \quad j = t + 1, \ldots, p \tag{4.21}$$
$$z_t^* = -\sum_{j=1}^{t-1} z_j^* - \sum_{j=t+1}^{p} z_j^* \tag{4.22}$$

Then x^* minimizes $f_1(x)$, $z_1^*, \ldots, z_p^*$ is a maximum feasible solution to the dual problem, and

$$f_1(x^*) = \sum_{j=1}^{p} c_j z_j^*$$

Proof: Recall that $c_1 < c_2 < \ldots < c_p$; thus

$$f(x^*) = f(c_{jt}) = \sum_{j=1}^{t-1} C_j(c_t - c_j) + \sum_{j=t+1}^{p} C_j(c_j - c_t)$$

Also, it follows from (4.22) that

$$\sum_{j=1}^{p} z_j^* = 0$$

so that

$$\begin{aligned}\sum_{j=1}^{p} c_j z_j^* &= \sum_{j=1}^{p} c_j z_j^* - c_t \sum_{j=1}^{p} z_j^* \\ &= \sum_{j=1}^{t-1} z_j^*(c_j - c_t) + \sum_{j=t+1}^{p} z_j^*(c_j - c_t) \\ &= \sum_{j=1}^{t-1} C_j(c_t - c_j) + \sum_{j=t+1}^{p} C_j(c_j - c_t) \\ &= f(c_t)\end{aligned}$$

Consequently the conclusion will follow from Property 2 if $|z_j^*| \leq C_j$ for $j = 1, \ldots, p$; (4.20) and (4.21) imply that $|z_j^*| = C_j$ for all j, except $j = t$, so it only remains to show that $|z_t^*| \leq C_t$. Using (4.22) and the definition of the z_j^* gives

$$\begin{aligned} z_t^* &= \sum_{j=1}^{t-1} C_j - \sum_{j=t+1}^{p} C_j \\ &= \sum_{j=1}^{t-1} C_j - \sum_{j=t}^{p} C_j + C_t \end{aligned}$$

So (4.18) now implies that

$$z_t^* < 0 + C_t = C_t \tag{4.23}$$

Also,

$$\begin{aligned} z_t^* &= \sum_{j=1}^{t-1} C_j - \sum_{j=t+1}^{p} C_j \\ &= \sum_{j=1}^{t} C_j - \sum_{j=t+1}^{p} C_j - C_t \end{aligned}$$

so (4.19) now implies that

$$z_t^* \geq 0 - C_t = -C_t \tag{4.24}$$

Combining (4.23) and (4.24) gives $-C_t \leq z_t^* < C_t$, which implies that $|z_t^*| \leq C_t$, and the proof is complete.

A few comments on Property 3 are in order. First, (4.18) states that the slope of $f_1(x)$ is negative for $c_{t-1} \leq x \leq c_t$, and (4.19) states that the slope is nonnegative for $c_t \leq x \leq c_{t+1}$; so (4.18) and (4.19) just state that $f_1(x)$ is minimized at the point where its slope changes from negative to nonnegative. This fact agrees with previous observations. Second, there is a useful, equivalent way of writing (4.18) and (4.19); rearranging (4.18) after adding and subtracting $\sum_{j=1}^{t-1} C_j$ gives

$$2\sum_{j=1}^{t-1} C_j < \sum_{j=1}^{p} C_j = \sum_{i=1}^{m} w_i$$

or

$$\sum_{j=1}^{t-1} C_j < \tfrac{1}{2}\sum_{i=1}^{m} w_i \tag{4.25}$$

Likewise, adding and subtracting $\sum_{j=1}^{t} C_j$ to (4.19) and rearranging terms gives

$$2\sum_{j=1}^{t} C_j \geq \sum_{j=1}^{p} C_j = \sum_{i=1}^{m} w_i$$

or

$$\sum_{j=1}^{t} C_j \geq \tfrac{1}{2}\sum_{i=1}^{m} w_i \tag{4.26}$$

Thus, (4.25) and (4.26) can be obtained from (4.18) and (4.19); on reversing the process, (4.18) and (4.19) can be obtained from (4.25) and (4.26); so conditions (4.18) and (4.19) are equivalent to conditions (4.25) and (4.26). The latter conditions are sometimes referred to as *median conditions*. Put in words, the median conditions state that $f_1(x)$ is minimized at the point c_t for which the partial sum of the C_j first exceeds half the total sum. There is certainly always an index t for which the median conditions hold; so $f_1(x)$ always has a minimum. Furthermore, the dual variables as defined by (4.20) and (4.21) can be given an interpretation in terms of $f_1(x)$; z_j^* gives the increase in the minimum value of $f_1(x)$ if c_t is increased by one unit; it is left as an exercise to determine why this is so.

Since $f_1(x)$ and $f_2(y)$ have the same structure, all the preceding analysis applies to $f_2(y)$ with obvious changes; the point d_s minimizes $f_2(y)$, where s is such that

$$\sum_{i=1}^{s-1} D_i < \tfrac{1}{2}\sum_{i=1}^{m} w_i$$

and

$$\sum_{i=1}^{s} D_i \geq \tfrac{1}{2}\sum_{i=1}^{m} w_i$$

4.4 Squared Euclidean-Distance Location Problems

Our attention now shifts to a new distance measure, the Euclidean distance. However, rather than treat initially the Euclidean-distance problem, we examine the single-facility location problem where cost is proportional to the square of the Euclidean distance. There are at least two reasons for studying the squared distance (gravity) problem. First, there exist location problems in which costs increase quadratically, instead of linearly, in terms of the Euclidean distance between the existing and new facilities. Second, the study of the gravity problem lays some of the groundwork for the Euclidean-distance problem to be examined subsequently.

The gravity problem can be formulated as

$$\underset{x,y}{\text{minimize}}\, f(x, y) = \sum_{i=1}^{m} w_i[(x - a_i)^2 + (y - b_i)^2] \tag{4.27}$$

Any point (x^*, y^*) that minimizes (4.27) must satisfy the conditions

$$\left(\frac{\partial f(x^*, y^*)}{\partial x^*}, \frac{\partial f(x^*, y^*)}{\partial y^*}\right) = (0, 0) \tag{4.28}$$

Computing the partial derivatives of (4.27) with respect to x and y and then setting them to zero gives the following unique solution:

$$x^* = \frac{\sum_{i=1}^{m} w_i a_i}{\sum_{i=1}^{m} w_i} \tag{4.29}$$

$$y^* = \frac{\sum_{i=1}^{m} w_i b_i}{\sum_{i=1}^{m} w_i} \tag{4.30}$$

The coordinates x^* and y^* of the new facility may thus be interpreted as weighted averages of the x and y coordinates of the existing facilities, and are, in fact, the coordinates that minimize (4.27). Conditions (4.28) can be shown to be both necessary and sufficient for a minimum. Thus, the gravity problem has a simple solution. The solution is sometimes referred to as the *centroid* or *center-of-gravity* solution. Hence, the title *gravity* problem. (According to Kelly [15], an interest in the single-facility gravity problem dates back at least to the work of Lagrange in the nineteenth century.)

Recall the example involving existing facilities at (4, 2), (8, 5), (11, 8), and (13, 2) with weights $\frac{1}{6}$, $\frac{1}{3}$, $\frac{1}{3}$, and $\frac{1}{6}$ respectively. Suppose that the cost of item movement between the new facility and each existing facility equals the product of the value of the associated weight and the square of the Euclidean distance between the new and existing facilities. From (4.29)

$$x^* = \frac{[\frac{1}{6}(4) + \frac{1}{3}(8) + \frac{1}{3}(11) + \frac{1}{6}(13)]}{1.0}$$

or $\qquad x^* = 9.167$

and from (4.30)

$$y^* = \frac{[\frac{1}{6}(2) + \frac{1}{3}(5) + \frac{1}{3}(8) + \frac{1}{6}(2)]}{1.0}$$

or

$$y^* = 5$$

Therefore, the optimum location of the new facility under the assumptions of the gravity problem is also an optimum solution to the rectilinear problem. [Note that the optimum location of the new facility in the rent-a-car example is (12.12, 9.77) under the gravity-problem assumptions, as compared to (8, 16) under the rectilinear-problem assumptions.]

For the reasons stated earlier for the rectilinear problem, contour lines for the gravity problem are of interest. Contour lines for the gravity problem can be obtained quite easily. Consider the contour lines given in Figure 4.7 for two simple cases of the gravity problem. In the first case there exists a single facility. In the second case there is equal item movement between the new facility and each of the two existing facilities. Consequently, it is easy to see that the contour lines will be concentric circles centered on the optimum location.

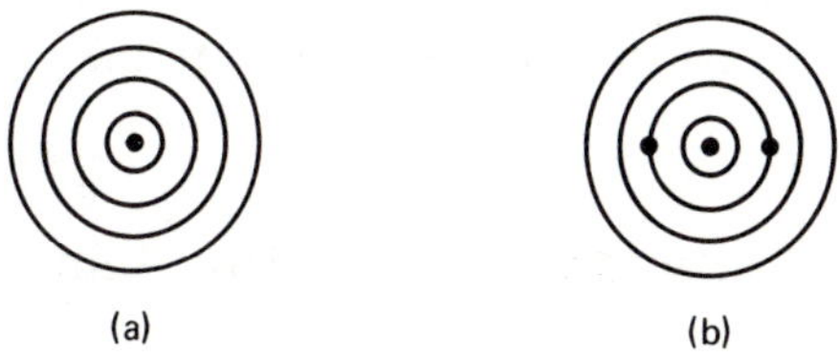

Figure 4.7. Contour lines for two simple gravity problems.

Now, what do you think the contour lines will look like when we have any number of existing facilities with unequal item movement? If you suspect the contour lines will still be concentric circles centered on the optimum location, your intuition is remarkable, for that is the case. To see why this is true, notice that from (4.27) we want to determine the set of all points (x, y) such that

$$k = \sum_{i=1}^{m} w_i[(x - a_i)^2 + (y - b_i)^2]$$

where k is a constant. Consequently, on expanding the squared terms we find that

$$\begin{aligned} k = x^2 \sum_{i=1}^{m} w_i - 2x \sum_{i=1}^{m} w_i a_i + \sum_{i=1}^{m} w_i a_i^2 \\ + y^2 \sum_{i=1}^{m} w_i - 2y \sum_{i=1}^{m} w_i b_i + \sum_{i=1}^{m} w_i b_i^2 \end{aligned} \tag{4.31}$$

If we let

$$W = \sum_{i=1}^{m} w_i$$

divide (4.31) by W, and employ the relations (4.29) and (4.30), we find that

$$\frac{k}{W} = x^2 - 2xx^* + \sum_{i=1}^{m} \frac{w_i a_i^2}{W} + y^2 - 2yy^* + \sum_{i=1}^{m} \frac{w_i b_i^2}{W} \tag{4.32}$$

On adding $(x^*)^2$ and $(y^*)^2$ to both sides of (4.32) and simplifying, we obtain the equation for a circle,

$$r^2 = (x - x^*)^2 + (y - y^*)^2 \tag{4.33}$$

centered on the point (x^*, y^*) with radius

$$r = \left[\frac{k}{W} + (x^*)^2 + (y^*)^2 - \sum_{i=1}^{m} \frac{w_i(a_i^2 + b_i^2)}{W}\right]^{1/2} \tag{4.34}$$

This is an interesting and, to us, a nonintuitive result. Based on this result, if you are unable to locate the new facility at the optimum location (x^*, y^*) and must evaluate alternative sites, you should always choose the one that has the smallest straight-line distance to the point (x^*, y^*).

4.5 Euclidean-Distance Location Problems

Thus far, we have considered the rectilinear problem and the gravity problem. We conclude the treatment of single-facility location problems by considering the Euclidean-distance problem. The Euclidean problem may be stated as

$$\underset{x,y}{\text{minimize}}\, f(x, y) = \sum_{i=1}^{m} w_i[(x - a_i)^2 + (y - b_i)^2]^{1/2} \tag{4.35}$$

The Euclidean problem is variously referred to as the Steiner–Weber problem or the general Fermat problem, and has an extraordinary longevity. In fact, a version of the problem for the case $m = 3$ and $w_i = 1$, for $i = 1, 2, 3$, was posed, purely as a problem in geometry, by Fermat early in the seventeenth century, and was solved by Toricelli prior to 1640. The problem was studied by Steiner, a Swiss mathematician, in the nineteenth century, and by Weber, a German economist, early in the twentieth century. A duality result, related to the duality discussion in Section 4.3, was obtained by Fasbender in 1846. Interestingly enough, however, it was not until the work by Kuhn, in 1963, that the problem could be considered to be essentially completely solved.

Euclidean-distance problems arise in a number of different contexts. As an example, consider the following example cited by Noble [23]. Ten towns are located at random in a large level area devoted to wheat ranching. A new cooperative power plant is to be located to serve each of the towns electing to join the cooperative. Separate electric cables are to be run from the power plant to each town. The sole criterion in locating the plant is to minimize the total required length of electrical cable. In this case $m = 10$ and $w_i = 1$ for all values of i.

The approach that immediately comes to mind in solving the Euclidean-distance problem is again to compute the partial derivatives of (4.35) and set them to zero. Assuming $(x, y) \neq (a_i, b_i)$, $i = 1, \ldots, m$, the partial derivatives are

$$\frac{\partial f(x, y)}{\partial x} = \sum_{i=1}^{m} \frac{w_i(x - a_i)}{[(x - a_i)^2 + (y - b_i)^2]^{1/2}} \tag{4.36}$$

and

$$\frac{\partial f(x, y)}{\partial y} = \sum_{i=1}^{m} \frac{w_i(y - b_i)}{[(x - a_i)^2 + (y - b_i)^2]^{1/2}} \tag{4.37}$$

Notice that if, for any i, $(x, y) = (a_i, b_i)$, then (4.36) and (4.37) are undefined. Consequently, we see that difficulties arise when the location for the new facility coincides (mathematically) with the location of some existing facility. If there were some guarantee that any optimum location of the new facility would never be the same as the location of an existing facility, then (4.36) and (4.37) would still give necessary and sufficient conditions for a least-cost location of the new facility. Unfortunately, there is no such guarantee available. Consequently, a modification of the partial derivative approach is required. The modification, due to Kuhn [17], is based on the two-tuple $R(x, y)$, which is defined as follows, if $(x, y) \neq (a_i, b_i)$, $i = 1, \ldots, m$:

$$R(x, y) = \left(\frac{\partial f(x, y)}{\partial x}, \frac{\partial f(x, y)}{\partial y}\right)$$

And if $(x, y) = (a_k, b_k)$, $k = 1, \ldots, m$,

$$R(x, y) = R(a_k, b_k) = \begin{cases} \left(\dfrac{u_k - w_k}{u_k} s_k, \dfrac{u_k - w_k}{u_k} t_k\right), & \text{if } u_k > w_k \\ (0, 0), & \text{if } u_k \leq w_k \end{cases}$$

where

$$s_k = \sum_{\substack{i=1 \\ \neq k}}^{m} \frac{w_i(a_k - a_i)}{[(a_i - a_k)^2 + (b_i - b_k)^2]^{1/2}}$$

$$t_k = \sum_{\substack{i=1 \\ \neq k}}^{m} \frac{w_i(b_k - b_i)}{[(a_i - a_k)^2 + (b_i - b_k)^2]^{1/2}}$$

and

$$u_k = (s_k{}^2 + t_k{}^2)^{1/2}$$

The two-tuple $R(x, y)$ is defined for all points in the plane. Kuhn [17] establishes that a necessary and sufficient condition for (x^*, y^*) to be a least-cost new facility location is that $R(x^*, y^*) = (0, 0)$. Consequently, the location of some existing facility, (a_k, b_k), will be the optimum location for the new facility if and only if $u_k \leq w_k$. Thus, one should compute the value of u_k and compare it with the value of w_k if it is suspected that the optimum new-facility location coincides with the location of existing facility k.

Although we have available necessary and sufficient conditions for an optimum solution to the Euclidean problem, we still do not have a way of determining (x^*, y^*). The two-tuple $R(x, y)$, referred to subsequently as Kuhn's modified gradient, can also be manipulated to provide the basis for a computational procedure for finding the location (x^*, y^*). Notice that, on setting (4.36) equal to zero, we obtain the expression

$$x \sum_{i=1}^{m} \frac{w_i}{[(x - a_i)^2 + (y - b_i)^2]^{1/2}} = \sum_{i=1}^{m} \frac{w_i a_i}{[(x - a_i)^2 + (y - b_i)^2]^{1/2}} \tag{4.38}$$

If we let

$$g_i(x, y) = \frac{w_i}{[(x - a_i)^2 + (y - b_i)^2]^{1/2}}, \quad i = 1, \ldots, m$$

then (4.38) can be given as

$$x = \frac{\sum_{i=1}^{m} a_i g_i(x, y)}{\sum_{i=1}^{m} g_i(x, y)}$$

Likewise, from (4.37) we obtain

$$y = \frac{\sum_{i=1}^{m} b_i g_i(x, y)}{\sum_{i=1}^{m} g_i(x, y)}$$

So long as $g_i(x, y)$ is defined, we can employ the following iterative procedure:

$$x^{(k)} = \frac{\sum_{i=1}^{m} a_i g_i(x^{(k-1)}, y^{(k-1)})}{\sum_{i=1}^{m} g_i(x^{(k-1)}, y^{(k-1)})} \tag{4.39}$$

and

$$y^{(k)} = \frac{\sum_{i=1}^{m} b_i g_i(x^{(k-1)}, y^{(k-1)})}{\sum_{i=1}^{m} g_i(x^{(k-1)}, y^{(k-1)})} \tag{4.40}$$

The superscripts denote the iteration number. Thus, a starting value $(x^{(0)}, y^{(0)})$ is required to determine $(x^{(1)}, y^{(1)})$. The value of $(x^{(1)}, y^{(1)})$ is used to determine the value of $(x^{(2)}, y^{(2)})$, and so forth. The iterative procedure continues until no appreciable improvement occurs in the estimate of the optimum location for the new facility, or until a location is found that satisfies Kuhn's modified gradient condition. The iterative procedure is guaranteed to converge to the optimum location; a discussion of convergence for this, as well as for more general versions of the problem, is given by Katz [14], Kuhn [18], and Weiszfeld [28].

Typically, the gravity solution is used as the starting value for the iterative procedure. As an illustration of the iterative procedure, let $\mathbf{P}_1 = (0, 0)$, $\mathbf{P}_2 = (0, 10)$, $\mathbf{P}_3 = (5, 0)$, $\mathbf{P}_4 = (12, 6)$, and let all the w_i values be equal. The gravity solution is easily shown to be (4.25, 4.00). Using the iterative procedure given by (4.39) and (4.40), we obtain

$$
\begin{aligned}
(x^{(1)}, y^{(1)}) &= (4.023, 3.111), & (x^{(6)}, y^{(6)}) &= (3.971, 2.074) \\
(x^{(2)}, y^{(2)}) &= (3.949, 2.627), & (x^{(7)}, y^{(7)}) &= (3.981, 2.045) \\
(x^{(3)}, y^{(3)}) &= (3.935, 2.358), & (x^{(8)}, y^{(8)}) &= (3.987, 2.028) \\
(x^{(4)}, y^{(4)}) &= (3.944, 2.209), & (x^{(9)}, y^{(9)}) &= (3.992, 2.017) \\
(x^{(5)}, y^{(5)}) &= (3.958, 2.124), & (x^{(10)}, y^{(10)}) &= (3.995, 2.011)
\end{aligned}
$$

Using Kuhn's modified gradient, it can be verified that the point $(x^*, y^*) = (4.0, 2.0)$ is a least-cost new-facility location. Notice how close the last five estimates are to the least-cost new-facility location.

It is of interest to notice that the previous example can also be solved geometrically. Specifically, for the case of four existing facilities with equal weights, suppose that the four points $\mathbf{P}_1$, $\mathbf{P}_2$, $\mathbf{P}_3$, and $\mathbf{P}_4$ can be arranged in pairs (A, B) and (C, D) so that the straight lines AB and CD intersect at a point X^* which lies between A and B and between C and D, as shown in Figure 4.8. The point X^* is the optimum location for the new facility. However, if one point, say D, lies inside the triangle ABC formed by the remain-

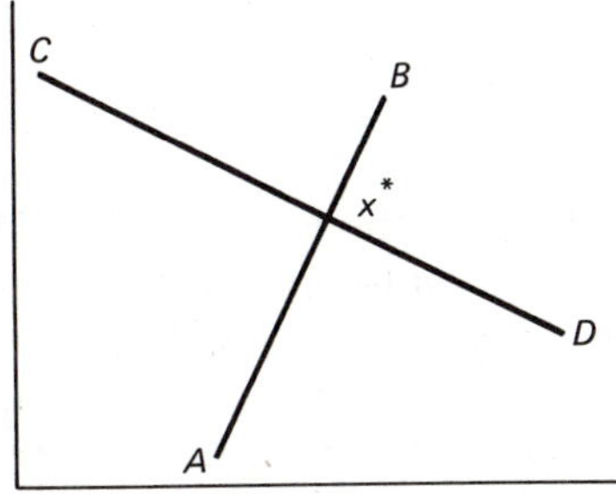

Figure 4.8. Graphical solution procedure for the Euclidean problem.

ing points, the optimum location for the new facility coincides with the location of existing facility D. Using the geometrical approach on the previous example, it is easy to show that $(x^*, y^*) = (4.0, 2.0)$.

For the case of three existing facilities (A, B, and C) with equal weights, if the triangle ABC formed by the existing facilities has all angles less than 120°, then the optimum location for the new facility is the point from which straight lines drawn to each existing facility form three angles of 120°. If the triangle ABC contains an angle greater than or equal to 120°, then the optimum location for the new facility is at that vertex [5, 8, 19].

An alternative iterative solution procedure can be employed to solve the Euclidean problem without employing Kuhn's modified gradient procedure. The procedure is almost identical to that given by (4.39) and (4.40), with the exception that $g_i(x, y)$ is defined as

$$g_i(x, y) = \frac{w_i}{[(x - a_i)^2 + (y - b_i)^2 + \epsilon]^{1/2}}, \quad i = 1, \ldots, m \tag{4.41}$$

where ϵ is an arbitrarily small, positive-valued constant. Notice that (4.41) is always defined. Furthermore, as the value of ϵ approaches zero, the new function approaches the original function. We have found that the use of (4.41) in (4.39) and (4.40) produces a very efficient solution procedure for the Euclidean problem. This procedure is called HAP (hyperboloid approximation procedure) and is discussed in greater detail in Chapter 5.

It is interesting to notice that HAP also can be used to solve rectilinear location problems. To see why this is true, consider three facilities with locations (0, 2), (0, 5), and (0, 8) and weights of $\frac{1}{3}$, $\frac{1}{3}$, and $\frac{1}{3}$. For this case,

$$f(x, y) = \tfrac{1}{3}[x^2 + (y - 2)^2]^{1/2} + \tfrac{1}{3}[x^2 + (y - 5)^2]^{1/2} + \tfrac{1}{3}[x^2 + (y - 8)^2]^{1/2} \tag{4.42}$$

For the expression (4.42), it should be obvious that for any $x \neq 0$, $f(x, y) > f(0, y)$. Therefore, it is only necessary to search along the line $x = 0$ to find a least-cost location for the new facility. Consequently, (4.42) reduces to

$$f(0, y) = \tfrac{1}{3}|y - 2| + \tfrac{1}{3}|y - 5| + \tfrac{1}{3}|y - 8| \tag{4.43}$$

which is identical to the previous expression (4.12). Thus, we see that a rectilinear location problem can be given as the sum of two Euclidean problems, $f(0, y)$ and $f(x, 0)$. Of course, we would not use HAP in solving single-facility rectilinear location problems since the optimum solution can be obtained directly. However, this is not necessarily the case for multiple new facilities, as discussed in Chapter 5.

The iterative procedure is commonly initiated using either the rectilinear solution or the centroid solution. In some cases, it might occur that the rectilinear solution is sufficiently close to the optimum Euclidean solution such that further search is not justified. Using the triangle inequality, the following upper and lower bounds can be obtained for the optimum Euclidean solution [25]. (The development of the lower bounds is given as an exercise at the end of the chapter.)

$$E(x^*, y^*) \geq E(x^0, y^0) \geq [R^2(x^*) + R^2(y^*)]^{1/2} \tag{4.44}$$

where (x^0, y^0) = optimum Euclidean solution

(x^*, y^*) = optimum rectilinear solution

$E(x, y)$ = value of the objective function for the Euclidean problem given the solution (x, y); see (4.35)

$R(x) = f_1(x)$ for rectilinear problem; see (4.6)

$R(y) = f_2(y)$ for rectilinear problem; see (4.7)

As an illustration of the use of the triangle inequality, consider the example involving four existing facilities, treated in Section 4.2, where $x^* = 8$, $y^* = 5$ is an optimum rectilinear solution. For that example,

$$\begin{aligned}
R(x^*) &= \tfrac{1}{6}(4) + \tfrac{1}{3}(0) + \tfrac{1}{3}(3) + \tfrac{1}{6}(5) \\
&= 2.5 \\
R(y^*) &= \tfrac{1}{6}(3) + \tfrac{1}{3}(0) + \tfrac{1}{3}(3) + \tfrac{1}{6}(3) \\
&= 2.0 \\
E(x^*, y^*) &= \tfrac{1}{6}[(8-4)^2 + (5-2)^2]^{1/2} + \tfrac{1}{3}[(8-8)^2 + (5-5)^2]^{1/2} \\
&\quad + \tfrac{1}{3}[(8-11)^2 + (5-8)^2]^{1/2} + \tfrac{1}{6}[(8-13)^2 + (5-2)^2]^{1/2} \\
&= 3.22
\end{aligned}$$

Thus, $$3.22 \geq E(x^0, y^0) \geq [(2.5)^2 + (2.0)^2]^{1/2} = 3.2$$

and, for this example, the maximum possible cost penalty associated with locating the new facility at the point (8, 5) is small. In fact, a computation of Kuhn's modified gradient shows that $u_2 = 0.063 < w_2 = 0.333$. Thus, $R(a_2, b_2) = (0, 0)$, and $x^* = 8$, $y^* = 5$ for the Euclidean problem. Furthermore, when HAP is used with $\epsilon = 10^{-12}$, it is found that $x^* = 8$ and $y^* = 5$.

We have previously emphasized the importance of contour lines in our discussion of rectilinear and gravity problems. Contour lines are equally important in the case of the Euclidean problem. Unfortunately, exact methods for constructing contour lines are not available for the Euclidean problem, except for the simplest cases where there are only one or two existing facilities, as illustrated in Figure 4.9. The contour lines for case (a) are

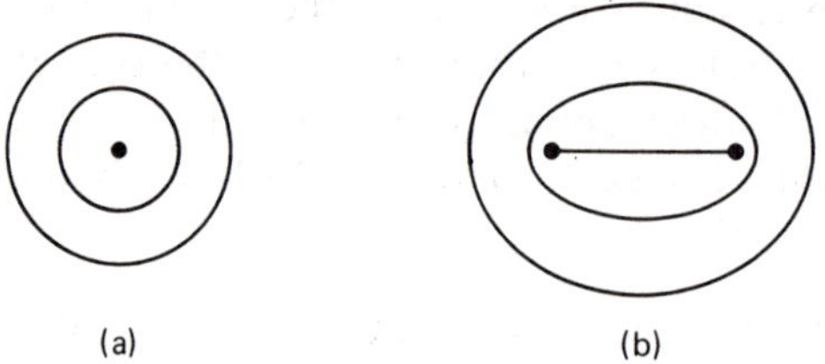

Figure 4.9. Contour lines for two simple Euclidean location problems.

for a single existing facility and for case (b) are for two existing facilities, each having equal item movement with the new facility.

It is relatively simple to obtain approximate contour lines by evaluating the cost function over, say, a rectangular grid of points covering the ranges of (x, y) values of interest. The contour lines can then be drawn by interpolating between grid points. Alternatively, one can assign a given value k to $f(x, y)$ in (4.35), pick a value of x, and search over y for the two values that yield the value k. The process is continued for successive values of x until a family of points is obtained for the contour line having value k. Such

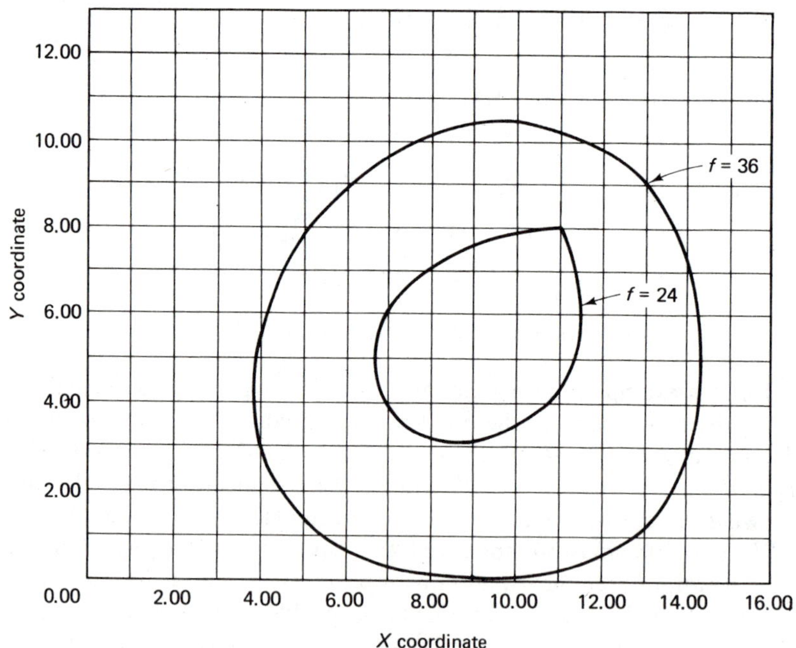

Figure 4.10. Contour lines for Euclidean example problem.

an approach was used to obtain the contour lines shown in Figure 4.10 for the example problem involving four existing facilities, where

$$f(x, y) = \tfrac{1}{6}[(x - 4)^2 + (y - 2)^2]^{1/2} + \tfrac{1}{3}[(x - 8)^2 + (y - 5)^2]^{1/2} + \tfrac{1}{3}[(x - 11)^2 + (y - 8)^2]^{1/2} + \tfrac{1}{6}[(x - 13)^2 + (y - 2)^2]^{1/2}$$

The optimum location for the new facility is $(x^*, y^*) = (8, 5)$ and $f(x^*, y^*) = 3.22$.

In addition to analytic solutions to the Euclidean problem, analog solutions have been obtained [2, 12, 24]. A very simple mechanical analog is illustrated in Figure 4.11. The mechanical analog consists of a board having holes placed appropriately to represent the locations of existing facilities. There are m strings attached at a common knot. A string passes through each hole. A weight of size w_i is attached to the end of each string (notice that this is the motivation for the use of the term "weight"). The system is allowed to reach an equilibrium condition. In the absence of friction, the knot will rest at the optimum location for the new facility. The mechanical analog is simple to construct and will yield reasonable approximations to the Euclidean problem. The solution obtained can be used as the initial solution for either HAP or Kuhn's modified gradient procedure if further refinement is desired. Alternatively, an electrical analog computer solution can be obtained [7, 13]. Using the analog computer with a plotter, contour lines can be constructed.

The mechanical analog can be used to demonstrate an important result, called the majority theorem [30], which states that if at least one half (a majority) the cumulative weight is associated with an existing facility the

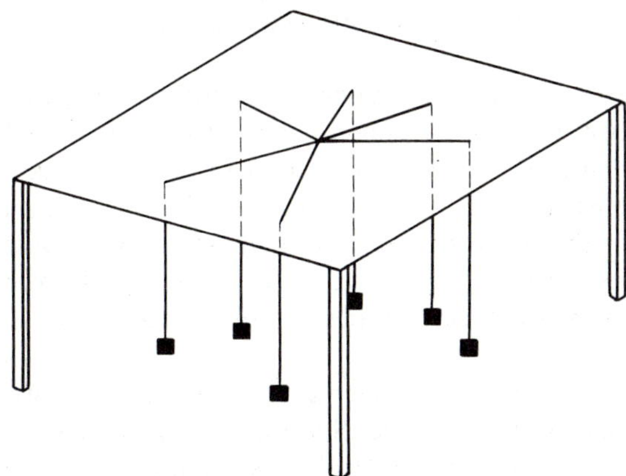

Figure 4.11. Mechanical analog model of the Euclidean problem.

optimum location for the new facility will coincide with the location of the existing facility. The theorem holds for both rectilinear and Euclidean location problems involving a single new facility.

4.6 Summary

In this chapter we have presented analytical approaches for solving some single-facility location problems. In particular, we concentrated on those single-facility location problems in which the cost of item movement was either a linear function of rectilinear or Euclidean distance or a quadratic function of Euclidean distance. Recognizing that other considerations might necessitate a modification in the solution obtained, contour lines were constructed to assist the layout analyst in choosing an appropriate location for the new facility. In a sense, the contour lines provide a means of imputing the cost of other considerations. For example, if the minimum-cost solution obtained by using the models in this chapter had a material-handling cost of \$100 and the preferred location (based on other considerations) had a material-handling cost of \$200, we know that satisfying the other considerations must be worth at least \$100.

Based on the discussion of contour lines, it is easy to assess the importance of locating at the optimum location. Typically, the total cost surface, $f(x, y)$, is relatively flat in the region near the optimum location. Consequently, if you place the new facility "close" to the optimum, a very good solution is normally obtained. Therefore, when using the iterative solution procedure to solve the Euclidean problem, we recommend that you not attempt to obtain *the* optimum solution; rather, you should strive to obtain a very good solution by choosing an appropriate stopping criterion for the procedure.

Remember, the solution obtained from the model is a benchmark solution. Consequently, if the optimum location for the new facility is the same as some existing facility, we do not expect to see you stack the new facility on top of the existing facility. Rather, we suggest that the new facility be located next to the existing facility, if practical to do so. Otherwise, contour lines can be used to guide your search for a feasible location. (Of course, if you really want to stack facilities on top of each other, we cannot stop you.)

In our discussion of single-facility location problems we have assumed that certain data were available. Specifically, the values of (a_i, b_i) and w_i were assumed to be known for each existing facility. In some cases these values might change over time. In others, the values might not be known at all. If the values change over time, one might be interested in minimizing the discounted cost of item movement. If the values of (a_i, b_i) and w_i are not known, we suggest that a number of different estimates of these values be made and that the location problem be solved for each combination of estimates. Based on the solutions obtained and the associated contour lines, you can choose that location which best satisfies your particular objectives.

Too often, analytical approaches are not used because of the "lack of accurate data." This seems to us to be a very poor excuse for at least two reasons. First, the models are normally quite insensitive to errors in estimating the values of the parameters of the model. Second, what alternative approach will be used to make a decision? Undoubtedly, the alternative approach will be based (at least implicitly) on subjective estimates of the values of parameters. If this is the case, why not make these estimates explicit and use the model?

If you still are unwilling to assign values to the parameters and, yet, a decision is made concerning the location of the new facility, then we can impute values or ranges of values to parameters. As an illustration, suppose that the assumptions of the gravity problem are met, and there exist three facilities located at $\mathbf{P}_1 = (0, 0)$, $\mathbf{P}_2 = (2, 6)$, and $\mathbf{P}_3 = (8, 4)$. You are told that it is impossible to estimate the values of w_i, but the decision has been made to locate the new facility at (4, 4). From (4.29) and (4.30), for (x^*, y^*) to be equal to (4, 4) it must be true that

$$4 = 0w_1 + 2w_2 + 8w_3$$
$$4 = 0w_1 + 6w_2 + 4w_3$$

where we have chosen our time scale such that

$$1 = w_1 + w_2 + w_3$$

Consequently, we have three equations in three unknowns, which can be solved to obtain

$$w_1 = 0.2$$
$$w_2 = 0.4$$
$$w_3 = 0.4$$

You are now able to ask, "Do we really expect there to be twice as much item movement to the second existing facility as there is to the first, and the same amount of item movement between the new facility and the second and third existing facilities?" If the answer is "No!", then values should be specified that are reasonable for the location problem in question. Similar approaches can be taken when the rectilinear and the Euclidean models are appropriate. However, the analysis is not as straightforward as for the gravity problem. This type of analysis is pursued further in the problems.

REFERENCES

1. BINDSCHEDLER, A. E., and J. M. MOORE, "Optimum Location of New Machines in Existing Plant Layouts," *The Journal of Industrial Engineering*, Vol. 12, No. 1, 1961, pp. 41–48.
2. BURSTALL, R. M., and R. A. LEAVER, "Evaluation of Transport Costs for Alternative Factory Sites—A Case Study," *Operational Research Quarterly*, Vol. 13, No. 4, 1962.
3. CHARNES, A., J. E. QUON, and S. J. WERSAN, "Location-Allocation Problems in the L_1 Metric," *Joint CORS-ORSA 25th National Meeting*, Montreal, Quebec, May 1964.
4. COOPER, L., "Location-Allocation Problems," *Operations Research*, Vol. 11, No. 3, 1963, pp. 331–344.
5. COOPER, L., "Heuristic Methods for Location-Allocation Problems," *SIAM Review*, Vol. 6, No. 1, 1964, pp. 37–52.
6. COOPER, L., "Solutions of Generalized Locational Equilibrium Models," *Journal of Regional Science*, Vol. 7, No. 1, 1967, pp. 1–18.
7. EILON, S., and D. P. DEZIEL, "Siting a Distribution Center," *Management Science*, Vol. 12, No. 6, 1966, pp. 245–254.
8. EILON, S., C. D. T. WATSON-GANDY, and N. CHRISTOFIDES, *Distribution Management: Mathematical Modelling and Practical Analysis*, Hafner Publishing Company, Inc., New York, 1971.
9. FRANCIS, R. L., "A Note on the Optimum Location of New Machines in Existing Plant Layouts," *The Journal of Industrial Engineering*, Vol. 14, No. 1, 1963, pp. 57–59.
10. FRANCIS, R. L., "Some Aspects of a Minimax Location Problem," *Operations Research*, Vol. 15, No. 6, 1967, pp. 1163–1168.
11. HAKIMI, S. L., "Optimum Location of Switching Centers and the Absolute Centers and Medians of a Graph," *Operations Research*, Vol. 12, No. 3, 1964, pp. 450–459.
12. HALEY, K. B., "The Siting of Depots," *International Journal of Production Research*, Vol. 2, 1963, pp. 41–46.
13. HITCHINGS, G. G., "Analogue Techniques for the Optimal Location of a Main Facility in Relation to Ancillary Facilities," *International Journal of Production Research*, Vol. 7, No. 3, 1967, pp. 189–197.
14. KATZ, N. I., "On the Convergence of a Numerical Scheme for Solving Some Locational Equilibrium Problems," *Journal of the Society of Industrial and Applied Mathematics*, Vol. 17, No. 6, 1969, pp. 1224–1231.
15. KELLY, L. M., "Optimal Distance Configurations," *Recent Progress in Combinatorics*, Proceedings of the Third Waterloo Conference on Combinatorics, W. T. Tuttle, editor, Academic Press, Inc., New York, 1969, pp. 111–122.

16. KUHN, H. W., "Locational Problems and Mathematical Programming," *Separatum-Colloquium on the Application of Mathematics to Economics*, 1965, pp. 235–242.

17. KUHN, H. W., "On a Pair of Dual Non-Linear Problems," *Non-Linear Programming*, Chapter 3, J. Abadie, ed., John Wiley & Sons, Inc., New York, 1967.

18. KUHN, H. W., "A Note on Fermat's Problem", *Mathematical Programming*, Vol. 4, No. 1, 1973, pp. 98–107.

19. KUHN, H. W., and E. KUENNE, "An Efficient Algorithm for the Numerical Solution of the Generalized Weber Problem in Spatial Economics," *Journal of Regional Science*, Vol. 4, No. 2, Winter 1962, pp. 21–33.

20. LEVY, J., "An Extended Theorem for Location on a Network," *Operational Research Quarterly*, Vol. 18, No. 4, 1967, pp. 433–444.

21. MCHOSE, A. H., "A Quadratic Formulation of the Activity Location Problem," *The Journal of Industrial Engineering*, Vol. 12, No. 5, 1961, pp. 334–338.

22. MIEHLE, W., "Link Length Minimization in Networks," *Operations Research*, Vol. 6, No. 2, 1958, pp. 232–243.

23. NOBEL, B., "Optimum Location Problems," *Applications of Undergraduate Mathematics in Engineering*, Chapter 2, The Macmillan Company, New York, 1967.

24. PALERMO, F. P., "A Network Minimization Problem," *IBM Journal of Research and Development*, Vol. 5, No. 4, 1961, pp. 335–337.

25. PRITSKER, A. A. B., and P. M. GHARE, "Locating New Facilities with Respect to Existing Facilities," *AIIE Transactions*, Vol. 2, No. 4, 1970, pp. 290–297.

26. TIDEMAN, M., "Comment on 'A Network Minimization Problem'," *IBM Journal of Research and Development*, Vol. 6, No. 2, 1962, p. 259.

27. VERGIN, R. C., and J. D. ROGERS, "An Algorithm and Computational Procedure for Locating Economic Activities," *Management Science*, Vol. 13, No. 6, 1967, pp. 240–254.

28. WEISZFELD, E., "Sur le point pour lequel la somme des distances de *n* points donnés est minimum," *Tohoku Mathematics Journal*, Vol. 43, 1936, pp. 355–386.

29. WHITE, J. A., "A Quadratic Facility Location Problem," *AIIE Transactions*, Vol. 3, No. 2, 1971, pp. 156–157.

30. WITZGALL, C., "Optimal Location of a Central Facility: Mathematical Models and Concepts," National Bureau of Standards Report 8388, Gaithersberg, Maryland, 1965.

PROBLEMS

4.1. Go through the complete procedure for constructing contour lines for the rectilinear-distance problem for which the data are as follows: $\mathbf{P}_1 = (4, 4)$, $\mathbf{P}_2 = (4, 11)$, $\mathbf{P}_3 = (7, 2)$, $\mathbf{P}_4 = (11, 11)$, $\mathbf{P}_5 = (14, 7)$; $w_1 = 3$, $w_2 = 2$, $w_3 = 2$, $w_4 = 4$, $w_5 = 1$. Also find all least-cost new-facility locations.

4.2. A conveyor line is being planned to run into a warehouse; it will begin at the point (0, 5) and run parallel to the x axis in an increasing direction of x [i.e., the conveyor line runs to the right of the point (0, 5)]. Items entering the warehouse on the line are picked up at the end of the line and transported directly to one of the truck docks at the points $\mathbf{P}_1 = (7, 10)$, $\mathbf{P}_2 = (15, 7)$, $\mathbf{P}_3 = (15, 3)$, and $\mathbf{P}_4 = (12, 0)$ (coordinates are given in tens of feet). The distance items travel between the line end and any one of the docks may be accurately approximated by the rectilinear distance between the end of the line and the dock. From data on expected usage rates of docks, labor costs, and material transport times, it is expected that the total annual cost per foot for transporting items between the end of the line and the docks at points $\mathbf{P}_1$, $\mathbf{P}_2$, $\mathbf{P}_3$, and $\mathbf{P}_4$ will be \$80, \$20, \$30, and \$70, respectively. The equivalent annual cost per foot of conveyor is \$90.00.

(a) Construct a model representing the total annual cost for transporting items from the production area along the line and from the line to the docks.

(b) Using the model developed in part (a), find the conveyor line length to minimize the total cost of item transport.

(c) What should be the conveyor line length to minimize the total cost if the equivalent annual cost per foot of conveyor line is \$30.00?

(d) How much would the equivalent annual cost per foot of conveyor line have to be in order for it to be cheaper to have no line at all?

4.3. Existing facilities are located as follows: $\mathbf{P}_1 = (4, 4)$, $\mathbf{P}_2 = (4, 10)$, $\mathbf{P}_3 = (6, 5)$, $\mathbf{P}_4 = (10, 5)$, $\mathbf{P}_5 = (10, 9)$, and $\mathbf{P}_6 = (12, 3)$. It is desired to locate one new machine with respect to the existing facilities. Travel between facilities is along a rectilinear aisle structure. The amount of item movement between the new facility and each existing facility is given as $w_1 = 4$, $w_2 = 4$, $w_3 = 2$, $w_4 = 3$, $w_5 = 5$, and $w_6 = 6$. Where should the new facility be located to minimize distance traveled? Plot contour lines and rank the following potential locations: $A = (4, 6)$, $B = (8, 8)$, $C = (8, 5)$, $D = (10, 3)$.

4.4. Four hospitals located within a city are cooperating to establish a centralized blood-bank facility that will serve the hospitals. The new facility is to be located such that distance traveled is minimized. The hospitals are located as follows: $\mathbf{P}_1 = (5, 10)$, $\mathbf{P}_2 = (7, 6)$, $\mathbf{P}_3 = (4, 2)$, and $\mathbf{P}_4 = (16, 3)$. The number of deliveries to be made per year between the blood-bank facility and each hospital is estimated to be 450, 1,200, 300, and 1,500, respectively. Assuming rectilinear travel, determine the optimum location. What assumption must be made concerning the deliveries of blood? Is it reasonable to expect delivery trips will always be to only one hospital when a delivery is made?

4.5. A large corporation has employed a campus plan layout for its corporate headquarters. A district heating system is to be installed, which will heat each of four buildings. Considering the cost of installation and the heat losses, it is agreed that the cost for the system is proportional to the square of the Euclidean distance between the heating facility and each building. The constant of proportionality is the Btu's per hour required at each building. The buildings to be served by the heating system are located as follows: $\mathbf{P}_1 = (18, 5)$, $\mathbf{P}_2 = (13, 9)$, $\mathbf{P}_3 = (23, 11)$, and $\mathbf{P}_4 = (7, 11)$. The Btu requirements per hour are 12,000, 5,000, 4,000, and 15,000, respectively. Find the least-cost location for the central heating facility.

4.6. An airport terminal is served by Pokey, Early, Delayed, and Undecided airlines. The airlines unload baggage from arriving airplanes at the points indicated in Figure P4.6.

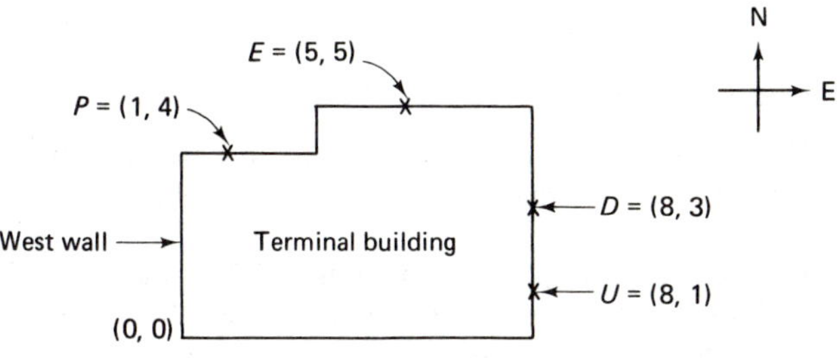

Figure P4.6

The number of arriving flights per day for each air line is 36, 22, 28, and 18, for Pokey, Early, Delayed, and Undecided, respectively. If a separate trip is made from the baggage receiving area to a passenger pickup point for each arriving flight, and rectilinear distance is assumed, where should the pickup point be located? Suppose that the pickup point must be outside and on either the south or west side of the terminal; where should the pickup point be located? Suppose that a conveyorized system is to be installed with separate conveyor belts joining each airline and the pickup point. If the conveyor layout follows a rectilinear layout pattern, what is the optimum location for the pickup point?

4.7. Mrs. Black has to pick up several items for her daughter's party at various locations. She wants to locate a parking space such that she can get to all the stores in a minimum amount of time without reparking. Each block is square —100 feet on a side. Streets running north to south are numbered consecutively. Those running east to west are lettered consecutively. The bakery is at 6th and E; she must walk half as fast as normal, so that she won't drop the cake or pastries. At 10th and D is the grocery store. The dress shop is at 12th and G. Mrs. Black picks up her daughter, Kathy, from the beauty parlor at 10th and G, and they walk twice as fast as normal back to the car, so that no one will see her with her hair up. It is assumed she must stay on the sidewalks that enclose each block—distance used crossing streets is considered negligible. Determine the location of the parking space that satisfied her objective.

4.8. A portable lunch wagon is to be located along the main aisle of an industrial plant. The vendor wishes to locate the wagon so that the total walking distance for his customers is minimized. On a coordinate system the main aisle runs parallel to the x axis at $y = 10$. Work stations are located at the following coordinate points. Travel from the work stations to the main aisle is rectilinear.

Station	*Location*	*Station*	*Location*
1	(4, 4)	6	(12, 12)
2	(4, 12)	7	(10, 14)
3	(6, 4)	8	(12, 6)
4	(6, 12)	9	(8, 14)
5	(8, 8)	10	(10, 4)

(a) If one customer is located at each station, where should the wagon be located along the main aisle to minimize cumulative customer walking distance?

(b) If one customer is located at stations 1, 2, 3, 4, and 5, and three customers are located at stations 6, 7, 8, 9, and 10, where should the wagon be located along the main aisle to minimize cumulative customer walking distance?

4.9. For the rectilinear-distance problem, it is known that some point c_t minimizes $f_1(x)$. Explain why $\sum_{j=1}^{t-1} C_j \leq \frac{1}{2} \sum_{i=1}^{m} w_i$ and $\sum_{j=1}^{t} C_j \geq \frac{1}{2} \sum_{i=1}^{m} w_i$; i.e., explain why the median conditions are necessary conditions for a minimum. [Property 3 of Section 4.3 and the subsequent discussion establish that if c_t is a point for which the median conditions hold, then c_t minimizes $f_1(x)$; what is being asked for here is the converse.]

4.10. For the rectilinear-distance location problem, explain why, for the median conditions

$$\sum_{j=1}^{t-1} C_j < \frac{1}{2} \sum_{i=1}^{m} w_i, \qquad \sum_{j=1}^{t} C_j = \frac{1}{2} \sum_{i=1}^{m} w_i$$

any point between c_t and c_{t+1} minimizes $f_1(x)$.

4.11. Assuming Euclidean distances, what is the optimum location for a single new facility if $\mathbf{P}_1 = (0, 9)$, $\mathbf{P}_2 = (1, 2)$, $\mathbf{P}_3 = (7, 10)$, $\mathbf{P}_4 = (8, 3)$, and $w_1 = w_2 = w_3 = w_4$?

4.12. Plot the contour line passing through the point (1, 0) for the problem

$$f(x, y) = 3|x - 1| + 2|y - 1| + 1|y - 2|$$

What is the value of $f(x, y)$ for this contour line?

4.13. Suggest an iterative procedure for solving the problem

$$\text{minimize} \sum_{i=1}^{m} w_i[(x - a_i)^2 + (y - b_i)^2]^k, \quad \text{where } k > 0$$

The procedure should agree with the one given in the case for $k = \frac{1}{2}$. What reasons can you suggest for considering values of k other than $\frac{1}{2}$?

4.14. Given existing facilities at points $\mathbf{P}_1 = (0, 0)$ and $\mathbf{P}_2 = (6, 0)$, with $w_1 = w_2 = 1$, plot several contour lines for the (a) rectilinear-distance model, (b) squared-distance model, and (c) Euclidean-distance model.

4.15. For Problem 4.14, which points will minimize the total cost function for each of the three cases?

4.16. Suppose that two people go to a shopping center. They have agreed beforehand that each will visit one store alone and then return to their parked car. The two stores are at opposite ends of the shopping center, and the car can be parked at any spot along a straight sidewalk running between the two stores. Where do you think they would park their car and why? Does your answer suggest that there may be quantifiable approaches to location problems other than minimum total cost approaches; if so, what are they?

4.17. Using the contour-line approach for evaluating locations for a new machine, determine the optimal location based on the following data:

Locations of Existing Machines			*Possible Locations for New Machines*		
i	x_i	y_i	j	x_j	y_j
1	8	7	1	8	10
2	13	2	2	16	17
3	13	13	3	2	16
4	18	10	4	16	3
5	2	10	5	16	8
6	8	16			
7	18	20			

The pallet-load movement per day between the new machine and the ith existing machine, w_i, is given to be

i	w_i
1	100
2	100
3	150
4	100
5	200
6	250
7	100

Rank the alternative locations in order of preference using contour lines. Rectilinear travel is used.

4.18. The ABC Auto Parts Company has 11 retail sales stores in the city of Greenville. The ABC Company needs a new warehouse facility to service its retail stores. The locations of the stores and the expected deliveries per week from the warehouse to each store are

Store	*Location (miles)*	*Expected Deliveries*
1	(1, 0)	1
2	(1, 3)	1
3	(2, 0)	1
4	(2, −1)	1
5	(1, −2)	3
6	(−1, 0)	5
7	(−1, 1)	6
8	(−1, 3)	2
9	(−2, −2)	4
10	(−3, −1)	1
11	(−3, 1)	2

Assume that travel within the city of Greenville is rectilinear and that after each delivery the delivery truck must return to the warehouse.

(a) If there are no restrictions on the warehouse location, where should it be located?

(b) If the ABC Company is going to open a new retail outlet at the warehouse location and this outlet must be at least 1 rectilinear mile away from an existing store, where should it be located?

4.19. A plumbing contractor wishes to minimize the cost of pipe for a water waste disposal system in a building under construction. The locations of all the facilities that require water waste removal have already been determined. However, all the individual drain pipes must be connected at some point along the inside basement wall of the building; i.e., $y = 0$. From the connection one main drain pipe is used to connect with the sewer system. Assume the (x, y, z) coordinates of the connection points within the building are $\mathbf{P}_1 = (8, 20, -8)$, $\mathbf{P}_2 = (12, 10, 2)$, $\mathbf{P}_3 = (25, 35, 2)$, $\mathbf{P}_4 = (40, 20, 2)$, $\mathbf{P}_5 = (25, 35, 12)$, $\mathbf{P}_6 = (30, 40, 12)$, $\mathbf{P}_7 = (15, 25, 22)$, and $\mathbf{P}_8 = (25, 35, 22)$. The main drain must be connected to the sewer line at $\mathbf{P}_0 = (20, -50, -10)$. The cost per foot of drain line from the connection point to each facility is $w_1 = \$4$, $w_2 = \$5$, $w_3 = \$3$, $w_4 = \$1.50$, $w_5 = \$3$, $w_6 = \$1.50$, $w_7 = \$2$, and $w_8 = \$3$, respectively. The cost per foot of pipe for the main drain is \$15. Pipe lines will be laid out rectilinearly. Determine the optimum location for the main drain connection. What is the optimum location if the main drain connection is constrained to have $x \leq 1$ and $z \leq -8$?

4.20. As Minister of Defense for the planet Xerxes in the galaxy Euclid, you are seriously concerned about your four most aggressive rivals and plan to place one outpost in space fully equipped with an army of laser ships to be dispatched, as necessary, to attack any of the four rival planets. The positions in space of the four rivals, relative to an arbitrarily selected origin, are $R_1 = (2, 2, 2)$, $R_2 = (12, 6, 8)$, $R_3 = (10, 0, -4)$, and $R_4 = (6, 20, -10)$, measured in millions of miles. The critical travel time is a function of the square of the Euclidean distance traveled (due to the repelling forces generated from the four planets). You have assessed the probability of an attack originating from each planet and assigned the following values: $p_1 = 0.4$, $p_2 = 0.6$, $p_3 = 0.4$, and $p_4 = 0.9$. (Notice that attacks are not mutually exclusive events.) Fortunately, due to the magnitude of the diameters of the orbits of all planets and the outpost (order of 10 million light-years), the entire coordinate system is considered to orbit, such that any changes in the relative positions of the planets due to orbiting can be ignored. Where should the outpost be located to minimize the expected distance traveled?

4.21. One new facility is to be located with respect to three existing facilities, and straight-line movement is used. Annual item movement between the new facility and each existing facility is given to be $w_1 = 5$, $w_2 = 2$, and $w_3 = 10$. The coordinate locations of the existing facilities are $\mathbf{P}_1 = (2, 6)$, $\mathbf{P}_2 = (4, 0)$, and $\mathbf{P}_3 = (8, 3)$.

(a) Give upper and lower bounds on the straight-line distance traveled per year. Use (4.44).

(b) Determine the optimum location for the new facility, assuming cost is proportional to the square of the straight-line distance traveled.

(c) Determine the optimum location for the Euclidean problem.

4.22. Are the following statements true or false? If false, give a reason for the statement being false.

(a) Let $\mathbf{P}_1 = (0, 0)$, $\mathbf{P}_2 = (0, 10)$, $\mathbf{P}_3 = (x, 5)$, and $w_1 = w_2 = w_3$. With rectilinear distance, the optimum location for the new facility is not dependent on the value of x.

(b) Under some conditions the rectilinear, Euclidean, and squared Euclidean problems will yield the same location for a given problem.

(c) The partial derivatives $\partial f/\partial x_j$ are always defined for both the Euclidean and squared Euclidean problem.

(d) The single-facility rectilinear problem is easiest solved using linear programming.

(e) The rectilinear, Euclidean, and squared Euclidean problems can have multiple optimum solutions.

(f) In a single-facility location problem (of the type treated in this chapter) involving existing facilities at $\mathbf{P}_1 = (10, 10)$, $\mathbf{P}_2 = (12, 15)$, and $\mathbf{P}_3 = (16, 18)$, the location (16, 10) can never be an optimum location.

4.23. Let four existing facilities be located at $\mathbf{P}_1 = (0, 0)$, $P_2 = (5, 0)$, $P_3 = (5, 5)$, and $\mathbf{P}_4 = (0, 5)$ with $w_1 = 2$, $w_2 = 1$, $w_3 = 2$, and $w_4 = 2$. Determine the optimum location for a single new facility when cost is proportional to (a) rectilinear distance, (b) squared Euclidean distance, and (c) Euclidean distance. Compare the minimum total cost obtained in part (c) with the upper bound from (4.44), using the solution obtained in part (a).

4.24. Let four existing facilities be located at $\mathbf{P}_1 = (0, 0)$, $\mathbf{P}_2 = (6, 0)$, $\mathbf{P}_3 = (6, 16)$, and $\mathbf{P}_4 = (12, 16)$ with $w_1 = w_2 = w_3 = w_4 = 2$. Determine the optimum location for a single new facility when cost is proportional to (a) rectilinear distance, (b) squared Euclidean distance, and (c) Euclidean distance.

Compare the minimum total cost obtained in part (c) with the upper bound from (4.44) using the following rectilinear solutions: (6, 0), (6, 8), and (6, 16).

4.25. A city wishes to locate a fire station so as to minimize the cost of fire damage. After some analysis, five major areas have been identified in the city as being primary risk areas. The feeling is that fire damage increases in proportion to the square of the time required to answer an alarm. It is also felt that time and distance are directly proportional. For purposes of the analysis straight-line distances are assumed. Weighting factors have been obtained such that the fire loss at area k is equal to $w_k d_k^2$, where w_k is the weighting factor and d_k is the Euclidean distance from the fire station to area k. The coordinate locations of the areas, along with the weight factors, are

k	x_k	y_k	w_k
1	4	12	4
2	8	2	2
3	14	10	2
4	2	3	8
5	14	4	9

What is the optimum location for the fire station?

4.26. Four machines are located at the coordinate points (0, 4), (0, 8), (4, 0), and (8, 0). Thirty per cent of the item movement between the existing machines and a new machine is between the new machine and each of the existing machines at (0, 4) and (4, 0). Twenty per cent is between the new machine and each of the existing machines at (0, 8) and (8, 0). Rectilinear movement is used. Construct the contour line passing through the positive quadrant and beginning at the point (0, 12). If the machine locations are interpreted as docks and the contour line is interpreted as the periphery of the warehouse, what warehousing interpretation might be given to the percentage weights? Explain.

4.27. Given three existing facilities located at $\mathbf{P}_1 = (0, 0)$, $\mathbf{P}_2 = (0, 4)$, and $\mathbf{P}_3 = (4, 2)$ with $w_1 = w_2 = w_3 = 1$, determine the upper and lower bounds on the minimum value of the objective function for the single-facility Euclidean problem using (4.44).

4.28. Let $z_1, \ldots, z_p$ be a feasible solution to the dual of the rectilinear problem. The following conditions will be called complementary slackness conditions: $z_j(c_j - x) = |z_j||c_j - x|$ and $(|z_j| - C_j)|x - c_j| = 0$ for $j = 1, \ldots, p$.

(a) Given a feasible solution to the dual problem $z_1, \ldots, z_p$ and a point x such that the complementary slackness conditions hold, prove that x minimizes $f_1(x)$, and $z_1, \ldots, z_p$ is a maximum feasible solution to the dual problem.

(b) Suppose that x minimizes $f_1(x)$ and $z_1, \ldots, z_p$ is a maximum feasible solution to the dual problem; prove that the complementary slackness conditions hold.

(c) Given that $c_1 < c_2 < \cdots < c_p$, that x minimizes $f_1(x)$, and that $z_1, \ldots, z_p$ is a maximum feasible solution to the dual problem, prove that at most one of the following inequalities can hold strictly:

$$|z_j| \leq C_j, j = 1, \ldots, p$$

4.29. Solve the Euclidean-distance problem given the following data:

(a) $\mathbf{P}_1 = (0, 0)$, $\mathbf{P}_2 = (2, 0)$, $\mathbf{P}_3 = (0, 2)$, $\mathbf{w}_1 = \frac{1}{5}$, $\mathbf{w}_2 = \frac{3}{5}$, $\mathbf{w}_3 = \frac{1}{5}$.

(b) $\mathbf{P}_1 = (1, 0)$, $\mathbf{P}_2 = (2, 1)$, $\mathbf{P}_3 = (1, 2)$, $\mathbf{P}_4 = (0, 1)$, $\mathbf{w}_i = 1$ for $i = 1, \ldots, 4$.

Be sure to justify your answers.

4.30. A new cooperative power plant is to be located to serve each of 10 towns located in a large level area devoted to wheat ranching. Separate electric cables are to be run from the power plant to each town. The locations of the 10 towns are $\mathbf{P}_1 = (0, 0)$, $\mathbf{P}_2 = (1, 3)$, $\mathbf{P}_3 = (2, 5)$, $\mathbf{P}_4 = (3, 9)$, $\mathbf{P}_5 = (4, 2)$, $\mathbf{P}_6 = (5, 4)$, $\mathbf{P}_7 = (6, 6)$, $\mathbf{P}_8 = (7, 1)$, $\mathbf{P}_9 = (8, 5)$, and $\mathbf{P}_{10} = (10, 8)$. Determine the location for the power plant that will minimize the total required length of electrical cable.

4.31. The Wei-Louse Army has combat outposts located at $\mathbf{P}_1 = (2, 7)$, $\mathbf{P}_2 = (3, 8)$, $\mathbf{P}_3 = (6, 9)$, and $\mathbf{P}_4 = (9, 8)$. The military commander, General B. Trayed, has decided to construct a temporary base from which helicopters will be dispatched to provide air support and supplies for the outposts. The airbase will receive its supplies by air from two supply points, $\mathbf{P}_5 = (0, 4)$ and $\mathbf{P}_6 = (8, 1)$. The expected number of flights per day between the new base and the point $\mathbf{P}_j$ is given by w_j, where $w_1 = 10$, $w_2 = 12$, $w_3 = 14$, $w_4 = 10$, $w_5 = 26$, and $w_6 = 34$. The local air traffic controller, Private Ama Kamikazee, indicates straight-line paths are to be used for all flights in order to confuse the enemy. Determine the location for the base such that the expected distance flown per day is minimized.

4.32. The Royal Canadian Hillies is going to locate a brandy carryout in the Yukon. A number of Saint Bernard dogs, trained by Sergeant Preston, will be used to rescue lost tourists looking for the South Pole. Based on past experience, it is anticipated that emergency rescue missions will have to be dispatched to the points $Q_1 = (18, 2)$, $Q_2 = (4, 0)$, $Q_3 = (6, 20)$, and $Q_4 = (12, 18)$ with weekly frequencies $w_1 = 6$, $w_2 = 2$, $w_3 = 7$, and $w_4 = 4$. Sergeant Preston has trained his dogs to travel along a straight-line path in carrying out the rescue mission.

(a) Determine the location of the brandy carryout that minimizes the distance traveled by the dogs.

(b) Assume the local SPCA has protested that the suffering of St. Bernards in carrying out rescue missions is proportional to the square of the distance traveled per mission. What location will minimize the total amount of suffering?

4.33. Suppose that an item is to be located in three-space instead of two-space. Suggest contexts in which (a) the location models considered in this chapter would extend in a logical manner to their three-space counterpart, and (b) would not extend in a logical manner to their three-space counterpart.

4.34. Existing facilities are located as follows:

$$\begin{array}{ll} \mathbf{P}_1 = (0, 0) & \mathbf{P}_6 = (7, 8) \\ \mathbf{P}_2 = (2, 8) & \mathbf{P}_7 = (9, 0) \\ \mathbf{P}_3 = (4, 0) & \mathbf{P}_8 = (9, 4) \\ \mathbf{P}_4 = (4, 4) & \mathbf{P}_9 = (13, 2) \\ \mathbf{P}_5 = (7, 2) & \mathbf{P}_{10} = (13, 8) \end{array}$$

There is equal item movement between the single new facility and each existing facility.

(a) Assuming rectilinear travel and assuming the new facility must be at least 3 rectilinear distance units away from an existing facility, what is the optimum location for the new facility? What is the associated value of $f(x, y)$?

(b) Assuming cost is proportional to the square of the straight-line distance between the new facility and each existing facility, where should the new facility be located?

4.35. Define the following approximating function for the Euclidean problem:

$$\text{minimize } g(x, y) = \sum_{i=1}^{m} w_i E_i(\epsilon)$$

where

$$E_i(\epsilon) = [(x - a_i)^2 + (y - b_i)^2 + (\epsilon)]^{1/2}$$

and ϵ is some arbitrarily small, positive constant. Let

$$g_x = \frac{\partial g(x, y)}{\partial x} \qquad g_{xx} = \frac{\partial^2 g(x, y)}{\partial x^2}$$

$$g_y = \frac{\partial g(x, y)}{\partial y} \qquad g_{yy} = \frac{\partial^2 g(x, y)}{\partial y^2}$$

$$g_{xy} = \frac{\partial^2 g(x, y)}{\partial x \partial y} = g_{yx},$$

Newton's method can be employed to solve the set of equations

$$g_x = 0$$

$$g_y = 0$$

Let (x_0, y_0) be an approximation of the optimum location for the new facility, (x^*, y^*). Successive approximations of (x^*, y^*) can be generated from the recursion formulas

$$x_{k+1} = x_k - \left[\frac{g_x g_{yy} - g_y g_{xy}}{g_{xx} g_{yy} - (g_{xy})^2}\right]_k$$

and

$$y_{k+1} = y_k - \left[\frac{g_y g_{xx} - g_x g_{xy}}{g_{xx} g_{yy} - (g_{xy})^2}\right]_k$$

where all functions involved are to be evaluated at (x_k, y_k). Use Newton's method to solve the Euclidean-distance problems in Problems 4.30, 4.31, and 4.32. Compare the solution accuracies and number of iterations required with those obtained using the iterative procedure described in the chapter.

4.36. Newton's method, as described in Problem 4.31, is not guaranteed to converge to the optimum solution. As an illustration, consider the following data for a single-facility location problem:

$$\mathbf{P}_1 = (0, 0), \qquad \mathbf{P}_2 = (4, 0), \qquad \mathbf{P}_3 = (2, 4)$$
$$\mathbf{w}_1 = 2, \qquad \mathbf{w}_2 = 3, \qquad \mathbf{w}_3 = 4$$

Based on Euclidean distances, perform four iterations using Newton's method and compare the locations with those obtained after performing four iterations using (4.39) and (4.40). Let $(x^0, y^0) = (3, 3)$.

4.37. Write a computer program to solve the Euclidean problem using the HAP iterative procedure.

4.38. Solve the Euclidean problem in Problem 4.36 using the HAP iterative procedure.

4.39. Plot the contour lines for $f(x, y) = 28$ given
(a) $f(x, y) = 3y + 2x + 4|y - 3| + 2|x - 2| - |x - 4|$.
(b) $f(x, y) = 3|y| + 2|x| + 4|y - 3| + 2|x - 2| - |x - 4|$.

4.40. A new facility is to be placed in an existing layout. The new facility interacts with four existing facilities located at (0, 0), (2, 4), (4, 2), and (5, 5). Rectilinear travel is used. The new facility is located at the point (2, 2). What can be said about the values of w_i in order for the location to minimize travel to and from the existing facilities?

4.41. Solve Problem 4.40 for the case of Euclidean distances and the case of the gravity problem.

4.42. Given $a_{1j}, a_{2j} \geq 0$, $j = 1, \ldots, n$; the following inequality is known as the triangle inequality:

$$\left[\sum_{j=1}^{n} (a_{1j} + a_{2j})^2\right]^{1/2} \leq \left[\sum_{j=1}^{n} a_{1j}^2\right]^{1/2} + \left[\sum_{j=1}^{n} a_{2j}^2\right]^{1/2}$$

By induction, given $a_{ij} \geq 0$,

$$\left[\sum_{j=1}^{n} \left(\sum_{i=1}^{m} a_{ij}\right)^2\right]^{1/2} \leq \sum_{i=1}^{m} \left[\sum_{j=1}^{n} a_{ij}^2\right]^{1/2}$$

Use the latter inequality to derive the lower bound for the optimum Euclidean solution given in (4.44).

4.43. Plot contour lines for the rent-a-car example problem for the cases of (a) rectilinear distance, (b) Euclidean distance, and (c) squared Euclidean distance.

4.44. A single-facility location problem is given in which the distance is *Euclidean*; the existing facility locations are given by

$$\mathbf{P}_1 = (0, 2), \quad \mathbf{P}_2 = (2, 7), \quad \mathbf{P}_3 = (3, 5), \quad \mathbf{P}_4 = (8, 9), \quad \mathbf{P}_5 = (5, 9),$$
$$\mathbf{P}_6 = (4, 4)$$

with corresponding weights

$$w_1 = 5, \quad w_2 = 4, \quad w_3 = 6, \quad w_4 = 2, \quad w_5 = 1, \quad w_6 = 3$$

Find a new-facility location for which the transportation cost will be at most 2% greater than the minimum transportation cost and, in particular, be certain to *give a rigorous justification* for your choice of the new-facility location found.

4.45. (a) A single decision is to be made that will be influenced by events occurring at known points in time $a_1, a_2, \ldots, a_m$. In particular, a cost per unit time of u_i will be incurred if the decision is made prior to a_i, and of v_i if after a_i. Thus, the cost incurred, if the decision is made at a point in time x, with respect to event i is $g_i(x)$, where

$$\begin{aligned} g_i(x) &= u_i(a_i - x), \quad \text{if } x \leq a_i \\ &= v_i(x - a_i), \quad \text{if } x \geq a_i \end{aligned}$$

The total cost incurred if the decision is made at time x is given by

$$f(x) = \sum_{i=1}^{m} g_i(x)$$

Set up a linear-programming problem which, if solved, would determine the point in time to make the decision that would minimize the total cost.

(b) The dual to the linear-programming problem is given by

$$\text{maximize} \quad \sum_{i=1}^{m} a_i y_i$$

subject to

$$\sum_{i=1}^{m} y_i = 0$$
$$-y_i \leq v_i, \quad i = 1, \ldots, m$$
$$y_i \leq u_i, \quad i = 1, \ldots, m$$

If there is a point in time that minimizes $f(x)$, what can you conclude, using duality theory and examining the dual problem, about the relationship of v_i and u_i, for $i = 1, \ldots, m$?

(c) Suppose that the a_i are ordered so that they are strictly increasing, i.e., $a_1 < a_2 < \cdots < a_m$. For $a_s \leq x \leq a_{s+1}$, show that

$$f(x) = M_s(x) + K_s$$

where
$$M_s = \sum_{i=1}^{s} v_i - \sum_{i=s+1}^{m} u_i$$
$$K_s = -\sum_{i=1}^{s} a_i v_i + \sum_{i=s+1}^{m} a_i u_i$$

(d) Suppose it is known that $f(x)$ is convex. Show that the point a_s will minimize $f(x)$ if the index s is such that

$$\sum_{i=1}^{s-1} (u_i + v_i) \leq \sum_{i=1}^{m} u_i$$
$$\sum_{i=1}^{s} (u_i + v_i) \geq \sum_{i=1}^{m} u_i$$

MULTIFACILITY LOCATION PROBLEMS

5.1 Introduction

In our previous discussion of facility location problems we treated the case of a single new facility to be located relative to a number of existing facilities. The discussion concentrated on rectilinear- and Euclidean-distance location problems, as well as the gravity problem. In this chapter, the analysis of Chapter 4 is extended to include the problem of locating multiple new facilities with respect to multiple existing facilities. Thus the single-facility location problem can be considered to be a special case of the multifacility location problem treated here. As you might expect, applications of multifacility location problems occur in the same contexts as discussed in Chapter 4.

We formulate the multifacility location problem as follows. Let m existing facilities be located at known distinct points $\mathbf{P}_1, \ldots, \mathbf{P}_m$ and let n new facilities be located at points $\mathbf{X}_1, \ldots, \mathbf{X}_n$ in the plane. Let $d(\mathbf{X}_j, \mathbf{P}_i)$ represent the distance between the locations of new facility j and existing facility i and $d(\mathbf{X}_j, \mathbf{X}_k)$ be the distance between the locations of new facilities j and k. Let the annual cost per unit distance between new facility j and existing facility i be denoted w_{ji}, with v_{jk} being the corresponding annual cost per unit distance between new facilities j and k. The total annual trans-

5

portation cost associated with new facilities located at $\mathbf{X}_1, \ldots, \mathbf{X}_n$ is given by

$$f(\mathbf{X}_1, \ldots, \mathbf{X}_n) = \sum_{1 \leq j < k \leq n} v_{jk}\, d(\mathbf{X}_j, \mathbf{X}_k) + \sum_{j=1}^{n} \sum_{i=1}^{m} w_{ji}\, d(\mathbf{X}_j, \mathbf{P}_i) \qquad (5.1)$$

Properly defining v_{jk} as the annual cost due to item movement *between* new facilities j and k, it is only necessary to sum over those values of j which are less than k and over those values of k from 2 to n. The multifacility location problem can be stated as the selection of locations $\mathbf{X}_1^*, \ldots, \mathbf{X}_n^*$ of the new facilities such that total annual cost is minimized.

Because n new facilities are to be located, where n is at least two, as might be expected the multifacility location problem is more difficult to solve than the one-facility location problem, and somewhat less geometric insight is available; the construction of contour lines is no longer possible except for certain special cases when $n = 2$.

Notice that it is the costs proportional to the distance between new facilities which distinguish the multifacility location problem from the one-facility location problem. In fact, when all the terms v_{jk} are zero, then (5.1) may be written

$$f(\mathbf{X}_1, \ldots, \mathbf{X}_n) = \sum_{j=1}^{n} f_j(\mathbf{X}_j) \tag{5.2}$$

where

$$f_j(\mathbf{X}_j) = \sum_{i=1}^{m} w_{ji}\, d(\mathbf{X}_j, \mathbf{P}_i) \tag{5.3}$$

Notice that (5.3) just defines a one-facility total cost expression so that (5.2) is the sum of n different one-facility cost expressions. Since the location of one new facility has no effect upon the cost of locating other new facilities for the special case where all v_{jk} are zero,

$$\min f(\mathbf{X}_1, \ldots, \mathbf{X}_n) = \sum_{j=1}^{n} \min f_j(\mathbf{X}_j)$$

That is, least-cost locations of the new facilities may be found by solving n one-facility location problems independently. Thus it is the terms in (5.1) involving the costs due to distances between new facilities that give the multifacility location problem its special interest.

At this point it is convenient to establish several definitions. New facilities j and k will be said to *have an exchange* when v_{jk} is positive, and to *have no exchange* when v_{jk} is zero. Thus in the case when new facilities *have no exchanges* the multifacility problem reduces to n one-facility problems. It will always be assumed subsequently that each new facility j has an exchange with at least one other new facility.

Not only shall we assume that there is an exchange between new facilities, but also it will be assumed that there is an exchange between new and existing facilities. As motivation for the latter assumption, consider a situation in which there exists a collection of new facilities that have exchanges only among those new facilities within the collection. Where should facilities within the collection be located?

We can make the preceding discussion more explicit by saying that new facility j_1 is *chained* if there is a sequence of distinct new facilities $j_1, \ldots, j_p$ and an existing facility i_j such that new facilities j_t and j_{t+1} have an exchange for $t = 1, \ldots, p - 1$, and new facility j_p and existing facility i_j have an exchange. New facilities that are not chained will be called *unchained.* Any multifacility problem that includes unchained new facilities is poorly formulated, in the sense that for any least-cost solution to the problem, if all unchained new facilities are located at the same point, this point can be *any* point in the plane; the details as to why this is so are developed in one of the problems. By giving all the unchained new facilities the same location, no costs are incurred due to the distances between unchained new facilities, and so unchained new facilities make no contribution to the total cost expression (5.1). Since a multifacility problem which includes unchained new facilities is poorly formulated, it will be assumed subsequently that all new facilities are chained.

5.2 Rectilinear-Distance Location Problems

For the case when distances are rectilinear, when $\mathbf{X}_j = (x_j, y_j)$ and $\mathbf{P}_i = (a_i, b_i)$, then

$$d(\mathbf{X}_j, \mathbf{X}_k) = |x_j - x_k| + |y_j - y_k| \tag{5.4}$$

and

$$d(\mathbf{X}_j, \mathbf{P}_i) = |x_j - a_i| + |y_j - b_i| \tag{5.5}$$

On substituting (5.4) and (5.5) into (5.1) and rearranging terms, we obtain

$$f(\mathbf{X}_1, \ldots, \mathbf{X}_n) = f_1(x_1 \ldots, x_n) + f_2(y_1, \ldots, y_n) \tag{5.6}$$

where

$$f_1(x_1, \ldots, x_n) = \sum_{1 \leq j < k \leq n} v_{jk}|x_j - x_k| + \sum_{j=1}^{n} \sum_{i=1}^{m} w_{ji}|x_j - a_i| \tag{5.7}$$

and

$$f_2(y_1, \ldots, y_n) = \sum_{1 \leq j < k \leq n} v_{jk}|y_j - y_k| + \sum_{j=1}^{n} \sum_{i=1}^{m} w_{ji}|y_j - b_i| \tag{5.8}$$

Subsequent reference to the total cost expressions f_1 and f_2 will be to the expressions defined by (5.7) and (5.8), respectively, unless otherwise indicated. The expressions f_1 and f_2 give the total cost incurred due to "travel" in the x and y directions, respectively.

From (5.6), it follows that

$$\min f(\mathbf{X}_1, \ldots, \mathbf{X}_n) = \min f_1(x_1, \ldots, x_n) + \min f_2(y_1, \ldots, y_n)$$

Thus, just as for the one new facility case, optimum x coordinates of the new facilities can be found independently of optimum y coordinates. Furthermore it is again the case that f_1 and f_2 have the same form, so any procedure developed for minimizing f_1 will also apply to f_2 on replacing x_j by y_j and a_i by b_i.

The procedure used to minimize f_1 depends on the idea of transforming it to an equivalent linear-programming problem; any optimum solution to the linear-programming problem will provide optimum x coordinates of the new facilities. An exactly analogous procedure is used to minimize f_2.

The determination of an equivalent linear-programming problem will now be considered. The following fact is needed: given numbers a, b, p, and q, if $a - b - p + q = 0$, $p \geq 0$, $q \geq 0$, and $pq = 0$, then $|a - b| = p + q$; the justification for this fact is left as an exercise. Thus minimizing f_1 is equivalent to the following problem:

$$\text{minimize} \quad \sum_{1 \leq j < k \leq n} v_{jk}(p_{jk} + q_{jk}) + \sum_{j=1}^{n} \sum_{i=1}^{m} w_{ji}(r_{ji} + s_{ji})$$

subject to

$$\begin{aligned}
x_j - x_k - p_{jk} + q_{jk} \qquad\qquad &= 0, \quad 1 \leq j < k \leq n \\
x_j \qquad\qquad - r_{ji} + s_{ji} &= a_i, \quad i = 1, \ldots, m \text{ and } j = 1, \ldots, n \\
p_{jk}, q_{jk} &\geq 0, \quad 1 \leq j < k \leq n \\
r_{ji}, s_{ji} &\geq 0, \quad i = 1, \ldots, m \text{ and } j = 1, \ldots, n \\
x_j \text{ unrestricted}, &\quad j = 1, \ldots, n \\
p_{jk} q_{jk} &= 0, \quad 1 \leq j < k \leq n \\
r_{ji} s_{ji} &= 0, \quad i = 1, \ldots, m \text{ and } j = 1, \ldots, n
\end{aligned}$$

Notice that with the exception of the last two sets of multiplicative constraints, the problem is a linear-programming problem; call the above problem P0, and call P1 the linear-programming problem obtained by deleting the multiplicative constraints from P0. Since P1 is a less constrained problem than P0, the minimum value of its objective function will be at least as small as the minimum value of the objective function of P0. If a minimum feasible solution to P1 satisfies *all* the constraints of P0 (i.e., the multiplicative constraints as well as the other) it is therefore a minimum feasible solution to P0.

In solving P1 by linear programming the theory of linear programming guarantees that some basic feasible solution will be a minimum feasible solution. For any basic feasible solution, if p_{jk} is in the basic feasible solution, q_{jk} will not be, and vice versa; likewise, if r_{ji} is in the basic feasible solution, s_{ji} will not be, and vice versa. Since variables not in the basic feasible solution are zero, the multiplicative constraints will therefore be satisfied for every basic feasible solution. To see why both p_{jk} and q_{jk} or both r_{ji} and s_{ji} would not be in the same basic feasible solution, suppose that the first two sets of equality constraints of P1 are written in matrix form; then the column of the matrix corresponding to p_{jk} is -1 times the column corresponding to q_{jk}, so that the two columns are linearly dependent. Likewise, the two columns corresponding to r_{ji} and s_{ji} are linearly dependent. But a basis consists of linearly independent vectors. If both p_{jk} and q_{jk} or if both r_{ji} and s_{ji} were in the same basic feasible solution, the corresponding columns making up the basis would be linearly dependent, which cannot be. Thus P0 and P1 are equivalent optimization problems, and we can employ linear programming to solve the rectilinear multifacility location problem.

At this point it is useful to consider an example problem. A small manufacturing shop is planning on adding two new machine tools to handle their expanded product line. The firm has a number of existing machines. However, only three of the existing machines will have a material-handling relationship with the new machines. The existing machines, $\mathbf{P}_1$, $\mathbf{P}_2$, and $\mathbf{P}_3$, have locations (10, 15), (20, 25) and (40, 5), respectively. It is anticipated that there will be 2 pallet loads transported per day between the new machines, located at $\mathbf{X}_1$ and $\mathbf{X}_2$, respectively. There will be 2 pallet loads transported per day between $\mathbf{X}_1$ and $\mathbf{P}_1$, 1 pallet load per day between $\mathbf{X}_1$ and $\mathbf{P}_2$, 4 pallet loads per day between $\mathbf{X}_2$ and $\mathbf{P}_1$, and 5 pallet loads per day between $\mathbf{X}_2$ and $\mathbf{P}_3$. It is desired that the new machines be located in order that total distance traveled per day be minimized. All item movement is along a set of rectilinear aisles.

Mathematically, the problem is

$$\begin{aligned}\text{minimize}\quad f(\mathbf{X}_1, \mathbf{X}_2) = {} & 2d(\mathbf{X}_1, \mathbf{X}_2) + 2d(\mathbf{X}_1, \mathbf{P}_1) + 1d(\mathbf{X}_1, \mathbf{P}_2) + 0d(\mathbf{X}_1, \mathbf{P}_3) \\ & + 4d(\mathbf{X}_2, \mathbf{P}_1) + 0d(\mathbf{X}_2, \mathbf{P}_2) + 5d(\mathbf{X}_2, \mathbf{P}_3)\end{aligned}$$

or

$$\text{minimize}\quad f(\mathbf{X}_1, \mathbf{X}_2) = f_1(x_1, x_2) + f_2(y_1, y_2)$$

where

$$\begin{aligned}f_1(x_1, x_2) = {} & 2|x_1 - x_2| + 2|x_1 - 10| + 1|x_1 - 20| \\ & + 4|x_2 - 10| + 5|x_2 - 40|\end{aligned} \tag{5.9}$$

and

$$\begin{aligned}f_2(y_1, y_2) = {} & 2|y_1 - y_2| + 2|y_1 - 15| + 1|y_1 - 25| \\ & + 4|y_2 - 15| + 5|y_2 - 5|\end{aligned} \tag{5.10}$$

Notice that it is unnecessary to include terms involving w_{13} and w_{22} since they are zero. The linear program equivalent to minimizing (5.9) is

$$\begin{aligned}&\text{minimize} \\ &z_1 = 2(p_{12} + q_{12}) + 2(r_{11} + s_{11}) + 1(r_{12} + s_{12}) + 4(r_{21} + s_{21}) + 5(r_{23} + s_{23}) \\ &\text{subject to} \\ &x_1 - x_2 - p_{12} + q_{12} = 0 \\ &x_1 - r_{11} + s_{11} = 10 \\ &x_1 - r_{12} + s_{12} = 20 \\ &x_2 - r_{21} + s_{21} = 10 \\ &x_2 - r_{23} + s_{23} = 40 \\ &p_{12}, q_{12}, r_{ji}, s_{ji} \geq 0 \\ &x_1, x_2 \text{ unrestricted}\end{aligned} \tag{5.11}$$

This problem can be even further simplified. Since x_1 and x_2 are unrestricted in sign, the second and fourth constraints may be deleted from the problem, solved for x_1 and x_2 respectively, and the values obtained for x_1 and x_2 substituted into the other constraints to give the following equivalent problem, which will be called the *reduced problem:*

minimize

$$z_1 = 2(p_{12}+q_{12}) + 2(r_{11}+s_{11}) + 1(r_{12}+s_{12}) + 4(r_{21}+s_{21}) + 5(r_{23}+s_{23})$$

subject to

$$\begin{aligned}
-p_{12}+q_{12}+r_{11}-s_{11} - r_{21}+s_{21} &= 0\\
r_{11}-s_{11} - r_{12}+s_{12} &= 10\\
r_{21}-s_{21} - r_{23}+s_{23} &= 30\\
p_{12}, q_{12}, r_{ji}, s_{ji} &\geq 0
\end{aligned}$$

The question arises as to why the second and fourth constraints were solved for x_1 and x_2 respectively. Actually, either the second or third constraint could have been solved for x_1, and either the fourth or fifth constraint could have been solved for x_2. The second constraint was selected since it had the smallest value on the right-hand side of the second and third constraints, and the fourth constraint was chosen since it had the smallest value on the right-hand side of the fourth and fifth constraints, respectively. This approach guarantees that the values on the right-hand side of all constraints except the first one involving r_{ji} and s_{ji}, will be nonnegative.

It can be verified that one minimum feasible solution to the latter linear program is given by $s_{12} = 10$, $s_{23} = 30$, with all other variables zero. Optimum x coordinates may now be computed using the second and fourth constraints of the original linear program; $x_1 = 10 + r_{11} - s_{11} = 10$, $x_2 = 10 + r_{21} - s_{21} = 10$. The minimum value of z_1 is 160, so the minimum value of $f_1(x_1, x_2)$ is also 160. The reduced problem has an alternative minimum feasible solution resulting in optimum x coordinates of $x_1 = 20$, $x_2 = 20$; one of the chapter problems consists of establishing this fact. One of the properties of linear-programming problems is that any convex combination of minimum feasible solutions is also a minimum feasible solution; the application of this property to the special case of P0, the original linear program equivalent to (5.9), implies that any point on the line segment joining the points (10, 10) and (20, 20) will also minimize (5.9).

In an exactly analogous manner, when the problem of minimizing f_2 is converted to a linear-programming problem, optimum y coordinates are found to be $y_1 = y_2 = 15$, and the minimum value of f_2 is computed to be 60. Thus the minimum total cost is given by $160 + 60 = 220$, and (x_1, y_1),

(x_2, y_2) give minimum-cost new-facility locations, where $10 \leq x_1 = x_2 \leq 20$, and $y_1 = y_2 = 15$. Whether or not this optimum solution is a realistic one depends, of course, on the problem context. In cases where the solution is not entirely realistic it may still be of value in providing a benchmark solution, in the sense that it is a starting point for finding a more realistic solution. For the example problem, we would want to locate the two new machines as close together as possible.

Several observations made in solving the example problem apply in general when solving P1. When minimizing f_1, constraints corresponding to terms of f_1 for which v_{jk} or w_{ji} is zero need not be included in the equivalent linear program; their inclusion is optional. Furthermore, just as in the example, any constraint of P1 involving a single x_j variable may be deleted from the problem, solved for x_j, and the resultant value substituted into the other constraints to eliminate x_j from all constraints. Alternatively, the original problem could be set up so that the locations of all existing facilities are in the first quadrant; then there will be least-cost locations of all new facilities in the first quadrant, so that all x_j and y_j may be constrained to be nonnegative. Also, the example problem illustrates a fact that is true in general; the rectilinear-distance multifacility location problem always has a minimum-cost solution where the x coordinate of each new facility is equal to the x coordinate of some existing facility, and the y coordinate of each new facility is equal to the y coordinate of some existing facility. Note that this need not imply that the optimum location of each new facility will be identical with the location of some existing facility, although this can, of course, be the case.

It may be of interest to attach additional linear constraints in the x's and y's to the rectilinear-distance multifacility location problem. Such constraints could be used, for example, to restrict the new facilities to certain regions of the plane. One procedure for attaching additional constraints in the x's and y's is as follows: Call P2 the linear program equivalent to minimizing f_2 as defined by (5.8). Call P3 the linear program having an objective function that is the sum of the objective functions of P1 and P2, and having as constraints all the constraints of P1 and P2; P3 is the linear program equivalent to minimizing f as defined by (5.6). The additional linear constraints are simply attached to the constraints of P3, and the desired equivalent linear program is obtained. Naturally, for this case the previous procedure of eliminating the x's and y's from the constraints cannot always be carried out. However, it is possible either to locate the axis so that it is reasonable to require each x_j and each y_j to be nonnegative, or to replace each x_j and each y_j by the difference of two nonnegative variables.

At this point, it is appropriate to consider alternative approaches for solving the multifacility rectilinear location problem. A straightforward linear-programming solution of P1 can be time consuming when a large number of new and existing facilities are involved. Specifically, from (5.11) observe that there can be as many as $n^2 + 2mn$ variables and $n(n - 1)/2 + mn$ constraints involved in determining the x coordinates for the new facilities. Of course, depending on the values of v_{jk} and w_{ji}, the actual number of variables and constraints can be considerably less than this maximum number. However, it is apparent that, for large-sized problems, more efficient solution procedures are desired.

Depending on the characteristics of a particular multifacility rectilinear location problem, the optimum solution can sometimes be obtained in an iterative fashion by solving single-facility location problems. The following are some useful properties of an optimum solution to the multifacility rectilinear location problem:

1. An optimum x coordinate for each new facility coincides with the x coordinate of some existing facility and an optimum y coordinate for each new facility coincides with the y coordinate of some existing facility.
2. The optimum x coordinate (y coordinate) for each new facility is greater than or equal to the minimum and less than or equal to the maximum values of the x coordinates (y coordinates) obtained by letting v_{jk} equal zero and solving n single-facility location problems.
3. When located optimally, each new facility is located at a "median location" with respect to all other facilities, both new and existing (see Chapter 4 for the definition of a "median location").
4. If each new facility is located at a median location with respect to all other facilities and no two new facilities have the same location for either coordinate, then an optimal solution has been obtained.

By way of illustrating the approaches that can be taken to solve some multifacility rectilinear location problems, consider an example problem involving two new facilities and three existing facilities with

$$\mathbf{P}_1 = (0, 0), \quad \mathbf{P}_2 = (2,4), \quad \mathbf{P}_3 = (4, 2)$$

$$v_{12} = 2, \quad w_{11} = 2, \quad w_{12} = 1, \quad w_{13} = 2, \quad w_{21} = 1, \quad w_{22} = 2, \quad w_{23} = 3$$

To begin, suppose that $v_{12} = 0$. In this case there is no exchange between the new facilities. Consequently, two single-facility problems are to be solved. On solving the single-facility location problems it is found that $x_1 = 2, y_1 = 2$, $2 \leq x_2 \leq 4$, $y_2 = 2$. Now, suppose that $v_{12} = 2$; what will happen? By the second property, $2 \leq y_1^* \leq 2$ and $2 \leq y_2^* \leq 2$. Consequently, $y_1^* = y_2^* = 2$. Furthermore, since x_2 can equal 2 without any increase in total cost, $2 \leq x_1^* \leq 2$ and $2 \leq x_2^* \leq 2$. Consequently, $x_1^* = x_2^* = 2$.

Next, consider an example for which $n = 2$ and $m = 4$, with $v_{12} = 6$ and

$$\mathbf{W} = \begin{pmatrix} 5 & 3 & 0 & 0 \\ 0 & 1 & 8 & 4 \end{pmatrix}$$

$$\mathbf{P}_1 = (0, 2), \quad \mathbf{P}_2 = (4, 0), \quad \mathbf{P}_3 = (6, 8), \quad \mathbf{P}_4 = (10, 4)$$

Solving the single-facility location problems by assuming that $v_{12} = 0$, it is found that $x_1 = 0$, $y_1 = 2$, $x_2 = 6$, and $y_2 = 8$. Now, suppose that $v_{12} = 6$; what will happen? In this case it is not immediately clear what will happen. The approach taken is to determine facility locations one at a time. To illustrate the approach, we fix the second new facility at (6, 8) and determine the single-facility location for the first new facility. In this case we find that $x_1 = 4$ and $y_1 = 2$. Now, the first new facility is fixed at (4, 2) and the single-facility location is determined for the second new facility. It is found that $x_2 = 6$ and $y_2 = 4$. Since the location for the second new facility has changed from (6, 8) to (6, 4), the second new facility is fixed at (6, 4) and the single-facility location is determined for the first new facility. In this case, the first new facility does not change locations; that is, $x_1 = 4$ and $y_1 = 2$. Consequently, by the third and fourth properties listed previously, since each new facility is at a median location with respect to all other facilities and the two new facilities do not have the same location for either coordinate, an optimal solution has been obtained: $x_1^* = 4$, $y_1^* = 2$, $x_2^* = 6$, and $y_2^* = 4$.

The preceding procedure is guaranteed to yield an optimum solution only if all new facilities have different values for both coordinates. It is possible for the median conditions to be satisfied at nonoptimum locations. However, such can only occur when two or more new facilities have the same value for one of the coordinates [23]. As an illustration, consider an example problem for which $n = 2$ and $m = 2$, with $v_{12} = 10$ and

$$\mathbf{W} = \begin{pmatrix} 8 & 2 \\ 1 & 4 \end{pmatrix}$$

$$\mathbf{P}_1 = (0, 0), \qquad \mathbf{P}_2 = (4, 4)$$

If we assume that $v_{12} = 0$ and solve two single-facility location problems, it is found that $\mathbf{X}_1 = (0, 0)$ and $\mathbf{X}_2 = (4, 4)$. Next, letting $v_{12} = 10$ and fixing the second new facility at (4, 4), we find that $\mathbf{X}_1 = (4, 4)$. With the first new facility fixed at (4, 4), it is found that $\mathbf{X}_2 = (4, 4)$. Notice that both new facilities satisfy the median conditions at (4, 4) even though the optimum locations are $\mathbf{X}_1^* = (0, 0)$ and $\mathbf{X}_2^* = (0, 0)$.

The "clustering" of new facilities can lead to nonoptimum solutions when one-at-a-time search procedures are used to solve multifacility location problems. However, clustering does not produce significant computational difficulties when the linear-programming approach is used. In the following section we present a dual formulation of the linear-programming problem P1, which can provide a more efficient solution to the multifacility location problem.

5.3 Duality Approach for Solving the Rectilinear Problem*

Since the problem of minimizing f_1 as defined by (5.7) is equivalent to a linear-programming problem, it has a dual, just as any linear-programming problem has a dual. Also as in linear programming, the solution to the dual problem provides a good deal of insight into the primal problem (minimize f_1); furthermore the dual problem provides an approach to minimizing f_1 that takes advantage of the special structure of the linear program P1 which is equivalent to minimizing f_1. Just as previously, since f_1 and f_2 as defined by (5.7) and (5.8) have the same structure, all the results obtained for f_1 will apply to f_2 on replacing a_i by b_i and x_j by y_j.

To motivate the derivation of the dual, consider first the derivation of the dual of the first linear-program example equivalent to minimizing f_1 as defined by (5.9). Associating the dual variables $\bar{z}_{12}$, $\bar{u}_{11}$, $\bar{u}_{12}$, $\bar{u}_{21}$, and $\bar{u}_{23}$ with the first through fifth constraints, respectively, gives the following dual problem:

$$
\begin{array}{llllll}
\text{maximize} & & 10\bar{u}_{11} + 20\bar{u}_{12} + 10\bar{u}_{21} + 40\bar{u}_{23} & & & \\
\text{subject to} & \bar{z}_{12} + \bar{u}_{11} + \bar{u}_{12} & & & & = 0 \\
& -\bar{z}_{12} & + \bar{u}_{21} + \bar{u}_{23} & & & = 0 \\
& -\bar{z}_{12} & & & & \leq 2 \\
& \bar{z}_{12} & & & & \leq 2 \\
& -\bar{u}_{11} & & & & \leq 2 \\
& \bar{u}_{11} & & & & \leq 2 \\
& -\bar{u}_{12} & & & & \leq 1 \\
& \bar{u}_{12} & & & & \leq 1 \\
& -\bar{u}_{21} & & & & \leq 4 \\
& \bar{u}_{21} & & & & \leq 4 \\
& -\bar{u}_{23} & & & & \leq 5 \\
& \bar{u}_{23} & & & & \leq 5
\end{array}
$$

---

*A background in duality theory of linear programming is required for the section. For an efficient means of solving the dual problem as a network problem, see [1].

The first two equality constraints are due to the fact that x_1 and x_2 are unrestricted in sign. Note that, in terms of the original notation, this problem may be rewritten as

$$\begin{aligned}
\text{maximize} \quad & \sum_{i=1}^{3} \sum_{j=1}^{2} a_i \bar{u}_{ji} \\
\text{subject to} \quad & \bar{z}_{12} + \sum_{i=1}^{3} \bar{u}_{1i} = 0 \\
& -\bar{z}_{12} + \sum_{i=1}^{3} \bar{u}_{2i} = 0 \\
& |\bar{z}_{12}| \leq v_{12} \\
& |\bar{u}_{ji}| \leq w_{ji}, \quad i = 1, \ldots, 3 \text{ and } j = 1, 2
\end{aligned}$$

The two problems are the same because the conditions $w_{13} = 0$ and $w_{22} = 0$ and the constraints $|\bar{u}_{ji}| \leq w_{ji}$ force $\bar{u}_{13}$ and $\bar{u}_{22}$ to be zero.

Using the same approach as for the example to obtain the dual to the linear-programming problem P1 equivalent to the general version of f_1 defined by (5.7) gives the following dual problem:

$$\text{maximize} \quad \sum_{i=1}^{m} \sum_{j=1}^{n} a_i \bar{u}_{ji}$$

subject to

$$\begin{aligned}
-\sum_{j=1}^{t-1} \bar{z}_{jt} + \sum_{k=t+1}^{n} \bar{z}_{tk} + \sum_{i=1}^{m} \bar{u}_{ti} = 0, \quad & t = 1, \ldots, n \\
|\bar{z}_{jk}| \leq v_{jk}, \quad & 1 \leq j < k \leq n \\
|\bar{u}_{ji}| \leq w_{ji}, \quad & i = 1, \ldots, m \text{ and } j = 1, \ldots, n
\end{aligned}$$

In this dual problem sums extending from 1 to 0 and $n + 1$ to n are defined to be zero.

Subsequently, it will be useful to refer to this problem, the dual of P1, as D1. The dual program D1 is, of course, a linear problem, although because of its special structure, not a usual looking one. Notice that D1 involves n constraints, plus $n(n - 1)/2 + mn$ upper bound constraints on the mn dual variables. Consequently, a bounded variable linear-programming algorithm can be used to solve D1. Additionally, further solution efficiencies can be obtained by converting D1 to a network flow problem. For a discussion of the latter approach, see [1].

Not only does the dual formulation provide additional solution efficiencies over the primal formulation, but also it provides considerable insight into the properties of an optimum solution to the primal formulation. To facilitate a consideration of these properties, we shall denote a feasible solution to D1 by $(\bar{\mathbf{Z}}, \bar{\mathbf{U}})$ and the objective function of D1 by $g(\bar{\mathbf{U}})$. The following properties result from the duality theory of linear programming.

Property 1: If $x_1, \ldots, x_n$ minimizes f_1, then there is a maximum feasible solution $(\bar{\mathbf{Z}}, \bar{\mathbf{U}})$ to D1, and $f_1(x_1, \ldots, x_n) = g(\bar{\mathbf{U}})$.

Property 2: If $(\bar{\mathbf{Z}}, \bar{\mathbf{U}})$ is a maximum feasible solution to D1, then there are numbers $x_1, \ldots, x_n$ that minimize f_1, and $f_1(x_1, \ldots, x_n) = g(\bar{\mathbf{Z}}, \bar{\mathbf{U}})$.

An important theorem of calculus, the extreme value theorem, or Weierstrass theorem, together with the properties of D1 (the set of all feasible solutions to D1 is closed, bounded, and nonempty, and the objective function is continuous), guarantees that D1 always has at least one maximum feasible solution; thus Property 2 assures that f_1 can always be minimized. Since f_1 and f_2 have the same structure, the dual of the problem of minimizing f_2 is obtained simply by replacing the a_i in D1 by the b_i; thus f_2 can also always be minimized, and so f, the sum of the expressions f_1 and f_2, can always be minimized.

The following property also follows from the duality theory of linear programming, but it is instructive to include a proof of it.

Property 3: For any feasible solution $(\bar{\mathbf{Z}}, \bar{\mathbf{U}})$ to D1 and any x coordinates $x_1, \ldots, x_n$,

$$g(\bar{\mathbf{U}}) \leq f_1(x_1, \ldots, x_n)$$

Proof: For the given hypotheses,

$$\begin{aligned} g(\bar{\mathbf{U}}) &= \sum_{i=1}^{m}\sum_{t=1}^{n} a_i \bar{u}_{ti} - \sum_{t=1}^{n} x_t(0) \\ &= \sum_{t=1}^{n}\sum_{i=1}^{m} a_i \bar{u}_{ti} - \sum_{t=1}^{n} x_t\left(-\sum_{j=1}^{t-1} \bar{z}_{jt} + \sum_{k=t+1}^{n} \bar{z}_{tk} + \sum_{i=1}^{m} \bar{u}_{ti}\right) \\ &= \sum_{t=1}^{n}\sum_{i=1}^{m} \bar{u}_{ti}(a_i - x_t) + \sum_{t=1}^{n}\sum_{j=1}^{t-1} x_t \bar{z}_{jt} - \sum_{t=1}^{n}\sum_{k=t+1}^{n} x_t \bar{z}_{tk} \end{aligned}$$

But
$$\sum_{t=1}^{n}\sum_{j=1}^{t-1} x_t \bar{z}_{jt} = \sum_{1 \leq j < k \leq n} x_k \bar{z}_{jk}$$

and
$$\sum_{t=1}^{n}\sum_{k=t+1}^{n} x_t \bar{z}_{tk} = \sum_{1 \leq j < k \leq n} x_j \bar{z}_{jk}$$

Thus
$$g(\bar{\mathbf{U}}) = \sum_{t=1}^{n}\sum_{i=1}^{m} \bar{u}_{ti}(a_i - x_t) + \sum_{1 \leq j < k \leq n} (x_k - x_j)\bar{z}_{jk} \tag{5.12}$$

Now for $t = 1, \ldots, n$ and $i = 1, \ldots, m$,

$$\bar{u}_{ti}(a_i - x_t) \leq |\bar{u}_{ti}(a_i - x_t)| = |\bar{u}_{ti}||x_t - a_i| \tag{5.13}$$

Also, for $1 \leq j < k \leq n$,

$$(x_k - x_j)\bar{z}_{jk} \leq |(x_k - x_j)\bar{z}_{jk}| = |\bar{z}_{jk}||x_j - x_k| \tag{5.14}$$

Furthermore, for $t = 1, \ldots, n$ and $i = 1, \ldots, m$,

$$|\bar{u}_{ti}||x_t - a_i| \leq w_{ti}|x_t - a_i| \tag{5.15}$$

and for $1 \leq j < k \leq n$,

$$|\bar{z}_{jk}||x_j - x_k| \leq v_{jk}|x_j - x_k| \tag{5.16}$$

From (5.13) and (5.15) it follows that

$$\sum_{t=1}^{n}\sum_{i=1}^{m} \bar{u}_{ti}(a_i - x_t) \leq \sum_{t=1}^{n}\sum_{i=1}^{m} w_{ti}|x_t - a_i| \tag{5.17}$$

while from (5.14) and (5.16) it follows that

$$\sum_{1 \leq j < k \leq n} (x_k - x_j)\bar{z}_{jk} \leq \sum_{1 \leq j < k \leq n} v_{jk}|x_j - x_k| \tag{5.18}$$

It now follows, on using identity (5.12) and adding (5.17) and (5.18), that

$$g(\bar{\mathbf{U}}) \leq f_1(x_1, \ldots, x_n)$$

The preceding properties can be combined to give Property 4. Expressions (5.19) through (5.22) in Property 4 will be referred to as the *complementary slackness conditions.*

Property 4: Let $x_1, \ldots, x_n$ be the x coordinates of the new facilities and let $(\bar{\mathbf{Z}}, \bar{\mathbf{U}})$ be a feasible solution to D1. Then $x_1, \ldots, x_n$ minimizes f_1 and $(\bar{\mathbf{Z}}, \bar{\mathbf{U}})$ is a maximum feasible solution to the dual problem D1 if and only if

$$\bar{u}_{ti}(a_i - x_t) = |\bar{u}_{ti}||x_t - a_i|, \quad i = 1, \ldots m \text{ and } t = 1, \ldots, n \tag{5.19}$$

$$\bar{z}_{jk}(x_k - x_j) = |\bar{z}_{jk}||x_j - x_k|, \quad 1 \leq j < k \leq n \tag{5.20}$$

$$(|\bar{u}_{ti}| - w_{ti})|x_t - a_i| = 0, \quad i = 1, \ldots, m \text{ and } t = 1, \ldots, n \tag{5.21}$$

$$(|\bar{z}_{jk}| - v_{jk})|x_j - x_k| = 0, \quad 1 \leq j < k \leq n \tag{5.22}$$

Proof: As in linear programming, it follows from Property 3 that $x_1, \ldots, x_n$ minimizes f_1 and $(\bar{\mathbf{Z}}, \bar{\mathbf{U}})$ is a maximum feasible solution to D1 if $g(\bar{\mathbf{U}}) = f_1(x_1, \ldots, x_n)$. If the complementary slackness conditions hold, an inspection of the proof of Property 3 establishes that inequalities (5.13), (5.14) (5.15), and (5.16) become equalities, so that the same approach as in the proof of Property 3 establishes that $g(\bar{\mathbf{U}}) = f_1(x_1, \ldots, x_n)$.

To complete the proof, let $x_1, \ldots, x_n$ minimize f_1 and let $(\bar{\mathbf{Z}}, \bar{\mathbf{U}})$ be a maximum feasible solution to D1. Then by either Property 1 or Property 2,

$f_1(x_1, \ldots, x_n) = g(\bar{\mathbf{U}})$. It remains to show that the complementary slackness conditions hold. In the proof of Property 3 it was established that conditions (5.19) through (5.22) still hold when each equality sign is replaced by an equal to or less than sign; if at least one of the inequalities held strictly, the same approach as in the proof of Property 3 would imply that $g(\bar{\mathbf{U}}) < f_1(x_1, \ldots, x_n)$, which would give a contradiction. Thus the complementary slackness conditions hold.

The complementary slackness conditions, especially (5.21) and (5.22), have several interesting implications. From (5.21), if $x_t \neq a_i$, then $|\bar{u}_{ti}| = w_{ti}$, and if $|\bar{u}_{ti}| < w_{ti}$ then $x_t = a_i$; thus, if the a_i are distinct, at most one of the inequalities $|\bar{u}_{t1}| \leq w_{t1}, |\bar{u}_{t2}| \leq w_{t2}, \ldots, |\bar{u}_{tm}| \leq w_{tm}$ can hold. Likewise, from (5.21), if $x_j \neq x_k$ then $|\bar{z}_{jk}| = v_{jk}$, and if $|\bar{z}_{jk}| < v_{jk}$, then $x_j = x_k$. The major computational usefulness of the complementary slackness conditions appears to be that if $(\bar{\mathbf{Z}}, \bar{\mathbf{U}})$ is any maximum feasible solution for which $|u_{ti}| < w_{ti}$, then for any optimum x coordinates it follows that $x_t = a_i$; that is, the value of at least one of the optimum x coordinates of new facility t is known; likewise, if $|\bar{z}_{jk}| < v_{jk}$, then $x_j = x_k$; so if the value of x_j is known, then the value of x_k is also known.

5.4 Squared Euclidean-Distance Location Problems

We now focus our attention on the multifacility extension of the squared Euclidean-distance problem described in Chapter 4. The motivation for studying the squared distance multifacility problem is the same as given in Chapter 4 for the squared distance single-facility problem, with the obvious difference being that more than one new facility is to be located.

Again supposing new facilities to be located at points $(x_1, y_1), \ldots, (x_n, y_n)$, and existing facilities to be located at points $(a_1, b_1), \ldots, (a_m, b_m)$, the squared distance multifacility location problem consists of finding locations of the new facilities that will minimize the following total cost expression:

$$f((x_1, y_1), \ldots, (x_n, y_n)) = \sum_{1 \leq j < k \leq n} v_{jk}[(x_j - x_k)^2 + (y_j - y_k)^2] + \sum_{j=1}^{n} \sum_{i=1}^{m} w_{ji}[(x_j - a_i)^2 + (y_j - b_i)^2] \tag{5.23}$$

The approach to finding new-facility locations that minimize (5.23) is the same for the single-facility problem; partial derivatives of (5.23) with respect to each variable are computed and set equal to zero. The result of the partial derivative computations is two sets of linear equations, one involving the x coordinates of the new facilities and the other involving the y coordinates of the new facilities. It can be shown when each new facility has an exchange with at least one existing facility that each set of equations has a unique solution; the unique solutions give the unique x coordinates and the unique y coordinates of the new facilities that minimize the squared distance cost expression (5.23).

To compute the partial derivatives, it is convenient to define a new quantity $\hat{v}_{jk}$ where

$$\hat{v}_{jk} = \begin{cases} v_{jk}, & k > j \\ v_{kj}, & k \leq j \end{cases}$$

Computing the partial derivative of (5.23) with respect to x_j gives, for $j = 1, \ldots, n$,

$$\frac{\partial f}{\partial x_j} = 2 \sum_{k=1}^{n} \hat{v}_{jk}(x_j - x_k) + 2 \sum_{i=1}^{m} w_{ji}(x_j - a_i) \tag{5.24}$$

Setting (5.24) equal to zero, dividing by 2, and collecting terms gives, for $j = 1, \ldots, n$,

$$x_j\left(\sum_{k=1}^{n} \hat{v}_{jk} + \sum_{i=1}^{m} w_{ji}\right) - \sum_{k=1}^{n} \hat{v}_{jk} x_k = \sum_{i=1}^{m} w_{ji} a_i \tag{5.25}$$

Note that (5.25), when considered for $j = 1, \ldots, n$, is a system of n linear equations in n variables. When the system is solved, it gives least-cost x coordinates of the new facilities. It is also interesting to note that (5.25) can be solved for x_j to obtain

$$x_j = \frac{\sum_{k=1}^{n} \hat{v}_{jk} x_k + \sum_{i=1}^{m} w_{ji} a_i}{\sum_{k=1}^{n} \hat{v}_{jk} + \sum_{i=1}^{m} w_{ji}} \tag{5.26}$$

From (5.26) we see that a necessary and sufficient condition for new facility j to be located optimally is that its location be at the weighted average position or gravity location with respect to all other facilities, both new and existing.

To find the least-cost y coordinates of the new facilities, the partial derivatives of expression (5.23) are computed with respect to each y_j. The resultant system of linear equations is given, for $j = 1, \ldots, n$, by

$$y_j\left(\sum_{k=1}^{n} \hat{v}_{jk} + \sum_{i=1}^{m} w_{ji}\right) - \sum_{k=1}^{n} \hat{v}_{jk} y_k = \sum_{i=1}^{m} w_{ji} b_i \tag{5.27}$$

Note that the coefficients of the variables in (5.25) and (5.27) are identical. Thus, if (5.25) is solved by computing the inverse of the matrix of coefficients of the variables, this same inverse may also be used in solving (5.27). In words, the values of **x** and **y** are to be determined where $\mathbf{Ax} = \mathbf{a}$ and $\mathbf{Ay} = \mathbf{b}$ with **A** an n by n matrix, and **x**, **y**, **a**, and **b** are n by 1 column vectors defined as

$$\mathbf{x} = \begin{pmatrix} x_1 \\ x_2 \\ \cdot \\ \cdot \\ \cdot \\ x_n \end{pmatrix}, \quad \mathbf{y} = \begin{pmatrix} y_1 \\ y_2 \\ \cdot \\ \cdot \\ \cdot \\ y_n \end{pmatrix}, \quad \mathbf{a} = \begin{pmatrix} \sum_{i=1}^m w_{1i} a_i \\ \sum_{i=1}^m w_{2i} a_i \\ \cdot \\ \cdot \\ \cdot \\ \sum_{i=1}^m w_{ni} a_i \end{pmatrix}, \quad \mathbf{b} = \begin{pmatrix} \sum_{i=1}^m w_{1i} b_i \\ \sum_{i=1}^m w_{2i} b_i \\ \cdot \\ \cdot \\ \cdot \\ \sum_{i=1}^m w_{ni} b_i \end{pmatrix}$$

$$\mathbf{A} = \begin{pmatrix} \sum_{k=1}^n \hat{v}_{1k} + \sum_{i=1}^m w_{1i} & -\hat{v}_{12} & \cdots & -\hat{v}_{1n} \\ -\hat{v}_{21} & \sum_{k=1}^n \hat{v}_{2k} + \sum_{i=1}^m w_{2i} & \cdots & -\hat{v}_{2n} \\ \cdot & \cdot & & \cdot \\ \cdot & \cdot & & \cdot \\ \cdot & \cdot & & \cdot \\ -\hat{v}_{n1} & -\hat{v}_{n2} & \cdots & \sum_{k=1}^n \hat{v}_{nk} + \sum_{i=1}^m w_{ni} \end{pmatrix}$$

To illustrate the solution procedure, consider finding locations of new facilities 1 and 2 to minimize (5.23) when $m = 3$, $n = 2$, $v_{12} = 2$,

$$\mathbf{W} = (w_{ji}) = \begin{pmatrix} 2 & 1 & 0 \\ 4 & 0 & 5 \end{pmatrix}$$

$(a_1, b_1) = (10, 15)$, $(a_2, b_2) = (20, 25)$, and $(a_3, b_3) = (40, 5)$. Note that the data for this example are the same as for the example of the rectilinear location problem examined previously. Given these data, notice that **A**, **a**, and **b** become

$$\mathbf{A} = \begin{pmatrix} 2+2+1+0 & -2 \\ -2 & 2+4+0+5 \end{pmatrix}, \qquad \mathbf{a} = \begin{pmatrix} 2(10) + 1(20) + 0(40) \\ 4(10) + 0(20) + 5(40) \end{pmatrix}$$

$$\mathbf{b} = \begin{pmatrix} 2(15) + 1(25) + 0(5) \\ 4(15) + 0(25) + 5(5) \end{pmatrix}$$

or, in matrix notation,

$$\begin{pmatrix} 5 & -2 \\ -2 & 11 \end{pmatrix} \begin{pmatrix} x_1 \\ x_2 \end{pmatrix} = \begin{pmatrix} 40 \\ 240 \end{pmatrix}$$

and

$$\begin{pmatrix} 5 & -2 \\ -2 & 11 \end{pmatrix} \begin{pmatrix} y_1 \\ y_2 \end{pmatrix} = \begin{pmatrix} 55 \\ 85 \end{pmatrix}$$

In equation form, from (5.25) the following system of equations is obtained

$$5x_1 - 2x_2 = 40$$
$$-2x_1 + 11x_2 = 240$$

and, from (5.27),

$$5y_1 - 2y_2 = 55$$
$$-2y_1 + 11y_2 = 85$$

Solving for the least-cost coordinates gives

$$(x_1, y_1) \doteq (18.039, 15.196)$$
$$(x_2, y_2) \doteq (25.098, 10.490)$$

It is interesting to note that while the data for this problem are the same as for the rectilinear-distance example problem, the least-cost new-facility coordinates certainly are not. Due to the differences in the structure of the rectilinear and squared distance models, substantially different answers may be obtained. Therefore, it is quite important to be sure beforehand which of the models is most appropriate to apply to any particular location problem being studied.

5.5 Euclidean-Distance Location Problems

In the previous treatment of the rectilinear-distance problem and the squared Euclidean-distance problem we found that multifacility problems were not substantially more difficult to solve than the corresponding single-facility versions. Such is not the case for the Euclidean-distance problem. The major difficulty is caused by the partial derivatives not always being defined. While a similar difficulty occurred with the single-facility Euclidean distance problem, the difficulties are much more severe for the multifacility problem, as there can be many more variables involved.

The multifacility Euclidean problem is formulated as follows:

$$\text{minimize} \quad f(\mathbf{X}_1, \ldots, \mathbf{X}_n) = \sum_{1 \le j < k \le n} v_{jk}[(x_j - x_k)^2 + (y_j - y_k)^2]^{1/2} + \sum_{j=1}^{n} \sum_{i=1}^{m} w_{ji}[(x_j - a_i)^2 + (y_j - b_i)^2]^{1/2} \tag{5.28}$$

The notation employed is the same as introduced for the multifacility rectilinear-distance problem. The necessary conditions for optimal locations of new facilities are that the partial derivatives of $f(\mathbf{X}_1, \ldots, \mathbf{X}_n)$ with respect to $\mathbf{X}_1, \ldots, \mathbf{X}_n$ be zero (or change sign) at the optimum locations.

Taking the partial derivatives of $f(\mathbf{X}_1, \ldots, \mathbf{X}_n)$ with respect to x_j and y_j, respectively, gives

$$\frac{\partial f}{\partial x_j} = \sum_{\substack{k=1 \\ \neq j}}^{n} \frac{\hat{v}_{jk}(x_j - x_k)}{D_{jk}} + \sum_{i=1}^{m} \frac{w_{ji}(x_j - a_i)}{E_{ji}}, \quad j = 1, \ldots, n \tag{5.29}$$

and

$$\frac{\partial f}{\partial y_j} = \sum_{\substack{k=1 \\ \neq j}}^{n} \frac{\hat{v}_{jk}(y_j - y_k)}{D_{jk}} + \sum_{i=1}^{m} \frac{w_{ji}(y_j - b_i)}{E_{ji}}, \quad j = 1, \ldots, n \tag{5.30}$$

where

$$D_{jk} = [(x_j - x_k)^2 + (y_j - y_k)^2]^{1/2} \tag{5.31}$$

and

$$E_{ji} = [(x_j - a_i)^2 + (y_j - b_i)^2]^{1/2} \tag{5.32}$$

Notice that if either new facilities j and k have the same location ($D_{jk} = 0$) or new facility j and existing facility i have the same location ($E_{ji} = 0$) both $\partial f/\partial x_j$ and $\partial f/\partial y_j$ are undefined.

Geometrically, each cost term included in (5.28) represents the equation of a right circular cone. To see why this is true suppose that existing facility i is located at the origin and w_{ji} is positive-valued. By letting new facility j move along the x axis a distance r_{ji}, the weighted distance function, f_{ji}, between new facility j and existing facility i can be described by the linear relationship

$$f_{ji} = w_{ji} r_{ji} \tag{5-33}$$

Since new facility j is not restricted to move in any specified direction, r_{ji} can be interpreted as describing the locus of points equidistant from existing facility i. As shown in Figure 5.1, the locus of points a distance r_{ji} from new facility i is a circle of radius r_{ji}. Consequently, f_{ji} can be given as

$$f_{ji} = w_{ji}(x_j^2 + y_j^2)^{1/2} \tag{5.34}$$

Thus (5.34) defines a right circular cone generated by revolving the straight line given by (5.33) about the f axis. If existing facility i has the location (a_i, b_i), then (5.34) becomes

$$f_{ji} = w_{ji}[(x_j - a_i)^2 + (y_j - b_i)^2]^{1/2} \tag{5.35}$$

Similarly, if new facility j is located at (x_j, y_j), new facility k is located at (x_k, y_k), and v_{jk} is positive-valued, then, relative to new facility j, the weighted distance function between new facilities j and k generates a right circular cone centered at the point (x_j, y_j). Consequently, (5.28) represents the sum of cones.

The sharp points of the cones involved in the summation result in the undefined derivatives and produce the *knife-edged surface* referred to by Vergin and Rogers [26]. Since a cone is a limiting form of a hyperboloid,

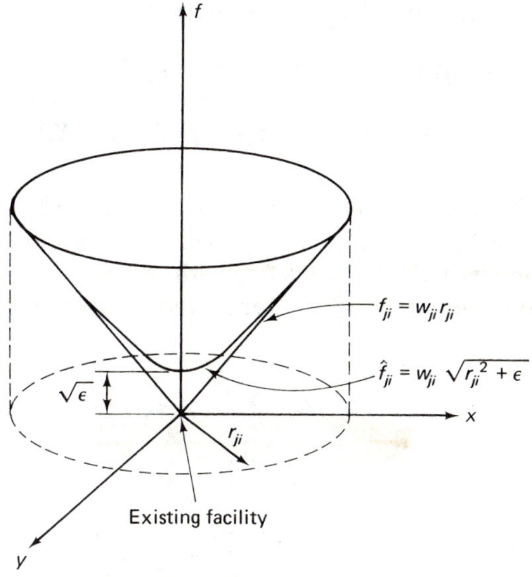

Figure 5.1. Geometrical interpretation of Euclidean location problem.

if the cones are replaced by hyperboloids, a smooth approximating function, $\hat{f}$, is obtained. Furthermore, since hyperboloids are strict convex functions and $\hat{f}$ is the sum of hyperboloids, $\hat{f}$ is a strict convex function.

As noted in Figure 5.1, the equation for a hyperbola lying in the first quadrant of the $\hat{f}x$ plane is given by

$$\hat{f}_{ji} = w_{ji}(r_{ji}^2 + \epsilon)^{1/2} \tag{5.36}$$

where ϵ is a positive-valued constant. The hyperboloid centered on the point (a_i, b_i) in the xy plane can be expressed as

$$\hat{f}_{ji} = w_{ji}[(x_j - a_i)^2 + (y_j - b_i)^2 + \epsilon]^{1/2} \tag{5.37}$$

From Figure 5.1 it is seen that the addition of the constant ϵ essentially results in the replacement of the point of a cone by a smooth hyperbolic surface. Consequently, by introducing ϵ the partial derivatives exist everywhere. Furthermore, the smaller the value of ϵ the closer the hyperboloid approximates the cone.

Letting

$$\hat{D}_{jk} = [(x_j - x_k)^2 + (y_j - y_k)^2 + \epsilon]^{1/2} \tag{5.38}$$

and

$$\hat{E}_{ji} = [(x_j - a_i)^2 + (y_j - b_i)^2 + \epsilon]^{1/2} \tag{5.39}$$

the new optimization problem can be given as

$$\text{minimize} \quad \hat{f}(X_1, \ldots, X_n) = \sum_{1 \le j < k \le n} v_{jk} \hat{D}_{jk} + \sum_{j=1}^{n} \sum_{i=1}^{m} w_{ji} \hat{E}_{ji} \tag{5.40}$$

where

$$\lim_{\epsilon \to 0} \hat{f}(X_1, \ldots, X_n) = f(X_1, \ldots, X_n)$$

Therefore, solving (5.40) using a very small value of ϵ yields a solution which is approximately the same as that obtained by solving (5.28).

Taking the partial derivatives of $\hat{f}$ with respect to x_j and y_j, setting the partials equal to zero, and solving for x_j and y_j gives the following iterative expressions:

$$x_j^{(h+1)} = \frac{\displaystyle\sum_{\substack{k=1 \\ \neq j}}^{n} \frac{\hat{v}_{jk} x_k^{(h)}}{\hat{D}_{jk}^{(h)}} + \sum_{i=1}^{m} \frac{w_{ji} a_i}{\hat{E}_{ji}^{(h)}}}{\displaystyle\sum_{\substack{k=1 \\ \neq j}}^{n} \frac{\hat{v}_{jk}}{\hat{D}_{jk}^{(h)}} + \sum_{i=1}^{m} \frac{w_{ji}}{\hat{E}_{ji}^{(h)}}} \tag{5.41}$$

$$y_j^{(h+1)} = \frac{\displaystyle\sum_{\substack{k=1 \\ \neq j}}^{n} \frac{\hat{v}_{jk} y_k^{(h)}}{\hat{D}_{jk}^{(h)}} + \sum_{i=1}^{m} \frac{w_{ji} b_i}{\hat{E}_{ji}^{(h)}}}{\displaystyle\sum_{\substack{k=1 \\ \neq j}}^{n} \frac{\hat{v}_{jk}}{\hat{D}_{jk}^{(h)}} + \sum_{i=1}^{m} \frac{w_{ji}}{\hat{E}_{ji}^{(h)}}} \tag{5.42}$$

where the superscripts denote the iteration number.

The iterative expressions (5.41) and (5.42) are subsequently referred to as the hyperboloid approximation procedure, or HAP [10]. In the case of a single new facility, (5.41) and (5.42) reduce to expressions similar to those given in Chapter 4. It can be established that if a convergent solution is obtained it will be an optimum solution. Even though it has not yet been shown that (5.41) and (5.42) will always yields a convergent solution, considerable computational experience with HAP has failed to produce a case in which the procedure has failed to converge.

In using HAP to solve multifacility location problems, it has been observed that the larger the value of ϵ the faster the convergence to the optimum value of the approximating function. However, the accuracy of the approximation decreases with increasing values of ϵ. Consequently, in solving location problems using HAP a large value of ϵ is used initially; the solution obtained is used as the starting solution, using a smaller value of ϵ; and the process is continued by successively reducing the value of ϵ until no significant decrease in the value of either (x_j, y_j) or $\hat{f}$ results.

Interestingly, HAP can be used to solve the rectilinear location problem by defining two approximating functions as follows:

$$\hat{f}_1(x) = \sum_{1 \le j < k \le n} v_{jk}[(x_j - x_k)^2 + \epsilon]^{1/2} + \sum_{j=1}^{n} \sum_{i=1}^{m} w_{ji}[(x_j - a_i)^2 + \epsilon]^{1/2} \tag{5.43}$$

and

$$\hat{f}_2(y) = \sum_{1\le j<k\le n} v_{jk}[(y_j - y_k)^2 + \epsilon]^{1/2} + \sum_{j=1}^{n}\sum_{i=1}^{m} w_{ji}[(y_j - b_i)^2 + \epsilon]^{1/2} \quad (5.44)$$

The HAP iterations for the rectilinear problem are given by (5.41) and (5.42), where $\hat{D}_{jk}$ and $\hat{E}_{ji}$ are replaced in (5.41) by

$$\hat{D}_{jk}^{(h)}(x) = [(x_j^{(h)} - x_k^{(h)})^2 + \epsilon]^{1/2} \quad (5.45)$$

and

$$\hat{E}_{ji}^{(h)}(x) = [(x_j^{(h)} - a_i)^2 + \epsilon]^{1/2} \quad (5.46)$$

In (5.42), $\hat{D}_{jk}^{(h)}$ and $\hat{E}_{ji}^{(h)}$ are replaced by

$$\hat{D}_{jk}^{(h)}(y) = [(y_j^{(h)} - y_k^{(h)})^2 + \epsilon]^{1/2} \quad (5.47)$$

and

$$\hat{E}_{ji}^{(h)}(y) = [(y_j^{(h)} - b_i)^2 + \epsilon]^{1/2} \quad (5.48)$$

The HAP approach can also be used to handle situations involving a mixture of rectilinear and Euclidean distances. The development of the iterative expressions for this case is left as an exercise. For further discussion of HAP, see [10].

Comparing the treatment of multifacility location problems with the discussion of single-facility location problems given in Chapter 4, it remains to consider multifacility extensions of the triangle inequality result and the graphical procedure for solving Euclidean problems. Applying the triangle inequality to the multifacility problem, the following upper and lower bounds are obtained for the optimum Euclidean solution [23].

$$E(\mathbf{X}^*, \mathbf{Y}^*) \ge E(\mathbf{X}^0, \mathbf{Y}^0) \ge [R^2(\mathbf{X}^*) + R^2(\mathbf{Y}^*)]^{1/2} \quad (5.49)$$

where $(\mathbf{X}^0, \mathbf{Y}^0)$ = optimum Euclidean solution
$(\mathbf{X}^*, \mathbf{Y}^*)$ = optimum rectilinear solution
$E(\mathbf{X}, \mathbf{Y})$ = value of the objective function for the Euclidean problem given the solution $(\mathbf{X}, \mathbf{Y})$, see (5.28)
$R(\mathbf{X}) = f_1(\mathbf{X})$ for rectilinear problem, see (5.7)
$R(\mathbf{Y}) = f_2(\mathbf{Y})$ for rectilinear problem, see (5.8)

For the example problem considered in the previous section,

$$(\mathbf{X}^*, \mathbf{Y}^*) = \begin{pmatrix} 10 & 15 \\ 10 & 15 \end{pmatrix}$$

$$R(\mathbf{X}^*) = 160$$

$$R(\mathbf{Y}^*) = 60$$

$$E(\mathbf{X}^*, \mathbf{Y}^*) = 172.28$$

Therefore,

$$172.28 \ge E(\mathbf{X}^0, \mathbf{Y}^0) \ge 170.88$$

demonstrating that in this case the rectilinear solution is a very good approximation to the optimum solution for the Euclidean problem.

In certain special cases the multifacility Euclidean problem can be solved graphically. As an example, suppose that there are five existing facilities located as shown in Figure 5.2. Furthermore, two new facilities are to be located, with $v_{12} = 1$ and

$$\mathbf{W} = \begin{pmatrix} 1 & 1 & 1 & 0 & 0 \\ 0 & 0 & 1 & 1 & 1 \end{pmatrix}$$

If we assume $\mathbf{X}_1^*$, the optimum location for the first new facility, lies outside the triangle $\mathbf{P}_3\mathbf{P}_4\mathbf{P}_5$, then $\mathbf{X}_2^*$, the optimum location for the second new facility, will be at the intersection of the lines $\mathbf{X}_1^*\mathbf{P}_4$ and $\mathbf{P}_3\mathbf{P}_5$. By a similar argument, $\mathbf{X}_1^*$ is the intersection of the lines $\mathbf{P}_1\mathbf{P}_3$ and $\mathbf{X}_2^*\mathbf{P}_2$. Thus $\mathbf{X}_1^*$ will be along the line $\mathbf{P}_1\mathbf{P}_3$ and $\mathbf{X}_2^*$ will be along the line $\mathbf{P}_3\mathbf{P}_5$. But since $\mathbf{X}_1^*$ is also along the line $\mathbf{X}_2^*\mathbf{P}_2$ and $\mathbf{X}_2^*$ is along the line $\mathbf{X}_1^*\mathbf{P}_4$, then $\mathbf{X}_1^*$ is the intersection of lines $\mathbf{P}_1\mathbf{P}_3$ and $\mathbf{P}_2\mathbf{P}_4$ and $\mathbf{X}_2^*$ is the intersection of the lines $\mathbf{P}_3\mathbf{P}_5$ and $\mathbf{P}_2\mathbf{P}_4$, as shown in Figure 5.2. To establish that an optimum solution is obtained, notice that the locations satisfy the conditions for the graphical solution procedure described in Chapter 4 and illustrated in Figure 4.8.

The graphical procedure is limited by its assumptions: (1) there is equal item movement between facilities (new and existing), (2) a total of four facilities (new and existing) have an interchange with each new facility, and (3) existing facilities are located such that the graphical procedure works. Thus the procedure is presented primarily because of its geometrical interest, rather than its potential for application.

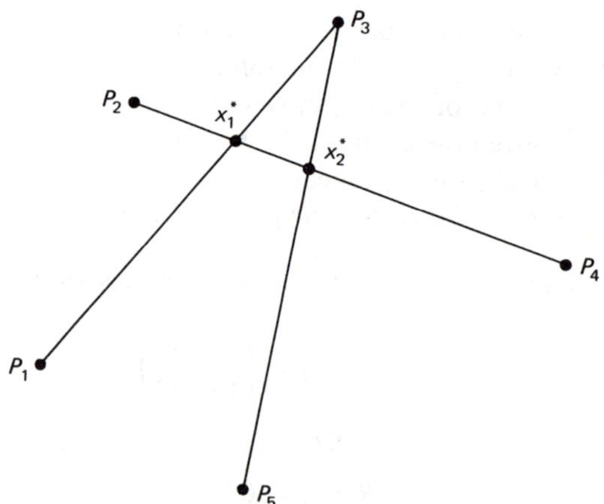

Figure 5.2. Graphical solution procedure.

5.6 Location-Allocation Problems

The multifacility location problems treated previously considered the situation in which the values of the weights v_{jk} and w_{ji} were not dependent on the locations of the new facilities. In this section we shall treat a class of location problems, called location-allocation problems, in which all v_{jk} values are zero and the w_{ji} values are decision variables, as are the locations for the new facilities. The treatment of the location-allocation problem in this section concentrates on the continuous solution-space problems examined by Cooper [3–6]. In Chapter 9 we examine discrete solution-space formulations of the location-allocation problem.

A very general statement of the location-allocation problem would involve a determination of the number of new facilities, as well as their locations, and the allocation of item movement between the new and existing facilities. A common example of a location-allocation problem involves the location of distribution centers that will receive products from production facilities and distribute products to retail or wholesale outlets. Another example of a location-allocation problem arises when a number of branch banks are to be located in a metropolitan area. Instead of branch banks or warehouses, the new facilities could easily be, say, hospitals or grocery stores. In the case of branch banks, hospitals, and grocery stores, the existing facilities would include the residences of the consumers.

A mathematical formulation of the location-allocation problem can be given as follows for the case of Euclidean distances.

$$\begin{aligned} \text{minimize} \quad & \Psi = \sum_{j=1}^{n} \sum_{i=1}^{m} z_{ji} w_{ji} [(x_j - a_i)^2 + (y_j - b_i)^2]^{1/2} + g(n) \\ \text{subject to} \quad & \sum_{j=1}^{n} z_{ji} = 1, \quad i = 1, \ldots, m \\ & n = 1, 2, \ldots, m \end{aligned} \tag{5.50}$$

where Ψ = total cost per unit time

w_{ji} = cost per unit time per unit distance if new facility j interacts with existing facility i

$z_{ji} = \begin{cases} 1, & \text{if new facility } j \text{ interacts with existing facility } i \\ 0, & \text{otherwise} \end{cases}$

(x_j, y_j) = coordinate location of new facility j

(a_i, b_i) = coordinate location of existing facility i

$g(n)$ = cost per unit time of providing n new facilities

The decision variables in (5.50) are n, z_{ji}, x_j, and y_j. Each constraint ensures that each existing facility interacts with only one new facility. Since no capacity constraints are given, it is assumed that a new facility is capable of handling all interchanges with existing facilities. Additionally, it is normally assumed that $w_{1i} = w_{2i} = \cdots = w_{ni}$, since identical new facilities are to be located. Unfortunately, an optimum solution to (5.50) is not easily obtained. The difficulty is presented by the decision variables n and z_{ji}. One approach that has been used to solve the problem is to fix the value of n and consider all possible combinations of z_{ji}. For each combination of z_{ji} an "optimum" solution is obtained using the single-facility Euclidean-distance solution procedure given in Chapter 4. Next the minimum-cost combination of z_{ji} is determined for the particular value of n. Finally, an optimum solution is obtained by performing a search over n.

Our procedure is not feasible for moderate- to large-sized problems, since the number of combinations of z_{ji} that must be considered for each value of n is given by the Stirling number of the second kind,

$$S(n, m) = \sum_{k=0}^{n} \frac{(-1)^k (n-k)^m}{k!\,(n-k)!} \tag{5.51}$$

If there are two new facilities and four existing facilities, there are seven combinations to be considered. The combinations are given in Table 5.1.

Table 5.1. COMBINATIONS OF z_{ji} FOR TWO NEW FACILITIES AND FOUR EXISTING FACILITIES

Combination	*New Facility*	*Existing Facility*
1	1	1, 2, 3
	2	4
2	1	1, 2, 4
	2	3
3	1	1, 3, 4
	2	2
4	1	2, 3, 4
	2	1
5	1	1, 2
	2	3, 4
6	1	1, 3
	2	2, 4
7	1	1, 4
	2	2, 3

Recall that the computation of the number of combinations is based on the assumption that the new facilities are identical: otherwise, even more combinations would have to be considered.

As an example of the application of the solution procedure just described, Cooper [3] considered a problem involving two facilities and seven existing facilities located as follows: (15, 15), (5, 10), (10, 27), (16, 8), (25, 14), (31, 23), and (22, 29). All w_{ji} values are equal to 1.0. Sixty-three allocation combinations are considered in determining the optimum allocation of the two new facilities and their locations. For one combination, a new facility is located at the point (15.420, 12.053) and allocated to existing facilities 1, 2, 4, and 5. The remaining new facility is located at the point (22.000, 29.000) and allocated to existing facilities 3, 6, and 7. To complete the location-allocation problem, the optimum number of new facilities must be determined by searching over n.

Since considerable computational effort can be required to solve the location-allocation problem exactly, a number of heuristic methods have been employed to obtain "good" solutions to the problem. For a discussion of several heuristic methods for solving location-allocation problems, see Cooper [4, 5]. Additionally, Kuenne and Soland [19] present a branch and bound algorithm for determining the optimum allocations and locations of n new facilities.

Location-allocation problems arise in a number of contexts and take on a wide variety of formulations. The distance metric employed can be either rectilinear, Euclidean, or some other appropriate measure.

In addition to the formulation of the location-allocation problem given in (5.50), a number of other formulations have been considered. As an example, when the w_{ji} values represent the shipment of products, there might be capacity constraints on the amount of product flowing to or from a particular facility. In such a case the problem is a capacitated location-allocation problem. Another variation of the location-allocation problem arises when one must also determine the amount of product to be shipped between new and existing facilities. Such a variation is referred to as a transportation location problem [7]. A third variation of the location-allocation problem (which has received considerable attention) involves a discrete solution space. Such problems are termed plant location and warehouse location problems, and are treated in Chapter 10.

5.7 Summary

In this chapter we presented multifacility extensions to the single-facility location problems described in the preceding chapter. We concentrated on those multifacility location problems in which the cost of item movement was either a linear function of rectilinear or Euclidean distance or a quadratic function of Euclidean distance. As in Chapter 4, our discussion concentrated on the case of a continuous solution space and both static and deterministic values of all parameters in the model. With the exception of our discussion of location-allocation problems, the level of interaction between new and existing facilities was a parameter of the model. In the case of location-allocation problems, the level of interaction was a decision variable, as were the number and locations of the new facilities.

Again, we emphasize that the solution obtained from the model is a *benchmark* or *ideal* solution. It is not unlikely that the solution obtained from the model will indicate either new facilities or new and existing facilities should have the same location. At this point practical limitations and modifying considerations should enter into the analysis. Furthermore, the "optimum location" obtained from the model is an optimum location on the basis of the structure of the model. If either the objective function does not include all the important criteria or the constraints do not account for all limitations on the solution, a subsequent evaluation of alternative solutions is required.

Our treatment in this chapter of multifacility location problems has not been exhaustive. A number of important variations of multifacility location problems remain to be considered. In Chapters 6, 8, and 10 multifacility location problems involving a discrete solution space are considered. Minimax versions of the problems treated in this chapter are considered in Chapter 9. Also, in Chapter 7 an interesting variation of a multifacility location problem is presented for the case of a continuous solution space.

REFERENCES

1. Cabot, A. V., R. L. Francis, and M. A. Stary, "A Network Flow Solution to a Rectilinear Distance Facility Location Problem," *AIIE Transactions*, Vol. 2, No. 2, 1970, pp. 132–141.

2. Cockayne, E. J., and Z. A. Melzak, "Steiner's Problem for Set-Terminals," *Quarterly of Applied Mathematics*, Vol. 36, No. 2, 1969, pp. 213–217.

3. Cooper, L., "Location-Allocation Problems," *Operations Research*, Vol. 11, No. 3, 1963, pp. 331–344.

4. Cooper, L., "Heuristic Methods for Location-Allocation Problems," *SIAM Review*, Vol. 6, No. 1, 1964, pp. 37–52.

5. Cooper, L., "Solution of Generalized Locational Equilibrium Models," *Journal of Regional Science*, Vol. 7, No. 1, 1967, pp. 1–18.

6. COOPER, L., "An Extension of the Generalized Weber Problem," *Journal of Regional Science*, Vol. 8, No. 2, 1969, pp. 181–197.
7. COOPER, L., "The Transportation-Location Problem," *Operations Research*, Vol. 20, No. 1, 1972, pp. 94–108.
8. DONATH, W. E., "Statistical Properties of the Placement of a Graph," *Journal of the Society of Industrial and Applied Mathematics*, Vol. 16, No. 2, 1968, pp. 439–457.
9. EYSTER, J. W., and J. A. WHITE, "Some Properties of the Squared Euclidean Distance Location Problem," *AIIE Transactions*, Vol. 5, No. 3, 1973, pp. 275–280.
10. EYSTER, J. W., J. A. WHITE, and W. W. WIERWILLE, "On Solving Multifacility Location Problems Using a Hyperboloid Approximation Procedure," *AIIE Transactions*, Vol. 5, No. 1, 1973, pp. 1–6.
11. FRANCIS, R. L., "A Note on the Optimum Location of New Machines in Existing Plant Layouts," *The Journal of Industrial Engineering*, Vol. 14, No. 1, 1963, pp. 57–59.
12. FRANCIS, R. L., "On the Location of Multiple New Facilities with Respect to Existing Facilities," *The Journal of Industrial Engineering*, Vol. 15, No. 2, 1964, pp. 106–107.
13. FRANCIS, R. L., and A. V. CABOT, "Properties of a Multifacility Location Problem Involving Euclidean Distances," *Naval Research Logistics Quarterly*, Vol. 19, No. 2, 1972, pp. 335–353.
14. GILBERT, E. N., and H. O. POLLAK, "Steiner Minimal Trees," *Journal of the Society of Industrial and Applied Mathematics*, Vol. 16, No. 1, 1968, pp. 1–29.
15. GOLDMAN, A. J., "Optimal Locations for Centers in a Network," *Transportation Science*, Vol. 3, No. 4, 1969, pp. 352–360.
16. GOLDMAN, A. J., and C. J. WITZGALL, "A Localization Theorem for Optimal Facility Placement," *Transportation Science*, Vol. 4, No. 4, 1970, pp. 406–408.
17. GOLDSTONE, L. A., "A Further Note on Warehouse Location," *Management Science*, Vol. 15, No. 4, 1968, pp. 132–133.
18. HANAN, M., "On Steiner's Problem with Rectilinear Distance," *Journal of the Society of Industrial and Applied Mathematics*, Vol. 14, No. 2, 1966, pp. 255–265.
19. KUENNE, R. E., and R. M. SOLAND, *The Multisource Weber Problem: Exact Solutions by Branch and Bound*, IDA Economic Papers (H. Williams, ed.), Program Analysis Division, Arlington, Va., Apr. 1971.
20. LEVY, J., "An Extended Theorem for Location on a Network," *Operational Research Quarterly*, Vol. 18, No. 4, 1967, pp. 433–442.
21. LOVE, R. F., "Locating Facilities in 3-Dimensional Space by Convex Programming," *Naval Research Logistics Quarterly*, Vol. 16, No. 4, 1969, pp. 503–516.
22. MELZAK, Z. A., "On the Problem of Steiner," *Canadian Mathematical Bulletin*, Vol. 4, No. 2, 1961, pp. 143–148.

23. PRITSKER, A. A. B., and P. M. GHARE, "Locating New Facilities with Respect to Existing Facilities," *AIIE Transactions*, Vol. 2, No. 4, 1970, pp. 290–297.

24. RAO, M. R., "On the Direct Search Approach to the Rectilinear Facilities Location Problem," *AIIE Transactions,* Vol. 5, No. 3, 1973, pp. 256–264.

25. STARY, M. A., "Some Aspects of a Two-Machine Location Problem," unpublished masters thesis, The Ohio State University, Columbus, Ohio, 1968.

26. VERGIN, R. C., and J. D. ROGERS, "An Algorithm and Computational Procedure for Locating Economic Facilities," *Management Science*, Vol. 13, No. 6, 1967, pp. 3240–3254.

27. WENDELL, R. E., and A. P. HURTER, JR., "Location Theory, Dominance, and Convexity," *Operations Research*, Vol. 21, No. 1, 1973, pp. 314–320.

28. WESOLOWSKY, G. O., and R. F. LOVE, "A Nonlinear Approximation Method for Solving a Generalized Rectangular Distance Weber Problem," *Management Science*, Vol. 18, No. 11, 1972, pp. 656–663.

29. WESOLOWSKY, G. O., and R. F. LOVE, "The Optimal Location of New Facilities Using Rectangular Distances," *Operations Research*, Vol. 19, No. 1, 1971, pp. 124–130.

30. WHITE, J. A., "A Note on the Quadratic Facility Location Problem," *AIIE Transactions*, Vol. 3, No. 2, 1971, pp. 156–157.

PROBLEMS

5.1. Given the assumptions of Section 5.1, but not the assumption that all new facilities are chained, explain why if there is one chained new facility there are at least two, and if there is at least one unchained new facility there are at least two. Also explain why chained and unchained new facilities have no exchanges.

5.2. Suppose a multifacility problem is given that has both chained and unchained facilities. What is the smallest value that n can take on?

5.3. Suppose a multifacility location problem is given that includes both chained and unchained new facilities; let the chained new facilities be numbered $1, 2, \ldots, p$, and let the unchained new facilities be numbered $p + 1, \ldots, n$. Explain why (5.1) may be written as

$$\sum_{1 \le j < k \le p} v_{jk}\, d(\mathbf{X}_j, \mathbf{X}_k) + \sum_{j=1}^{p} \sum_{i=1}^{m} w_{ji}\, d(\mathbf{X}_j, \mathbf{P}_i) + \sum_{p+1 \le j < k \le n} v_{jk}\, d(\mathbf{X}_j, \mathbf{X}_k)$$

Then make use of this expression to give a rigorous explanation of why a multifacility problem that includes unchained new facilities is poorly formulated.

5.4. A small machine shop has five existing machines located at coordinate locations $\mathbf{P}_1 = (8, 20)$, $\mathbf{P}_2 = (10, 10)$, $\mathbf{P}_3 = (16, 30)$, $\mathbf{P}_4 = (30, 10)$, and $\mathbf{P}_5 = (40, 20)$. Two new machines are to be located in the shop. Item movement is rectilinear. It is anticipated that there will be four trips per day between the new machines. The number of trips per day between each new machine and each existing machine is

$$\mathbf{W} = \begin{pmatrix} 8 & 6 & 5 & 4 & 3 \\ 2 & 3 & 4 & 6 & 7 \end{pmatrix}$$

Determine the optimum locations for the new machines.

5.5. In Problem 5.4, determine the optimum locations when cost is proportional to the square of the Euclidean distance between new machines and between new and existing machines.

5.6. Three existing facilities are located at $\mathbf{P}_1 = (3, 4)$, $\mathbf{P}_2 = (8, 7)$, and $\mathbf{P}_3 = (15, 2)$. Two new facilities are to be located with respect to the existing machines. The cost data for the location problem are

$$v_{12} = 3$$

$$\mathbf{W} = \begin{pmatrix} 2 & 6 & 0 \\ 4 & 5 & 1 \end{pmatrix}$$

Determine the optimum locations for the new facilities, assuming cost is proportional to (a) rectilinear distance, (b) squared Euclidean distance, and (c) Euclidean distance.

5.7. Two new facilities are to be located with respect to three existing facilities located at $\mathbf{P}_1 = (8, 15)$, $\mathbf{P}_2 = (10, 20)$, and $\mathbf{P}_3 = (30, 10)$. The cost data for the location problem are

$$v_{12} = 8$$

$$\mathbf{W} = \begin{pmatrix} 6 & 3 & 5 \\ 0 & 7 & 2 \end{pmatrix}$$

Determine the optimum locations for the new facilities, assuming cost is proportional to (a) rectilinear distance, (b) squared Euclidean distance, and (c) Euclidean distance.

5.8. The Columbia Broadcasting Company (CBC) has decided to transmit its radio and television programs from its central studios by laser beams. The beams are unidirectional, and transmission power varies as the square of the distance transmitted. CBC has four central studios located at Los Angeles, Chicago, Houston, and New York. However, owing to the curvature of the earth (a laser beam travels only in straight lines), it is not possible to construct broadcasting towers high enough for Los Angeles to be "visible" from New York. Thus CBC has decided to install two transmitting stations. The first will be in a position to communicate with Los Angeles, Chicago, Houston, and the second station. The second station communicates with all others except Los Angeles. The transmission loads from each station are unequal due to their location characteristics. All space programs originate at Houston; financial and cultural news comes from Los Angeles and New York. Crime news originates from Chicago. The coordinate locations of the existing studios are $\mathbf{P}_1 = (1, 2)$, $\mathbf{P}_2 = (2, 0)$, $\mathbf{P}_3 = (3, 3)$, and $\mathbf{P}_4 = (4, 2)$. Other data are

$$v_{12} = 8$$

$$\mathbf{W} = \begin{pmatrix} 4 & 2 & 2 & 0 \\ 0 & 2 & 2 & 4 \end{pmatrix}$$

Determine the optimum locations for the transmitting stations.

5.9. Find an alternative minimum feasible solution to the reduced linear-programming problem equivalent to minimizing f_1 as defined by (5.9), which results in optimum x coordinates of $x_1^* = 20$, $x_2^* = 20$.

5.10. Set up and solve the reduced form of the linear-programming problem equivalent to minimizing the example of f_2 defined by (5.10).

5.11. Construct and solve an example multifacility rectilinear-distance problem using linear programming directly.

5.12. Construct an example multifacility rectilinear-distance location problem having additional linear constraints on the x's and y's which you believe is a realistic one.

5.13. Given numbers a, b, p, and q such that

$$a - b = p - q \tag{i}$$

$$p \geq 0, \qquad q \geq 0 \tag{ii}$$

$$pq = 0 \tag{iii}$$

show that $|a - b| = p + q$.
[*Hint:* Consider three cases; $a - b > 0$, $a - b = 0$, and $a - b < 0$. As a start, here is the proof for the first case. From (ii), $p \geq 0$ and $q \geq 0$; either $p > 0$ or $q > 0$, for otherwise (i) implies that $a - b = 0$, which cannot be. If $q > 0$, (iii) implies that $p = 0$, so (i) implies that $a - b = -q < 0$, which cannot be; thus $q = 0$. Since $a - b > 0$, $|a - b| = a - b$, and $a - b = p$ by (i); since $q = 0$, $|a - b| = a - b = p + q$.]

5.14. As well as attaching additional linear constraints to the x's and y's of the linear-programming problem equivalent to the multifacility rectilinear-distance problem, suppose that linear constraints involving other variables are attached as well. In addition to not making any physical sense in most cases, doing so will usually destroy the equivalence of the resultant linear program to the actual location problem you wish to solve. Explain why.

5.15. Consider the problem of plotting contour lines of the function $f_1(x_1, x_2) = v_{12}|x_1 - x_2| + \sum_{j=1}^{p} C_j|x_1 - c_j| + \sum_{i=1}^{q} D_i|x_2 - d_i|$, where x_1 and x_2 are identified with the x and y axes, respectively; $c_1 < \cdots < c_p$, $d_1 < \cdots < d_q$, and $v_{12} > 0$. This problem is identical with that of plotting contour lines of the one-facility rectilinear-distance problem with the exception of the inclusion of the term $v_{12}|x_1 - x_2|$. Show that for any contour line passing through that part of a region (s, t) lying on or above the line $x_2 = x_1$ the slope is given by $-(M_t - v_{12})/(N_s + v_{12})$, and that for any contour line passing through that part of region (s, t) lying on or below the line $x_2 = x_1$ the slope is given by $-(M_t + v_{12})/(N_s - v_{12})$.

5.16. Using the procedure of Problem 5.15, construct contour lines for the example function f_1 defined by (5.9), and for the example function f_2 defined by (5.10).

5.17. Actions are to be taken at points $x_1, \ldots, x_n$ in time, which must be determined. These actions will be affected by events occurring at known points in time $a_1, \ldots, a_m$, and by whether or not one action precedes another. More exactly, a cost per time unit w''_{ji} is incurred if $x_j \leq a_i$, a cost per time unit w'_{ji} is incurred if $x_j > a_i$, a cost per time unit v''_{jk} is incurred if $x_j \leq x_k$, and a cost per time unit v'_{jk} is incurred if $x_j > x_k$. Construct a linear-programming problem that, if solved, would determine the points in time at which the n actions should be taken so as to minimize the total cost incurred.

5.18. Let $x_1, \ldots, x_n$ be the x coordinates of the new facilities. Prove that $x_1, \ldots, x_n$ minimizes f_1 if and only if there exists a feasible solution $(\bar{\mathbf{Z}}, \bar{\mathbf{U}})$ to D1 such that the complementary slackness conditions hold.

5.19. Three new facilities are to be located relative to six existing facilities. The cost of item movement is proportional to the square of the straight-line distance between facilities. The weighting factors are

$$\begin{array}{llllll} v_{12} = 6, & v_{13} = 1, & v_{23} = 4, & w_{11} = 4, & w_{12} = 0, & w_{13} = 2 \\ w_{14} = 0, & w_{15} = 6, & w_{16} = 5, & w_{21} = 3, & w_{22} = 6, & w_{23} = 0 \\ w_{24} = 0, & w_{25} = 8, & w_{26} = 1, & w_{31} = 0, & w_{32} = 0, & w_{33} = 8 \\ w_{34} = 10, & w_{35} = 2, & w_{36} = 6, & & & \end{array}$$

The coordinate locations of the existing facilities are $\mathbf{P}_1 = (0, 12)$, $\mathbf{P}_2 = (2, 1)$, $\mathbf{P}_3 = (10, 2)$, $\mathbf{P}_4 = (6, 12)$, $\mathbf{P}_5 = (20, 10)$, and $\mathbf{P}_6 = (5, 20)$. Determine the optimum locations for the new facilities.

5.20. The following data are given for the function $f_1(x_1, x_2, x_3)$ defined by Equation (5.7):

$$v_{12} = 8, \qquad v_{13} = 5, \qquad v_{23} = 1$$

$$\mathbf{P}_1 = (2, 0), \qquad \mathbf{P}_2 = (4, 0), \qquad \mathbf{P}_3 = (6, 0)$$

$$\mathbf{W} = (w_{ji}) = \begin{pmatrix} 3 & 4 & 2 \\ 1 & 2 & 3 \\ 2 & 1 & 1 \end{pmatrix}$$

(a) Construct the dual of the problem equivalent to minimizing $f_1(x_1, x_2, x_3)$.

(b) Solve the problem of minimizing $f_1(x_1, x_2, x_3)$, and give a justification for the solutions.

5.21. Let $(\bar{\mathbf{Z}}, \bar{\mathbf{U}})$ be a feasible solution to D1. Prove that $(\bar{\mathbf{Z}}, \bar{\mathbf{U}})$ is a maximum feasible solution to D1 if and only if there exist x coordinates $x_1, \ldots, x_n$ of the new facilities such that the complementary slackness conditions hold.

5.22. The following data are given for a multifacility rectilinear-distance location problem:

$$v_{12} = 4$$

$$\mathbf{W} = \begin{pmatrix} 4 & 0 & 5 \\ 2 & 1 & 0 \end{pmatrix} = (w_{ji})$$

$$\mathbf{P}_1 = (10, 15), \qquad \mathbf{P}_2 = (20, 25), \qquad \mathbf{P}_3 = (40, 5)$$

(a) Find optimum x and y coordinates of new facilities 1 and 2 by using linear programming directly.

(b) Find optimum x and y coordinates of new facilities 1 and 2 by plotting contour lines of f_1 and f_2.

5.23. The following data are given for a multifacility rectilinear-distance location problem: $m = 3$, $n = 5$, $\mathbf{P}_1 = (1, 0)$, $\mathbf{P}_2 = (2, 0)$, $\mathbf{P}_3 = (3, 0)$, $v_{12} = 2$, $v_{13} = 2$, $v_{14} = 2$, $v_{15} = 2$, $v_{23} = 20$, $v_{24} = 1$, $v_{25} = 0$, $v_{34} = 0$, $v_{35} = 0$, $v_{45} = 40$, and

$$\mathbf{W}' = (w_{ij}) = \begin{pmatrix} 10 & 4 & 4 & 4 & 4 \\ 0 & 1 & 1 & 5 & 5 \\ 0 & 4 & 4 & 4 & 4 \end{pmatrix}$$

(a) Show that $(x_1, y_1) = (1, 0)$ and $(x_j, y_j) = (2, 0)$ for $j = 2, \ldots, 5$ is a median location.

(b) Show that the solution in part (a) is not an optimum solution to the rectilinear problem.

5.24. For the data in Problem 5.23, assume Euclidean distance and determine the optimum locations for the five new facilities.

5.25. Given the data of Problem 5.22, construct and solve the analogous squared distance problem.

5.26. A minimax version of the multifacility Euclidean-distance location problem may be formulated as follows. Locations of existing facilities at points $(a_1, b_1), \ldots, (a_m, b_m)$ are known; locations of new facilities at points $(x_1, y_1), \ldots, (x_n, y_n)$ are to be determined. Nonnegative constants v_{jk} for $1 \leq j < k \leq n$, and w_{ji}, for $i = 1, \ldots, m$ and $j = 1, \ldots, n$, are given. The function $g((x_1, y_1), \ldots, (x_n, y_n))$ is defined by

$$g((x_1, y_1), \ldots, (x_n, y_n)) = \max_{1 \leq j < k \leq n} v_{jk}[(x_j - x_k)^2 + (y_j - y_k)^2]^{1/2}$$

The function $h((x_1, y_1), \ldots, (x_n, y_n))$ is defined by

$$h((x_1, y_1), \ldots, (x_n, y_n)) = \max_{\substack{1 \leq i \leq m \\ 1 \leq j \leq n}} w_{ji}[(x_j - a_i)^2 + (y_j - b_i)^2]^{1/2}$$

and the function $f((x_1, y_1), \ldots, (x_n, y_n))$ is defined by

$$\begin{aligned} &f((x_1, y_1), \ldots, (x_n, y_n)) \\ &\qquad = \max\,[g((x_1, y_1), \ldots, (x_n, y_n)), h((x_1, y_1), \ldots, (x_n, y_n))] \end{aligned}$$

The minimax location problem is to find locations of the new facilities that minimize the function $f(x_1, y_1), \ldots, (x_n, y_n))$. Identify some possible applications for which you believe the minimax location model would be more appropriate than the multifacility location models considered in Chapter 5.

5.27. The general HAP formulation can be given as

$$\hat{f}(\mathbf{X}_1, \ldots, \mathbf{X}_n) = \sum_{1 \leq j < k \leq n} v_{jk}\, \hat{d}(\mathbf{X}_j, \mathbf{X}_k) + \sum_{j=1}^{n} \sum_{i=1}^{m} w_{ji}\, \hat{d}(\mathbf{X}_j, \mathbf{P}_i)$$

where

$$\hat{d}(\mathbf{X}_j, \mathbf{X}_k) = \begin{cases} [(x_j - x_k)^2 + (y_j - y_k)^2 + \epsilon]^{1/2}, & \text{if Euclidean} \\ [(x_j - x_k)^2 + \epsilon]^{1/2} + [(y_j - y_k)^2 + \epsilon]^{1/2}, & \text{if rectilinear} \end{cases}$$

$$\hat{d}(\mathbf{X}_j, \mathbf{P}_i) = \begin{cases} [(x_j - a_i)^2 + (y_j - b_i)^2 + \epsilon]^{1/2}, & \text{if Euclidean} \\ [(x_j - a_i)^2 + \epsilon]^{1/2} + [(y_j - b_i)^2 + \epsilon]^{1/2}, & \text{if rectilinear} \end{cases}$$

Obtain the iterative expressions that would hold regardless of the distance measure used.

5.28. Consider a multifacility rectilinear location problem for which $n = 2$ and $m = 4$, with

$$v_{12} = 6$$

$$\mathbf{W} = \begin{pmatrix} 5 & 3 & 0 & 0 \\ 0 & 1 & 8 & 4 \end{pmatrix}$$

$$\mathbf{P}_1 = (0, 2), \quad \mathbf{P}_2 = (4, 0), \quad \mathbf{P}_3 = (6, 8), \quad \mathbf{P}_4 = (10, 4)$$

Is the optimum solution to this problem given by $(x_1, y_1) = (4, 2)$ and $(x_2, y_2) = (6, 4)$? Justify your answer.

5.29. Three existing facilities are located at $\mathbf{P}_1 = (3, 4)$, $\mathbf{P}_2 = (8, 7)$, and $\mathbf{P}_3 = (15, 2)$. Two new facilities are to be located with respect to the existing machines. The cost data for the location problem are

$$v_{12} = 3$$

$$\mathbf{W} = \begin{pmatrix} 2 & 6 & 0 \\ 4 & 5 & 1 \end{pmatrix}$$

Determine the optimum locations for the new facilities, assuming cost is proportional to (a) rectilinear distance, and (b) squared Euclidean distance.

5.30. Three new facilities are to be located relative to five existing facilities. The data for the problem are

$$v_{12} = 0, \qquad v_{13} = 2, \qquad v_{23} = 1$$

$$\mathbf{W} = \begin{pmatrix} 6 & 1 & 2 & 0 & 0 \\ 0 & 0 & 1 & 3 & 4 \\ 0 & 5 & 2 & 0 & 2 \end{pmatrix}$$

$$\mathbf{P}_1 = (0, 0), \quad \mathbf{P}_2 = (2, 8), \quad \mathbf{P}_3 = (5, 4), \quad \mathbf{P}_4 = (7, 6), \quad \mathbf{P}_5 = (8, 2)$$

(a) Determine the optimum locations for the new facilities, assuming (1) rectilinear travel, and (2) straight-line travel.
(b) Determine the optimum locations for the new facilities, assuming cost is proportional to the square of the straight-line distance.
(c) Using (5.49), compute the upper and lower bounds on the minimum-cost solution to the Euclidean location problem.

5.31. Three existing facilities are located at the points $\mathbf{P}_1 = (0, 0)$, $\mathbf{P}_2 = (20, 15)$, and $\mathbf{P}_3 = (40, 15)$. Each new facility costs \$1,200 per month. The cost of item movement is \$5 per unit of distance. The number of trips per month between some new facility and each existing facility equals ten. Determine the optimum number, location, and allocation of new facilities, assuming at least one new facility is required and item movement occurs between an existing facility and only one new facility. Base your calculations on (a) rectilinear distance and (b) Euclidean distance.

5.32. Four plants are located at the points (0, 0), (10, 10), (0, 50), and (20, 40). The shipment of goods from the plants to a distribution center per month, measured in truckloads, is 100, 60, 120, and 50 loads, respectively. The cost of transporting a truckload of product from the plant to a distribution center is \$20 per unit of distance. The monthly cost of owning and operating a distribution center is \$40,000. Determine the optimum number, location, and allocation of distribution centers, assuming at least one distribution center is required and a plant ships to only one distribution center (use a Euclidean distance approximation).

5.33. Five special-purpose machines are located in a plant at the points (0, 0), (0, 10), (20, 0), (40, 10), and (10, 10). The machines require maintenance at expected frequencies of 10, 6, 12, 10, and 5 times per month, respectively. Due to the nature of the maintenance, all maintenance must be performed at the maintenance center. The cost of transporting the machines to the maintenance center is \$5 per unit of distance. The monthly cost of owning and operating a maintenance center is \$5,000. (a) Determine the optimum number, location, and allocation of maintenance centers. Assume a machine is serviced by only one maintenance center and travel is rectilinear. (b) What values of travel cost result in one maintenance center being the optimum number?

DISCRETE LOCATION AND LAYOUT PROBLEMS

6.1 Introduction

The location problems treated in the two previous chapters have been ones where the facilities could be located anywhere in the plane, and where the facilities were idealized as points. In Chapter 4 we pointed out that these assumptions might not be valid and suggested the use of contour lines as an aid in resolving the problem. However, as noted in Chapter 5, the contour-line approach is not feasible when we are faced with the problem of locating multiple facilities. Furthermore, in summarizing the discussion in Chapter 5, we cited the need to restrict the solution by not allowing multiple facilities at the same location. In this chapter we consider the problem of locating multiple new facilities among a discrete number of possible locations, or sites. In our subsequent discussion the new facilities may be allowed to have a positive area. We shall assume no interchanges between new facilities. Thus the problems considered are, roughly speaking, discrete analogs of the problems considered in Chapter 4.

As an illustration of the type of problem considered initially in this chapter, suppose that three new facilities, *a*, *b*, and *c*, are to be located in the Atlanta metropolitan area. There exist six facilities in the Atlanta area (labeled *p* through *u*) that have a material-handling relationship with at least

6

one of the new facilities. The existing facilities are located at (0, 0), (0, 1), (0, 3), (1, 1), (2, 2), and (4, 0), respectively. A preliminary analysis indicates that there are five possible locations (labeled v through z) with coordinate locations (1, 0), (1, 2), (2, 0), (4, 1), and (4, 3), respectively, for the new facilities. However, zoning restrictions prohibit the location of new facility a at location v and space limitations prevent new facility b being located at location w. There are no interchanges among the three new facilities. The matrix $\mathbf{W} = (w_{ik})$, with w_{ik} being the trips per day made between new facility i and existing facility k, is

w_{ik}	p	q	r	s	t	u
a	4	0	1	2	0	2
b	1	2	3	0	2	1
c	0	1	4	0	2	3

All travel is assumed to occur along a rectilinear set of streets. How should the new facilities be assigned among the locations in order to minimize distance traveled per day? We shall see subsequently that facilities a, b, and

c can be assigned to locations x, v, and w, respectively, and distance will be minimized.

6.2 Assignment Model

The simplest discrete location problem is perhaps one in which only one new facility is to be located, and a single choice must be made from among a finite number of sites, say n. Given that the total annual cost is known for locating the new facility at each possible site, the obvious solution to the problem is to locate the new facility at the site having the smallest cost among the n sites considered.

Depending upon the problem context, however, it may not be as simple to solve the problem as to formulate it. In particular, it may not be an easy matter to determine the cost of locating the new facility at a given site. For the example problems considered in this chapter, the costs can be readily determined. It is important to be aware of the fact that this will not always be the case, especially when a problem such as choosing a least-cost site for a new plant is considered.

When two or more new facilities are to be located, and n possible sites are available, the location problem becomes substantially more difficult. Indeed, if there are m new facilities, with each requiring a site from among n sites, and m is no greater than n, there are $\frac{n!}{(n-m)!}$ possible different assignments of new facilities to sites. It is for solving problems such as these, where a number of new facilities are to be located, that the assignment model is useful.

Suppose for the moment that n new facilities are to be located, one at each site, and that there are exactly n sites available. Let x_{ij} be a variable that is one if new facility i is located at site j, and zero otherwise. Thus

$$x_{ij} = 0 \quad \text{or} \quad 1, \quad \text{for } i = 1, \ldots, n \text{ and } j = 1, \ldots, n \tag{6.1}$$

Since some new facility must be located at each site, it follows that

$$\sum_{i=1}^{n} x_{ij} = 1, \quad \text{for } j = 1, \ldots, n \tag{6.2}$$

In other words, condition (6.2) states that exactly one new facility will be located at each site j. Likewise, each new facility i must be located at exactly one site, which leads to the condition

$$\sum_{j=1}^{n} x_{ij} = 1, \quad \text{for } i = 1, \ldots, n \tag{6.3}$$

It remains to compute the total cost; suppose that c_{ij} is the cost if new facility

i is located at site j. Then, given variables x_{ij} satisfying (6.1), (6.2), and (6.3), that is, given an assignment of new facilities to sites, the term $\sum_{j=1}^{n} c_{ij}x_{ij}$ represents the cost due to locating new facility i, so that the total cost due to locating all new facilities is given by

$$\sum_{i=1}^{n}\sum_{j=1}^{n} c_{ij}x_{ij} \tag{6.4}$$

The assignment problem may now be stated as

$$\begin{aligned} &\text{minimize} \quad f(\mathbf{x}) = \sum_{i=1}^{n}\sum_{j=1}^{n} c_{ij}x_{ij} \\ &\text{subject to} \quad \sum_{i=1}^{n} x_{ij} = 1, \quad j = 1, \ldots, n \\ &\qquad\qquad \sum_{j=1}^{n} x_{ij} = 1, \quad i = 1, \ldots, n \\ &\qquad\qquad x_{ij} = 0 \quad \text{or} \quad 1 \quad \text{for all } i \text{ and } j \end{aligned} \tag{6.5}$$

As an example of the use of the assignment model in a context similar to those considered in Chapter 4, suppose that there are p existing facilities, that d_{kj} represents an appropriately determined distance between existing facility k and site j, and that w_{ik} represents a (nonnegative) total cost per unit distance for some given time period incurred in transporting items between new facility i and existing facility k. Then, for a given assignment, $w_{ik}\sum_{j=1}^{n} d_{kj}\, x_{ij}$ represents the total cost for a given time period incurred in transporting items between new facility i and existing facility k. The total cost of transporting all items is then given by

$$\sum_{i=1}^{n}\sum_{k=1}^{p} w_{ik}\sum_{j=1}^{n} d_{kj}x_{ij} = \sum_{i=1}^{n}\sum_{j=1}^{n} c_{ij}x_{ij} \tag{6.6}$$

The right side of the total cost expression (6.6) is obtained by rearranging the order of summation on the left side and defining the term c_{ij} as follows:

$$c_{ij} = \sum_{k=1}^{p} w_{ik}d_{kj}, \quad i = 1, \ldots, n \text{ and } j = 1, \ldots, n \tag{6.7}$$

With the definition of c_{ij} given by (6.7) the problem of locating new facilities with respect to existing facilities may now be solved as an assignment problem. It is interesting to note that the expression for c_{ij} is directly analogous to the basic location model of Chapter 4, inasmuch as c_{ij} is a sum of weighted distances.

Expression (6.7) is simply a sum of nonnegative numbers times distances, and represents the cost when new facility i is located at site j. Furthermore, if the n by n, n by p, and p by n matrices **C**, **W**, and **D** are defined, respectively, by $\mathbf{C} = (c_{ij})$, $\mathbf{W} = (w_{ik})$, and $\mathbf{D} = (d_{kj})$, then expression (6.7) may be written in matrix notation as

$$\mathbf{C} = \mathbf{WD} \tag{6.8}$$

In cases where there are a large number of sites and existing facilities, it may be advantageous to use expression (6.8) and a computer program for multiplying matrices to compute the terms c_{ij}.

In addition to costs, as given by (6.7), incurred when new facility i is assigned to site j, other costs, such as site preparation costs, or site purchase costs, may be incurred as well. If c''_{ij} denotes the sum of these other costs, and c'_{ij} represents the costs defined by (6.7), then the cost terms c_{ij} to use in solving the assignment problem would be given by $c_{ij} = c'_{ij} + c''_{ij}$. Naturally, c'_{ij} and c''_{ij} must have the same dimensions.

There is one other point to consider in formulating location problems to be solved by the assignment model. In many situations there are fewer new facilities to be located than there are sites available. Such situations are handled as follows: suppose that q new facilities are to be located, where $q \leq n$. Then number the actual new facilities to be located 1 through q and let c_{ij} be defined as usual for $i = 1, \ldots, q$ and $j = 1, \ldots, n$. For $i = q + 1, \ldots, n, j = 1, \ldots, n$, take c_{ij} to be zero. Conditions (6.2), (6.3), and (6.4) remain unchanged. This approach may be thought of, in linear-programming terms, as including slack variables with associated costs of zero. In a physical context, we are creating "dummy" new facilities to take up the unused sites without incurring any cost.

Now consider the example problem described earlier. The matrix **D** is given by

$$\mathbf{D} = (d_{kj}) = \begin{array}{c} \\ p \\ q \\ r \\ s \\ t \\ u \end{array} \begin{array}{c} \begin{matrix} v & w & x & y & z \end{matrix} \\ \begin{pmatrix} 1 & 3 & 2 & 5 & 7 \\ 2 & 2 & 3 & 4 & 6 \\ 4 & 2 & 5 & 6 & 4 \\ 1 & 1 & 2 & 3 & 5 \\ 3 & 1 & 2 & 3 & 3 \\ 3 & 5 & 2 & 1 & 3 \end{pmatrix} \end{array}$$

Recalling the **W** matrix defined earlier, the cost matrix is obtained as follows:

$$\mathbf{C} = \mathbf{WD} = \begin{array}{c} \\ a \\ b \\ c \end{array} \begin{array}{c} \begin{matrix} v & w & x & y & z \end{matrix} \\ \begin{pmatrix} 16 & 26 & 21 & 34 & 48 \\ 26 & 20 & 29 & 38 & 40 \\ 33 & 27 & 33 & 37 & 37 \end{pmatrix} \end{array}$$

Remember facilities a and b are not allowed at locations v and w, respectively. To avoid the possibility of these allocations, we let c_{11} and c_{22} take on very large positive values. Thus, on defining two dummy facilities, d and e, the cost matrix becomes

$$\mathbf{C} = \begin{pmatrix} \infty & 26 & 21 & 34 & 48 \\ 26 & \infty & 29 & 38 & 40 \\ 33 & 27 & 33 & 37 & 37 \\ 0 & 0 & 0 & 0 & 0 \\ 0 & 0 & 0 & 0 & 0 \end{pmatrix}$$

This particular assignment problem can be solved by inspection to yield an assignment of facilities a, b, and c to sites x, v, and w, respectively. Of course, not all assignment problems are as easily solved as the example problem. Methods for solving assignment problems are presented in most introductory operations research texts.

6.3 Generalized Assignment Model

In contrast to the situation where a new facility can take up only one site, the situation will now be considered where a new facility can take up a number of sites. To provide a context for the discussion, warehouse-layout problems will be considered. Other similar problems should suggest themselves, however.

Suppose that a region $\mathbf{L}$, which might be that part of the floor of a warehouse where items can be stored, is subdivided into n grid squares of equal size, numbered in any convenient manner from 1 to n. Suppose that m items are to be stored in the warehouse, and let the total number of grid squares which item i will take up be denoted by A_i; for example, if item 3 takes up 4 grid squares, then $A_3 = 4$. Let the warehouse have p docks, the locations of which are known, and denote by d_{kj} an appropriately determined distance between dock k and the center of grid square j. Denote by $\mathbf{S}_i$ the numbers of grid squares that are taken up by item i; for example, if item i takes up 4 grid squares, which have the numbers 12, 13, 29, and 30, then $\mathbf{S}_i = \{12, 13, 29, 30\}$. It is intuitively useful to think of $\mathbf{S}_i$ as specifying the location, or storage region, of item i, and of A_i as specifying the area taken up by item i. Given the assumption, for a given storage region for an item, that an item is equally likely to travel between dock k and any grid square taken up by item i, the average distance item i travels between dock k and its storage region is given by

$$\sum_{j \in \mathbf{S}_i} \frac{1}{A_i} d_{kj} \tag{6.9}$$

That is, the average distance is the sum of the distances between dock k and all the grid squares taken up by item i divided by A_i. Finally, let w_{ik} represent a known total cost per unit of average distance incurred in transporting item i between dock k and its storage region for a given time period. Typically, if items are stored on pallets, for example, w_{ik} would be directly proportional to the number of pallet loads of item i moving between dock k and the storage region of item i for some given time period. Thus

$$w_{ik} \sum_{j \in \mathbf{S}_i} \frac{1}{A_i} d_{kj} \tag{6.10}$$

is the total average cost of transporting item i between dock k and the storage region of item i. Dimensionally, w_{ik} equals (cost per average distance) (trips of item i to dock k per time period). Furthermore, expression (6.9) has dimensions of (average distance per trip between dock k and region j). Thus it follows that the dimensions of (6.10) are (cost per time period), and the total average cost per time period due to transporting items to and from storage is given by

$$F(\mathbf{S}_1, \ldots, \mathbf{S}_m) = \sum_{i=1}^{m} \sum_{k=1}^{p} w_{ik} \left(\sum_{j \in \mathbf{S}_i} \frac{1}{A_i} d_{kj} \right) \tag{6.11}$$

The warehouse-layout problem consists of finding storage regions for each item (given that no more than one item can take up any grid square) which will minimize the total cost expression (6.11). Since sets $\mathbf{S}_1, \ldots, \mathbf{S}_m$ determine the storage regions of the items, it is useful to think of the collection of all sets, denoted by $\{\mathbf{S}_1, \ldots, \mathbf{S}_m\}$, as representing a layout. The problem is then to find a layout to minimize the total cost.

To convert the layout problem into a generalized assignment problem, define the variables x_{ij}, for $i = 1, \ldots, m$ and $j = 1, \ldots, n$ as

$$x_{ij} = \begin{cases} 1, & \text{if item } i \text{ takes up grid square } j \\ 0, & \text{if item } i \text{ does not take up grid square } j \end{cases}$$

It will be assumed for the time being that the total number of grid squares to be taken up by items is the same as the total number of grid squares; that is, $\sum_{i=1}^{m} A_i = n$. Since item i takes up a total of A_i grid squares, it follows that

$$\sum_{j=1}^{n} x_{ij} = A_i \quad \text{for } i = 1, \ldots, m \tag{6.12}$$

Since each grid square is also taken up by one item, it also follows that

$$\sum_{i=1}^{m} x_{ij} = 1, \quad \text{for } j = 1, \ldots, n \tag{6.13}$$

Given the definition of the variables x_{ij}, $\mathbf{S}_i$ is the collection of all grid squares j for which $x_{ij} = 1$; that is, $\mathbf{S}_i = \{j : x_{ij} = 1\}$, so expression (6.10) is equivalent to the expression

$$w_{ik} \sum_{j=1}^{n} \frac{1}{A_i} d_{kj} x_{ij}$$

and the total cost expression (6.11) is equivalent to the expression

$$\sum_{i=1}^{m} \sum_{k=1}^{p} w_{ik} \left(\sum_{j=1}^{n} \frac{1}{A_i} d_{kj} x_{ij} \right) \tag{6.14}$$

If the term c_{ij} is defined by

$$c_{ij} = \frac{1}{A_i} \sum_{k=1}^{p} w_{ik} d_{kj}, \quad \text{for } i = 1, \ldots, m \text{ and } j = 1, \ldots, n \tag{6.15}$$

then the total cost expression (6.14) may be rewritten, on rearranging the order of summation, as

$$\sum_{i=1}^{m} \sum_{j=1}^{n} c_{ij} x_{ij} \tag{6.16}$$

The generalized assignment-problem version of the layout problem may now be stated:

$$\begin{aligned} \text{minimize} \quad & \sum_{i=1}^{m} \sum_{j=1}^{n} c_{ij} x_{ij} \\ \text{subject to} \quad & \sum_{j=1}^{n} x_{ij} = A_i, \quad i = 1, \ldots, m \\ & \sum_{i=1}^{m} x_{ij} = 1, \quad j = 1, \ldots, n \\ & x_{ij} \in \{0, 1\}, \quad i = 1, \ldots, m \\ & \qquad\qquad\qquad\; j = 1, \ldots, n \end{aligned} \tag{6.17}$$

The motivation for the terminology "generalized assignment problem" should now be clear: taking each A_i to be 1 gives the assignment problem. When some A_i are positive integers greater than 1 a generalization of the assignment problem is obtained. The generalized assignment problem is itself a special case of the transportation problem. In fact, the common way to solve the generalized assignment problem is as a transportation problem.

Prior to considering an example, it is useful to consider how to handle the situation where $\sum_{i=1}^{m} A_i < n$; that is, the total number of grid squares to be taken up by items is less than the total number of grid squares available. In this case a "dummy item" $m + 1$ is defined to take up $A_{m+1} = n - \sum_{i=1}^{m} A_i$

grid squares, and the constraints in (6.17) are changed to

$$\sum_{j=1}^{n} x_{ij} = A_i, \quad \text{for } i = 1, \ldots, m+1 \tag{6.18}$$

$$\sum_{i=1}^{m+1} x_{ij} = 1, \quad \text{for } j = 1, \ldots, n \tag{6.19}$$

and $$x_{ij} = 0 \quad \text{or} \quad 1, \quad \text{for } i = 1, \ldots, m+1 \text{ and } j = 1, \ldots, n \tag{6.20}$$

The objective function remains unchanged, so that the generalized assignment problem now consists of minimizing (6.16) subject to constraints (6.18), (6.19), and (6.20). In the transportation problem context, the inclusion of the dummy item is equivalent to the inclusion of a dummy destination: in a general linear-programming context the inclusion of a dummy item is equivalent to the inclusion of slack variables and one redundant constraint. When the cost coefficients c_{ij} for the dummy item case are computed using (6.15), then another way of obtaining the same terms c_{ij} is to define the terms $w_{m+1,1}, \ldots, w_{m+1,p}$ to be zero and use the equation

$$c_{ij} = \frac{1}{A_i} \sum_{k=1}^{p} w_{ik} d_{kj}, \quad \text{for } i = 1, \ldots, m+1 \text{ and } j = 1, \ldots, n \tag{6.21}$$

The use of Equation (6.21) for computing c_{ij} is of some intuitive appeal since it is based on the fact that the dummy item has no exchange with any dock; that is, all $w_{m+1,k} = 0$.

Figure 6.1 illustrates the solution of a layout problem using the generalized assignment model. The rectangular region shown may be considered to be a warehouse, and is subdivided into 288 grid squares; it has dimensions of 12 grid squares by 24 grid squares with the size of each grid square being of unit length. Supposing the rectangle to be located in the first quadrant, with its lower left-hand corner at the origin, the docks are located at the points $\mathbf{P}_1 = (7, 0)$, $\mathbf{P}_2 = (12, 0)$, and $\mathbf{P}_3 = (17, 0)$. Three items are to be laid out: items 1, 2, and 3 will take up $A_1 = 45$, $A_2 = 156$, and $A_3 = 45$ grid squares, respectively. A dummy item 4 takes up the remaining $A_4 = 288 - 246 = 42$ grid squares. The terms w_{ik} are given in the matrix

$$\mathbf{W} = \begin{pmatrix} 45 & 312 & 45 \\ 90 & 156 & 90 \\ 45 & 312 & 45 \\ 0 & 0 & 0 \end{pmatrix}$$

The distance d_{kj} between dock k and the center of grid square j for this example was computed as the rectilinear distance, with the length of the side of a grid square representing one unit of distance. For this example

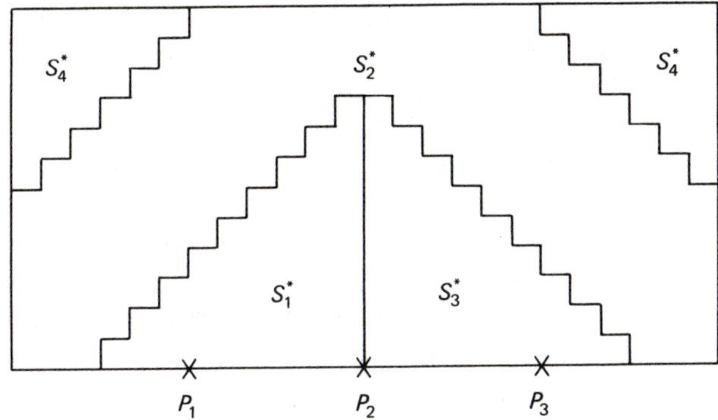

Figure 6.1. Generalized assignment solution of a layout problem.

problem there are 4(288) = 1,152 variables, so it is infeasible, of course, to solve the problem by hand, or even to illustrate the values of all the variables in the text. The solution is easy to illustrate, however, as is seen in Figure 6.1. A least-cost layout is shown with items 1, 2, 3, and 4 taking up the grid squares indicated by $\mathbf{S}_1^*$, $\mathbf{S}_2^*$, $\mathbf{S}_3^*$, and $\mathbf{S}_4^*$, respectively. Notice, as would be expected, that the unused grid squares, represented by $\mathbf{S}_4^*$, are in the corners of the warehouse farthest from the docks. For this example problem, denoting the total cost of movement of item i by $F(\mathbf{S}_i^*)$, it was found that $F(\mathbf{S}_1^*) = 2{,}715$, $F(\mathbf{S}_2^*) = 4{,}486$, $F(\mathbf{S}_3^*) = 2{,}715$, and, of course, $F(\mathbf{S}_4^*) = 0$. The least total cost is then given by $F(\mathbf{S}_1^*, \mathbf{S}_2^*, \mathbf{S}_3^*, \mathbf{S}_4^*) = 9{,}916$. Suppose, in the previous example, that the A_i had the following values: $A_1 = 44$, $A_2 = 156$, $A_3 = 44$, and $A_4 = 44$. Also suppose that the length of each side of every grid square was 1 foot. An alternative choice of dimensions of grid squares could be made for which the length of each grid square would be 2 feet; the rectangular region would then be subdivided into only 72 grid squares, and would have dimensions of 6 grid squares by 12 grid squares. Furthermore the values of the A_i would be given by $A_1 = 11$, $A_2 = 39$, $A_3 = 11$, and $A_4 = 11$. If the problem were then set up as a generalized assignment problem, there would be 4(72) = 288 variables, as opposed to 1,152 variables in the previous problem, and 4 + 72 = 76 constraints instead of 4 + 288 = 292 constraints. Thus, if it can be justified physically, there is a very good computational reason for choosing the dimensions of every grid square to be as large as possible. In many warehouse contexts, items are all stored on pallets, each of which has the same dimensions; in such a case the dimensions of each grid square would have to be at least as large as those of each pallet. Computational savings can be obtained, although perhaps at the risk of loss of accuracy in the final answer, by making the dimensions of each grid square to be an integer multiple of those of each pallet. Clearly, the choice of the dimensions of each grid square depends to some extent upon judgment.

One additional point concerning the use of the generalized assignment model can be made: when any generalized assignment problem is solved using a transportation algorithm, degeneracy occurs. Consequently, there will be one or more variables in any basic feasible solution having values of zero. While degeneracy causes no theoretical difficulties, it does make such problems, even when they have relatively few variables, somewhat awkward to solve by hand. Thus a computer program of a transportation algorithm would commonly be employed to solve the generalized assignment problem. Also, a computer program for the multiplication of matrices might be used to compute the cost coefficients defined by (6.15).

6.4 Locating One Item

Whereas the generalized assignment model provides a quite general means of solving warehouse-layout problems, more efficient solution procedures can be developed for solving problems having special structure. Perhaps the simplest problem to consider, in terms of special structure, is the one in which only one item is to be located. Suppose, with the single exception that only one item is to be located, that the problem of interest is the same as that considered in the previous section. Suppose that the total number of grid squares the item requires is given by A, and that $\mathbf{S}$ represents the collection of numbers of grid squares which the item is assigned. Then the average distance an item moves between storage and dock k is given by

$$\sum_{j \in \mathbf{S}} \frac{1}{A} d_{kj} \tag{6.22}$$

Given that the total cost of movement of the item between storage and dock k is directly proportional to the average distance, with w_k being the constant of proportionality, the total cost of moving the item between storage and the docks will be given by

$$F(\mathbf{S}) = \sum_{k=1}^{p} w_k \left(\sum_{j \in \mathbf{S}} \frac{1}{A} d_{kj} \right) \tag{6.23}$$

which, on defining

$$f_j = \sum_{k=1}^{p} w_k d_{kj}, \quad j = 1, \ldots, n$$

and rearranging the order of summation, may be written as

$$F(\mathbf{S}) = \sum_{j \in \mathbf{S}} \frac{1}{A} f_j \tag{6.24}$$

The problem of interest may now be stated as follows: find A grid squares, with $\mathbf{S}^*$ being the set of the numbers of the grid squares, for which $F(\mathbf{S}^*) \leq F(\mathbf{S})$ where $\mathbf{S}$ represents a set consisting of the numbers of any other A grid squares, and where $F(\mathbf{S})$ is defined by (6.24). This problem is a very simple one, as the following roughly equivalent restatement illustrates: given n numbers $f_1, \ldots, f_n$ from which A must be selected, chose those A numbers whose sum is the smallest. Intuitively speaking, this latter problem would be solved simply by choosing A of the f_j having the smallest values. For example, if $n = 6$, $A = 3$, $f_1 = 10$, $f_2 = 3$, $f_3 = 6$, $f_4 = 6$, $f_5 = 4$, and $f_6 = 7$, then either the numbers f_2, f_3, and f_5, or f_2, f_4, and f_5, would be chosen, giving either $\mathbf{S}^* = \{2, 3, 5\}$ or $\mathbf{S}^* = \{2, 4, 5\}$. For the general problem, the solution procedure would be as follows: let $j_1, j_2, \ldots, j_n$ be any permutation of 1, 2, $\ldots, n$ for which $f_{j_1} \leq f_{j_2} \leq \cdots \leq f_{j_n}$: then a least-cost solution is given by $\mathbf{S}^* = \{j_1, j_2, \ldots, j_A\}$. With reference to the numerical example, $f_2 \leq f_5 \leq f_3 \leq f_4 \leq f_6 \leq f_1$, so one least-cost solution is given by $\mathbf{S}^* = \{2, 5, 3\}$.

For purposes of subsequent development, and in order to give a rigorous justification for the solution procedure, it is useful to establish some terminology. We prove subsequently that the procedure does indeed give a least-cost solution. Let $\mathbf{L}$ denote the collection of integers $1, 2, \ldots, n$, and let $H(\mathbf{L}: A)$ denote the collection of all subsets of $\mathbf{L}$ consisting of A integers; any element $\mathbf{S}$ in $H(\mathbf{L}: A)$ specifies the location of an item. Then the problem may be restated as follows: find an element $\mathbf{S}^*$ in $H(\mathbf{L}: A)$ for which $F(\mathbf{S}^*) \leq F(\mathbf{S})$ for any other element $\mathbf{S}$ in $H(\mathbf{L}: A)$. The solution procedure we have described is based on the following important result (which is proved in the next section):

Property 1: Suppose that there exists a number k for which $\mathbf{S}^*$ is an element of $H(\mathbf{L}: A)$, where $\mathbf{S}^*$ has the following property: $f_j \leq k$ for every element j in $\mathbf{S}^*$, and $f_j \geq k$ for every element j not in $\mathbf{S}^*$; then $F(\mathbf{S}^*) \leq F(\mathbf{S})$, where $\mathbf{S}$ is any other element in $H(\mathbf{L}: A)$.

As a layout example of the foregoing solution procedure, consider the rectangle shown in Figure 6.2, placed in the first quadrant with its lower left-hand corner at the origin. The rectangle has dimensions of 7 units in the x direction and 18 units in the y direction, has been subdivided into $7(18) = 126$ grid squares, and may be considered to represent a warehouse. Docks are located at the points $\mathbf{P}_1 = (0, 5)$ and $\mathbf{P}_2 = (0, 13)$, with $w_1 = w_2 = 1$, and it is given that $A = 70$. The distance between dock k and the center of grid square j, d_{kj}, is the rectilinear distance, with the length of the side of a grid square representing 1 unit of distance. After computing $f_j = d_{1j} + d_{2j}$ for each grid square j, the least-cost solution $\mathbf{S}^*$ is as shown in Figure 6.2. The least-cost solution $\mathbf{S}^*$ has the property that $f_j \leq 18$ for every element j in $\mathbf{S}^*$, and $f_j \geq 18$ for every element j not in $\mathbf{S}^*$ (in fact, $f_j \geq 19$ for every

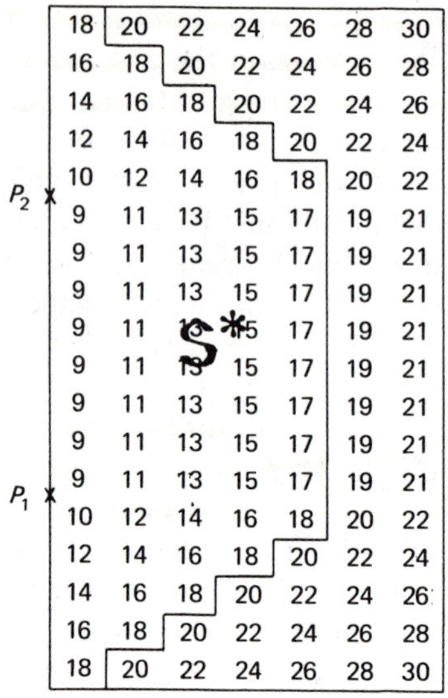

	18	20	22	24	26	28	30
	16	18	20	22	24	26	28
	14	16	18	20	22	24	26
	12	14	16	18	20	22	24
	10	12	14	16	18	20	22
P_2 ×	9	11	13	15	17	19	21
	9	11	13	15	17	19	21
	9	11	13	15	17	19	21
	9	11	S* 13	15	17	19	21
	9	11	13	15	17	19	21
	9	11	13	15	17	19	21
	9	11	13	15	17	19	21
P_1 ×	9	11	13	15	17	19	21
	10	12	14	16	18	20	22
	12	14	16	18	20	22	24
	14	16	18	20	22	24	26
	16	18	20	22	24	26	28
	18	20	22	24	26	28	30

Figure 6.2. Warehouse layout involving two docks and a single item.

element j not in **S***). Thus Property 1 guarantees that **S*** is a least-cost solution.

Property 1 gives sufficient conditions for **S*** to minimize $F(\mathbf{S})$; that is, it states conditions which, if satisfied, guarantee that **S*** minimizes $F(\mathbf{S})$. It is useful to know that the sufficient conditions are also necessary; that is, any **S*** in $H(\mathbf{L}:A)$ which minimizes $F(\mathbf{S})$ satisfies the conditions of Property 1. The necessary conditions can be stated formally as

Property 2: Let **S*** be an element of $H(\mathbf{L}:A)$ and suppose that $F(\mathbf{S}^*) \leq F(\mathbf{S})$ for every element **S** in $H(\mathbf{L}:A)$. Then **S*** has the following property: there exists a number k such that $f_j \leq k$ for every element j in **S***, and $f_j \geq k$ for every element j not in **S***.

Properties 1 and 2 may be summarized to yield the following necessary and sufficient conditions:

Property 3: Let **S*** be an element of $H(\mathbf{L}:A)$. Then

$$F(\mathbf{S}^*) \leq F(\mathbf{S})$$

for every element **S** in $H(\mathbf{L}:A)$ if and only if **S*** has the following property:

there exists a number k such that $f_j \leq k$ for every element j in $\mathbf{S}^*$, and $f_j \geq k$ for every element j not in $\mathbf{S}^*$.

6.5 Proofs of Properties 1 and 2*

In this section Properties 1 and 2 are proved.

Property 1: Suppose that there exists a number k for which $\mathbf{S}^*$ is an element of $H(\mathbf{L}:A)$, where $\mathbf{S}^*$ has the following property: $f_j \leq k$ for every element j in $\mathbf{S}^*$, and $f_j \geq k$ for every element j not in $\mathbf{S}^*$; then $F(\mathbf{S}^*) \leq F(\mathbf{S})$, where $\mathbf{S}$ is any other element in $H(\mathbf{L}:A)$.

Proof: Let $\mathbf{S}$ be any element in $H(\mathbf{L}:A)$. Denote by $\mathbf{SS}^*$ all those elements common to both $\mathbf{S}$ and $\mathbf{S}^*$, denote by $\mathbf{T}^*$ those elements of $\mathbf{S}^*$ not in $\mathbf{SS}^*$, and denote by $\mathbf{T}$ those elements of $\mathbf{S}$ not in $\mathbf{SS}^*$. Then

$$\sum_{j \in \mathbf{S}} f_j = \sum_{j \in \mathbf{T}} f_j + \sum_{j \in \mathbf{SS}^*} f_j \tag{6.25}$$

and

$$\sum_{j \in \mathbf{S}^*} f_j = \sum_{j \in \mathbf{T}^*} f_j + \sum_{j \in \mathbf{SS}^*} f_j \tag{6.26}$$

Suppose that $\mathbf{SS}^*$ consists of p elements, so that both $\mathbf{T}$ and $\mathbf{T}^*$ consist of $A - p$ elements. Since every element of $\mathbf{T}^*$ is an element of $\mathbf{S}^*$, the assumed property of $\mathbf{S}^*$ implies that $f_j \leq k$ for every element in $\mathbf{T}^*$, so

$$\sum_{j \in \mathbf{T}^*} f_j \leq \sum_{j \in \mathbf{T}^*} k = k(A - p) \tag{6.27}$$

Since every element of $\mathbf{T}$ is not an element of $\mathbf{S}^*$, the assumed property of $\mathbf{S}^*$ implies that $f_j \geq k$ for every element in $\mathbf{T}$; so

$$\sum_{j \in \mathbf{T}} f_j \geq \sum_{j \in \mathbf{T}} k = k(A - p) \tag{6.28}$$

From (6.27) and (6.28),

$$\sum_{j \in \mathbf{T}^*} f_j \leq \sum_{j \in \mathbf{T}} f_j \tag{6.29}$$

So (6.25), (6.26), and (6.29) imply that

$$\sum_{j \in \mathbf{S}^*} f_j \leq \sum_{j \in \mathbf{S}} f_j \tag{6.30}$$

Dividing inequality (6.30) by A gives $F(\mathbf{S}^*) \leq F(\mathbf{S})$.

---

*This section can be omitted without loss of continuity.

Our proof may be easily illustrated with reference to the previous numerical example, where $\mathbf{L} = \{1, 2, \ldots, 6\}$. Take $\mathbf{S}^* = \{2, 5, 3\}$ and $\mathbf{S} = \{4, 5, 6\}$. Then $\mathbf{SS}^* = \{5\}$, $\mathbf{T}^* = \{2, 3\}$, and $\mathbf{T} = \{4, 6\}$; also, $k = f_3 = 6$. Furthermore,

$$\sum_{j \in \mathbf{S}} f_j = (f_4 + f_6) + f_5 = (6 + 7) + 4 \geq (6 + 6) + 4 = 16$$

and
$$\sum_{j \in \mathbf{S}^*} f_j = (f_2 + f_3) + f_5 = (3 + 6) + 4 \leq (6 + 6) + 4 = 16$$

so that

$$\sum_{j \in \mathbf{S}^*} \tfrac{1}{3} f_j \leq \tfrac{16}{3} \leq \sum_{j \in \mathbf{S}} \tfrac{1}{3} f_j$$

Given the general situation with n numbers $f_1, \ldots, f_n$, and a permutation $j_1, j_2, \ldots, j_n$ of $1, \ldots, n$ so that $f_{j_1} \leq f_{j_2} \leq \cdots \leq f_{j_n}$, the number k is just the term f_{j_A}; for if $\mathbf{S}^* = \{j_1, j_2, \ldots, j_A\}$, then $f_i \leq f_{j_A}$ for any element i in $\mathbf{S}^*$, and $f_i \geq f_{j_A}$ for any element i not in $\mathbf{S}^*$.

Property 2: Let $\mathbf{S}^*$ be an element of $H(\mathbf{L}:A)$, and suppose that

$$F(\mathbf{S}^*) \leq F(\mathbf{S}) \tag{6.31}$$

for every element $\mathbf{S}$ in $H(\mathbf{L}:A)$. Then $\mathbf{S}^*$ has the following property: there exists a number k such that $f_j \leq k$ for every element j in $\mathbf{S}^*$, and $f_j \geq k$ for every element j not in $\mathbf{S}^*$.

Proof: Let

$$k = \max_{j \in \mathbf{S}^*} f_j = f_q$$

Then, by the definition of k, $f_j \leq k$ for every element j in $\mathbf{S}^*$. To show that $f_j \geq k$ for every element j not in $\mathbf{S}^*$, suppose the opposite; that is, there is at least one element r not in $\mathbf{S}^*$ for which $f_r < k$. Then let $\mathbf{S} = (\mathbf{S}^* - \{q\}) \cup \{r\}$; that is, let $\mathbf{S}$ be constructed from $\mathbf{S}^*$ by deleting the element q and including the element r. Certainly, $\mathbf{S}$ is an element of $H(\mathbf{L}:A)$, since $\mathbf{S}$ consists of A integers. Furthermore, a direct computation establishes that

$$F(\mathbf{S}) = F(\mathbf{S}^*) - \frac{1}{A}(f_q - f_r)$$

$$= F(\mathbf{S}^*) - \frac{1}{A}(k - f_r) < F(\mathbf{S}^*)$$

which contradicts (6.31). Thus there is no element r not in $\mathbf{S}^*$ for which $f_r < k$; so $f_j \geq k$ for every element j not in $\mathbf{S}^*$.

6.6 Locating *m* Items: The Factoring Case

Again we consider the problem treated in Section 6.3; i.e., finding a layout to minimize the expression

$$F(\mathbf{S}_1, \ldots, \mathbf{S}_m) = \sum_{i=1}^{m} \sum_{k=1}^{p} w_{ik} \left(\sum_{j \in \mathbf{S}_i} \frac{1}{A_i} d_{kj} \right) \tag{6.32}$$

In this section, an assumption, which is realistic in many cases, will be made that will permit a least-cost layout to be found by a simpler procedure than the generalized assignment approach.

To state the assumption, it is useful to make the following definition: the matrix $\mathbf{W} = (w_{ik})$, which has m rows and p columns, will be said to *factor* if and only if there exist numbers $u_1, \ldots, u_m$ and $v_1, \ldots, v_p$ such that

$$w_{ik} = u_i v_k, \quad \text{for } i = 1, \ldots, m \text{ and } k = 1, \ldots, p \tag{6.33}$$

Subsequently, it will be assumed that the matrix **W** factors: it should be noted that the factoring assumption is always satisfied when there is only one dock.

Given the factoring assumption, $F(\mathbf{S}_1, \ldots, \mathbf{S}_m)$ can be rewritten, on rearranging the order of summation, as

$$F(\mathbf{S}_1, \ldots, \mathbf{S}_m) = \sum_{i=1}^{m} \frac{u_i}{A_i} \sum_{j \in \mathbf{S}_i} f_j \tag{6.34}$$

where

$$f_j = \sum_{k=1}^{p} v_k d_{kj}, \quad \text{for } j = 1, \ldots, n \tag{6.35}$$

As will be seen, expression (6.34) for $F(\mathbf{S}_1, \ldots, \mathbf{S}_m)$ will permit a solution procedure for finding a least-cost layout that is very similar to the solution procedure of Section 6.4.

Some insight into the factoring assumption can be obtained by observing that an equivalent condition for the matrix **W** to factor is that

$$w_{ik} = c_i w_k, \quad \text{for } i = 1, \ldots, m \text{ and } k = 1, \ldots, p \tag{6.36}$$

where

$$c_i = \sum_{k=1}^{p} w_{ik}, \quad \text{for } i = 1, \ldots, m \tag{6.37}$$

and

$$w_k = \frac{\sum_{i=1}^{m} w_{ik}}{\sum_{k=1}^{p} \sum_{i=1}^{m} w_{ik}}, \quad \text{for } k = 1, \ldots, p \tag{6.38}$$

To see that expressions (6.33) and (6.36) are equivalent, first suppose that

(6.36) holds: then just defining $u_i = c_i$ and $v_k = w_k$ gives (6.33); so if (6.36) holds, then the matrix **W** factors. Next suppose that the matrix **W** factors: then

$$\sum_{k=1}^{p} w_{ik} = u_i \sum_{k=1}^{p} v_k \tag{6.39}$$

$$\sum_{i=1}^{m} w_{ik} = v_k \sum_{i=1}^{m} u_i \tag{6.40}$$

and

$$\sum_{k=1}^{p} \sum_{i=1}^{m} w_{ik} = \left(\sum_{i=1}^{m} u_i\right)\left(\sum_{k=1}^{p} v_k\right) \tag{6.41}$$

Solving (6.39) for u_i and (6.40) for v_k, and then using (6.41), gives

$$u_i v_k = \frac{\left(\sum_{k=1}^{p} w_{ik}\right)\left(\sum_{i=1}^{m} w_{ik}\right)}{\sum_{k=1}^{p} \sum_{i=1}^{m} w_{ik}} = c_i w_k$$

Since the matrix **W** factors, $w_{ik} = u_i v_k$, and so $w_{ik} = c_i w_k$. Thus the factoring assumption is equivalent to condition (6.36), with c_i and w_k defined by (6.37) and (6.38), respectively; so u_i in (6.34) and v_k in (6.35) may be replaced by c_i and w_k, respectively.

Expression (6.36) may be given a useful physical interpretation. Suppose, in a warehouse context, that w_{ik} represents the total number of pallets of item i traveling in and out of storage from dock k per time period. Then c_i represents the total number of pallets of item i traveling in and out of storage per time period, while w_k represents the percentage of all pallets traveling in and out of storage from dock k. *Thus the factoring assumption is satisfied whenever the total number of pallets of item i traveling in and out of storage from dock k per time period can be obtained by multiplying the total number of pallets of item i traveling in and out of storage per time period by the percentage of all pallets traveling in and out of storage from dock k.*

To motivate the procedure to be developed for finding a least-cost layout, consider the following two sequences of numbers: (1, 3, 4, 6, 8) and (5, 6, 2, 9, 7). If the second sequence is to be rearranged such that vector multiplication will maximize the scalar product, what arrangement should be used? Hopefully, it is intuitive that the maximum scalar product occurs with the following arrangement: (1, 3, 4, 6, 8) and (2, 5, 6, 7, 9), since the largest numbers are multiplied together, the second largest numbers are multiplied together, and so on. Consequently,

$$8(9) + 6(7) + 4(6) + 3(5) + 1(2) = 155$$

is the maximum scalar product. Next, consider the situation in which the scalar product is to be minimized. In this case the following arrangement will minimize the scalar product: (1, 3, 4, 6, 8) and (9, 7, 6, 5, 2). Thus,

$$1(9) + 3(7) + 4(6) + 6(5) + 8(2) = 100$$

is the minimum scalar product. (We also use this result in Chapter 8.)

Our discussion can be related to a layout problem by letting the first of the sequences represent the average distance between a grid square and the docks and letting the second sequence represent the numbers $c_1/A_1, \ldots, c_m/A_m$.

Recalling that u_i can be replaced by c_i and v_k by w_k, the items will be numbered so that

$$\frac{c_1}{A_1} \geq \frac{c_2}{A_2} \geq \cdots \geq \frac{c_m}{A_m} \tag{6.42}$$

The problem of interest is to find a layout $\{\mathbf{S}_1^*, \ldots, \mathbf{S}_m^*\}$ such that $F(\mathbf{S}_1^*, \ldots, \mathbf{S}_m^*) \leq F(\mathbf{S}_1, \ldots, \mathbf{S}_m)$, where $\{\mathbf{S}_1, \ldots, \mathbf{S}_m\}$ is any other layout. It will be useful to denote the class of all layouts by $H_m(\mathbf{L}: A)$. The procedure for finding a least-cost layout may now be stated. For $j = 1, \ldots, n$, compute

$$f_j = \sum_{k=1}^{p} w_k d_{kj} \tag{6.43}$$

and let $j_1, j_2, \ldots, j_n$ be a permutation of $1, 2, \ldots, n$ such that $f_{j_1} \leq f_{j_2} \leq \cdots \leq f_{j_n}$. A least-cost layout $\{\mathbf{S}_1^*, \ldots, \mathbf{S}_m^*\}$ in $H_m(\mathbf{L}: A)$ is then given, on defining $B_i = \sum_{h=1}^{i} A_h$, for $i = 1, \ldots, m$, by

$$\begin{aligned}
\mathbf{S}_1^* &= \{j_1, \ldots, j_{B_1}\} \\
\mathbf{S}_2^* &= \{j_{B_1+1}, \ldots, j_{B_2}\} \\
\mathbf{S}_m^* &= \{j_{B_{m-1}+1}, \ldots, j_{B_m}\}
\end{aligned}$$

As a layout example of the solution procedure, the example illustrated in Figure 6.2 will be considered again, but with the following changes: there are $m = 2$ items, $A_1 = 30$, $A_2 = 84$, and

$$\mathbf{W} = (w_{ik}) = \begin{pmatrix} 2 & 2 \\ 4 & 4 \end{pmatrix}$$

Thus $c_1 = 4$, $c_2 = 8$, $w_1 = \frac{1}{2}$, and $w_2 = \frac{1}{2}$, so the total cost expression (6.34) becomes

$$F(\mathbf{S}_1^*, \mathbf{S}_2^*) = \frac{4}{30} \sum_{j \in \mathbf{S}_1^*} f_j + \frac{8}{84} \sum_{j \in \mathbf{S}_2^*} f_j$$

where
$$f_j = \tfrac{1}{2} d_{1j} + \tfrac{1}{2} d_{2j}, \quad \text{for } j = 1, \ldots, 126$$

A least-cost layout $\{\mathbf{S}_1^*, \mathbf{S}_2^*\}$ is shown in Figure 6.3. $\mathbf{S}_1^*$ has the property that $f_j \leq 6.5$ for every element j in $\mathbf{S}_1^*$, and $\mathbf{S}_2^*$ has the property that $7 \leq f_j \leq 12$ for every element j in $\mathbf{S}_2^*$. For every element j neither in $\mathbf{S}_1^*$ nor $\mathbf{S}_2^*$, $f_j \geq 13$; so $\mathbf{S}_1^*$ and $\mathbf{S}_2^*$ have indeed been constructed as the preceding procedure dictates.

	9	10	11	12	13	14	15
	8	9	10	11	12	13	14
	7	8	9	10	11	12	13
	6	7	8	9	10	11	12
P_2 ×	5	6	7	8	9	10	11
	1.5	5.5	6.5	7.5	8.5	9.5	10.5
	4.5	5.5	6.5	7.5	8.5	9.5	10.5
	4.5	5.5	6.5	7.5	8.5	9.5	10.5
	4.5	S_1^*	6.5	7.5	8.5	S_2^*	10.5
	4.5		6.5	7.5	8.5		10.5
	4.5	5.5	6.5	7.5	8.5	9.5	10.5
	4.5	5.5	6.5	7.5	8.5	9.5	10.5
	4.5	5.5	6.5	7.5	8.5	9.5	10.5
P_1 ×	5	6	7	8	9	10	11
	6	7	8	9	10	11	12
	7	8	9	10	11	12	13
	8	9	10	11	12	13	14
	9	10	11	12	13	14	15

Figure 6.3. Warehouse layout involving two docks and two items with factoring.

Some comments on the general solution procedure are pertinent at this point. Note that f_j as defined by (6.43) may be interpreted as an average distance between grid square j and docks. Thus, given that the items are numbered as specified by (6.42), the solution procedure states that item 1 takes up the A_1 grid squares having the closest average distances to the dock, item 2 takes up the A_2 grid squares having the next closest average distances, and so forth. The inequalities stated in (6.42) may be motivated as follows: an item taking up a large number of grid squares and having a small total number of pallets traveling in and out of storage per time period should take up grid squares having greater average distances from the dock than an item taking up a small number of grid squares and having a large number of pallets traveling in and out of storage per time period. With reference to the example illustrated in Figure 6.3, note that, even though $c_2 = 8 > c_1 = 4$, item 2 takes up grid squares having greater average distances from the docks than item 1. To find a least-cost layout, it is not enough simply to examine the values of c_1 and c_2; the ratios c_1/A_1 and c_2/A_2 must be considered.

6.7 Development of the Solution Procedure*

Prior to developing a rigorous justification for the solution procedure for finding a least total cost layout, it is convenient to introduce some notation. Given any layout $\{\mathbf{S}_1, \ldots, \mathbf{S}_m\}$, the union of $\mathbf{S}_1, \mathbf{S}_2, \ldots, \mathbf{S}_q$, denoted by $\bigcup_{i=1}^{q} \mathbf{S}_i$, for $q = 1, 2, \ldots, m$, is defined to be the collection of all grid square numbers in at least one of the sets $\mathbf{S}_1, \mathbf{S}_2, \ldots, \mathbf{S}_q$. For example, given a layout $\{\mathbf{S}_1, \mathbf{S}_2, \mathbf{S}_3\}$, where $\mathbf{S}_1 = \{1, 3, 5\}$, $\mathbf{S}_2 = \{6, 13, 23, 24\}$, and $\mathbf{S}_3 = \{10, 12\}$, then $\bigcup_{i=1}^{2} \mathbf{S}_i = \{1, 3, 5, 6, 13, 23, 24\}$, and $\bigcup_{i=1}^{3} \mathbf{S}_i = \{1, 3, 5, 6, 13, 23, 24, 10, 12\}$.

The following property develops a rigorous justification for the procedure for finding a least total cost layout when the factoring assumption holds.

Property 4: Let $\{\mathbf{S}_1^*, \ldots, \mathbf{S}_m^*\}$ be a layout in $H_m(\mathbf{L}:A)$. Given that

$$\frac{c_1}{A_1} \geq \cdots \geq \frac{c_m}{A_m} > 0 \tag{6.44}$$

then

$$F(\mathbf{S}_1^*, \ldots, \mathbf{S}_m^*) \leq F(\mathbf{S}_1, \ldots, \mathbf{S}_m)$$

where $\{\mathbf{S}_1, \ldots, \mathbf{S}_m\}$ is any other layout in $H_m(\mathbf{L}:A)$, if $\{\mathbf{S}_1^*, \ldots, \mathbf{S}_m^*\}$ has the following property: there exist numbers $k_1 \leq k_2 \leq \cdots \leq k_m$ such that, for $q = 1, \ldots, m$,

$$f_j \leq k_q, \quad \text{for every element } j \text{ in } \bigcup_{i=1}^{q} \mathbf{S}_i^* \tag{6.45}$$

and

$$f_j \geq k_q, \quad \text{for every element } j \text{ not in } \bigcup_{i=1}^{q} \mathbf{S}_i^* \tag{6.46}$$

Proof: The proof will be developed for the case $m = 2$; this case includes all the essential ideas of the proof. A general proof can be readily developed by following the approach to be presented.

Let $\{\mathbf{S}_1^*, \mathbf{S}_2^*\}$ be a layout having the assumed properties, and let $\{\mathbf{S}_1, \mathbf{S}_2\}$ be any other layout. Then

$$\begin{aligned} F(\mathbf{S}_1, \mathbf{S}_2) &= \frac{c_1}{A_1} \sum_{j \in \mathbf{S}_1} f_j + \frac{c_2}{A_2} \sum_{j \in \mathbf{S}_2} f_j \\ &\quad - \frac{c_2}{A_2} \sum_{j \in \mathbf{S}_1} f_j + \frac{c_2}{A_2} \sum_{j \in \mathbf{S}_1} f_j \\ &= \left(\frac{c_1}{A_1} - \frac{c_2}{A_2}\right) \sum_{j \in \mathbf{S}_1} f_j + \frac{c_2}{A_2} \sum_{j \in \mathbf{S}_1 \cup \mathbf{S}_2} f_j \end{aligned} \tag{6.47}$$

---

*The solution procedure for the m-item location problem is rigorously developed for the case of w_{ik} factoring. An understanding of set theory is assumed in this section. The section can be omitted without loss of continuity.

Likewise,

$$F(\mathbf{S}_1^*, \mathbf{S}_2^*) = \left(\frac{c_1}{A_1} - \frac{c_2}{A_2}\right) \sum_{j \in \mathbf{S}_1^*} f_j + \frac{c_2}{A_2} \sum_{j \in \mathbf{S}_1^* \cup \mathbf{S}_2^*} f_j \tag{6.48}$$

Now, given the assumed property of $\mathbf{S}_1^*$ and of $\mathbf{S}_1^* \cup \mathbf{S}_2^*$, exactly the same approach as that used to obtain inequality (6.30) in the proof of Property 1 establishes that

$$\sum_{j \in \mathbf{S}_1^*} f_j \leq \sum_{j \in \mathbf{S}_1} f_j \tag{6.49}$$

since $\mathbf{S}_1^*$ and $\mathbf{S}_1$ both consist of A_1 elements, and that

$$\sum_{j \in \mathbf{S}_1^* \cup \mathbf{S}_2^*} f_j \leq \sum_{j \in \mathbf{S}_1 \cup \mathbf{S}_2} f_j \tag{6.50}$$

since $\mathbf{S}_1^* \cup \mathbf{S}_2^*$ and $\mathbf{S}_1 \cup \mathbf{S}_2$ both consist of $B_2 = A_1 + A_2$ elements. Multiplying (6.49) by $(c_1/A_1) - (c_2/A_2) \geq 0$ and (6.50) by $c_2/A_2 > 0$ and adding the resulting inequalities gives

$$\left(\frac{c_1}{A_1} - \frac{c_2}{A_2}\right) \sum_{j \in \mathbf{S}_1^*} f_j + \frac{c_2}{A_2} \sum_{j \in \mathbf{S}_1^* \cup \mathbf{S}_2^*} f_j \leq \left(\frac{c_1}{A_1} - \frac{c_2}{A_2}\right) \sum_{j \in \mathbf{S}_1} f_j + \frac{c_2}{A_2} \sum_{j \in \mathbf{S}_1 \cup \mathbf{S}_2} f_j \tag{6.51}$$

Thus, on using identities (6.47) and (6.48) and inequality (6.51), it follows that $F(\mathbf{S}_1^*, \mathbf{S}_2^*) \leq F(\mathbf{S}_1, \mathbf{S}_2)$. Parenthetically, it should be pointed out that it is not necessary to assume that (6.45) and (6.46) hold for the case $q = 1$ when $c_1/A_1 = c_2/A_2$.

The least-cost layout shown in Figure 6.3 satisfies the conditions of Property 4; for that layout, $k_1 = 6.5$ and $k_2 = 12$.

Property 4 gives sufficient conditions for a least-cost layout; that is, any layout satisfying the conditions of Property 4 will be a least-cost layout. It is also of interest to determine whether or not the sufficient conditions are also necessary, that is, to determine whether or not any least-cost layout will satisfy the conditions of Property 4. As will be seen, the sufficient conditions, with slight modifications, are also necessary.

For any layout $\{\mathbf{S}_1, \ldots, \mathbf{S}_m\}$ in $H_m(\mathbf{L} : A)$, it is useful to define the terms

$$F(\mathbf{S}_i) = \frac{c_i}{A_i} \sum_{j \in \mathbf{S}_i} f_j, \quad \text{for } i = 1, \ldots, m$$

Given the factoring assumption, it is not difficult to verify that

$$\sum_{k=1}^{p} w_{ik} \sum_{j \in \mathbf{S}_i} \frac{1}{A_i} d_{kj} = F(\mathbf{S}_i), \quad \text{for } i = 1, \ldots, m \tag{6.52}$$

Thus $F(\mathbf{S}_i)$ is that part of the total cost expression due to item i, so that

$$F(\mathbf{S}_1, \ldots, \mathbf{S}_m) = \sum_{i=1}^{m} F(\mathbf{S}_i)$$

In obtaining necessary conditions, only the case where there are two items will be considered. This case includes all the features of the general case of m items, and provides more insight than the general case. The extension of the following properties to m items is left as an exercise.

In obtaining the necessary conditions, it will be useful to consider two separate situations, the first of which is given by the following property.

Property 5A: Let $\{\mathbf{S}_1^*, \mathbf{S}_2^*\}$ be a least-cost layout in $H_2(\mathbf{L} : A)$. Given that

$$\frac{c_1}{A_1} = \frac{c_2}{A_2} > 0 \tag{6.53}$$

then $\{\mathbf{S}_1^*, \mathbf{S}_2^*\}$ has the following property: there exists a number k_2 such that

$$f_j \leq k_2, \quad \text{for every element } j \text{ in } \bigcup_{i=1}^{2} \mathbf{S}_i^*$$

and

$$f_j \geq k_2, \quad \text{for every element } j \text{ not in } \bigcup_{i=1}^{2} \mathbf{S}_i^*$$

Furthermore, if $\{\mathbf{S}_1', \mathbf{S}_2'\}$ is any other layout for which

$$\bigcup_{i=1}^{2} \mathbf{S}_i' = \bigcup_{i=1}^{2} \mathbf{S}_i^* \tag{6.54}$$

then $\{\mathbf{S}_1', \mathbf{S}_2'\}$ is also a least-cost layout; that is,

$$F(\mathbf{S}_1^*, \mathbf{S}_2^*) = F(\mathbf{S}_1', \mathbf{S}_2') \tag{6.55}$$

Proof: It is first convenient to establish (6.55). Using identities (6.47) and (6.48) together with the fact that $c_1/A_1 - c_2/A_2 = 0$ gives

$$F(\mathbf{S}_1', \mathbf{S}_2') = \frac{c_2}{A_2} \sum_{j \in \mathbf{S}_1' \cup \mathbf{S}_2'} f_j \tag{6.56}$$

and

$$F(\mathbf{S}_1^*. \mathbf{S}_2^*) = \frac{c_2}{A_2} \sum_{j \in \mathbf{S}_1^* \cup \mathbf{S}_2^*} f_j \tag{6.57}$$

But now (6.55) follows from (6.54), (6.56), and (6.57).

To complete the proof, let $\{\mathbf{S}_1^*, \mathbf{S}_2^*\}$ be a least-cost layout, and let $\{\mathbf{S}_1, \mathbf{S}_2\}$ be any other layout, so that

$$F(\mathbf{S}_1^*, \mathbf{S}_2^*) \leq F(\mathbf{S}_1, \mathbf{S}_2)$$

or, on again using identities (6.47) and (6.48),

$$\frac{c_2}{A_2} \sum_{j \in \mathbf{S}_1^* \cup \mathbf{S}_2^*} f_j \leq \frac{c_2}{A_2} \sum_{j \in \mathbf{S}_1 \cup \mathbf{S}_2} f_j$$

But now $\mathbf{S}_1^* \cup \mathbf{S}_2^*$ and $\mathbf{S}_1 \cup \mathbf{S}_2$ both have $A_1 + A_2$ elements; so (6.45) and (6.46) follow on using exactly the same approach as in the proof of Property 2.

Note that when condition (6.53) holds, nothing can be said about $\mathbf{S}_1^*$ and $\mathbf{S}_2^*$ separately; only a statement about their union can be made. The reason for this is given in equation (6.55), which states that there are alternative least-cost layouts, and that in fact any layout $\{\mathbf{S}_1', \mathbf{S}_2'\}$ for which (6.53) holds will be a least-cost layout.

When condition (6.53) is changed slightly, statements can be made, however, about $\mathbf{S}_1^*$ and $\mathbf{S}_2^*$ individually, where $\{\mathbf{S}_1^*, \mathbf{S}_2^*\}$ is any least-cost layout, as the following property establishes.

Property 5B: Let $\{\mathbf{S}_1^*, \mathbf{S}_2^*\}$ be a least-cost layout in $H_2(\mathbf{L} : A)$. Given that

$$\frac{c_1}{A_1} > \frac{c_2}{A_2} > 0$$

then $\{\mathbf{S}_1^*, \mathbf{S}_2^*\}$ has the following property: there exist numbers $k_1 \leq k_2$ such that, for $q = 1, 2$,

$$f_j \leq k_q, \quad \text{for every element } j \text{ in } \bigcup_{i=1}^{q} \mathbf{S}_i^* \tag{6.58}$$

and

$$f_j \geq k_q, \quad \text{for every element } j \text{ not in } \bigcup_{i=1}^{q} \mathbf{S}_i^* \tag{6.59}$$

Proof: Consider first the case $q = 1$. Let

$$k_1 = \max_{j \in \mathbf{S}_1^*} f_j$$

Then certainly (6.58) holds. It remains to establish (6.59). Suppose that there is at least one element r not in $\mathbf{S}_1^*$ for which $f_j < k_1$; then either r is in $\mathbf{S}_2^*$, or r is in the complement of $\mathbf{S}_1^* \cup \mathbf{S}_2^*$. Consider the case where r is in $\mathbf{S}_2^*$. Let $f_t = k_1$, and define the layout $\{\mathbf{S}_1, \mathbf{S}_2\}$ as follows: $\mathbf{S}_1 = (\mathbf{S}_1^* - \{t\}) \cup \{r\}$, $\mathbf{S}_2 = (\mathbf{S}_2^* - \{r\}) \cup \{t\}$. That is, roughly speaking, $\mathbf{S}_1$ and $\mathbf{S}_2$ are obtained from $\mathbf{S}_1^*$ and $\mathbf{S}_2^*$ by interchanging grid squares r and t. Then a direct computation establishes that

$$\begin{aligned} F(\mathbf{S}_1, \mathbf{S}_2) &= F(\mathbf{S}_1^*) + \frac{c_1}{A_1}(f_r - f_t) + F(\mathbf{S}_2^*) + \frac{c_2}{A_2}(f_t - f_r) \\ &= F(\mathbf{S}_1^*, \mathbf{S}_2^*) + \left(\frac{c_1}{A_1} - \frac{c_2}{A_2}\right)(f_r - k_1) \\ &< F(\mathbf{S}_1^*, \mathbf{S}_2^*) \end{aligned}$$

contradicting the fact that $\{\mathbf{S}_1^*, \mathbf{S}_2^*\}$ is a least-cost layout. Thus r is not in $\mathbf{S}_2^*$, so it follows that $k_1 \leq f_j$ for every element j in $\mathbf{S}_2^*$. In particular, if

$$k_2 = \max_{j \in \mathbf{S}_2^*} f_j$$

it follows that $k_1 \leq k_2$. Next suppose that r is in the complement of $\mathbf{S}_1^* \cup \mathbf{S}_2^*$, and define the layout $\{\mathbf{S}_1, \mathbf{S}_2\}$ by $\mathbf{S}_1 = (\mathbf{S}_1^* - \{t\}) \cup \{r\}$, $\mathbf{S}_2 = \mathbf{S}_2^*$. Then again a direct computation establishes that

$$\begin{aligned} F(\mathbf{S}_1, \mathbf{S}_2) &= F(\mathbf{S}_1^*) + \frac{c_1}{A_1}(f_r - f_t) + F(\mathbf{S}_2^*) \\ &= F(\mathbf{S}_1^*, \mathbf{S}_2^*) + \frac{c_1}{A_1}(f_r - k_1) \\ &< F(\mathbf{S}_1^*, \mathbf{S}_2^*) \end{aligned}$$

which again contradicts the fact that $\{\mathbf{S}_1^*, \mathbf{S}_2^*\}$ is a least-cost layout. Thus, for every element j not in $\mathbf{S}_1^*$, $k_1 \leq f_j$; so (6.59) is established.

It remains to establish (6.58) and (6.59) for the case $q = 2$. Let

$$k_2 = \max_{j \in \mathbf{S}_2^*} f_j = f_v$$

Then certainly $f_j \leq k_2$ for every element j in $\mathbf{S}_2^*$. Furthermore, since $k_1 \leq k_2$ and $f_j \leq k_1$ for every element j in $\mathbf{S}_1^*$, $f_j \leq k_2$ for every element j in $\mathbf{S}_1^*$ so that (6.58) is established. To establish (6.59), suppose that there is at least one element r not in $\bigcup_{i=1}^{2} \mathbf{S}_i^*$ for which $f_r < k_2 = f_v$, and define the layout $\{\mathbf{S}_1, \mathbf{S}_2\}$ as follows: $\mathbf{S}_1 = \mathbf{S}_1^*$, $\mathbf{S}_2 = (\mathbf{S}_2 - \{v\}) \cup \{r\}$. Then a computation establishes that

$$\begin{aligned} F(\mathbf{S}_1, \mathbf{S}_2) &= F(\mathbf{S}_1^*) + F(\mathbf{S}_2^*) + \frac{c_2}{A_2}(f_r - f_v) \\ &= F(\mathbf{S}_1^*, \mathbf{S}_2^*) + \frac{c_2}{A_2}(f_r - k_2) \\ &< F(\mathbf{S}_1^*, \mathbf{S}_2^*) \end{aligned}$$

which again contradicts the fact that $\{\mathbf{S}_1^*, \mathbf{S}_2^*\}$ is a least-cost layout. Thus (6.59) is established, which completes the proof.

Although the proof of Property 5-B may at first glance seem involved, it is really rather simple. The underlying idea in the proof is that if either (6.58) or (6.59) does not hold it is possible to construct a layout which has a lower cost than the given least-cost layout; thus (6.58) and (6.59) must hold, as it is of course impossible for any layout to have a cost less than that of the given least-cost layout.

Properties 4 and 5 can now be combined to give necessary and sufficient conditions for a layout to be a least-cost layout.

Property 6: Let $\{\mathbf{S}_1^*, \mathbf{S}_2^*\}$ be a layout in $H_2(\mathbf{L}:A)$. Given that

$$\frac{c_1}{A_1} = \frac{c_2}{A_2} > 0$$

then $\{\mathbf{S}_1^*, \mathbf{S}_2^*\}$ is a least-cost layout if and only if it has the following property: there exists a number k_2 such that

$$f_j \leq k_2, \quad \text{for every element } j \text{ in } \bigcup_{i=1}^{2} \mathbf{S}_i^*$$

and

$$f_j \geq k_2, \quad \text{for every element } j \text{ not in } \bigcup_{i=1}^{2} \mathbf{S}_i^*$$

Given that

$$\frac{c_1}{A_1} > \frac{c_2}{A_2} > 0$$

$\{\mathbf{S}_1^*, \mathbf{S}_2^*\}$ is a least-cost layout if and only if it has the following property: there exists numbers $k_1 \leq k_2$ such that, for $q = 1, 2$,

$$f_j \leq k_q, \quad \text{for every element } j \text{ in } \bigcup_{i=1}^{q} \mathbf{S}_i^*$$

and

$$f_j \geq k_q, \quad \text{for every element } j \text{ not in } \bigcup_{i=1}^{q} \mathbf{S}_i^*$$

6.8 Summary

In this chapter the problem treated was that of allocating m facilities among n possible locations. It was assumed that there was no interchange among new facilities. The problem was generalized to include the situation in which facilities (items) required more than one location (grid square). The resulting generalized assignment problem was applied to a warehouse-layout problem. A number of properties were developed for the warehouse-layout problem in order to obtain efficient solutions to the problem.

Property 1, a basic tool for finding least-cost layouts for the problems considered in Sections 6.4 and 6.6, is a special case of the Neyman–Pearson lemma [11] a fundamental lemma in statistics that has a number of applications not of a statistics nature. The solution procedure for finding a least-cost layout developed immediately prior to the procedure given in Property 4 is based on an ordering procedure first published by Hardy, Littlewood, and Polya [7]. The solution procedure given in Property 4 is based on a procedure examined in the next chapter. Condition (6.44) given in Property 4 also occurs in a similar form in sequencing problems [13].

There are minimax approaches to the location problems considered in this chapter. Specifically, Garfinkel [4] has developed an algorithm that may be used to solve minimax assignment problems; Garfinkel and Rao [5] have developed an algorithm for solving minimax transportation problems (which include the minimax version of the generalized assignment problem as a special case). A number of minimax problems are considered in Chapter 9.

REFERENCES

1. DuPont, A., "Necessary Conditions for Optimum Solutions to Special... Assignment Problems," unpublished master's thesis, The Ohio State University, Columbus, Ohio, 1968.
2. Edmonds, J., and D. R. Fulkerson, "Bottle Neck Extrema," Rand Corporation, Memorandum–5375-PR, Santa Monica, Calif., 1968; subsequently published in *Journal of Combinatorial Theory*, Vol. 8, No. 4, 1970, pp. 229–306.
3. Francis, R. L., "Sufficient Conditions for Some Optimum-Property Facility Designs," *Operations Research*, Vol. 15, No. 3, 1967, pp. 448–466.
4. Garfinkel, R. S., "An Improved Algorithm for the Bottleneck Assignment Problem," *Operations Research*, Vol. 19, No. 7, 1971, pp. 1747–1750.
5. Garfinkel, R. S., and M. R. Rao, "The Bottleneck Transportation Problem," *Naval Research Logistics Quarterly*, Vol. 18, No. 4, 1971, pp. 465–472.
6. Goldman, A. J., "Optimal Locations for Centers in a Network," *Transportation Science*, Vol. 3, No. 4, 1969, pp. 352–360.
7. Hardy, G. H., J. E. Littlewood, and G. Polya, *Inequalities*, Cambridge University Press, New York, 1952.
8. Koopmans, J. C., and M. Beckmann, "Assignment Problems and the Location of Economic Activities," *Econometrica*, Vol. 25, No. 1, 1957, pp. 53–76.
9. Mallette, A. J., and R. L. Francis, "A Generalized Assignment Approach to Optimal Facility Layout," *AIIE Transactions*, Vol. 4, No. 2, 1972, pp. 144–147.
10. Moore, J. M., "Optimal Locations for Multiple Machines," *The Journal of Industrial Engineering*, Vol. 12, No. 5, 1961, pp. 34–38.
11. Neyman, J., and E. S. Pearson, "On the Problem of the Most Efficient Tests of Statistical Hypotheses," *Philosophical Transactions of the Royal Society of London*. Series A, Vol. 231, 1933, pp. 289–337.
12. Roberts, S. D. and Reed, R., "On the Problem of Optimal Warehouse Bay Configurations," *AIIE Transactions*, Vol. 4, No. 3, 1972, pp. 178–185.

13. SMITH, W. E., "Various Optimizers for Single-Stage Production," *Naval Research Logistics Quarterly*, Vol. 3, No. 1, 1956.

PROBLEMS

6.1. Four new facilities are to be assigned to four sites. The following matrix $\mathbf{C} = (c_{ij})$ gives the cost of assigning new facility i to site j:

$$\mathbf{C} = \begin{pmatrix} 2 & 10 & 6 & 6 \\ 16 & 14 & 10 & 2 \\ 8 & 12 & 8 & 12 \\ 2 & 10 & 4 & 6 \end{pmatrix}$$

Find an assignment of new facilities to sites that will minimize the total cost.

6.2. Four sites are to be chosen from five that are available, and four new facilities are to be assigned to the four chosen sites. The following matrix $\mathbf{C} = (c_{ij})$ gives the cost of assigning new facility i to site j:

$$\mathbf{C} = \begin{pmatrix} 2 & 10 & 6 & 6 & 8 \\ 16 & 14 & 10 & 2 & 16 \\ 8 & 12 & 8 & 12 & 12 \\ 2 & 10 & 4 & 6 & 10 \end{pmatrix}$$

Determine which sites should be chosen and how new facilities should be assigned to sites so as to minimize the total cost.

6.3. Four machines are to be placed in a plant layout. Sites, numbered 1 through 4, available for the locations of the machines have the coordinates (5, 5), (7, 5), (5, 2), and (7, 2), respectively. Products will move between the new machines and existing machines, numbered 1 through 3, having the coordinates (2, 2), (5, 7), and (9, 4), respectively. The matrix $\mathbf{W} = (w_{ik})$ specifies total costs per unit of rectilinear distance for a given time period incurred in moving products between new and existing machines:

$$\mathbf{W} = \begin{pmatrix} 2 & 1 & 3 \\ 3 & 2 & 4 \\ 1 & 1 & 2 \\ 2 & 2 & 2 \end{pmatrix}$$

(a) Formulate completely a problem that, if solved, would determine an assignment of new machines to sites which would minimize the total cost of product movement.

(b) Solve the problem formulated in part (a).

6.4. For the example illustrated in Figure 6.3,
(a) Suppose that A is changed to 78. Determine $\mathbf{S}^*$ for this case.
(b) Find $\mathbf{S}^*$ when $A = 80$. Is there a unique solution for $\mathbf{S}^*$ in this case?
(c) Find $\mathbf{S}^*$ when $A = 88$. Is there a unique solution for $\mathbf{S}^*$ in this case?

6.5. A rectangular region consists of 42 grid squares, the length of each side of a grid square is 1, and the region has dimensions of 7 grid squares in the x direction, 6 grid squares in the y direction, and is located in the first quadrant with its lower left-hand corner at the origin. Docks are located at the points $\mathbf{P}_1 = (0, 3)$ and $\mathbf{P}_2 = (7, 3)$. One item is to be located in the region, and will take up 28 grid squares. Rectilinear distances between docks and grid squares are computed in the same manner as in the examples of the chapter. Given that $\mathbf{W} = (w_1, w_2) = (1, 1)$, find a least-cost location for the item.

6.6. A rectangular region consists of 15 grid squares, the length of each side of a grid square is 1, and the region has dimensions of 5 grid squares in the x direction and 3 in the y direction; the region is located in the first quadrant with its lower left-hand corner at the origin. Docks have the locations $\mathbf{P}_1 = (1, 0)$, $\mathbf{P}_2 = (4, 0)$, $\mathbf{P}_3 = (1, 3)$, and $\mathbf{P}_4 = (4, 3)$. Two items are to be located in the region; item 1 will take up $A_1 = 5$ grid squares, and item 2 will take up $A_2 = 10$ grid squares. The data w_{ik} are given by the matrix

$$\mathbf{W} = (w_{ik}) = \begin{pmatrix} 0 & 1 & 0 & 1 \\ 2 & 2 & 1 & 0 \end{pmatrix}$$

Grid squares are numbered from 1 to 15, beginning with the number 1 in the upper left-hand corner and proceeding sequentially from left to right. The distances between docks and grid squares are determined in the same manner as in the examples of the chapter.
(a) Set up the tableau for the generalized assignment problem that, if solved, would determine least-cost locations of items 1 and 2.
(b) Solve the problem set up in part (a).

6.7. Two new machines are to be located in an existing layout containing five machines at the locations $\mathbf{P}_1 = (1, 3)$, $\mathbf{P}_2 = (4, 3)$, $\mathbf{P}_3 = (8, 2)$, $\mathbf{P}_4 = (5, 1)$, and $\mathbf{P}_5 = (3, 4)$. Two sites are available for location of the new machines; the site locations are $\mathbf{Q}_1 = (3, 1)$ and $\mathbf{Q}_2 = (6, 3)$. The number of pallet loads moved per week between the new and existing machines is given by the following $\mathbf{W}$ matrix:

$$\mathbf{W} = \begin{pmatrix} 13 & 26 & 2 & 8 & 10 \\ 11 & 0 & 7 & 15 & 8 \end{pmatrix}$$

The cost to transport a pallet load by a fork lift truck is a linear function of distance. Thus,

$$c = 0.5 + 22d$$

where

$$c = \text{cost per trip}$$

$$d = \text{distance per trip}$$

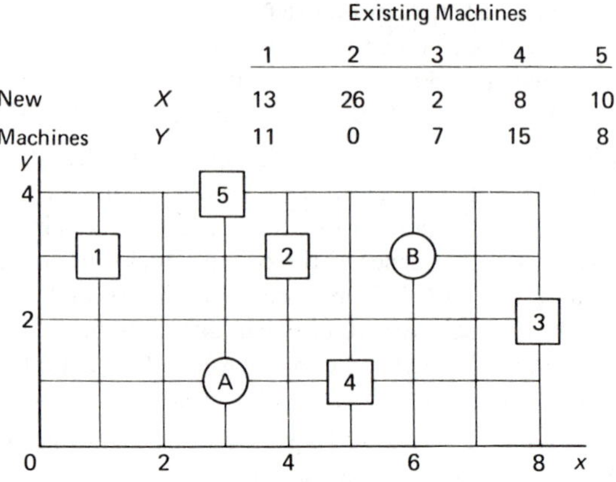

		Existing Machines				
		1	2	3	4	5
New	X	13	26	2	8	10
Machines	Y	11	0	7	15	8

Figure P6.7

Rectilinear item movement is used.

(a) Determine the optimum locations for the two new machines.

(b) What is the minimum weekly cost resulting from the solution in (a)?

6.8. A firm wishes to locate two new machines, *A* and *B*, in an existing layout. There are three possible locations to be considered, *E*, *F*, and *G*. There are three existing machines, *J*, *K*, and *L*, in the layout. Determine the locations of the new machines that will minimize item movement-distance per week. The coordinate locations of the existing machines and the possible locations for the new machines are given below, along with the amount of item movement between the existing machines and the new machines. All item movement between *J* and the new machines is straight-line movement. All other item movement is rectilinear. Space limitations prohibit locating *A* at *G*.

Location of Existing Machine

Machine	x	y
J	0	0
K	2	9
L	8	5

Possible Locations for New Machines

Location	x	y
E	0	7
F	4	3
G	6	8

Item movement per week between new machine and existing machines

New	Existing Machines		
Machines	*J*	*K*	*L*
A	4	6	6
B	6	8	5

6.9. Three new machines are to be located in an existing layout. There are four possible locations for the new machines. A material-handling relationship exists between the new machines and three existing machines as given below in pallet loads per day.

$$
\text{New Machines}\quad
\begin{array}{c}
 \\ A \\ B \\ C
\end{array}
\overset{\begin{array}{c}\text{Existing Machines}\\ J \quad K \quad L\end{array}}{\begin{pmatrix} 10 & 4 & 8 \\ 0 & 8 & 6 \\ 8 & 6 & 4 \end{pmatrix}}
$$

The distances between the existing machines and the possible locations are given in 10-foot increments as follows:

$$
\text{Possible Locations}\quad
\begin{array}{c}
 \\ E \\ F \\ G \\ H
\end{array}
\overset{\begin{array}{c}\text{Existing Machines}\\ J \quad K \quad L\end{array}}{\begin{pmatrix} 1 & 3 & 4 \\ 4 & 2 & 3 \\ 5 & 3 & 5 \\ 6 & 4 & 2 \end{pmatrix}}
$$

Space limitations prohibit locating B at H. Where should the new machines be located to minimize distance traveled per day?

6.10. Three new facilities A, B, and C are to be located in a machine shop. There are five location sites available, V, W, X, Y, and Z. Cost studies have yielded the following daily costs of assigning facilities to locations

$$
\begin{array}{c}
 \\ A \\ B \\ C
\end{array}
\overset{\begin{array}{ccccc}V & W & X & Y & Z\end{array}}{\begin{pmatrix} 14 & 6 & 12 & 12 & 11 \\ 19 & 7 & 14 & 16 & 10 \\ 13 & 10 & 19 & 16 & 9 \end{pmatrix}}
$$

Due to qualitative considerations, it is agreed that either A, B, or C must be located at V. Use the assignment algorithm to determine the optimum assignment of facilities to location sites.

6.11. Let $\mathbf{L}$ be the union of the first and fourth quadrants, and let $\mathbf{L}$ be subdivided into grid squares of unit length. Let "docks" have the locations $\mathbf{P}_1 = (0, 0)$, $\mathbf{P}_2 = (0, 2)$, and $\mathbf{P}_3 = (0, 4)$, and let d_{kj} denote the rectilinear distance between dock k and the center of grid square j. Given a value $A = 12$, let $H(\mathbf{L}: A)$ denote the collection of all sets of 12 distinct numbers of grid squares in $\mathbf{L}$. Finally, let $w_1 = 1$, $w_2 = 2$, $w_3 = 1$. Find a design $\mathbf{S}^*$ in $H(\mathbf{L}: A)$ such that

$$F(\mathbf{S}^*) \leq F(\mathbf{S}), \quad \text{for any other } \mathbf{S} \text{ in } H(\mathbf{L}: A)$$

where $F(\mathbf{S})$ is defined by

$$F(\mathbf{S}) = \sum_{k=1}^{3} w_k \left(\sum_{j \in \mathbf{S}} \frac{1}{A} d_{kj} \right)$$

6.12. A firm wishes to locate four new machines, A, B, C, and D, in an existing layout with the objective of minimizing annual cost. There are six possible locations to be considered, E, F, G, H, I, and J. There are three existing machines in the layout, L, M, and N. Items are transported with rectilinear travel using a fork truck with a variable operating cost of $0.05 per 1,000 feet traveled. The coordinate locations of the existing machines and the possible locations for the new machines are given below, along with the amount of item movement between the existing machines and the new machines. All movement is rectilinear. Space limitations prohibit locating A at E or I and locating B at G or J. There is no item movement between new machines.

Data

1. Location of existing machines (measured in 100-foot increments)

Machines	x	y
L	2	9
M	8	3
N	12	12

2. Possible locations for new machines (measured in 1,000-foot increments)

Location	x	y
E	0	7
F	4	3
G	6	8
H	10	2
I	14	8
J	5	16

3. Trips per week between new and existing machines

New	Existing L	M	N
A	60	60	80
B	80	50	60
C	20	50	40
D	80	20	60

6.13. Four warehouses are to be located. Six sites are available. The costs of assigning warehouses to sites are provided in the following matrix:

Warehouse	New Location A	B	C	D	E	F
1	8	14	10	6	12	9
2	11	17	15	10	8	7
3	13	14	8	12	15	10
4	12	11	9	13	14	15

Determine the locations to minimize the total cost.

6.14. As a facility design engineer, you have been asked to assign the locations for four new machines in a machine shop. Five possible new locations are to be considered with the associated costs given in the following matrix, C.

		New Location				
		A	*B*	*C*	*D*	*E*
Machine	1	4	8	12	10	6
	2	13	9	6	8	7
	3	16	9	17	15	10
	4	15	10	11	12	9

How would you assign the new machines to the new locations so as to minimize the total cost?

6.15. A rectangular region is partitioned into 4 by 6 grid squares, with the side having a length of 6 grid square widths being parallel to the x axis. The rectangle represents a warehouse that has two docks, one at the midpoint of each side of the rectangle having a length of 4 grid square widths. One type of item will be stored in the warehouse, and will take up 16 grid squares. All items enter at the dock on the left side of the warehouse as it appears in a drawing, and depart from the other dock. All travel between docks and storage follows a rectilinear pattern. All item movement into the warehouse is handled by warehouse personnel, but *only half* of item movement out of the warehouse is handled by warehouse personnel, while the other half is handled by customers (properly supervised). If an item is equally likely to travel between any grid square in the chosen storage region when entering, and equally likely to travel between any grid square in the chosen storage region when leaving, which grid squares should the items be stored in to minimize the cost of item movement incurred by *warehouse personnel only*?

6.16. Design a layout for a warehouse having two docks located at (0, 0) and (0, 60). Rectilinear travel is to be used within the warehouse; both docks are to be located along the same exterior wall of the warehouse; the warehouse is to have an area of 1,600 square feet; and the warehouse is to be made up of storage bays of size 20 by 20 feet. Determine the configuration for the warehouse design that minimizes the expected distance traveled per unit time when twice as many trips are made to the dock at (0, 0) as to the dock at (0, 60). The warehouse is to lie in the first and fourth quadrants.

6.17. Solve Problem 6.16 for

(a) A storage-bay size of 10 by 10 feet.

(b) A storage-bay size of 5 by 5 feet.

(c) What do you suspect the layout would be if the storage bays were infinitesimally small?

(d) Suppose that there are two existing facilities located at (0, 0) and (0, 60) and there is twice as much item movement to the facility at (0, 0) as to the facility at (0, 60). Construct the contour line lying in the first and fourth quadrants and enclosing an area of 1,600 square feet. Compare the contour line with the solution to part (c).

6.18. An alternative representation of the total cost expression (6.11) is given by

$$F(\mathbf{S}_1, \ldots, \mathbf{S}_m) = \sum_{i=1}^{m} F(\mathbf{S}_i)$$

where $$F(\mathbf{S}_i) = \sum_{j \in \mathbf{S}_i} c_{ij}, \quad i = 1, \ldots, m$$

and c_{ij} is defined by (6.15). Given a layout $\{\mathbf{S}_1, \ldots, \mathbf{S}_m\}$, suppose that p is the number of a grid square in $\mathbf{S}_q$ and r is the number of a grid square in $\mathbf{S}_t$. Let $\mathbf{S}'_q$ and $\mathbf{S}'_t$ be obtained from $\mathbf{S}_q$ and $\mathbf{S}_t$ by interchanging p and r; that is,

$$\mathbf{S}'_q = (\mathbf{S}_q - \{p\}) \cup \{r\} \tag{i}$$

and $$\mathbf{S}'_t = (\mathbf{S}_t - \{r\}) \cup \{p\} \tag{ii}$$

(a) Show that $F(\mathbf{S}_q) + F(\mathbf{S}_t) = F(\mathbf{S}'_q) + F(\mathbf{S}'_t)$ if and only if

$$c_{qp} + c_{tr} = c_{qr} + c_{tp} \tag{iii}$$

(b) Given a layout $\{\mathbf{S}_1, \ldots, \mathbf{S}_m\}$ with $\mathbf{S}'_q$ and $\mathbf{S}'_t$ defined by (i) and (ii), respectively, and with $\mathbf{S}'_i = \mathbf{S}_i$ otherwise, show that $F(\mathbf{S}_1, \ldots, \mathbf{S}_m) = F(\mathbf{S}'_1, \ldots, \mathbf{S}'_m)$ if and only if condition (iii) holds. Note that if $\{\mathbf{S}_1, \ldots, \mathbf{S}_m\}$ is a least-cost layout, then condition (iii) is a necessary and sufficient condition for $\{\mathbf{S}'_1, \ldots, \mathbf{S}'_m\}$ to also be a least-cost layout. Thus condition (iii) is useful for finding alternative least-cost layouts.

6.19. Suppose that the entire first quadrant is subdivided into grid squares, with the length of each side of every grid square being 1. One dock is located at the point (0, 10). One item is to be located in the first quadrant and will take up $A = 30$ grid squares. Given that f_j represents the rectilinear distance between grid square j and the dock, find $\mathbf{S}^*$ in $H(\mathbf{L}:A)$, where $H(\mathbf{L}:A)$ is the collection of all subsets of 30 integers representing 30 different grid squares in the first quadrant, for which $F(\mathbf{S}^*) \leq F(\mathbf{S})$ for any element $\mathbf{S}$ in $H(\mathbf{L}:A)$, and $F(\mathbf{S})$ is defined by (6.24). Also compute $F(\mathbf{S}^*)$. Based on the shape of $\mathbf{S}^*$, what would you guess $\mathbf{S}^*$ would look like for an arbitrary value of A? For this case the intuitive solution procedure involving ordering the f_j can no longer be applied, while Property 1 is still applicable.

6.20. Suppose that the entire first and second quadrants are subdivided into grid squares, with the length of each side of every grid square being 1. Everything else is given as in Problem 6.19, with the single exception that $A = 40$. Again find $\mathbf{S}^*$ and compute $F(\mathbf{S}^*)$. Based on the values of $F(\mathbf{S}^*)$ for this problem and Problem 6.19, what would you guess would be the case in general when the dock is not restricted to lie on the boundary of the warehouse?

6.21. Determine whether or not the following matrices factor:

$$\begin{pmatrix} 1 & 2 \\ 3 & 4 \end{pmatrix}, \quad \begin{pmatrix} 1 & 2 \\ 2 & 4 \end{pmatrix}, \quad \begin{pmatrix} 6 & 3 \\ 12 & 6 \end{pmatrix}$$

6.22. Develop a proof of Property 4 (a) for the case $m = 3$, and (b) for the general case.

6.23. An existing warehouse is 200 feet long and 100 feet wide with three docks located as shown in Figure P6.23. Two products are to be stored in the ware-

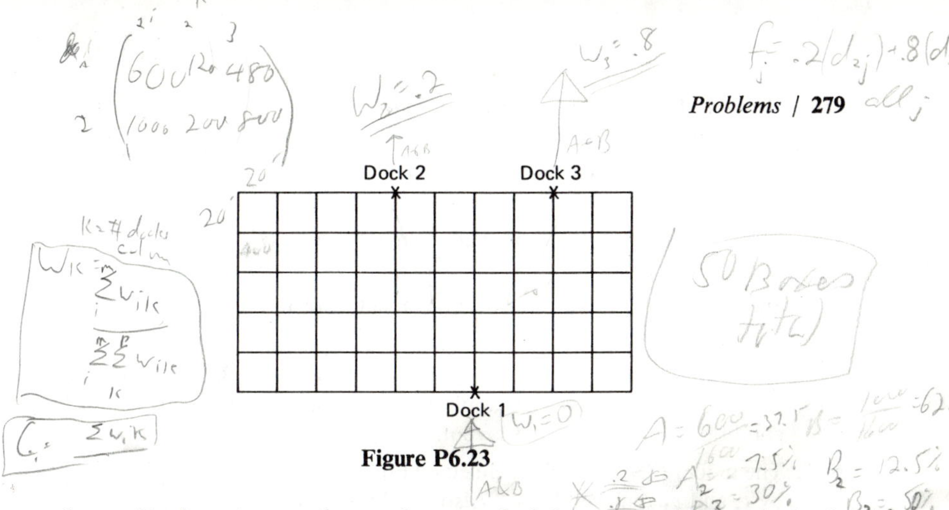

Figure P6.23

house. Product A enters the warehouse at dock 1 at a rate of 600 pallet loads per month, and is shipped from docks 2 and 3 at rates of 120 and 480 pallet loads per month, respectively. Product B enters the warehouse at dock 1 at a rate of 1,000 pallet loads per month and is shipped from docks 2 and 3 at rates of 200 and 800 pallet loads per month, respectively. Storage spaces of 8,000 and 12,000 square feet are required for A and B, respectively. The warehouse is arranged into bays of dimension 20 by 20 feet, and only one type of product can be stored in a given bay.

(a) Develop the warehouse layout that minimizes expected distance traveled per month and, for the design developed, determine the expected distance traveled per month. Assume rectilinear travel.

(b) Suppose that all shipment of product A from the warehouse is from dock 2; how would you solve the layout design problem? Do not solve, but specify very explicitly the approach you would take.

6.24. An existing warehouse of dimensions 120 by 160 feet is to be used for the storage of two products, A and B. The warehouse consists of storage bays of size 20 by 20 feet. The storage area required for product A equals 10,000 square feet and for product B equals 6,000 square feet. Item movement between storage and one of the three docks equals 2,000 loads per month for product A and 2,500 loads per month for product B. Rectilinear travel is used. The three docks are equally spaced along one of the 160-foot walls. Design the layout such that the average distance traveled per month is minimized, assuming (a) any unit of product is equally likely to travel to and from either dock, (b) units are twice as likely to travel to and from the centrally located dock as to either of the remaining two docks.

6.25. Given a layout $\{\mathbf{S}_1, \ldots, \mathbf{S}_m\}$, and $F(\mathbf{S}_i)$ for $i = 1, \ldots, m$ as defined by the expression immediately preceding (6.52), show that (6.52) holds.

6.26. Suppose that the factoring assumption is given; let $\{\mathbf{S}_1^*, \ldots, \mathbf{S}_m^*\}$ be a least-cost layout in $H_m(\mathbf{L} : A)$, and suppose that

$$\frac{c_1}{A_1} > \frac{c_2}{A_2} > \cdots > \frac{c_m}{A_m} > 0$$

Prove that $\{\mathbf{S}_1^*, \ldots, \mathbf{S}_m^*\}$ has the following property: there exist numbers

$k_1 \leq k_2 \leq \cdots \leq k_m$ such that, for $p = 1, \ldots, m$,

$$f_j \leq k_p, \quad \text{for every element } j \text{ in } \bigcup_{i=1}^{p} \mathbf{S}_i^*$$

and
$$f_j \geq k_p, \quad \text{for every element } j \text{ not in } \bigcup_{i=1}^{p} \mathbf{S}_i^*$$

Note that if $\{\mathbf{S}_1^*, \ldots, \mathbf{S}_m^*\}$ is a layout in $H_m(\mathbf{L}:A)$ satisfying these conditions Property 4 also applies, so that one has necessary and sufficient conditions for a least-cost layout.

6.27. The model $F(\mathbf{S}_1, \ldots, \mathbf{S}_m)$ defined by Equation (6.34) is given, with u_i replaced by c_i and v_k replaced by w_k. Also it is known that

$$\frac{c_1}{A_1} = \frac{c_2}{A_2} = \cdots = \frac{c_m}{A_m} = r \tag{i}$$

When (i) holds, it follows that if $\bar{c} = \sum_{j=1}^{m} c_i$ and $B_m = \sum_{i=1}^{m} A_i$ then

$$r = \frac{\bar{c}}{B_m} \tag{ii}$$

Given that (i) holds, if $\{\mathbf{S}_1, \ldots, \mathbf{S}_m\}$ is any layout in $H_m(\mathbf{L}:A)$, and $T_m = \bigcup_{i=1}^{m} \mathbf{S}_i$, show that

$$F(\mathbf{S}_1, \ldots, \mathbf{S}_m) = \bar{c} \sum_{j \in T_m} \frac{1}{B_m} f_j \tag{iii}$$

If $\{\mathbf{S}_1', \ldots, \mathbf{S}_m'\}$ is another layout in $H_m(\mathbf{L}:A)$ for which $T_m = \bigcup_{i=1}^{m} \mathbf{S}_1'$, what can one conclude about $F(\mathbf{S}_1, \ldots, \mathbf{S}_m)$ relative to $F(\mathbf{S}_1', \ldots, \mathbf{S}_m')$, and what physical meaning does the conclusion have?

6.28. The development of discrete layouts in this chapter assumed some portion of item i was to be transported to either dock in the case of a multidock design. Suppose that item movement is always from a grid square to the nearest dock. Let d_j be the distance from grid square j to the nearest dock. Develop an expression for the average distance an item travels if assigned to grid square j. Develop an expression comparable to (6.11) for the nearest dock assumption. Show that Property 3 holds and that the factoring assumption always holds under the nearest dock assumption. For the case of travel to the nearest dock, obtain the warehouse designs comparable to those given in Figures 6.2 and 6.3.

6.29. Consider the warehouse shown in Figure P6.29. Storage bays are of size $20' \times 20'$. Docks $\mathbf{P}_1$ and $\mathbf{P}_2$ are for truck delivery; docks $\mathbf{P}_3$ and $\mathbf{P}_4$ are for rail delivery. Randomized storage and retrieval policies are used. Sixty percent of all item movement in and out of storage is to either $\mathbf{P}_1$ or $\mathbf{P}_2$, with each dock equally likely to be used. Forty percent of all item movement in and out of storage is equally divided between docks $\mathbf{P}_3$ and $\mathbf{P}_4$. Three products, A, B, and

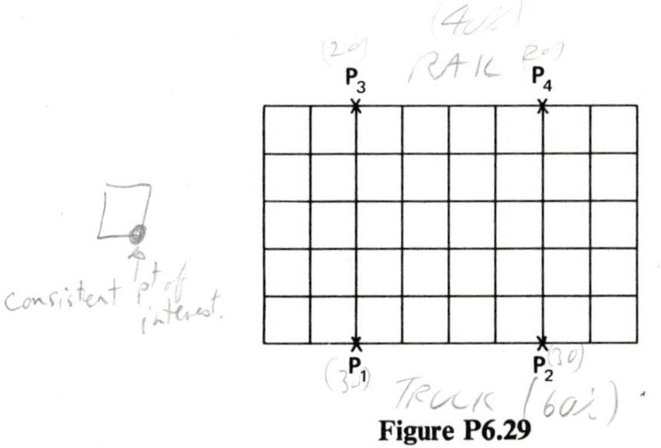

Figure P6.29

C, are to be stored in the warehouse with only one-type product stored in a given storage bay. Product A requires 4,000 square feet of storage space and enters and leaves storage at a rate of 800 loads per month; product B requires 5,000 square feet of storage space and leaves storage at a rate of 900 loads per month; product C requires 3,500 square feet of storage space and enters and leaves storage at a rate of 750 loads per month. Assuming rectilinear travel, design the warehouse layout which minimizes expected distance traveled.

CONTINUOUS FACILITY DESIGN AND LAYOUT PROBLEMS

7.1 Introduction

In Chapter 6 we considered a number of multifacility location problems involving a finite solution space. In the case of the warehouse-layout design problems the facilities were items and the location sites were grid squares or storage bays. Instead of point locations we were concerned with area locations. In this chapter continuous formulations are developed for the warehouse-layout problems considered in Chapter 6. Thus the results in this chapter can be thought of as continuous analogs of those developed in Chapter 6.

As an overview of the chapter, we shall find that if a warehouse is to be designed for the storage of a single item then the warehouse layout that minimizes expected distance traveled per unit time is obtained by constructing a contour line which encloses the appropriate area. If multiple items are to be stored and the factoring assumption holds, the items should be placed in the warehouse in the same order as given in Chapter 6. The storage area for each item is determined by the shape of a contour line.

In addition to developing the preceding ideas formally, the major con-

7

tributions of this chapter are the development of a single-integral expression for computing the expected distance traveled per unit time within the warehouse and the consideration of rectangular warehouse layouts. Many of the ideas presented are quite simple; however, the mathematics required to develop formally these ideas, at times, become somewhat involved. We have found it instructive to remember that integrals are used in this chapter as replacements for the summations in the previous chapters.

As an illustration of the warehouse layout designs suggested by the contour-line approach, consider Figure 7.1. Layout designs (a) and (b) suggest that the single dock should be located in the middle of the warehouse to minimize expected distance traveled per unit time. If travel is rectilinear, the diamond-shaped layout is obtained; if travel is Euclidean, a circular design is obtained. If the dock must be located along the periphery of the warehouse, designs (c) and (d) are obtained for the cases of rectilinear and Euclidean travel, respectively. If two docks are located as shown and travel is rectilinear, the layout design obtained is given by design (e). Finally, if

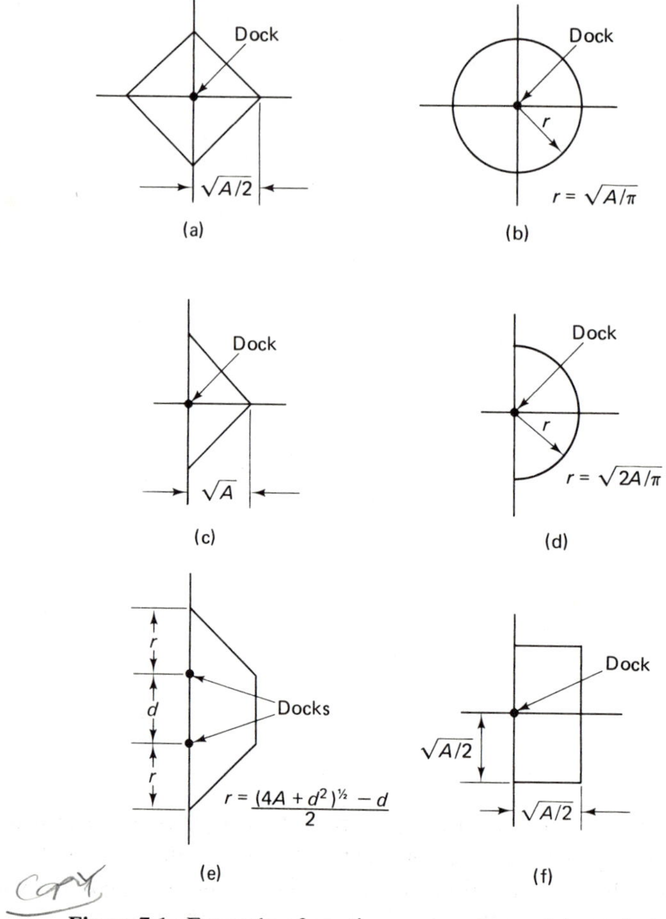

Figure 7.1. Example of continuous warehouse layout designs.

the design must be rectangular with the dock located along the periphery of the warehouse, the design that minimizes expected distance traveled per unit time is given by design (f) for the case of rectilinear travel.

The designs suggested by Figure 7.1 reinforce our recommendation in each of the preceding chapters that the results obtained from the model be interpreted as *benchmark* or *ideal* solutions. We certainly do not suggest that you run out and recommend triangular-shaped warehouses to your boss, unless, of course, you are not particularly happy with your present employment. Rather, what we suggest is that you compare alternative designs against the "least-cost" design obtained from this chapter. The majority of the models we present are based solely on the objective of minimizing some linear function of the expected distance traveled per unit time. Consequently, some of the designs we obtain might be very expensive to construct, since construction costs have not been included explicitly in the analysis. One final observation should be made concerning the proper application of the warehouse design results; the warehouse layout obtained represents the configuration required for the storage of items. By adding additional space for equipment storage, offices, and locker rooms, it is possible to obtain an overall configuration that can be constructed economically and still have the least-cost storage design.

7.2 One Set Designs

7.2.1 *Single-dock case*

The continuous formulation that we employ in this chapter is motivated by the special structure of the problems considered in Sections 6.4 and 6.6. Their structure suggests that it may be possible to find closed-form solutions for these problems. If such is the case, then instead of finding a least-cost solution by the use of an algorithm, the least-cost solution can be stated explicitly. In cases where a closed-form solution cannot easily be found, we shall see it will still be possible to state conditions which, if satisfied, guarantee that the solution will indeed be a least-cost solution.

In order to motivate the development, consider the following situation. Let **L** be the collection of all grid squares of unit dimensions in the union of the first and fourth quadrants, and let f_j represent the rectilinear distance between the center of grid square j and the origin.

The first problem of interest is to find 156 grid squares in **L**, represented by **S**′, so that

$$F(\mathbf{S}') = \sum_{j \in \mathbf{S}'} \tfrac{1}{156} f_j \leq \sum_{j \in \mathbf{S}} \tfrac{1}{156} f_j = F(\mathbf{S})$$

where **S** consists of any other 156 grid squares in **L**. (We have chosen a problem involving 156 grid squares because this number of grid squares makes the relationship between the discrete and continuous problems particularly apparent.) Property 1 of Chapter 6 can be used to find **S**′, as shown in Figure 7.2. A computation establishes that $F(\mathbf{S}') = \frac{1{,}300}{156} = 8.333$. On observing that **S**′ has an area of 156 units and a nearly triangular shape, one might guess that the second problem of interest, finding a closed-form solution, would have a solution such as is shown in Figure 7.3, a triangle **S*** with an area of $A = 156$. Since **S*** is a region in the plane instead of a

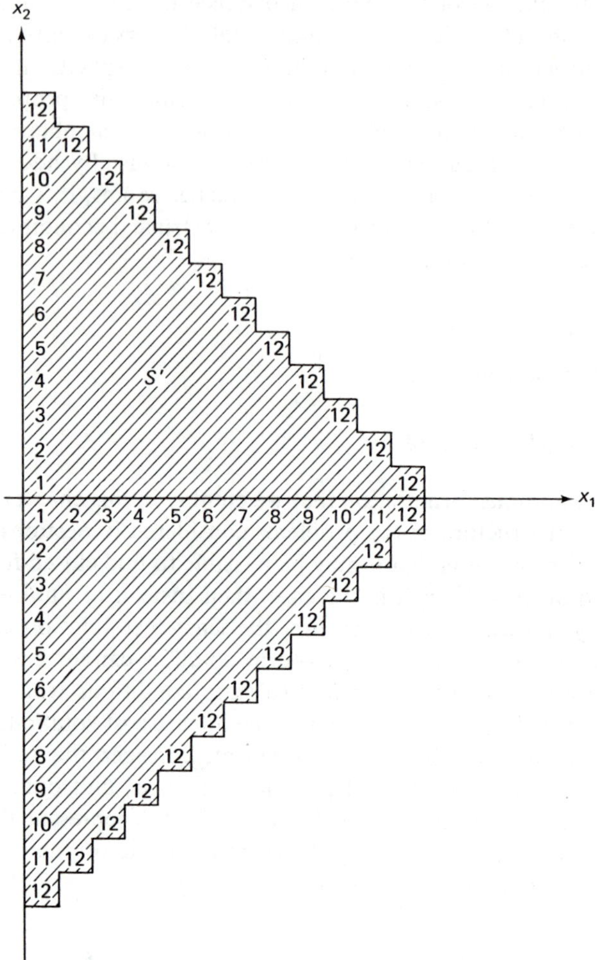

Figure 7.2. Discrete layout design for a single item and a single dock.

collection of grid squares, the average distance between **S*** and the origin is computed using integrals, and is as follows:

$$\iint_{S^*} \frac{1}{156}[|x_1| + |x_2|]\, dx_1 dx_2 = 8.327 \tag{7.1}$$

Notice the close agreement between the values of average distance found in the two different problems. As will be seen subsequently, Figure 7.3 does in fact give a closed-form solution. What remains to be done is to formulate

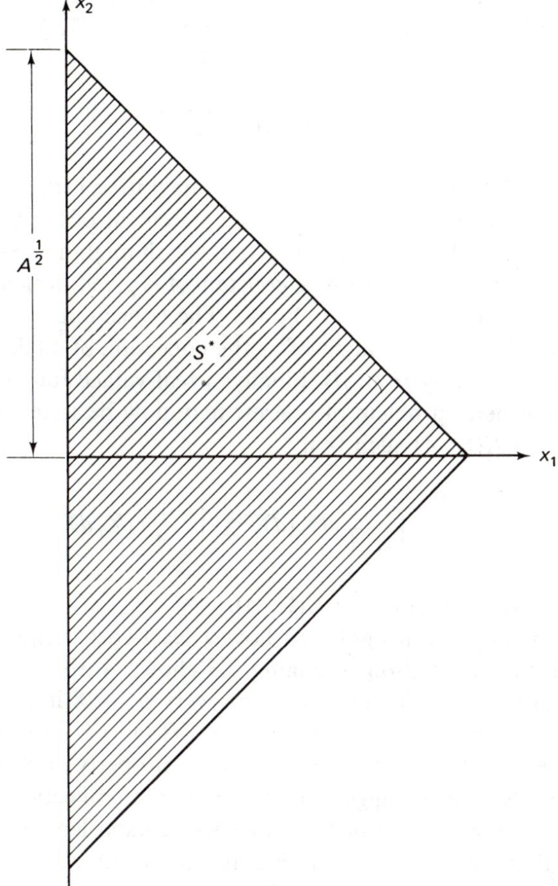

Figure 7.3. Continuous layout design for a single item and a single dock with rectilinear travel.

more precisely the problem for which Figure 7.3 gives a closed-form solution, and then develop the formulation of a more general class of similar problems. A physical interpretation of the solution shown in Figure 7.3 might be as follows: of all warehouse designs **S** of area A in **L** that have a single dock located at the origin, the warehouse design **S*** minimizes the average rectilinear distance that products travel between storage and the dock.

It is useful at this point to introduce some vector notation. If the vector, or two-tuple, is defined by $\mathbf{X} = (x_1, x_2)$, and the function $f(\mathbf{X})$ is defined by

$$f(\mathbf{X}) = |x_1| + |x_2| \tag{7.2}$$

then on taking $A = 156$ an alternative, more convenient single-integral representation of the double integral on the left side of (7.1) is

$$\int_{S^*} \frac{1}{A} f(\mathbf{X})\, d\mathbf{X} \tag{7.3}$$

where it is understood that the region of integration is the set $\mathbf{S}^*$ in the plane.

The vector notation now permits a statement of the *general* problem of interest. A region $\mathbf{L}$ in the plane is given, which might represent, for example, a plot of ground. Also, a positive constant A is given, which might represent, for example, the area of a warehouse. The collection of all subsets of $\mathbf{L}$ of area A, which may be thought of as the collection of all designs in $\mathbf{L}$ of area A, will be denoted by $H(\mathbf{L}: A)$. Finally, a real-valued function $f(\mathbf{X})$ is defined on $\mathbf{L}$. The general problem of interest may now be stated as follows: find a design $\mathbf{S}^*$ in $H(\mathbf{L}: A)$ such that

$$\int_{S^*} \frac{1}{A} f(\mathbf{X})\, d\mathbf{X} \leq \int_{S} \frac{1}{A} f(\mathbf{X})\, d\mathbf{X} \tag{7.4}$$

where $\mathbf{S}$ is any other design in $H(\mathbf{L}: A)$.

To obtain some insight into how a least-cost design might be found, note that an examination of Figure 7.3 shows, with $f(\mathbf{X})$ defined by (7.2), that $\mathbf{S}^*$ has the following property: for every point $\mathbf{X}$ in $\mathbf{S}^*, f(\mathbf{X}) \leq k$, where $k = (156)^{1/2}$; for every point $\mathbf{X}$ not in $\mathbf{S}^*, f(\mathbf{X}) \geq k$ [in fact, for every point $\mathbf{X}$ not in $\mathbf{S}^*, f(\mathbf{X}) > k$]. Thus $\mathbf{S}^*$ satisfies a property very similar to the one satisfied by $\mathbf{S}'$, which suggests that a proper modification of Property 1 of Chapter 6, a special case of the Neyman–Pearson lemma, could be used to establish that the set $\mathbf{S}^*$ shown in Figure 7.3 does in fact minimize the average rectilinear distance as defined by (7.1). Indeed, as we shall now see, a proper modification of Property 1 of Chapter 6 provides sufficient conditions for a solution to the general problem of interest. In going through the proof, you may find it useful to refer to Figure 7.4, which illustrates the concepts of the proof by the use of the design $\mathbf{S}^*$ shown in both Figures 7.3 and 7.4, and the use of an alternative design $\mathbf{S}$ shown in Figure 7.4. [Notice the analogy between Property 1 of Chapter 6 and Property 1 presented here.]

Property 1: Let $\mathbf{L}$ be a region in the plane, and let $f(\mathbf{X})$ be a real-valued function defined on $\mathbf{L}$. Let A be a positive constant, and denote by $H(\mathbf{L}: A)$ the collection of all sets of area A contained in $\mathbf{L}$. Suppose that there exists a number k for which $\mathbf{S}^*$ is a design in $H(\mathbf{L}: A)$, where $\mathbf{S}^*$ has the following property: $f(\mathbf{X}) \leq k$ for every point $\mathbf{X}$ in $\mathbf{S}^*$, and $f(\mathbf{X}) \geq k$ for every point

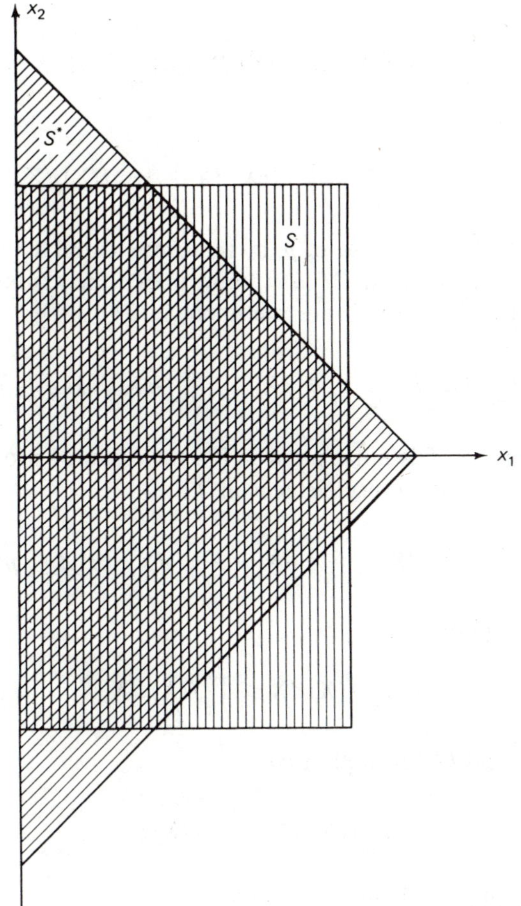

Figure 7.4. Illustration of two layout designs of area A contained in L.

X not in **S***; then

$$\int_{\mathbf{S}^*} \frac{1}{A} f(\mathbf{X})\, d\mathbf{X} \leq \int_{\mathbf{S}} \frac{1}{A} f(\mathbf{X})\, d\mathbf{X} \tag{7.5}$$

where **S** is any other design in $H(\mathbf{L}: A)$.

Proof: Let **S** be any design in $H(\mathbf{L}: A)$. Denote by **SS*** all those points common to both **S** and **S***, denote by **T*** all those points in **S*** which are not in **SS***, and denote by **T** all those points in **S** which are not in **SS***. Therefore,

$$\int_{\mathbf{S}} f(\mathbf{X})\,d\mathbf{X} = \int_{\mathbf{T}} f(\mathbf{X})\,d\mathbf{X} + \int_{\mathbf{SS}^*} f(\mathbf{X})\,d\mathbf{X} \tag{7.6}$$

and

$$\int_{\mathbf{S}^*} f(\mathbf{X})\,d\mathbf{X} = \int_{\mathbf{T}^*} f(\mathbf{X})\,d\mathbf{X} + \int_{\mathbf{SS}^*} f(\mathbf{X})\,d\mathbf{X} \tag{7.7}$$

Suppose that **SS*** has an area of p, so that both **T** and **T*** have an area of $A - p$. Since every point in **T*** is a point in **S***, the assumed property of **S*** implies that $f(\mathbf{X}) \leq k$ for every element in **T***; so

$$\int_{\mathbf{T}^*} f(\mathbf{X})\,d\mathbf{X} \leq \int_{\mathbf{T}^*} k\,d\mathbf{X} = k\int_{\mathbf{T}^*} d\mathbf{X} = k(A - p) \tag{7.8}$$

Since every point in **T** is not a point in **S***, the assumed property of **S*** implies that $f(\mathbf{X}) \geq k$ for every element in **T**; so

$$\int_{\mathbf{T}} f(\mathbf{X})\,d\mathbf{X} \geq \int_{\mathbf{T}} k\,d\mathbf{X} = k\int_{\mathbf{T}} d\mathbf{X} = k(A - p) \tag{7.9}$$

From (7.8) and (7.9),

$$\int_{\mathbf{T}^*} f(\mathbf{X})\,d\mathbf{X} \leq \int_{\mathbf{T}} f(\mathbf{X})\,d\mathbf{X} \tag{7.10}$$

so (7.6), (7.7), and (7.10) imply that

$$\int_{\mathbf{S}^*} f(\mathbf{X})\,d\mathbf{X} \leq \int_{\mathbf{S}} f(\mathbf{X})\,d\mathbf{X} \tag{7.11}$$

Dividing inequality (7.11) by A gives (7.5).

A design **S*** satisfying inequality (7.5) will be called a *least-cost design.* Property 1 states sufficient conditions for a design to be a least-cost design. To find a least-cost design satisfying the conditions of Property 1, the geometrical approach used may be stated roughly as follows. Construct a contour line of the function $f(\mathbf{X})$; for example, with reference to Figure 7.3, one contour line of the function is just the boundary of the set **S***, and is the collection of all points in **L** for which $f(\mathbf{X}) = k$, where, for this example, $k = (156)^{1/2}$. Next compute the area of the region "inside" the contour line. Again with reference to Figure 7.3, the area of the region inside the contour line is just k^2; since the area of **S*** must be $A = 156$, it follows that $k = (156)^{1/2}$. One final point may be made concerning the design shown in Figure 7.3: it illustrates the solution for any arbitrary positive value of A, and not just the special case $A = 156$. **S*** is the collection of all points

X in **L** for which $f(\mathbf{X}) \leq (A)^{1/2}$, where $f(\mathbf{X})$ is defined by (7.2). Using set notation, **S*** may be represented as

$$\mathbf{S}^* = \{\mathbf{X} \in \mathbf{L}: f(\mathbf{X}) \leq k\}$$

where, for this example, $k = (A)^{1/2}$. Thus **S*** satisfies the conditions stated in Property 1.

The warehouse design given in Figure 7.3 is the design for a new warehouse. In a number of practical situations a layout design is to be developed for an existing warehouse. In the case of an existing warehouse, the cost of constructing the facility has already occurred. Therefore, minimizing material-handling cost appears to be a reasonable objective. As an illustration of the approach suggested by Property 1 in designing the layout, we let the existing facility be denoted by **L**. Consequently, we wish to obtain a design **S*** of area A contained in **L**. Consider a warehouse of dimensions 200 by 150 feet having a single dock located as shown in Figure 7.5. Only one type of item is to be stored in the warehouse, and it requires 18,000 square feet of storage space. Assuming item movement is equally likely to occur between the dock and any point in the storage region **S**, and assuming travel is rectilinear, we construct the contour line having an area of 18,000 square feet, as shown in Figure 7.5.

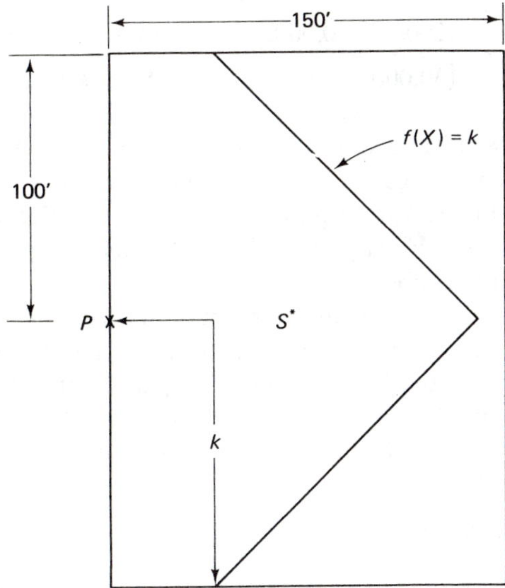

Figure 7.5. Layout design within an existing warehouse.

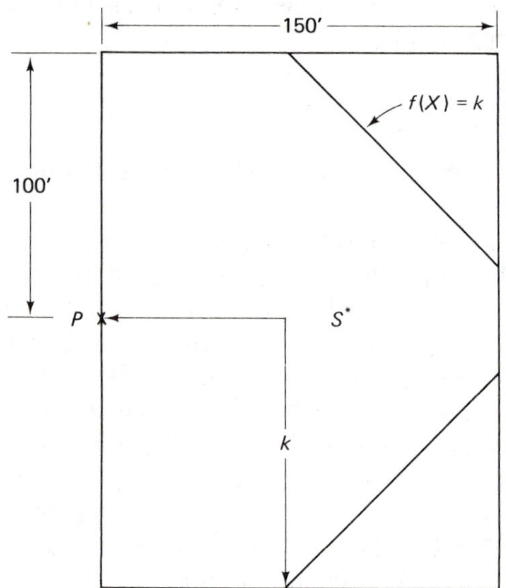

Figure 7.6. Layout design within an existing warehouse.

The area of **S*** in Figures 7.5 and 7.6 can be expressed as a function of k, the value of the contour line:

$$A = \begin{cases} 200k - 10{,}000, & 100 \le k \le 150 \\ 30{,}000 - (250 - k)^2, & 150 \le k \le 250 \end{cases}$$

Letting A equal 18,000 and solving for k gives a value of 140 feet and results in the design given in Figure 7.5. Notice that if $20{,}000 \le A \le 30{,}000$ then **S*** is as given in Figure 7.6. In particular, if A equals 27,500, then k equals 200 feet in Figure 7.6. Finally, if $A \le 10{,}000$, then A equals k^2, and a triangular storage area is obtained.

The space within the existing warehouse not used for item storage can be used for offices, equipment storage, locker rooms, future expansion of the storage area, and the like. We again emphasize that the storage-space configuration given by **S*** is a benchmark solution and should be used as a design aid in obtaining the final warehouse layout.

7.2.2 *Multidock case*

Since vector notation is now being used, it is also useful to have a vector notation to represent the distance between any two points X and Y in the plane. Given any two points $\mathbf{X} = (x_1, x_2)$ and $\mathbf{Y} = (y_1, y_2)$ in the plane, and any number p that is at least 1, the l_p distance between **X** and **Y** will be defined as

$$|\mathbf{X}-\mathbf{Y}|=[|x_1-y_1|^p+|x_2-y_2|^p]^{1/p} \tag{7.12}$$

To avoid any notational confusion, it should be pointed out that the vertical bars on the right side of (7.12) represent absolute values, while the vertical bars on the left side of (7.12) are simply part of the definition. Note that when $p = 1$ the l_p distance is the rectilinear distance, whereas when $p = 2$ the l_p distance is the Euclidean distance. Thus the l_p distance provides a convenient single means of representing several distances. In the sequel, the value of p of interest in a specific instance either will be stated explicity or will be clear from the context.

For all values of p greater than or equal to 1, it can be shown that the l_p distance has the following properties: $|\mathbf{X} - \mathbf{Y}| \geq 0$, for all **X** and **Y** in the plane; the equation $|\mathbf{X} - \mathbf{Y}| = 0$ is equivalent to $\mathbf{X} = \mathbf{Y}$; for any number b, $|b(\mathbf{X} - \mathbf{Y})| = \text{abs}(b)|\mathbf{X} - \mathbf{Y}|$, where abs($b$) denotes the absolute value of b, and **X** and **Y** are any two points in the plane; for any points **X**, **Y**, and **Z** in the plane, $|\mathbf{X}| - |\mathbf{Y}| \leq |\mathbf{X} - \mathbf{Z}| + |\mathbf{Z} - \mathbf{Y}|$. The latter inequality is referred to as the triangle inequality, since it includes as a special case, when $p = 2$, the fact that the length of the hypotenuse of a right triangle is equal to or less than the sum of the lengths of the other two sides of the triangle.

It will be convenient to develop the subsequent discussion in this section in a warehouse design context. Let **L** be a region in the plane in which the warehouse will be located. Let A be the area of the warehouse, and let $H(\mathbf{L}: A)$ denote the class of all warehouse designs in **L** of area A. Suppose that warehouse docks are located at known points $\mathbf{P}_1, \ldots, \mathbf{P}_n$ in the plane; then

$$\int_{\mathbf{S}} \frac{1}{A}|\mathbf{X}-\mathbf{P}_j|\,d\mathbf{X}, \quad j=1,\ldots,n$$

represents the average distance an item in the warehouse moves between storage and dock j. Also, let the total cost of movement between storage and dock j be directly proportional to the average distance. Let w_j be the product of cost per unit average distance and the number of trips made between storage and dock j per time period. Therefore, if we assume each point in **S** is equally likely to be chosen for movement to and from the dock, the average total cost of moving the item between storage and the docks will be given by

$$\sum_{j=1}^{n} w_j \int_{\mathbf{S}} \frac{1}{A}|\mathbf{X}-\mathbf{P}_j|\,d\mathbf{X} \tag{7.13}$$

which, on defining the function $f(\mathbf{X})$ by

$$f(\mathbf{X}) = \sum_{j=1}^{n} w_j|\mathbf{X} - P_j| \tag{7.14}$$

and rearranging the order of summation and integration, may be rewritten as

$$\int_S \frac{1}{A} f(\mathbf{X})\, d\mathbf{X} \tag{7.15}$$

Since cost expressions (7.13) and (7.15) are the same when $f(\mathbf{X})$ is defined by (7.14), and since Property 1 applies to cost expression (7.15), it also applies to expression (7.13).

Consider now an example of the application of Property 1 to find a design to minimize the total cost expression (7.13). Let $\mathbf{L}$ be the union of the first and fourth quadrants, suppose that there are two docks with locations $\mathbf{P}_1 = (0, 0)$, $\mathbf{P}_2 = (0, d)$, where $d \geq 0$, and that $w_1 = w_2 = w$. Then the total cost of item movement for a design S is obtained by multiplying each of the two integrals in the following expression by w:

$$\int_S \frac{1}{A}|\mathbf{X} - \mathbf{P}_1|\, d\mathbf{X} + \int_S \frac{1}{A}|\mathbf{X} - \mathbf{P}_2|\, d\mathbf{X} \tag{7.16}$$

Since w is just a positive constant, finding a design to minimize expression (7.16) is equivalent to finding a minimum total cost design. For this example it will be assumed that the distances in (7.16) are rectilinear, so that $p = 1$. To find a least-cost design, the geometric approach mentioned previously will be followed. A contour line of the function $f(\mathbf{X}) = |\mathbf{X} - \mathbf{P}_1| + |\mathbf{X} - \mathbf{P}_2|$ is plotted, and is shown in Figure 7.7. Note that the procedure for plotting contour lines developed in Chapter 4 can be used when the distances are rectilinear, since the function for which a contour line is being plotted is a special case of the function defined by (7.14), and the latter function was studied in Chapter 4. With reference to Figure 7.7, if the contour line represents the collection of all points $\mathbf{X}$ in $\mathbf{L}$ for which $f(\mathbf{X}) = k$, then the cross-hatched region $\mathbf{S}^*$ represents the collection of all points $\mathbf{X}$ in $\mathbf{L}$ for which $f(\mathbf{X}) \leq k$. To find the proper value for k, note that, again with reference to Figure 7.7, $k = d + 2c$, so that

$$c = \tfrac{1}{2}(k - d) \tag{7.17}$$

Furthermore, if $\mathbf{S}^*$ is to have an area of A, then

$$A = c^2 + cd \tag{7.18}$$

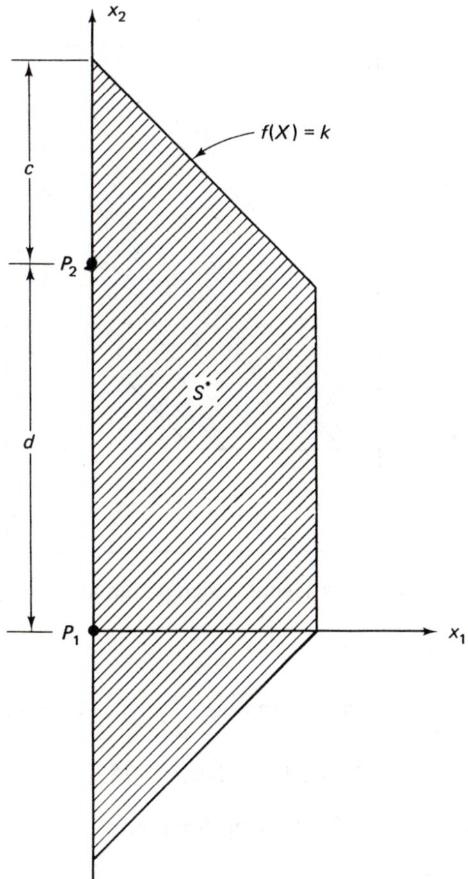

Figure 7.7. Multidock layout design.

So substituting (7.17) into (7.18) and solving for k gives $k = (4A + d^2)^{1/2}$. Thus, if the design $\mathbf{S}^*$ is defined by

$$\mathbf{S}^* = \{\mathbf{X} \in \mathbf{L}: |\mathbf{X} - \mathbf{P}_1| + |\mathbf{X} - \mathbf{P}_2| \leq (4A + d^2)^{1/2}\}$$

then $\mathbf{S}^*$ has an area of A and satisfies the conditions of Property 1, so that $\mathbf{S}^*$ is a least-cost design.

Recalling our earlier discussion of layout design within an existing warehouse, suppose in our earlier example that the existing warehouse has two docks, one for truck delivery and one for rail delivery, as shown in Figure 7.8. Letting $w_1 = w_2$ gives the contour line shown in Figure 7.8.

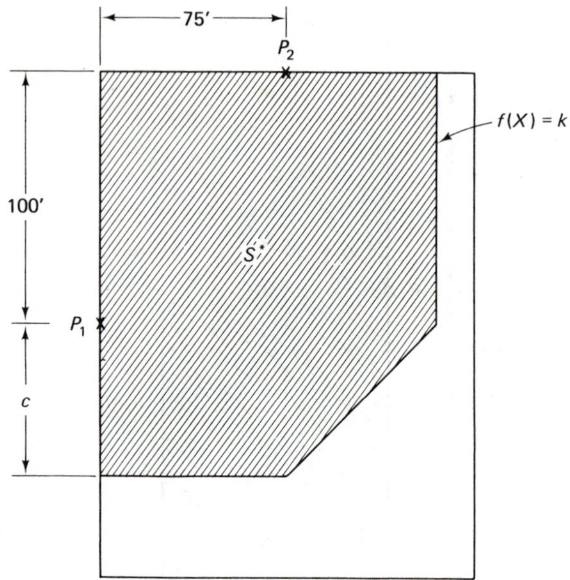

Figure 7.8. Two-dock layout design within an existing warehouse.

The area of **S*** is given by

$$A = \begin{cases} 7{,}500 + 175c + \dfrac{c^2}{2}, & 0 \leq c \leq 75 \\ 12{,}187.50 + 150c, & 75 \leq c \leq 100 \end{cases}$$

for $7{,}500 \leq A \leq 27{,}187.50$. Solving for c gives

$$c = \begin{cases} -175 + (15{,}625 + 2A)^{1/2}, & 7{,}500 \leq A \leq 23{,}437.50 \\ \dfrac{A - 12{,}187.50}{150}, & 23{,}437.50 \leq A \leq 27{,}187.50 \end{cases}$$

Therefore, if A equals 18,000, then c equals approximately 52.5 feet and **S*** is as depicted in Figure 7.8.

The storage areas shown in Figures 7.3, 7.5, 7.6, 7.7, and 7.8 are such that they can be denoted using set thoery by

$$\mathbf{S}^* = \{\mathbf{X} \in \mathbf{L}: f(\mathbf{X}) \leq k\}$$

where k is a constant chosen so that **S*** has an area of A, and $f(\mathbf{X})$ is a continuous function. In general, when least-cost designs can be denoted in this manner, it is possible, under weak assumptions, to develop a single-integral expression for the minimum total cost. The single-integral expression is useful, as it may be awkward to compute the minimum total cost directly using the double-integral expression; furthermore, most of the information needed for the single-integral expression is a by-product of information required to find the proper value of k.

7.2.3 *Single-integral expression for* $F(S^*)$

To develop single-integral expressions in general for minimum total cost designs, it is useful to state assumptions and establish some notation. Let $\mathbf{L}$ be a connected region in the plane, and let $f(\mathbf{X})$ be a continuous function defined for every point $\mathbf{X}$ in $\mathbf{L}$. It will be assumed that $f(\mathbf{X})$ has a minimum value on $\mathbf{L}$, denoted by f^*. For every number $z \geq f^*$ assume that the area of the set $\{\mathbf{X} \in \mathbf{L}: f(\mathbf{X}) \leq z\}$ exists, and denote the area of the set by $q(z)$. For the situation for which Figure 7.2 shows a least-cost design, f^* is just zero, and $q(z) = z^2$. For the situation for which Figure 7.7 shows a least-cost design, f^* is just d, and, on replacing k by z in (7.17), substituting into the right side of (7.18), and simplifying, it follows that $q(z) = \frac{1}{4}(z^2 - d^2)$. Notice that in both situations $q(z)$ is a strictly increasing function, and that its first derivative, denoted by $q'(z)$, also exists. For the general case $q(z)$ will always be a nondecreasing function; we shall assume that it is strictly increasing and that its first derivative exists. Since, by the assumption, $q(z)$ is strictly increasing, it will always have an inverse function, denoted by $r(t)$; recall that the inverse function has the property that $r(q(z)) = z$, and $q(r(t)) = t$. With reference to the two situations for which Figures 7.2 and 7.4 are applicable, in the first case $t = q(r(t)) = r(t)^2$, so that $r(t) = (t)^{1/2}$, while, for the second case, $t = q(r(t)) = \frac{1}{4}[r(t)^2 - d^2]$, so that $r(t) = (4t + d^2)^{1/2}$.

Having established some notation and assumptions, we now derive the single-integral expression in general for a least-cost design.

Property 2: Suppose there exists a number k such that S^* is a design in $H(\mathbf{L}: A)$, where $\mathbf{S}^* = \{\mathbf{X} \in \mathbf{L}: f(\mathbf{X}) \leq k\}$. Then $k = r(A)$, and

$$\int_{\mathbf{S}^*} \frac{1}{A} f(\mathbf{X})\, d\mathbf{X} = \frac{1}{A} \int_{f^*}^{r(A)} q'(z) z\, dz \tag{7.19}$$

*Proof:** Let $\bar{\mathbf{Y}}$ be a random variable with a density function $1/A$ on $\mathbf{S}^*$, and a density function of zero elsewhere. If $\bar{\mathbf{Z}}$ is defined by $\bar{\mathbf{Z}} = f(\bar{\mathbf{Y}})$, then $\bar{\mathbf{Z}}$ is also a random variable, and takes on values between f^* and k. If $E\{\bar{\mathbf{Z}}\}$

*This proof requires an understanding of probability theory. A reader who is unfamiliar with probability theory should not expect to understand this proof.

denotes the expected value of $\bar{\mathbf{Z}}$, the definition $\bar{\mathbf{Z}}$ implies that

$$E\{\bar{\mathbf{Z}}\} = \int_{\mathbf{S}^*} \frac{1}{A} f(\mathbf{X})\, d\mathbf{X}$$

For all z between f^* and k, denote the distribution function of $\bar{\mathbf{Z}}$ by $F(z)$, so that

$$F(z) = Pr\{\bar{\mathbf{Z}} \leq z\} = Pr\{f(\bar{\mathbf{Y}}) \leq z\}$$

Thus, if $\mathbf{S}(z) = \{\mathbf{X} \in \mathbf{L}: f(\mathbf{X}) \leq z\}$, then

$$F(z) = \int_{\mathbf{S}(z)} \frac{1}{A}\, d\mathbf{X} = \frac{1}{A} \int_{\mathbf{S}(z)} d\mathbf{X} = \frac{1}{A} q(z)$$

It now follows that

$$E\{\bar{Z}\} = \int_{f^*}^{k} F'(z) z\, dz = \frac{1}{A} \int_{f^*}^{k} q'(z) z\, dz \tag{7.20}$$

From the definition of $\mathbf{S}^*$, the area of $\mathbf{S}^*$ is $A = q(k)$, so that $r(A) = r(q(k)) = k$. Replacing k by $r(A)$ in (7.20) gives (7.19).

To illustrate the use of Property 2, consider first the situation for which Figure 7.3 shows a least-cost design; for this case,

$$\int_{\mathbf{S}^*} \frac{1}{A} f(\mathbf{X})\, d\mathbf{X} = \int_0^{A^{1/2}} \frac{1}{A} (2z) z\, dz = \frac{2}{3} A^{1/2}$$

For the situation for which Figure 7.7 shows a least-cost design,

$$\int_{\mathbf{S}^*} \frac{1}{A} f(\mathbf{X})\, d\mathbf{X} = \int_d^{(4A+d^2)^{1/2}} \frac{1}{A} \frac{1}{4} (2z) z\, dz$$

$$= \frac{1}{6A} [(4A + d^2)^{3/2} - d^3]$$

so multiplying the latter expression by w gives the expression for the cost.

7.2.4 *n docks versus 1 dock*

The warehouse example problems considered so far in this chapter have been ones in which the warehouse had either one dock or two docks. Furthermore, item movement was allowed from any point in the warehouse to any dock, with w_j being the positive constant of proportionality. A more general problem which might be considered is that of choosing between a warehouse with n docks, each with "weight" w_j, or a warehouse with one dock having a weight $\bar{w} = \sum_{j=1}^{n} w_j$, the sum of the weights associated with the n docks. It is an interesting fact that for a given design S there will always be a location of a single dock having a weight of $\bar{w}$ for which the total cost will be equal to or less than the cost for the same design with n docks. If the single dock location is a realistic one, and giving it a weight of $\bar{w}$ does not "overload" the dock, then the value of average total cost is less with only one dock.

To justify the discussion regarding having a single dock, it is necessary to develop an inequality. Let $\mathbf{P}_1, \ldots, \mathbf{P}_n$ be dock locations with associated nonnegative weights $w_1 \ldots, w_n$, let $\bar{w} = \sum_{j=1}^{n} w_j$, and define the dock location $\mathbf{P}$ by $\mathbf{P} = \sum_{j=1}^{n}(w_j/\bar{w})\mathbf{P}_j$; then for all points $\mathbf{X}$ in the plane

$$\sum_{j=1}^{n} w_j|\mathbf{X} - \mathbf{P}_j| \geq \bar{w}|\mathbf{X} - \mathbf{P}| \tag{7.21}$$

To establish inequality (7.21), note that $\bar{w}|\mathbf{X} - \mathbf{P}| = |\bar{w}(\mathbf{X} - \mathbf{P})|$, and that $\bar{w}(\mathbf{X} - \mathbf{P}) = \bar{w}\mathbf{X} - \sum_{j=1}^{n} w_j\mathbf{P}_j = \sum_{j=1}^{n} w_j(\mathbf{X} - \mathbf{P}_j)$, so that $\bar{w}|\mathbf{X} - \mathbf{P}| = |\sum_{j=1}^{n} w_j(\mathbf{X} - \mathbf{P}_j)|$. But now repeated use of the triangle inequality establishes that $|\sum_{j=1}^{n} w_j(\mathbf{X} - \mathbf{P})| \leq \sum_{j=1}^{n} |w_j(\mathbf{X} - \mathbf{P}_j)| = \sum_{j=1}^{n} w_j|\mathbf{X} - \mathbf{P}_j|$; so inequality (7.21) is established. Now given a design $\mathbf{S}$ with n docks, the total cost is given by

$$\sum_{j=1}^{n} w_j \int_{\mathrm{S}} \frac{1}{A}|\mathbf{X} - \mathbf{P}_j|\, d\mathbf{X} = \int_{\mathrm{S}} \frac{1}{A}\left(\sum_{j=1}^{n} w_j|\mathbf{X} - \mathbf{P}_j|\right) d\mathbf{X}$$

But, by inequality (7.21),

$$\int_{\mathrm{S}} \frac{1}{A}\left(\sum_{j=1}^{n} w_j|\mathbf{X} - \mathbf{P}_j|\right) d\mathbf{X} \geq \int_{\mathrm{S}} \frac{1}{A}(\bar{w}|\mathbf{X} - \mathbf{P}|)\, d\mathbf{X}$$

so that

$$\sum_{j=1}^{n} w_j \int_{\mathrm{S}} \frac{1}{A}|\mathbf{X} - \mathbf{P}_j|\, d\mathbf{X} \geq \bar{w} \int_{\mathrm{S}} \frac{1}{A}|\mathbf{X} - \mathbf{P}|\, d\mathbf{X} \tag{7.22}$$

Inequality (7.22) is the desired result, and states that the total cost for a design $\mathbf{S}$ with one dock $\mathbf{P}$ and a weight $\bar{w}$ is equal to or less than the total cost for a design $\mathbf{S}$ with n docks. Inequality (7.22) may at first be nonintuitive; if so, it should be remembered that, for the case of n docks, it is still being assumed that each dock services the entire warehouse, in the sense that the average distance is computed between each dock and the entire design $\mathbf{S}$. If the situation is allowed where one dock services only a portion of the warehouse, then a design with n docks may very well have a smaller cost than a design with only one dock.

To illustrate inequality (7.22), consider the example for which Figure 7.7 shows a least-cost design. Since $\mathbf{P}_1 = (0, 0)$, $\mathbf{P}_2 = (0, d)$, and $w_1 = w_2 = w$, $\bar{w} = 2w$ and $\mathbf{P} = (w_1/\bar{w})\mathbf{P}_1 + (w_2/\bar{w})\mathbf{P}_2 = (0, d/2)$; the point $\mathbf{P}$ lies at the midpoint of the line segment joining $\mathbf{P}_1$ and $\mathbf{P}_2$. The inequality establishes that the total cost for the design $\mathbf{S}^*$ shown in Figure 7.7, $(w/6A)[(4A + d^2)^{3/2} - d^3]$, is equal to or greater than

$$2w \int_{\mathbf{S}^*} \frac{1}{A} |\mathbf{X} - \mathbf{P}|\, d\mathbf{X}$$

Now if one has an option of changing the design as well, then the same approach as used in the example for which Figure 7.3 shows a least-cost design establishes that

$$2w \int_{\mathbf{S}^*} \frac{1}{A} |\mathbf{X} - \mathbf{P}|\, d\mathbf{X} \geq 2w \int_{\bar{\mathbf{S}}} \frac{1}{A} |\mathbf{X} - \mathbf{P}|\, d\mathbf{X} = \frac{4w}{3} A^{1/2}$$

where $\bar{\mathbf{S}} = \{\mathbf{X} \in \mathbf{L} : |\mathbf{X} - \mathbf{P}| \leq A^{1/2}\}$. Thus, when it is reasonable to change the design as well as have a single dock with a weight of $2w$, a lower cost will be incurred. Note that for a more general version of the example for which Figure 7.7 shows a least-cost design, where there are n docks instead of two and all docks are located on the boundary of $\mathbf{L}$, that the point $\mathbf{P}$ will also be located on the boundary of $\mathbf{L}$, so that exactly the same approach as used previously is still applicable. Thus there will be a design having one dock with a weight of $\bar{w}$ and a total cost of $(2\bar{w}/3)A^{1/2}$, and this total cost will be equal to or less than the total cost of any design having n docks, with respective weights $w_1, \ldots, w_n$.

7.3 Locating *m* Items: The Factoring Case

7.3.1 *Single-dock case*

The previous discussion has dealt with finding a single design or set in the plane; the set could be thought of, in a warehousing context, as specifying the location of one item. It is logical to consider next the problem of locating m items, where m is greater than 1.

It will again be convenient to develop the discussion in a warehousing context. Suppose that the warehouse is to include m items; the set of points in the plane that item i takes up will be represented by $\mathbf{S}_i$, and it will be assumed that $\mathbf{S}_i$ has a known area, represented by A_i. It will further be assumed that for any two items i and j the sets $\mathbf{S}_i$ and $\mathbf{S}_j$ do not overlap. Since the sets $\mathbf{S}_1, \ldots, \mathbf{S}_m$ specify the locations of all m items, it will be convenient to speak of the collection $\{\mathbf{S}_1, \ldots, \mathbf{S}_m\}$ as being a *layout*. It will be assumed for any layout that each set $\mathbf{S}_i$ is contained in a known region $\mathbf{L}$, which might represent, for example, the plot of ground on which the warehouse is to be placed. The collection of all layouts will be denoted by $H_m(\mathbf{L}: A)$. It will be assumed that the warehouse will have n docks, at known locations $\mathbf{P}_1, \ldots, \mathbf{P}_n$. Given any layout $\{\mathbf{S}_1, \ldots, \mathbf{S}_m\}$, under an equal-likelihood assumption, the average distance that item i travels to or from dock j may be represented by

$$\int_{\mathbf{S}_i} \frac{1}{A_i} |\mathbf{X} - \mathbf{P}_j| \, d\mathbf{X}$$

The cost of the travel of item i to and from dock j for a given time period will be assumed to be directly proportional to the average distance, with w_{ij} being the nonnegative constant of proportionality, so that the total cost of movement of item i to and from dock j for a given time period is given by

$$w_{ij} \int_{\mathbf{S}_i} \frac{1}{A_i} |\mathbf{X} - \mathbf{P}_j| \, d\mathbf{X}$$

The total cost of movement of all items is then

$$\sum_{i=1}^{m} \sum_{j=1}^{n} w_{ij} \int_{\mathbf{S}_i} \frac{1}{A_i} |\mathbf{X} - \mathbf{P}_j| \, d\mathbf{X} \tag{7.23}$$

As in the previous chapter, it will be assumed that the matrix $\mathbf{W} = (w_{ij})$ factors; the assumption is equivalent to the assumption that $w_{ij} = c_i w_j$, where

$$c_i = \sum_{j=1}^{n} w_{ij}, \quad \text{for } i = 1, \ldots, m \quad \text{and} \quad w_j = \frac{\sum_{i=1}^{m} w_{ij}}{\sum_{i=1}^{m} \sum_{j=1}^{n} w_{ij}}$$

Also, as previously, the factoring assumption has the same physical interpretation. That is, w_j is the fraction of total item movement between storage and dock j, and c_i is the product of the cost per unit average distance traveled and the number of trips between storage per time period for item i.

Given the factoring assumption, (7.23) may be rewritten, on interchanging the order of summation and integration, as

$$\sum_{i=1}^{m} \frac{c_i}{A_i} \int_{\mathbf{S}_i} f(\mathbf{X}) \, d\mathbf{X} \tag{7.24}$$

where

$$f(\mathbf{X}) = \sum_{j=1}^{n} w_j |\mathbf{X} - \mathbf{P}_j| \tag{7.25}$$

It will be convenient to represent expression (7.24) by $F(\mathbf{S}_1, \ldots, \mathbf{S}_m)$; expression (7.24) represents the total cost of item movement for a layout $\{\mathbf{S}_1, \ldots, \mathbf{S}_m\}$. The problem of interest is then to find a layout $\{\mathbf{S}_1^*, \ldots, \mathbf{S}_m^*\}$ for which $F(\mathbf{S}_1^*, \ldots, \mathbf{S}_m^*) \leq F(\mathbf{S}_1, \ldots, \mathbf{S}_m)$, where $\{\mathbf{S}_1, \ldots, \mathbf{S}_m\}$ is any other layout; that is, the problem of interest is to find a least-cost layout.

To find a least-cost layout, conditions very analogous to those given in Property 4 of Chapter 6 will be developed. To develop the conditions, it is useful first to establish some notation. Given any layout $\{\mathbf{S}_1, \ldots, \mathbf{S}_m\}$, the union of $\mathbf{S}_1, \ldots, \mathbf{S}_q$, denoted by $\bigcup_{i=1}^{q} \mathbf{S}_i$ for $q = 1, 2, \ldots, m$, is defined to be the collection of all points in at least one of the sets $\mathbf{S}_1, \ldots, \mathbf{S}_q$. In a warehousing context, $\bigcup_{i=1}^{q} \mathbf{S}_i$ simply specifies the location of all the items 1 through q, and will have an area of B_q, where B_q is defined by $B_q = A_1 + \ldots + A_q$, for $q = 1, \ldots, m$. It may be useful intuitively to think of $\bigcup_{i=1}^{m} \mathbf{S}_i$ as being a warehouse design, since it represents the points taken up by all the items, and will have an area which is the sum of the areas of the individual items; that is, it will have an area of B_m. Figure 7.9 illustrates a layout $\{\mathbf{S}_1^*, \mathbf{S}_2^*\}$, while the resultant design $\bigcup_{i=1}^{2} \mathbf{S}_i^*$, is shown in Figure 7.10.

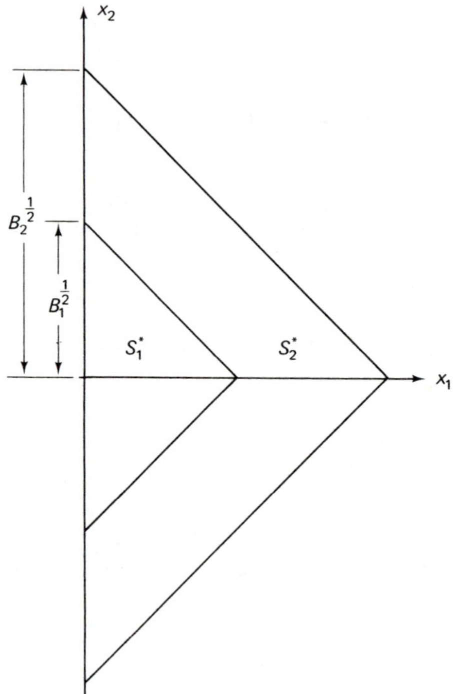

Figure 7.9. Multiitem, single-dock storage area layout.

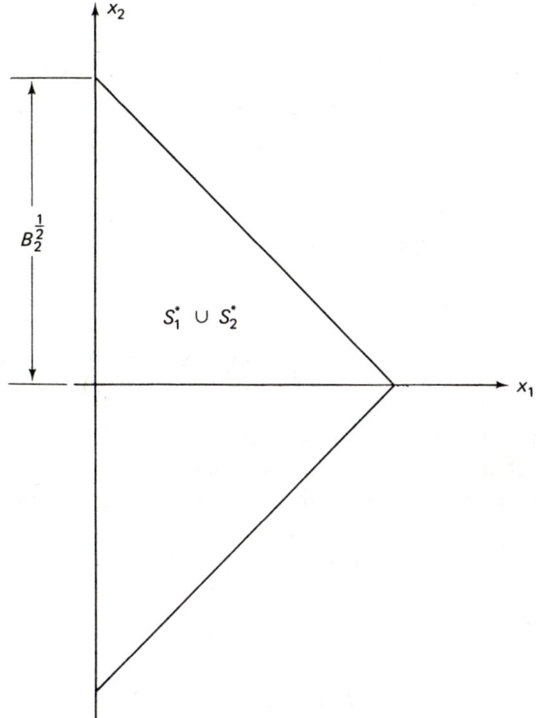

Figure 7.10. Multiitem, single-dock warehouse design.

Property 3: Let $\{\mathbf{S}_1^*, \ldots, \mathbf{S}_m^*\}$ be a layout in $H_m(\mathbf{L}:A)$. Given that

$$\frac{c_1}{A_1} \geq \cdots \geq \frac{c_m}{A_m} > 0$$

then
$$F(\mathbf{S}_1^*, \ldots, \mathbf{S}_m^*) \leq F(\mathbf{S}_1, \ldots, \mathbf{S}_m)$$

where $\{\mathbf{S}_1, \ldots, \mathbf{S}_m\}$ is any other layout in $H_m(\mathbf{L}: A)$, if $\{\mathbf{S}_1^*, \ldots, \mathbf{S}_m^*\}$ has the following property: there exist numbers $k_1 \leq \ldots \leq k_m$ such that, for $q = 1, \ldots, m$,

$$f(\mathbf{X}) \leq k_q, \quad \text{for every point } \mathbf{X} \text{ in } \textstyle\bigcup_{i=1}^{q} \mathbf{S}_i^*$$

and
$$f(\mathbf{X}) \geq k_q, \quad \text{for every point } \mathbf{X} \text{ not in } \textstyle\bigcup_{i=1}^{q} \mathbf{S}_i^*$$

Proof: Just as in Chapter 6, the proof will be developed for the case $m = 2$. Let $\{\mathbf{S}_1^*, \mathbf{S}_2^*\}$ be a layout having the assumed properties, and let

$\{\mathbf{S}_1, \mathbf{S}_2\}$ be any other layout. Then

$$\begin{aligned}F(\mathbf{S}_1, \mathbf{S}_2) &= \frac{c_1}{A_1}\int_{\mathbf{S}_1} f(\mathbf{X})\,d\mathbf{X} + \frac{c_2}{A_2}\int_{\mathbf{S}_2} f(\mathbf{X})\,d\mathbf{X} \\ &\quad -\frac{c_2}{A_2}\int_{\mathbf{S}_1} f(\mathbf{X})\,d\mathbf{X} + \frac{c_2}{A_2}\int_{\mathbf{S}_1} f(\mathbf{X})\,d\mathbf{X} \\ &= \left(\frac{c_1}{A_1} - \frac{c_2}{A_2}\right)\int_{\mathbf{S}_1} f(\mathbf{X})\,d\mathbf{X} + \frac{c_2}{A_2}\int_{\mathbf{S}_1\cup\mathbf{S}_2} f(\mathbf{X})\,d\mathbf{X}\end{aligned} \tag{7.26}$$

Likewise,

$$F(\mathbf{S}_1^*, \mathbf{S}_2^*) = \left(\frac{c_1}{A_1} - \frac{c_2}{A_2}\right)\int_{\mathbf{S}_1^*} f(\mathbf{X})\,d\mathbf{X} + \frac{c_2}{A_2}\int_{\mathbf{S}_1^*\cup\mathbf{S}_2^*} f(\mathbf{X})\,d\mathbf{X} \tag{7.27}$$

Now given the assumed property of $\mathbf{S}_1^*$, and of $\mathbf{S}_1^* \cup \mathbf{S}_2^*$, exactly the same approach as that used to obtain inequality (7.5) in the proof of Property 1 establishes, since $\mathbf{S}_1^*$ and $\mathbf{S}_1$ both have the same area, that

$$\int_{\mathbf{S}_1^*} f(\mathbf{X})\,d\mathbf{X} \le \int_{\mathbf{S}_1} f(\mathbf{X})\,d\mathbf{X} \tag{7.28}$$

and, since $\mathbf{S}_1^* \cup \mathbf{S}_2^*$ and $\mathbf{S}_1 \cup \mathbf{S}_2$ both have an area of $B_2 = A_1 + A_2$, that

$$\int_{\mathbf{S}_1^*\cup\mathbf{S}_2^*} f(\mathbf{X})\,d\mathbf{X} \le \int_{\mathbf{S}_1\cup\mathbf{S}_2} f(\mathbf{X})\,d\mathbf{X} \tag{7.29}$$

Multiplying (7.28) by $(c_1/A_1 - c_2/A_2) \ge 0$ and (7.29) by $c_2/A_2 > 0$ and adding the resultant inequalities gives

$$\begin{aligned}&\left(\frac{c_1}{A_1} - \frac{c_2}{A_2}\right)\int_{\mathbf{S}_1^*} f(\mathbf{X})\,d\mathbf{X} + \frac{c_2}{A_2}\int_{\mathbf{S}_1^*\cup\mathbf{S}_2^*} f(\mathbf{X})\,d\mathbf{X} \\ &\quad\le \left(\frac{c_1}{A_1} - \frac{c_2}{A_2}\right)\int_{\mathbf{S}_1} f(\mathbf{X})\,d\mathbf{X} + \frac{c_2}{A_2}\int_{\mathbf{S}_1\cup\mathbf{S}_2} f(\mathbf{X})\,d\mathbf{X}\end{aligned} \tag{7.30}$$

On using identities (7.26) and (7.27) and inequality (7.30), it follows that $F(\mathbf{S}_1^*, \mathbf{S}_2^*) \le F(\mathbf{S}_1, \mathbf{S}_2)$.

As an illustration of the use of Property 3, let $\mathbf{L}$ be the union of the first and fourth quadrants, suppose that a single dock is located at the origin, and let $f(\mathbf{X})$ be the rectilinear distance between the point $\mathbf{X}$ and the origin. Suppose that a layout involving two items, having areas A_1 and A_2, is to be found, and represent a layout by $\{\mathbf{S}_1, \mathbf{S}_2\}$. Then the average distance item i travels to or from storage is, for $i = 1, 2$,

$$\int_{\mathbf{S}_i} \frac{1}{A_i} f(\mathbf{X})\,d\mathbf{X}$$

So if c_i is the positive constant of proportionality converting average distance to a total cost per time period for movement of item i, then the total cost is given by

$$F(\mathbf{S}_1, \mathbf{S}_2) = c_1 \int_{\mathbf{S}_1} \frac{1}{A_1} f(\mathbf{X})\, d\mathbf{X} + c_2 \int_{\mathbf{S}_2} \frac{1}{A_2} f(\mathbf{X})\, d\mathbf{X} \tag{7.31}$$

A layout $\{\mathbf{S}_1^*, \mathbf{S}_2^*\}$ minimizing the total cost expression (7.31) is shown in Figure 7.9. Note that $\mathbf{S}_1^* = \{\mathbf{X} \in \mathbf{L}: f(\mathbf{X}) \le k_1 = B_1^{1/2}\}$, and $\mathbf{S}_1^* \cup \mathbf{S}_2^* = \{\mathbf{X} \in \mathbf{L}: f(\mathbf{X}) \le k_2 = B_2^{1/2}\}$; so both $\mathbf{S}_1^*$ and $\mathbf{S}_1^* \cup \mathbf{S}_2^*$ satisfy the conditions specified by Property 3. A computation to be completed subsequently establishes that, for the layout shown in Figure 7.9, the total cost is given by

$$F(\mathbf{S}_1^*, \mathbf{S}_2^*) = \frac{c_1}{A_1}\frac{2}{3} B_1^{3/2} + \frac{c_2}{A_2}\frac{2}{3}(B_2^{3/2} - B_1^{3/2}) \tag{7.32}$$

The layout $\{\mathbf{S}_1^*, \mathbf{S}_2^*\}$ illustrates a useful fact, which should be intuitive. Given any layout $\{\mathbf{S}_1^*, \ldots, \mathbf{S}_m^*\}$ in $H_m(\mathbf{L}: A)$, if there exist constants $k_1 < \ldots < k_m$ such that

$$\mathbf{S}_1^* = \{\mathbf{X} \in \mathbf{L}: f(\mathbf{X}) \le k_1\} \tag{7.33}$$

and

$$\mathbf{S}_i^* = \{\mathbf{X} \in \mathbf{L}: k_{i-1} \le f(\mathbf{X}) \le k_i\},\ i = 2, \ldots, m \tag{7.34}$$

then, for $q = 1, \ldots, m$,

$$f(\mathbf{X}) \le k_q, \quad \text{for every point } \mathbf{X} \text{ in } \textstyle\bigcup_{i=1}^{q} \mathbf{S}_i^*$$

and

$$f(\mathbf{X}) \ge k_q, \quad \text{for every point } \mathbf{X} \text{ not in } \textstyle\bigcup_{i=1}^{q} \mathbf{S}_i^*$$

Thus $\{\mathbf{S}_1^*, \ldots, \mathbf{S}_m^*\}$ satisfies the conditions of Property 3, and so is a least-cost layout. For the layout shown in Figure 7.9, the values of the constants k_1 and k_2 are given by $k_1 = B_1^{1/2}$ and $k_2 = B_2^{1/2}$.

7.3.2 *Multidock case*

As a multidock example of the use of Property 3, again let $\mathbf{L}$ be the union of the first and fourth quadrants, suppose that a warehouse is to have two docks located at the points $\mathbf{P}_1 = (0, 0)$ and $\mathbf{P}_2 = (0, d)$, where d is positive, that there are $m = 2$ items, $A_1 = 30$, $A_2 = 70$, and

$$\mathbf{W} = (w_{ij}) = \begin{pmatrix} 2 & 2 \\ 4 & 4 \end{pmatrix}$$

Note that the $\mathbf{W}$ matrix factors, so that $c_1 = 4$, $c_2 = 8$, $w_1 = \frac{1}{2}$, and $w_2 = \frac{1}{2}$.

Thus the function $f(\mathbf{X})$ defined by (7.25) becomes

$$f(\mathbf{X}) = \tfrac{1}{2}|\mathbf{X} - \mathbf{P}_1| + \tfrac{1}{2}|\mathbf{X} - \mathbf{P}_2| \tag{7.35}$$

where the distances are rectilinear, and all the data are now available for substitution into the total cost model $F(\mathbf{S}_1, \mathbf{S}_2)$ defined by (7.24). To find a least-cost layout $\{\mathbf{S}_1^*, \mathbf{S}_2^*\}$, the geometric approach used is quite similar to that used for finding least-cost designs discussed previously. A contour line of the function $f(\mathbf{X})$ defined by (7.35) is first constructed so that the region $\mathbf{S}_1^*$ "inside" the contour line has an area of A_1; if the contour line is the collection of all points in $\mathbf{L}$ for which $f(\mathbf{X}) = k_1$, then $\mathbf{S}_1^* = \{\mathbf{X} \in \mathbf{L}: f(\mathbf{X}) \leq k_1\}$. Due to the similarity of this example to the example illustrated in Figure 7.7, the analysis for that example is applicable after making a slight modification due to the difference in the definition of $f(\mathbf{X})$; the value of k_1 is given by $k_1 = [A_1 + (d^2/4)]^{1/2}$, where, for this case, $A_1 = 30$. Next, a contour line, the collection of all points in $\mathbf{L}$ for which $f(\mathbf{X}) = k_2$, is constructed so that the region $\mathbf{S}_1^* \bigcup \mathbf{S}_2^*$ inside the contour line has an area of $B_2 = A_1 + A_2$; in this case $k_2 = [B_2 + (d^2/4)]^{1/2}$, where $B_2 = 100$. The set $\mathbf{S}_2^*$ is defined by $\mathbf{S}_2^* = \{\mathbf{X} \in \mathbf{L}: k_1 \leq f(\mathbf{X}) \leq k_2\}$. The layout $\{\mathbf{S}_1^*, \mathbf{S}_2^*\}$ is illustrated in Figure 7.11; the total cost for this layout will be computed subsequently.

7.3.3 *Single-integral expression for* $F(S_1^*, \ldots, S_m^*)$

Just as in the previous section, it is both desirable and useful to develop single-integral expressions for least-cost layouts. The same notation developed prior to the statement of Property 2 will be used, and the same assumptions will be made. Furthermore, it is convenient to adopt one additional notational convention; both the terms $r(B_0)$ and k_0 will represent the minimum value of the function $f(\mathbf{X})$.

Property 4: Suppose that there exist numbers $k_0 < k_1 < \ldots < k_m$ such that $\{\mathbf{S}_1^*, \ldots, \mathbf{S}_m^*\}$ is a layout in $H_m(\mathbf{L}: A)$, where

$$\mathbf{S}_i^* = \{\mathbf{X} \in \mathbf{L}: k_{i-1} \leq f(\mathbf{X}) \leq k_i\}, \quad i = 1, \ldots, m$$

Then, for $i = 1, \ldots, m$, $k_i = r(B_i)$, and

$$\int_{\mathbf{S}_i^*} \frac{1}{A_i} f(\mathbf{X})\, d\mathbf{X} = \frac{1}{A_i} \int_{r(B_{i-1})}^{r(B_i)} q'(z) z\, dz \tag{7.36}$$

Proof: For $h = 1, \ldots, m$, let $\mathbf{T}_h = \bigcup_{i=1}^h \mathbf{S}_i^*$. Then it should be intuitive, and can be proved using set theory, that

$$\mathbf{T}_h = \{\mathbf{X} \in \mathbf{L}: f(\mathbf{X}) \leq k_h\} \tag{7.37}$$

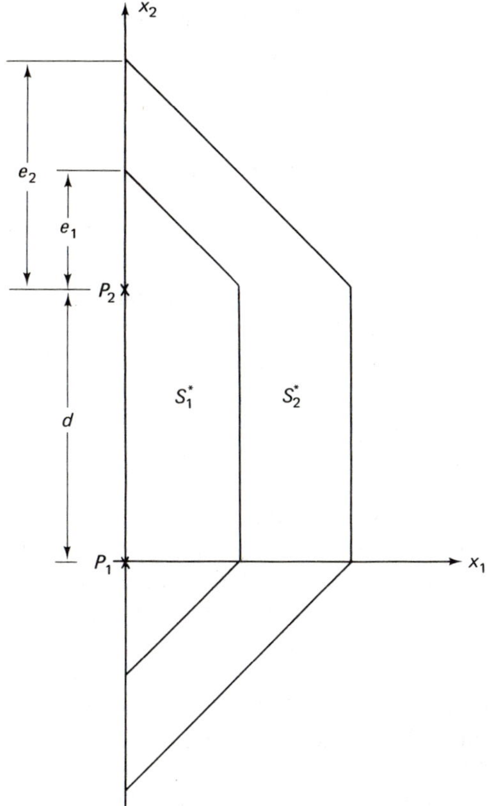

Figure 7.11. Multidock, multiitem storage area layout.

So the definition of the function $q(z)$ implies that the area of $\mathbf{T}_h$ is $q(k_h)$. But the area of $\mathbf{T}_h$ is also B_h, and so

$$r(B_h) = r\big(q(k_h)\big) = k_h \tag{7.38}$$

Now for the case $i = 1$, (7.36) follows from Property 2. For $i = 2, \ldots, m$, the same approach as used to establish (7.37) can be used to establish that $\mathbf{T}_i = \mathbf{T}_{i-1} \bigcup \mathbf{S}_i^*$, and so

$$\int_{\mathbf{S}_i^*} f(\mathbf{X})\, d\mathbf{X} = \int_{\mathbf{T}_i} f(\mathbf{X})\, d\mathbf{X} - \int_{\mathbf{T}_{i-1}} f(\mathbf{X})\, d\mathbf{X} \tag{7.39}$$

Again making use of Property 2,

$$\int_{\mathbf{T}_i} f(\mathbf{X})\, d\mathbf{X} = \int_{f^*}^{k_i} q'(z)z\, dz = \int_{f^*}^{k_{i-1}} q'(z)z\, dz + \int_{k_{i-1}}^{k_i} q'(z)z\, dz \tag{7.40}$$

and
$$\int_{\mathbf{T}_{i-1}} f(\mathbf{X})\, d\mathbf{X} = \int_{f^*}^{k_{i-1}} q'(z)z\, dz \tag{7.41}$$

Substituting (7.40) and (7.41) into (7.39) gives

$$\int_{\mathbf{S}_i^*} f(\mathbf{X})\, d\mathbf{X} = \int_{k_{i-1}}^{k_i} q'(z)z\, dz \tag{7.42}$$

Equation (7.36) now follows on substituting (7.38) into (7.42) and dividing by A_i.

Given a layout $\{\mathbf{S}_1^*, \ldots, \mathbf{S}_m^*\}$ with $\mathbf{S}_i^*$ defined by (7.33) which satisfies the conditions of Property 3, it now follows that the layout is a least-cost layout, and the minimum total cost is given by

$$F(\mathbf{S}_1^*, \ldots, \mathbf{S}_m^*) = \sum_{i=1}^{m} \frac{c_i}{A_i} \int_{r(B_{i-1})}^{r(B_i)} q'(z)z\, dz \tag{7.43}$$

To illustrate the use of Property 4, consider first the example for which Figure 7.9 shows a least-cost layout. From the discussion of the illustration of Property 2 using the example shown in Figure 7.3, we know that the function $q(z)$ is given by $q(z) = z^2$, and that $r(t) = t^{1/2}$. Substituting this information into (7.43) and carrying out the computation results in the value of $F(\mathbf{S}_1^*, \mathbf{S}_2^*)$ given by (7.32).

As a second illustration of the use of Property 4, consider the example for which Figure 7.11 shows a least-cost layout. From the discussion of the illustration of Property 2 using the example shown in Figure 7.7, we know that the function $q(z)$ is given by $q(z) = \frac{1}{4}(4z^2 - d^2)$ for $z \geq d/2$, that $r(t) = [t + (d^2/4)]^{1/2}$, and that $f^* = r(B_0) = d/2$. The difference in the function $q(z)$ for the examples shown in Figures 7.7 and 7.11 is due to the difference in the definition of $f(X)$ for the two examples. Substituting the information obtained into (7.43) gives the following expression for $F(\mathbf{S}_1^*, \mathbf{S}_2^*)$:

$$\frac{2c_1}{3A_1}\left[\left(B_1 + \frac{d^2}{4}\right)^{3/2} - \left(\frac{d}{2}\right)^3\right] + \frac{2c_2}{3A_2}\left[\left(B_2 + \frac{d^2}{4}\right)^{3/2} - \left(B_1 + \frac{d^2}{4}\right)^{3/2}\right]$$

Substituting $c_1 = 4$, $c_2 = 8$, $A_1 = B_1 = 30$, $A_2 = 70$, and $B_2 = 100$ into the expression obtained for $F(\mathbf{S}_1^*, \mathbf{S}_2^*)$ gives the numerical value of $F(\mathbf{S}_1^*, \mathbf{S}_2^*)$.

7.3.4 *n docks versus 1 dock*

As in the previous section, the question of choosing a layout having one dock versus choosing a layout having n docks is of interest, and a similar result can be obtained. Given the definition of $f(\mathbf{X})$ in (7.25), $f(\mathbf{X})$ is just the term on the left side of (7.21); furthermore, given the factoring assumption, $\bar{w} = 1$, so $\mathbf{P} = \sum_{j=1}^{n} w_j \mathbf{P}_j$, and the right side of (7.21) is just $|\mathbf{X} - \mathbf{P}|$. Inequality (7.21) thus becomes $f(\mathbf{X}) \geq |\mathbf{X} - \mathbf{P}|$, so, given any layout $\{\mathbf{S}_1, \ldots, \mathbf{S}_m\}$ in $H_m(\mathbf{L}: A)$, the use of the inequality implies that

$$\int_{\mathbf{S}_i} f(\mathbf{X})\, d\mathbf{X} \geq \int_{\mathbf{S}_i} |\mathbf{X} - \mathbf{P}|\, d\mathbf{X}, \quad i = 1, \ldots, m \tag{7.44}$$

Multiplying inequalities (7.44) by c_i/A_i and summing gives

$$\sum_{i=1}^{m} c_i \int_{\mathbf{S}_i} \frac{1}{A_i} f(\mathbf{X})\, d\mathbf{X} \geq \sum_{i=1}^{m} c_i \int_{\mathbf{S}_i} \frac{1}{A_i} |\mathbf{X} - \mathbf{P}|\, d\mathbf{X}$$

or

$$F(\mathbf{S}_1, \ldots, \mathbf{S}_m) \geq \sum_{i=1}^{m} c_i \int_{\mathbf{S}_i} \frac{1}{A_i} |\mathbf{X} - \mathbf{P}|\, d\mathbf{X} \tag{7.45}$$

Notice that the term on the right side of inequality (7.45) has a direct physical interpretation; if $\mathbf{P}$ is considered to be the location of a single dock, then the ith integral on the right side of (7.45) expresses the average distance that item i moves between storage and the dock. The term c_i may be interpreted, in a warehousing context, as the total number of pallet loads of item i going in and out of storage per time period, and so each term in the sum on the right side of (7.45) is directly proportional to the total cost of movement of item i per time period. Given a warehousing context, $F(\mathbf{S}_1, \ldots, \mathbf{S}_m)$ is also directly proportional to the total cost of item movement, so it will be convenient to refer to the expressions on both sides of inequality (7.45) as total cost expressions. Inequality (7.45) thus states that for any layout $\{\mathbf{S}_1, \ldots, \mathbf{S}_m\}$ there is a total cost expression involving one dock which has a value no greater than that of the total cost expression involving n docks. From the definition of c_i, $\sum_{i=1}^{m} c_i = \sum_{i=1}^{m} \sum_{j=1}^{n} w_{ij}$, so that, for the two expressions in (7.45), the sum of the dock "weights" is the same; thus, if the one dock is not overloaded, either choice of docks would result in the same total item "turnover" per time period.

To illustrate inequality (7.45), consider the example with two docks for which Figure 7.11 shows a least-cost design. Since $\mathbf{P}_1 = (0, 0)$, $\mathbf{P}_2 = (0, d)$, and $w_1 = w_2 = \frac{1}{2}$, $\mathbf{P} = \frac{1}{2}\mathbf{P}_1 + \frac{1}{2}\mathbf{P}_2 = (0, d/2)$. Recall for this example that $c_1 = 4$, $c_2 = 8$, $A_1 = 30$, and $A_2 = 70$. Thus the minimum total cost for the least-cost layout shown in Figure 7.11 is given by

$$F(\mathbf{S}_1^*, \mathbf{S}_2^*) = 4 \int_{\mathbf{S}_1^*} \tfrac{1}{30} f(\mathbf{X})\, d\mathbf{X} + 8 \int_{\mathbf{S}_2^*} \tfrac{1}{70} f(\mathbf{X})\, d\mathbf{X}$$

where $f(\mathbf{X})$ is defined by (7.35) for this example. From inequality (7.45),

$$F(\mathbf{S}_1^*, \mathbf{S}_2^*) \geq 4 \int_{\mathbf{S}_1^*} \tfrac{1}{30} |\mathbf{X} - \mathbf{P}|\, d\mathbf{X} + 8 \int_{\mathbf{S}_2^*} \tfrac{1}{70} |\mathbf{X} - \mathbf{P}|\, d\mathbf{X} \tag{7.46}$$

Given the option of changing the layout as well, the same approach as used in the example for which Figure 7.11 shows a least-cost layout establishes that the right side of (7.46) is equal to or greater than the total cost of the layout $\{\bar{\mathbf{S}}_1, \bar{\mathbf{S}}_2\}$ given by

$$4\int_{\bar{\mathbf{S}}_1} \tfrac{1}{30}|\mathbf{X}-\mathbf{P}|\,d\mathbf{X} + 8\int_{\bar{\mathbf{S}}_2} \tfrac{1}{70}|\mathbf{X}-\mathbf{P}|\,d\mathbf{X} = \tfrac{8}{90}(30^{1/2}) + \tfrac{8}{105}(100^{1/2} - 30^{1/2})$$

The layout $\{\bar{\mathbf{S}}_1, \bar{\mathbf{S}}_2\}$ is defined by $\bar{\mathbf{S}}_1 = \{\mathbf{X} \in \mathbf{L}: |\mathbf{X}-\mathbf{P}| \leq 30^{1/2}\}$, and $\bar{\mathbf{S}}_2 = \{\mathbf{X} \in \mathbf{L}: 30^{1/2} \leq |\mathbf{X}-\mathbf{P}| \leq 100^{1/2}\}$.

7.4 Rectangular Warehouse Design and Layout

It has been emphasized that many of the models developed in this text are design guides and should serve as benchmarks in designing layouts that are more acceptable from an operational viewpoint. The layouts presented thus far in this chapter are radically different from most warehouse layouts. There are a number of reasons for this. For example, we have restricted our choice of designs to those which minimize material-handling cost. No consideration was given to construction cost. An additional reason might be the resistance to change in individuals. The most popular warehouse configuration is a rectangular configuration. Since it is so popular, construction methods have been sufficiently standardized such that rectangular warehouses are also economical to build, as compared to other shapes.

Since rectangular warehouses are quite common, it appears worthwhile to treat such designs as a special topic. We shall consider now the problem of designing a new rectangular warehouse and the associated layout problem. It is assumed that the warehouse height and the area of the warehouse are predetermined quantities. Two basic types of cost are considered: costs due to item movement within the warehouse and costs due to the warehouse perimeter, such as perimeter construction and maintenance costs.

To begin, consider the rectangle **S** shown in Figure 7.12; the rectangle has *dimensions* a by b and area A, such that $ab = A$. The lower left-hand corner of **S** is denoted by the point (p, q). In set notation

$$\mathbf{S} = \{(x, y): p \leq x \leq p + a, q \leq y \leq q + b, ab = A\}$$

That is, **S** is the set of all points (x, y), where x is between p and $p + a$, y is between q and $q + b$, and $ab = A$. The set will be considered to be a rectangular warehouse design, with one dock at the point $(0, 0)$, the origin. Under the assumption that items are equally likely to move between any

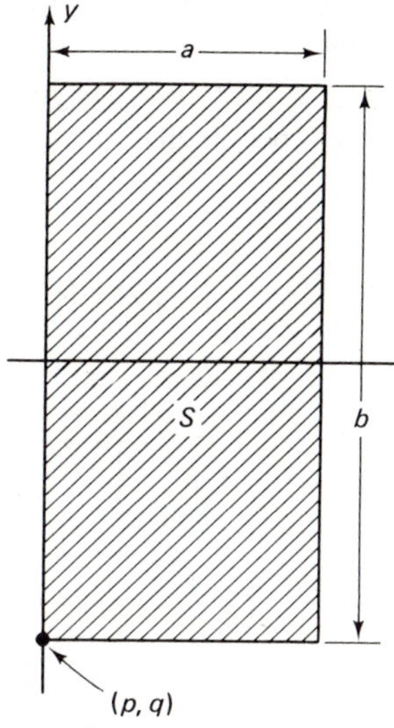

Figure 7.12. Rectangular warehouse design.

point in the warehouse and the dock, the average rectilinear distance an item travels in or out of storage may be represented by

$$\iint_S \frac{1}{A}(|x| + |y|)\, dx\, dy \tag{7.47}$$

(Notice that it is notationally convenient in this section to use iterated single integrals.)

On assuming that the total annual cost of item movement is directly proportional to the average rectilinear distance, where c is the positive constant of proportionality, the total annual cost of item movement is obtained by multiplying the average distance expression (7.47) by c. Let r be an annual total perimeter cost per foot for the warehouse design, representing such costs as perimeter construction and maintenance. Since the perimeter of the warehouse design is $2(a + b)$, the total annual perimeter cost will be $2r(a + b)$. Thus if the total annual cost of a rectangular warehouse design **S** represented by $FR(\mathbf{S})$, is defined to be the sum of item movement costs and perimeter cost, then

$$FR(\mathbf{S}) = c \iint_S \frac{1}{A}(|x| + |y|)\, dx\, dy + 2r(a + b) \tag{7.48}$$

The problem of interest is now to find a rectangular warehouse design of area A to minimize the total cost expression (7.48). To assure that the dock does not lie inside the warehouse, the condition $p \geq 0$ will be imposed when finding a least-cost design.

The first step in finding a least-cost design is to get expression (7.48) into a more tractable form; specifically, it will be useful to evaluate the double integral. Note that

$$\iint_{\mathbf{S}} \frac{1}{A}(|x| + |y|)\,dx\,dy = \iint_{\mathbf{S}} \frac{1}{A}|x|\,dx\,dy + \iint_{\mathbf{S}} \frac{1}{A}|y|\,dx\,dy$$

Furthermore,

$$\iint_{\mathbf{S}} \frac{1}{A}|x|\,dx\,dy = \int_{q}^{q+b} \int_{p}^{p+a} \frac{1}{ab}|x|\,dx\,dy$$

$$= \int_{p}^{p+a} \frac{1}{a}|x|\,dx \tag{7.49}$$

Likewise,

$$\iint_{\mathbf{S}} \frac{1}{A}|y|\,dx\,dy = \int_{q}^{q+b} \frac{1}{b}|y|\,dy \tag{7.50}$$

Thus, if $f_1(a, p)$ is defined to be the integral in (7.49), and $f_2(b, q)$ is defined to be the integral on the right side of (7.50), and the expression $f(a, b, p, q)$ is defined by

$$f(a, b, p, q) = c[f_1(a, p) + f_2(b, q)] + 2r(a + b) \tag{7.51}$$

then the total cost for the design **S** is given by $f(a, b, p, q)$. Thus the problem of finding a least-cost rectangular warehouse design may be reformulated as follows: minimize $f(a, b, p, q)$, subject to the conditions that $ab = A$ and $p \geq 0$.

One of the chapter problems involves establishing that

$$\begin{aligned} f_1(a, p) &= \frac{a}{4} + \frac{1}{a}\left(p + \frac{a}{2}\right)^2, \quad -a \leq p \leq 0 \\ f_1(a, p) &= \left|p + \frac{a}{2}\right|, \quad p \leq -a \quad \text{or} \quad p \geq 0 \end{aligned} \tag{7.52}$$

and that

$$f_2(b, q) = \frac{b}{4} + \frac{1}{b}\left(q + \frac{b}{2}\right)^2, \quad -b \leq q \leq 0 \tag{7.53}$$

$$f_2(b, q) = \left|q + \frac{b}{2}\right|, \quad q \leq -b \quad \text{or} \quad q \geq 0 \tag{7.54}$$

Now consider letting p and q vary, subject to the condition $p \geq 0$. From (7.52), when $p \geq 0$, then

$$f_1(a, p) = p + \frac{a}{2} \geq \frac{a}{2} = f_1(a, 0) \tag{7.55}$$

From (7.54), in a similar manner,

$$f_2(b, q) \geq \frac{b}{2} = f_2(b, 0) = f_2(b, -b), \quad q \leq -b \quad \text{or} \quad q \geq 0 \tag{7.56}$$

while, from (7.53), due to the squared term,

$$f_2(b, q) \geq \frac{b}{4} = f_2\left(b, -\frac{b}{2}\right) \tag{7.57}$$

From inequalities (7.55) and (7.57), it follows, when $p \geq 0$, that

$$\begin{aligned} f(a, b, p, q) &\geq c\left[f_1(a, 0) + f_2\left(b, -\frac{b}{2}\right)\right] + 2r(a + b) \\ &= c\left(\frac{a}{2} + \frac{b}{4}\right) + 2r(a + b) \\ &= \left(\frac{c}{2} + 2r\right)a + \left(\frac{c}{4} + 2r\right)b \\ &= f\left(a, b, 0, -\frac{b}{2}\right) \end{aligned} \tag{7.58}$$

Now since $ab = A$, $b = A/a$, and

$$f\left(a, b, 0, -\frac{b}{2}\right) = \left(\frac{c}{2} + 2r\right)a + \left(\frac{c}{4} + 2r\right)\frac{A}{a} \tag{7.59}$$

Computing the derivative of (7.59) with respect to a, setting the resultant expression to zero, and solving for a gives

$$a^* = \left(\frac{c + 8r}{2c + 8r}\right)^{1/2} A^{1/2} \tag{7.60}$$

Since the value of a given by (7.60) minimizes the right side of (7.58), on substituting (7.60) into (7.58) it follows, with b^* defined by $b^* = A/a$, that

$$f\left(a, b, 0, -\frac{b}{2}\right) \geq f\left(a^*, b^*, 0, -\frac{b^*}{2}\right) \tag{7.61}$$

Combining inequalities (7.58) and (7.61) gives

$$f(a, b, p, q) \geq f\left(a^*, b^*, 0, -\frac{b^*}{2}\right)$$

for any a, b, p, and q such that $ab = A$ and $p \geq 0$. From the definition of b^*,

$$b^* = \left(\frac{2c + 8r}{c + 8r}\right)^{1/2} A^{1/2} \tag{7.62}$$

A least-cost rectangular warehouse design has now been found; it has dimensions a^* and b^*, and the coordinates of the lower left-hand corner are given by $(0, -b^*/2)$. Using set notation, the least-cost rectangular warehouse design is given by

$$\mathbf{S}^* = \left\{(x, y): 0 \leq x \leq a^*, -\frac{b^*}{2} \leq y \leq \frac{b^*}{2}\right\}$$

The substitution of (7.60) and (7.62) into (7.51) establishes that the value of the least total cost is given by

$$FR(\mathbf{S}^*) = 2\left[\left(\frac{c}{2} + 2r\right)\left(\frac{c}{4} + 2r\right)\right]^{1/2} A^{1/2} \tag{7.63}$$

An example of the warehouse design obtained by taking r to be zero, so that perimeter costs are not considered, is shown in Figure 7.13.

It is left as an exercise to establish that the least-cost rectangular warehouse design obtained when the requirement is omitted that the dock not lie inside the warehouse, that is, $p \geq 0$, is a square of dimensions $A^{1/2}$ by $A^{1/2}$ with the dock at its center. The total cost for this design is $[(c/2) + 4r)]A^{1/2}$, and is less than the cost given by (7.63). A comparison of the two designs indicates, roughly speaking, how much it costs to require the dock not to be inside the warehouse.

Following the approaches of Chapter 6 and earlier sections in this chapter, it can be established that a rectangular warehouse containing n types of items should be designed such that items having the largest c/A values are located nearest the dock. Let the types of items be numbered so that

$$\frac{c_1}{A_1} \geq \frac{c_2}{A_2} \geq \cdots \geq \frac{c_n}{A_n} \tag{7.64}$$

For $j = 1, \ldots, n$, denote the union of $\mathbf{S}_1$ through $\mathbf{S}_j$ by $\bigcup_{i=1}^{j} \mathbf{S}_i$ as being

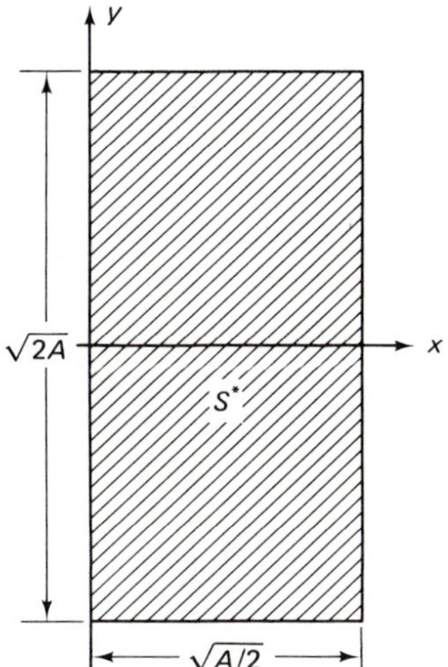

Figure 7.13. Optimum dimensions for a rectangular warehouse.

the total floor space taken up by item types 1 through j. For $j = 1, \ldots, n$, it will be assumed that $\bigcup_{i=1}^{j} \mathbf{S}_i$ is a rectangle, of area $\sum_{i=1}^{j} A_i$, lying to the right of the dock. We call this the *rectangularity assumption.* Any sets $\mathbf{S}_1, \ldots, \mathbf{S}_n$ numbered so that (7.64) holds and satisfying the rectangularity assumption we call an *ordered rectangular layout.*

Among all ordered rectangular layouts, the layout $\mathbf{S}_1^*, \ldots, \mathbf{S}_n^*$ shown in Figure 7.14 is a minimum-cost layout. Total cost, in this case, is the cost of item movement per time period and is expressed as

$$FR(\mathbf{S}_1, \ldots, \mathbf{S}_n) = \sum_{j=1}^{n} c_j \iint_{\mathbf{S}_j} \frac{1}{A_j}(|x| + |y|)\, dx\, dy$$

It can be established [3] that the total cost for the layout $\{\mathbf{S}_1^*, \ldots, \mathbf{S}_n^*\}$ given in Figure 7.14 is given by

$$FR(\mathbf{S}_1^*, \ldots, \mathbf{S}_n^*) = \sum_{j=1}^{n} \left(2^{1/2} \frac{c_j}{2A_j}\right)(B_j^{3/2} - B_{j-1}^{3/2}) \tag{7.65}$$

where
$$B_j = \sum_{k=1}^{j} A_k, \quad j = 1, \ldots, n$$

The term c_j is typically the product of the total cost per foot to move one unit of item type j in and out of storage and the expected number of

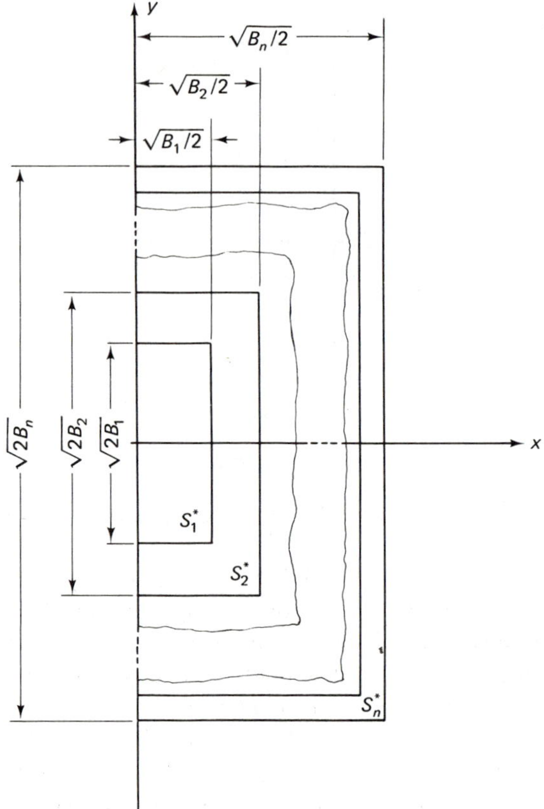

Figure 7.14. Optimum ordered rectangular layout.

items of type j moving in and out of storage per time period. If items are moved in and out of the warehouse on pallets, if a unit of item type j is interpreted as being a pallet load, and if the total cost per foot to move a pallet is the same for all types of items, then a simpler interpretation of c_j is the expected number of pallet loads of item type j going in and out of storage per time period.

In our previous discussion of warehouse design, the question of aisle space has been largely ignored. We have assumed, for example, that an orthogonal network of aisles existed in the warehouse, such that rectilinear distance was appropriate. Considering a rectangular warehouse and, more specifically, (7.48), let A_a denote the area of all warehouse aisle space and let A_s denote the area of all space actually taken up by items. Let r be the ratio of aisle space area to item space area,

$$r = \frac{A_a}{A_s}$$

Let A be the total warehouse area,

$$\begin{aligned} A &= A_a + A_s \\ &= (1 + r)A_s \end{aligned}$$

The position and arrangement of the aisles will have little effect upon the model, provided they constitute an orthogonal network and the equal-likelihood assumption is still approximately satisfied.

From (7.63), on replacing A by $(1 + r)A_s$, the total cost of the warehouse design S^* is given by

$$2[(\tfrac{1}{2}c + 2k)(\tfrac{1}{4}c + 2k)]^{1/2}(1 + r)^{1/2}(A_s)^{1/2} \tag{7.66}$$

The sensitivity of total cost to r is obvious from (7.66); that is, the minimum total cost increases as the square root of $1 + r$. Thus a value of r of 0.21 would cause total cost to be increased by

$$(1 + 0.21)^{1/2} - 1 = 0.10$$

or 10% greater than the total cost when there is no aisle space. Thus the minimum total cost is relatively insensitive to aisle space considerations.

7.5 Summary

In this chapter we considered continuous formulations of some warehouse design and layout problems. Our discussion was presented in the context of a location problem involving existing facilities (docks), which were idealized as points, and new facilities (items), which were characterized as areas. The location problem reduced to a determination of the location of the items relative to the docks. Associated with the determination was a decision concerning the configuration of the region required for each new facility (item).

The continuous formulation of facility layout and location problems introduced in this chapter can also be extended to include location problems in which the new facilities are idealized as points and the existing facilities are either points or areas [5]. We previously considered the situation in which both new and existing facilities were areas; such a situation was referred to as a plant layout problem and was treated in Chapters 2 and 3.

In a sense, the continuous warehouse design and layout problem treated in this chapter can be considered to be a special case of the more general problem of regional design. The regional design problem can be defined as the determination of optimal regions in the plane following some appropriate criteria. Thus the regional design problem involves a partitioning of a given region into subregions. Examples of regional design problems include the partitioning of a state into congressional districts, the partitioning of a county into school districts, the division of a given region into districts of telephone subscribers, the design of a source region for a business with several branch offices, the design of a parking lot based on the destinations of the people who park in the lot, and the design of a warehouse based on dock locations. For a discussion of regional design as it relates to facility layout and location, see [1].

We have previously emphasized that the warehouse designs and layouts developed in this chapter are intended as design aids, not final designs and layouts. The analysis only considered material-handling cost, and, in the discussion of rectangular warehouse design, a perimeter cost. Additionally, nonstorage areas were not considered explicitly in the analysis. Consequently, the results obtained from the models should be examined on the basis of practical limitations and modifying considerations. Where appropriate, changes should be made in the designs and a number of alternative solutions evaluated on the basis of the appropriate criteria.

REFERENCES

1. Corley, H. W., Jr., and S. D. Roberts, "A Partitioning Problem with Applications in Regional Design," *Operations Research*, Vol. 2, No. 5, 1972, pp. 1010–1019.

2. Francis, R. L., "On Some Optimum Facility Design Problems," unpublished Ph.D. dissertation, Northwestern University, Evanston, Ill., 1967; also Order No. 67–15, 232, University Microfilms, P.O. Box 1346, Ann Arbor, Mich. 48106.

3. Francis, R. L., "On Some Problems of Rectangular Warehouse Design and Layout," *The Journal of Industrial Engineering*, Vol. 18, No. 10, 1967, pp. 595–604.

4. Francis, R. L., "Sufficient Conditions for Some Optimum-Property Facility Designs," *Operations Research*, Vol. 15, No. 3, 1967, pp. 448–466.

5. WESOLOWSKY, G. O., and R. F. LOVE, "Location of Facilities with Rectangular Distances Among Point and Area Destinations," *Naval Research Logistics Quarterly*, Vol. 18, No. 1, 1971, pp. 83–90.

6. WHITE, J. A., "On the Optimum Design of Warehouses Having Radial Aisles," *AIIE Transactions*, Vol. 4, No. 4, 1973, pp. 333–336.

PROBLEMS

7.1. A warehouse is to be constructed for the storage of two products. The storage area required for the products equals 2,400 and 2,500 square feet, respectively. Item movement (measured in trips in and out of storage per week) equals 1,000 and 2,000, respectively, for the two products. Rectilinear item movement is to be used in the warehouse. The new warehouse will have three docks located on the same periphery of the warehouse. The docks will be separated by 60 feet. The probability an item will move to and from a particular dock is the same for each of the three docks.
(a) Determine the configuration, including dimensions, that minimizes expected distance traveled per week.
(b) Determine the value for expected distance traveled per week.

7.2. A warehouse having an area of 1,000 square feet is to be designed with a single dock. The warehouse is to be designed such that the cost of travel in moving items in and out of storage is minimized. Give the optimum dimensions for the warehouse based on
(a) Rectilinear travel with the dock located along an exterior wall.
(b) Rectilinear travel.
(c) Straight-line travel with the dock located along an exterior wall.
(d) Straight-line travel.
(e) Squared straight-line distance with the dock located along an exterior wall.
(f) Squared straight-line distance.
(g) Rectilinear travel in a rectangular-shaped warehouse with the dock located along an exterior wall.

7.3. An existing warehouse of dimensions 100 by 200 feet is to be used for the storage of two products, *A* and *B*. The storage area required for product *A* equals 9,000 square feet and for product *B* equals 6,000 square feet. Item movement between storage and one of the four docks equals 2,000 loads per month for product *A* and 2,500 loads per month for product *B*. Rectilinear item movement is used. Each dock is spaced at a midpoint of each of the four walls. Assuming any unit of product is equally likely to travel to either one of the four docks, determine the dimensions for the area of the warehouse to be used for the storage of each product in order to minimize the average distance traveled per month.

7.4. There currently exists a rectangular warehouse having dimensions of 100 by 150 feet with the longest dimension being parallel with the x axis. The warehouse has three doors at $\mathbf{X} = (0, 50)$, $\mathbf{Y} = (50, 0)$ and $\mathbf{Z} = (100, 0)$ when the warehouse is considered to lie in the first quadrant with the lower left-hand corner of the warehouse located at the origin. Three products are to be stored in the warehouse. Product A requires 4,000 square feet of floor space; product B requires 6,000 square feet of floor space; and product C requires 5,000 square feet of floor space. All three products are brought into the warehouse through the door at (0, 50). For each product 40% of the items leave the warehouse through the door at (50, 0), with the remaining 60% leaving through the door at (100, 0). On the average, 10 pallet loads of product A, 20 pallet loads of product B, and 40 pallet loads of product C enter the warehouse per day. Movement from storage to the doors at **Y** and **Z** also takes place in pallet load quantities at the same average daily rates. Design a continuous layout (with dimensions shown) to minimize the expected distance traveled per day. All travel within the warehouse is rectilinear, and the layout design must conform to the present warehouse design.

7.5. Design a continuous facility layout for a warehouse having two docks located at $\mathbf{P}_1 = (0, 0)$ and $\mathbf{P}_2 = (0, 30)$. Travel occurs along a set of rectilinear aisles with twice as many trips being made to the dock at $\mathbf{P}_1$ as to the dock at $\mathbf{P}_2$. The two docks are to be located along the periphery of the warehouse. The warehouse is to have an area of 1,000 square feet. Determine the configuration, including dimensions, and the expected distance traveled for the warehouse design.

7.6. Design a continuous facility layout for a warehouse that has two docks located at (0, 0) and (0, 60). Rectilinear travel is used within the warehouse. The docks are located along the same exterior wall of the warehouse. The warehouse is to have an area of 1,700 square feet. Determine the configuration, giving all dimensions, for the warehouse design that minimizes the expected distance traveled per unit time when twice as many trips are made to the dock at (0, 0) as to the dock at (0, 60). Also, determine the expected distance traveled per unit time for the design obtained. The warehouse design is to lie in the first and fourth quadrants.

7.7. Solve Problem 6.16 assuming a continuous layout is to be designed.

7.8. Solve Problem 6.23 assuming a continuous layout is to be designed.

7.9. Solve Problem 6.24 assuming a continuous layout is to be designed.

7.10. Solve Problem 7.2, giving dimensions for all storage areas, assuming two products, A and B, are to be stored in the warehouse. Storage area requirements are 400 and 600 square feet, respectively. Item movement equals 500 loads per week for product A and 1,000 loads per week for product B.

7.11. Given any two points **X** and **P** in the plane, denote by $g(\mathbf{X}, \mathbf{P})$ an *arbitrary* distance (not necessarily the l_p distance) between **X** and **P**. A famous theorem in mathematics states that $g(\mathbf{X}, \mathbf{P})$ is equal to or greater than the Euclidean distance between **X** and **P**; that is, the shortest distance between two points is a straight line. Make use of this theorem to prove, for any set **S** of area A, that

$$\int_{\mathbf{S}} \frac{1}{A} g(\mathbf{X}, \mathbf{P})\, d\mathbf{X} \geq \frac{2}{3}\left(\frac{A}{\pi}\right)^{1/2}$$

Can you suggest any practical use of this inequality?

7.12. Let **L** be the union of the first and fourth quadrants, let b be a known positive constant, define the points $\mathbf{P}_1 = (0, 0)$, $\mathbf{P}_2 = (0, b)$, $\mathbf{P}_3 = (0, 2b)$, denote the rectilinear distance **X** and $\mathbf{P}_j$ by $d(\mathbf{X}, \mathbf{P}_j)$, and define the function $f(\mathbf{X})$ by

$$f(\mathbf{X}) = d(\mathbf{X}, \mathbf{P}_1) + 2d(\mathbf{X}, \mathbf{P}_2) + d(\mathbf{X}, \mathbf{P}_3)$$

If $q(z)$ is the area of the set $\mathbf{S}(z)$, where $\mathbf{S}(z) = \{\mathbf{X} \in \mathbf{L}: f(\mathbf{X}) \leq z\}$, it is given that

$$q(z) = \frac{1}{2}\left(\frac{z}{2} - b\right)^2 \quad \text{for } 2b \leq z \leq 4b$$

$$q(z) = \frac{z^2}{16} - \frac{b^2}{2} \quad \text{for } 4b < z$$

The positive constant A is given, $H(\mathbf{L}: A)$ denotes the collection of all sets (or designs) **S** in **L** of area A, and for every design **S** in $H(\mathbf{L}: A)$ the expression $F(\mathbf{S})$ is defined by

$$F(\mathbf{S}) = \int_{\mathbf{S}} \frac{1}{A} f(\mathbf{X})\, d\mathbf{X}$$

(a) What physical interpretation can you give to the expression $F(\mathbf{S})$?

(b) Find a design $\mathbf{S}^*$ in $H(\mathbf{L}: A)$ that minimizes $F(\mathbf{S})$ when
   (1) $A \leq b^2/2$.
   (2) $A > b^2/2$.
   Why is it necessary to consider these two different cases?

(c) For the designs $\mathbf{S}^*$ found in part (b), develop single-integral expressions for $F(\mathbf{S}^*)$, using the function $q(z)$ as given. Do not evaluate the single integrals.

7.13. An existing warehouse of dimensions 120 by 240 feet is to be used for the storage of three products A, B, and C. The storage area required for product A equals 9,000 square feet, for product B, 6,000 square feet, and for product C, 12,000 square feet. Item movement between storage and one of the three docks equals 2,000 loads per month for product A, 2,500 loads per month for product B, and 4,000 loads per month for product C. Rectilinear item movement is used. The three docks are equally spaced along one of the 240-foot walls. Determine the dimensions for the area of the warehouse used to store each product such that the average annual distance traveled per month is minimized. Assume any unit of product is equally likely to travel to either dock.

7.14. Solve Problem 7.13 for the case of a new warehouse. Let the three docks be located at (0, 0), (50, 0), and (0, 50). Design a warehouse having an area of 30,000 square feet and lying in the first quadrant.

7.15. Suppose that a warehouse is to be located in a region **L**, which is the union of the first and fourth quadrants. The warehouse will have an area of A and will have one dock, located at the origin. Represent the Euclidean distance from any point **X** to the dock by $f(\mathbf{X})$.

(a) Find a warehouse design **S*** of area A in **L** that minimizes the average distance items will travel between the dock and storage.

(b) Compute the average distance items travel between the dock and storage for the warehouse design **S***.

7.16. Solve Problem 7.15 when **L** is the entire plane instead of the union of the first and fourth quadrants, and compute the ratio of the two least average distances.

7.17. Given a number $d > 0$, let $\mathbf{P}_1 = (0, 0)$ and $\mathbf{P}_2 = (d, 0)$. Let **L** consist of the "strip" in the first and fourth quadrants, parallel to the y axis, made up of all points lying between $\mathbf{P}_1$ and $\mathbf{P}_2$; that is, $\mathbf{L} = \{(x, y): 0 \leq x \leq d, -\infty < y < \infty\}$. Denote the rectilinear distance between **X** and $\mathbf{P}_i$ by $f_i(X)$, and for every design **S** in $H(\mathbf{L}: A)$, define an average distance $F(\mathbf{S})$ by

$$F(\mathbf{S}) = \sum_{i=1}^{2} \int_{\mathbf{S}} \frac{1}{A} f_i(\mathbf{X})\, d\mathbf{X}$$

(a) Find and draw a design **S*** in $H(\mathbf{L}: A)$ to minimize $F(\mathbf{S})$. [*Hint:* $q(z) = d(z - d)$.] Show that

$$F(\mathbf{S}^*) = \frac{A + 2d^2}{2d}$$

Note that this problem may be considered as that of finding a warehouse design where all items enter at one dock and depart from the other dock.

(b) Suppose that you have the option of choosing d as well. Find the value of d to minimize $F(\mathbf{S}^*)$. In the event that **S*** turns out to be a rectangle, determine its x and y dimensions.

7.18. Let **L** be the union of the first and fourth quadrants, let d be a positive number, and define the points

$$\mathbf{P}_1 = (0, 0), \qquad \mathbf{P}_2 = (0, d), \qquad \mathbf{P}_3 = (0, 2d)$$

and let

$$w_1 = 1, \qquad w_2 = 2, \qquad w_3 = 1.$$

Given that $f_i(\mathbf{X})$ represents the rectilinear distance between $\mathbf{X}$ and $\mathbf{P}_i$, multiplied by w_i, define a cost of item movement $F(\mathbf{S})$ for every design in $H(\mathbf{L}: A)$ by

$$F(\mathbf{S}) = \sum_{i=1}^{3} \int_{\mathbf{S}} \frac{1}{A} f_i(\mathbf{X})\, d\mathbf{X}$$

and define the function $f(\mathbf{X})$ by

$$f(\mathbf{X}) = \sum_{i=1}^{3} f_i(\mathbf{X})$$

so that

$$F(\mathbf{S}) = \int_{\mathbf{S}} \frac{1}{A} f(\mathbf{X})\, d\mathbf{X}.$$

(a) Plot contour lines of $f(\mathbf{X})$ in $\mathbf{L}$ for $f(\mathbf{X}) = 3d$ and $f(\mathbf{X}) = 5d$.

(b) Show, using trigonometric arguments, that the function $q(z)$ is given by

$$q(z) = \frac{1}{2}\left(\frac{z}{2} - d\right)^2, \quad 2d \leq z \leq 4d$$

$$q(z) = \frac{z^2}{16} - \frac{d^2}{2}, \quad 4d \leq z$$

(c) Show that

$$r(t) = 2(2t)^{1/2} + 2d, \quad 0 \leq t \leq \frac{d^2}{2}$$

$$r(t) = 4\left(t + \frac{d^2}{2}\right)^{1/2}, \quad \frac{d^2}{2} \leq t$$

(d) Find a design $\mathbf{S}^*$ to minimize $F(\mathbf{S})$ for (1) $A \leq d^2/2$ and (2) $A > d^2/2$.

(e) For $A \leq d^2/2$, show that

$$F(\mathbf{S}^*) = \frac{4}{3}(2A)^{1/2} + 2d$$

(f) For $A > d^2/2$, show that

$$F(\mathbf{S}^*) = \frac{8}{3A}\left(A + \frac{d^2}{2}\right)^{3/2} - \frac{d^3}{A}$$

7.19. Given a positive number a and any real number p, show that the value of the integral

$$\int_{p}^{a+p} \frac{1}{a} |x|\, dx$$

is given by $|p + (a/2)|$ when $p \geq 0$ or $p \leq -a$, and by $a/4 + (1/a)[p + (a/2)]^2$ when $-a \leq p \leq 0$. [*Hint:* When $p \geq 0$

$$\int_p^{a+p} \frac{1}{a}|x|\,dx = \int_p^{a+p} \frac{1}{a}x\,dx$$

When $p \leq -a$, then $a + p \leq 0$, and so

$$\int_p^{a+p} \frac{1}{a}|x|\,dx = \int_p^{a+p} \frac{1}{a}(-x)\,dx$$

Finally, when $-a \leq p \leq 0$, then $a + p \geq 0$, and so

$$\int_p^{a+p} \frac{1}{a}|x|\,dx = \int_p^{0} \frac{1}{a}(-x)\,dx + \int_0^{a+p} \frac{1}{a}x\,dx.]$$

7.20. Show that the least-cost rectangular warehouse design obtained when the requirement is omitted that the dock not lie inside the warehouse is a square of dimensions $A^{1/2}$ by $A^{1/2}$ with the dock at its center. Derive the total cost for the design.

7.21. Rework the example problem for which the cost of a design is given by Equation (7.36) when (a) **L** is the entire plane, (b) distance is Euclidean instead of rectilinear, (c) distance is Euclidean and **L** is the entire plane.

7.22. Solve Problem 7.21 for the case of m sets instead of 2 sets.

7.23. Rework the second example problem in the text that illustrates the use of Property 3. Consider the following cases: (a) **L** is the entire plane, (b) distances are Euclidean, (c) distances are Euclidean and **L** is the entire plane.

7.24. Solve Problem 7.23 for the case of m sets instead of 2 sets.

7.25. The function $f(\mathbf{X})$ is defined by

$$f(\mathbf{X}) = \sum_{j=1}^{n} w_j \|\mathbf{X} - \mathbf{P}_j\|$$

where the w_j are positive constants, and $\|\mathbf{X} - \mathbf{P}_j\|$ is the l_p distance between $\mathbf{X}$ and $\mathbf{P}_j$. Let t be an integer between 1 and $n - 1$, and define the terms

$$w^* = \sum_{j=t+1}^{n} w_j, \qquad \mathbf{P}^* = \sum_{j=t+1}^{n} \frac{w_j}{w^*}\mathbf{P}_j$$

(a) Prove that

$$f(\mathbf{X}) \geq \sum_{j=1}^{t} w_j \|\mathbf{X} - \mathbf{P}_j\| + w^* \|\mathbf{X} - \mathbf{P}^*\|$$

(b) Explain the implication the inequality of part (a) has when deciding upon the number of docks to be used in a warehouse.

7.26. A rectangular warehouse is to have an x dimension of 300 feet and a y dimension of 100 feet. The axis is situated so that the lower left-hand corner of the warehouse coincides with the origin. The warehouse has docks located at the points $\mathbf{P}_1 = (100, 100)$, $\mathbf{P}_2 = (150, 0)$, and $\mathbf{P}_3 = (200, 100)$. Two items will be located in the warehouse; each item will take up 10,000 square feet. The item–dock $\mathbf{W}$ matrix is as follows:

$$\mathbf{W} = \begin{pmatrix} 2{,}000 & 4{,}000 & 2{,}000 \\ 1{,}500 & 3{,}000 & 1{,}500 \end{pmatrix}$$

Let $\mathbf{L}$ represent the warehouse, $H_2(\mathbf{L}: A)$ the class of all layouts, $r(\mathbf{X}, \mathbf{P}_i)$ the rectilinear distance between $\mathbf{X}$ and $\mathbf{P}_i$, and define the total cost for a layout by

$$F(\mathbf{S}_1, \mathbf{S}_2) = \sum_{i=1}^{2} \sum_{j=1}^{3} w_{ij} \int_{\mathbf{S}_i} \frac{1}{A_i} r(\mathbf{X}, \mathbf{P}_i)\, d\mathbf{X}$$

(a) With c_i and w_j defined as usual based on the $\mathbf{W}$ matrix, show that

$$\sum_{j=1}^{3} w_{ij} \int_{\mathbf{S}_i} \frac{1}{A_i} r(\mathbf{X}, \mathbf{P}_i)\, d\mathbf{X} = c_i \int_{\mathbf{S}_i} \frac{1}{A_i} r(\mathbf{X})\, d\mathbf{X} \qquad (1)$$

where
$$f(\mathbf{X}) = \sum_{j=1}^{3} w_j r(\mathbf{X}, \mathbf{P}_i)$$

Also, give a physical interpretation to the left side of (1).

(b) Given that the function $q(z)$ is defined by

$$q(z) = 400z - 30{,}000, \quad 75 \le z \le 100$$
$$q(z) = 200z - 10{,}000, \quad 100 \le z$$

find a layout $\{\mathbf{S}_1^*, \mathbf{S}_2^*\}$ in the $H_2(\mathbf{L}: A)$ to minimize $F(\mathbf{S}_1, \mathbf{S}_2)$ and make a sketch of the layout.

(c) Compute the value of equation (1) for $i = 1$ and $i = 2$.

(d) Compute $F(\mathbf{S}_1^*, \mathbf{S}_2^*)$.

(e) Show that the function $q(z)$ is in fact as given in part (b).

7.27. Design a continuous facility layout for a warehouse that has two docks located at $\mathbf{P}_1 = (0, 50)$ and $\mathbf{P}_2 = (50, 0)$. The warehouse is to be located in the first quadrant. Only one product is to be stored in the warehouse. Rectilinear travel is to be used.

(a) Determine the configuration, giving all dimensions, for the warehouse design that minimizes the expected distance traveled per unit time when twice as many trips are made to the dock at $\mathbf{P}_1$ as to the dock at $\mathbf{P}_2$.

(b) Determine the expected distance traveled per unit time based on the optimum configuration.

7.28. Consider a rectangular warehouse design. Compute the ratio a^*/b^* and plot a family of curves indicating the sensitivity of the ratio to various values of c and r. Suppose that $r > 20c$; what range of values can be placed on the ratio? What physical interpretation can be given to the inequality $r > 20c$?

7.29. Solve Problem 7.1 for the case of a single dock located at the origin and **L** defined as the first and fourth quadrants. Develop solutions and determine the values of $F(\mathbf{S}^*)$ for

(a) Rectilinear travel.

(b) Euclidean travel.

(c) Rectilinear travel, rectangular design (no perimeter cost).

(d) Rectilinear travel, rectangular design, $r = 10c$.

7.30. Based on the "nearest-dock assumption" (see Problem 6.28), develop the warehouse design for the two-dock case given in Figure 7.7.

7.31. Suppose that a warehouse has radial aisles leading to a single dock, and travel from a storage point (x, y) is made rectilinearly to a radial aisle and then along the radial aisle to the dock. The radial aisles emanate from the dock similar to spokes in a wheel. Assuming travel is made to the adjacent radial aisle that would minimize distance traveled, approximate the expected distance traveled for the radial-aisle configurations shown in Figure P7.31. Compare your formulation and solutions with those given in [6].

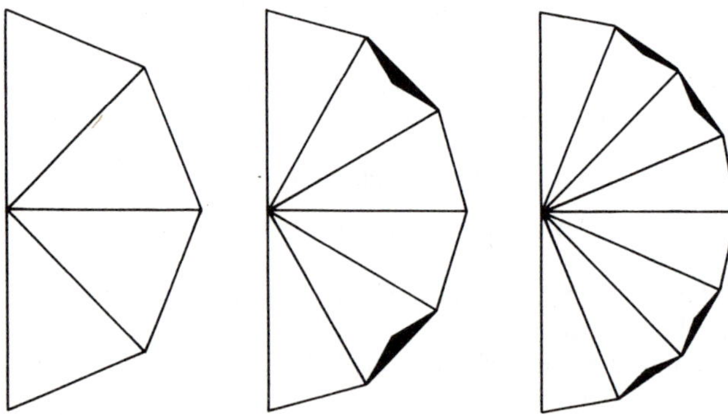

Figure P7.31

7.32. An identity useful for a general proof of Property 3 is known as summation by parts, and may be stated as follows: given numbers $a_1, \ldots, a_m$ and $b_1, \ldots, b_m$, if $S_j = b_1 + b_2 + \cdots + b_j$ for $j = 1, \ldots, m$, then $\sum_{i=1}^{m} a_i b_i = \sum_{j=1}^{m-1} (a_j - a_{j+1})S_j + a_m S_m$. Derive the summation by parts identity.

7.33. Prove Property 3 for (a) the case $m = 3$, and (b) the case where m is any positive integer.

7.34. Determine the value of $F(\mathbf{S}^*)$ for the example problem resulting in Figures 7.5, 7.6, and 7.8.

7.35. Formulate the location problem involving n new facilities given as points and m existing facilities, which can be points or areas. Assume rectilinear travel. Compare your formulation with that given by Wesolowsky and Love [5].

QUADRATIC ASSIGNMENT LOCATION PROBLEMS

8.1 Introduction

In Chapter 6 we considered the problem of assigning facilities to location sites. In that discussion, it was assumed there was no interchange between new facilities. We consider now the problem of assigning facilities to sites when there is an interchange between pairs of new facilities. As before, we consider a finite number of facilities and sites. Furthermore, exactly one facility is to be assigned to each site. The sites might be, for example, rooms in a plant, and the facilities might be departments to be assigned to the rooms. Alternatively, the facilities might be new machines to be located in a job shop, and the sites might be the possible locations of new facilities. As a third example, the sites might be actual geographic sites, and the facilities might be plants to be assigned to the sites.

The problem of assigning new facilities to sites when there is an interchange between new facilities is referred to as a quadratic assignment problem. The quadratic assignment problem is closely related to the multifacility location problem treated in Chapter 5. The major difference in the two problems concerns the number of location sites. In Chapter 5 we assumed an infinite solution space; whereas here we assume a finite number of location

8

sites. The location problem can be formulated as follows. Let c_{ikjh} denote the annual cost of having facility i located at site k and facility j located at site h. Also, let the decision variable x_{ik} equal one if facility i is located at site k and equal zero, otherwise. If there are n new facilities and sites, we wish to

$$\text{minimize} \qquad f(x) = \frac{1}{2} \sum_{i=1}^{n} \sum_{k=1}^{n} \sum_{j=1}^{n} \sum_{h=1}^{n} c_{ikjh} x_{ik} x_{jh}$$

$$\text{subject to} \qquad \sum_{i=1}^{n} x_{ik} = 1, \quad k = 1, \ldots, n$$

$$\sum_{k=1}^{n} x_{ik} = 1, \quad i = 1, \ldots, n$$

$$x_{ik} = 0 \text{ or } 1, \quad \text{for all i, } k$$

Notice that if facility i is located at site k and facility j is located at site h then x_{ik} and x_{jh} both equal 1 and the cost term c_{ikjh} is included in the total cost calculation. The first set of constraints ensures that exactly one facility is assigned to each site; the second set of constraints results in each facility being assigned to exactly one site. As with the assignment problem described in Chapter 6, we might have to add dummy facilities in order for there to be exactly n facilities to be assigned to the n sites.

We have previously discussed an important class of quadratic assignment problems, the plant layout problem. In Chapters 2 and 3 we presented various approaches that have been used to solve plant layout problems. Recall that both the ALDEP program and the CORELAP program assigned numerical values to the A, E, I, O, U, and X closeness ratings; these values are analogous to the c_{ikjh} values in the quadratic assignment formulation. The CRAFT program combined the flow data, cost data, and the distances between sites to obtain the matrix of c_{ikjh} values. (Recall that we made special mention of the use of CRAFT in solving quadratic assignment problems.)

In addition to the plant layout problem, a number of other location problems can be correctly formulated as quadratic assignment problems. The design of control panels to minimize the expected time required to execute a sequence of operations is one illustration of a quadratic assignment problem. If the sequence of operations involved the adjustment of control knob i located at position k, followed by the adjustment of control knob j located at position h, then c_{ikjh} would represent the time required to go from position k to position h.

The location of items in storage bins in a storeroom is another example of a quadratic assignment problem. The stockkeeper would want the items placed in such a way that the time required for him to fill a customer's order would be minimized. Thus, if an order involved the selection of item i located in bin k followed by the selection of item j located in bin h, the time required to walk from bin k to bin h would be given by c_{ikjh}. Additional illustrations of quadratic assignment problems are given as example problems in the subsequent discussion and as exercises at the end of the chapter.

Our discussion of the quadratic assignment problem begins with a brief summary of basic ideas, followed by a treatment of several heuristic solution procedures. Specifically, a steepest-descent pairwise-interchange procedure similar to that employed in CRAFT is presented, as are the heuristic procedures developed by Vollmann, Nugent, and Zartler [12] and Hillier [6]. Our treatment of the quadratic assignment problem concludes with a discussion of an exact solution procedure, based on the work of Gilmore [4] and Lawler [8].

The material presented in the chapter is arranged in an order of increasing difficulty of understanding, based on our experience in teaching the subject matter in both undergraduate and graduate level courses. Depending on the amount of exposure you have had to branch and bound methods, you may choose to omit the treatment of the exact solution procedure. Since you have been exposed previously to the CRAFT procedure, the discussion of the pairwise steepest-descent procedure should be easily understood.

The Vollmann, Nugent, Zartler procedure and the Hillier procedure are related to the CRAFT procedure and should be quite accessible.

8.2 Basic Ideas

In this section we lay the groundwork for the subsequent discussions concerning solution procedures for the quadratic assignment problem. Several basic ideas are presented as well as an example problem, which is used to illustrate the various solution procedures.

Given an assignment of facilities to sites, we find it instructive to let $a(i)$ denote the number of the site to which facility i is assigned, and let $\mathbf{a}$ be the assignment vector

$$\mathbf{a} = (a(1), a(2), \ldots, a(n))$$

Note that the ith component of the assignment vector $\mathbf{a}$ is the number of the site to which facility i has been assigned.

As an illustration of the use of the assignment vector, consider the problem of designing a complex of six novelty and craft shops in a resort area. The six shops are to be located in a building having the design shown in Figure 8.1(a). Each of the six sites is a candidate for the location of each shop. If the shops (facilities) are assigned to the sites according to the assignment vector

$$\mathbf{a} = (2, 4, 5, 3, 1, 6)$$

then the shops will be located in the building as shown in Figure 8.1(b). Thus, for the assignment, x_{12}, x_{24}, x_{35}, x_{43}, x_{51}, and x_{66} equal one and all other x_{ik} are zero.

Given an assignment of facilities to sites, the next step is to compute a total cost for the assignment. For all integer values of k and h between 1 and

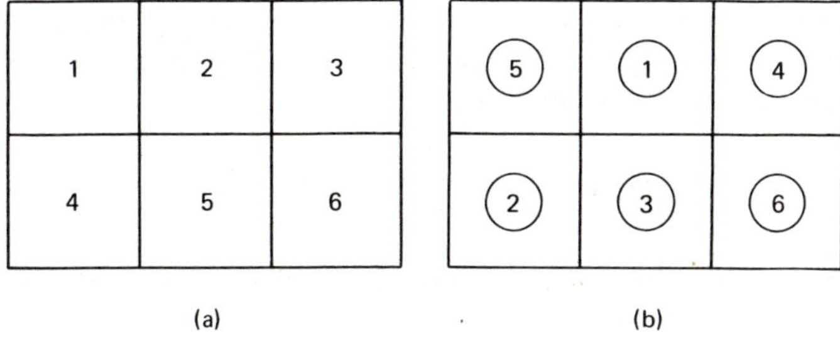

Figure 8.1. Example of (a) sites and (b) assignment of facilities to sites.

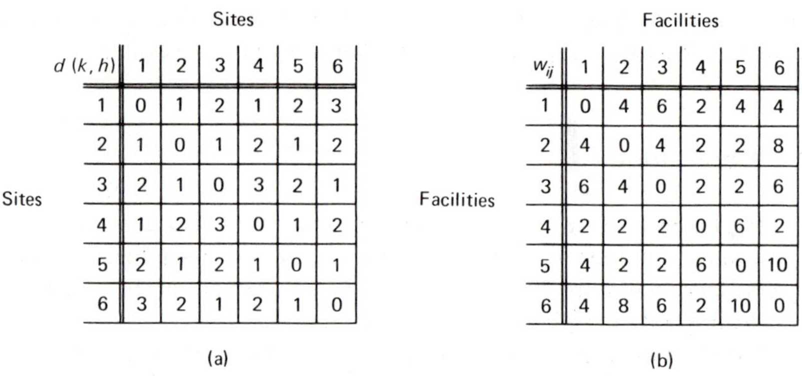

Sites × Sites

$d(k, h)$	1	2	3	4	5	6
1	0	1	2	1	2	3
2	1	0	1	2	1	2
3	2	1	0	3	2	1
4	1	2	3	0	1	2
5	2	1	2	1	0	1
6	3	2	1	2	1	0

(a)

Facilities × Facilities

w_{ij}	1	2	3	4	5	6
1	0	4	6	2	4	4
2	4	0	4	2	2	8
3	6	4	0	2	2	6
4	2	2	2	0	6	2
5	4	2	2	6	0	10
6	4	8	6	2	10	0

(b)

Figure 8.2. Example of (a) distance matrix and (b) weight matrix.

n, let $d(k, h)$ denote an appropriately determined distance between site k and site h. As a subsequent notational convenience, it will be useful to suppose that $d(k, h) = d(h, k)$ for all k and h. A site–distance matrix $\mathbf{D} = (d(k, h))$ for the example is shown in Figure 8.2(a); for the example, the distances between sites are assumed to be the rectilinear distances, measured in units of site widths, between the centers of sites. Suppose finally that w_{ij} is a constant of proportionality converting the distance between facilities i and j, for all $i < j$, into a cost, so that if facility i is at site $a(i)$ and facility j is at site $a(j)$ the total cost for facilities i and j is $w_{ij}\, d(a(i), a(j))$. The total cost for an assignment of facilities to sites is then

$$TC(\mathbf{a}) = \sum_{1 \le i < j \le n} w_{ij}\, d(a(i), a(j)) \tag{8.1}$$

The total cost for the assignment shown in Figure 8.1(b), using the w_{ij} terms above the diagonal in Figure 8.2(b), is 114.

Notice that the weight matrix $\mathbf{W} = (w_{ij})$ shown in Figure 8.2(b) includes entries on and below the diagonal, as well as above the diagonal. Also, notice that $w_{ij} = w_{ji}$ and that all the diagonal terms are zero. It will be assumed in general that any weight matrix $\mathbf{W}$ has the same property. The reason for the assumption is basically one of computational convenience; for example, the total cost of an assignment is also given by

$$TC(\mathbf{a}) = \tfrac{1}{2} \sum_{i=1}^{n} \sum_{j=1}^{n} w_{ij}\, d(a(i), a(j)) \tag{8.2}$$

To establish that expressions (8.1) and (8.2) are identical, note that (8.2) may be written as

$$\tfrac{1}{2}\left[\sum_{1 \le i < j \le n} w_{ij} d(a(i), a(j)) + \sum_{n \ge i > j \ge 1} w_{ij} d(a(i), a(j)) + \sum_{i=1}^{n} w_{ii} d(a(i), a(i))\right]$$

Now, the last sum in the latter expression is just zero, while the second sum may be written, using a change of indices, as

$$\sum_{n \geq e > f \geq 1} w_{ef} d(a(e), a(f)) \tag{8.3}$$

By assumption, $w_{ef} = w_{fe}$ and $d(a(e), a(f)) = d(a(f), a(e))$, so (8.3) is the same as

$$\sum_{n \geq e > f \geq 1} w_{fe}\, d(a(f), a(e)) = \sum_{1 \leq i < j \leq n} w_{ij}\, d(a(i), a(j)) \tag{8.4}$$

It now follows from the last three expressions that (8.1) and (8.2) give the same total cost for an assignment. Note that the right side of (8.4) is obtained from the left side of (8.4) just by replacing f by i and e by j, and observing that the set of summation indices $\{e, f: n \geq e > f \geq 1\}$ is the same as the set of summation indices $\{i, j: 1 \leq i < j \leq n\}$; the only difference between the left and right sides of (8.4) is the order of summation.

Another, and different, total cost expression will be useful subsequently; given an assignment of facilities to locations, the total cost for facility k is given, for $k = 1, \ldots, n$, by

$$p_k(\mathbf{a}) = \sum_{1 \leq i < k} w_{ik}\, d(a(i), a(k)) + \sum_{k < j \leq n} w_{kj}\, d(a(k), a(j)) \tag{8.5}$$

With reference to the example, (8.5), for $k = 3$, is

$$\begin{aligned} p_3(\mathbf{a}) = w_{13}\, d(a(1), a(3)) + w_{23}\, d(a(2), a(3)) + w_{34} d(a(3), a(4)) \\ + w_{35}\, d(a(3), a(5)) + w_{36}\, d(a(3), a(6)) \end{aligned} \tag{8.6}$$

Notice that (8.6) may also be written, using the symmetry of the **D** and **W** matrices, as

$$\begin{aligned} w_{31}\, d(a(3), a(1)) + w_{32}\, d(a(3), a(2)) + w_{33}\, d(a(3), a(3)) + w_{34}\, d(a(3), a(4)) \\ + w_{35}\, d(a(3), a(5)) + w_{36}\, d(a(3), a(6)) \end{aligned}$$

By simply generalizing the same approach, another equivalent expression for $p_k(\mathbf{a})$ is

$$p_k(\mathbf{a}) = \sum_{j=1}^{n} w_{kj}\, d(a(k), a(j)) \tag{8.7}$$

Given an assignment of facilities to sites, with a corresponding total cost $TC(\mathbf{a})$, it is natural to ask how an assignment with a lower cost might be found. One approach is simply to find two facilities, say u and v, that by interchanging their locations will yield an assignment with a lower total cost. The locations of facilities u and v could be interchanged to determine a new assignment, say $\mathbf{a}'$, with a corresponding total cost $TC(\mathbf{a}')$, and then $TC(\mathbf{a}) - TC(\mathbf{a}')$ could be computed to determine if the total cost has been decreased. This approach is unduly complicated, however, since both the costs $TC(\mathbf{a})$ and $TC(\mathbf{a}')$ include a number of common terms, that is, all those not involving facilities u or v. Thus, to compute $TC(\mathbf{a}) - TC(\mathbf{a}')$, it is only necessary to subtract those terms involving facilities u or v in $TC(\mathbf{a}')$ from those terms involving facilities u or v in $TC(\mathbf{a})$. The sum of all terms involving facilities u or v in $TC(\mathbf{a})$ is given by

$$\sum_{\substack{i=1\\ \neq v}}^{n} w_{iu}\, d(a(i), a(u)) + \sum_{\substack{i=1\\ \neq u}}^{n} w_{iv}\, d(a(i), a(v)) + w_{uv}\, d(a(u), a(v)) \tag{8.8}$$

When the sites at which facilities u and v are located are interchanged, facility u will be located at site $a(v)$, and facility v will be located at site $a(u)$, so that the sum of all terms involving facilities u or v in $TC(\mathbf{a}')$ is given by

$$\sum_{\substack{i=1\\ \neq v}}^{n} w_{iu}\, d(a(i), a(v)) + \sum_{\substack{i=1\\ \neq u}}^{n} w_{iv}\, d(a(i), a(u)) + w_{uv}\, d(a(v), a(u)) \tag{8.9}$$

Thus it is only necessary to subtract (8.9) from (8.8) to compute $TC(\mathbf{a}) - TC(\mathbf{a}')$. Prior to carrying out the computation, it is convenient to obtain alternative expressions for (8.8) and (8.9). Note that (8.8) may be written, on adding

$$0 = w_{vu}\, d(a(v), a(u)) + w_{uv}\, d(a(u), a(v)) - w_{vu}\, d(a(v), a(u)) - w_{uv}\, d(a(u), a(v))$$

as

$$\sum_{i=1}^{n} w_{iu}\, d(a(i), a(u)) + \sum_{i=1}^{n} w_{iv}\, d(a(i), a(v)) - w_{vu}\, d(a(v), a(u)) \tag{8.10}$$

Likewise, (8.9) may be written, on adding

$$0 = w_{vu}\, d(a(v), a(v)) + w_{uv}\, d(a(u), a(u))$$

as

$$\sum_{i=1}^{n} w_{iu}\, d(a(i), a(v)) + \sum_{i=1}^{n} w_{iv}\, d(a(i), a(u)) + w_{uv}\, d(a(v), a(u)) \tag{8.11}$$

Thus the change in cost obtained by interchanging the location of facilities u and v for a given assignment, which will be denoted by $DTC_{uv}(\mathbf{a})$, is readily obtained by subtracting (8.11) from (8.10), yielding

$$DTC_{uv}(\mathbf{a}) = \sum_{i=1}^{n} (w_{iu} - w_{iv})\,[d(a(i), a(u)) - d(a(i), a(v))] - 2w_{uv}\, d(a(u), a(v)) \tag{8.12}$$

Equation (8.12) will be illustrated for $u = 2$ and $v = 4$ in the example; $DTC_{24}(\mathbf{a})$ is equal to

$$\begin{aligned}
&(w_{12} - w_{14})[d(2, 4) - d(2, 3)] + (w_{22} - w_{24})[d(4, 4) - d(4, 3)] \\
&\quad + (w_{32} - w_{34})[d(5, 4) - d(5, 3)] + (w_{42} - w_{44})\,[d(3, 4) - d(3, 3)] \\
&\quad + (w_{52} - w_{54})[d(1, 4) - d(1, 3)] + (w_{62} - w_{64})\,[d(6, 4) - d(6, 3)] \\
&\quad - 2w_{24}\, d(4, 3) \\
&= 2(1) - 2(-3) + 2(-1) + 2(3) - 4(-1) + 6(1) - 2(2)(3) = 10
\end{aligned}$$

Thus, when the locations of facilities 2 and 4 are interchanged, giving the new assignment

$$\mathbf{a}' = (2, 3, 5, 4, 1, 6)$$

then the value of $TC(\mathbf{a}')$ is $TC(\mathbf{a}) - DTC_{24}(\mathbf{a}) = 114 - 10 = 104$.

Several points about the expression $DTC_{uv}(\mathbf{a})$ are worth making. Note that in $DTC_{uv}(\mathbf{a})$ the terms $w_{1u} - w_{1v}$, $w_{2u} - w_{2v}, \ldots, w_{nu} - w_{nv}$ are just the differences of the entries in columns u and v of the matrix $\mathbf{W}$, computed first for the first row, then for the second row, and so forth. If the computation of $DTC_{uv}(\mathbf{a})$ is to be carried out for a large number of different assignments, it may be convenient to construct a table of these differences, rather than recompute them using the data in the $\mathbf{W}$ matrix each time. The terms

$$d(a(i), a(u)) - d(a(i), a(v)), \quad i = 1, \ldots, n$$

are the differences between the entries in columns $a(u)$ and $a(v)$ of the $\mathbf{D}$ matrix, computed in the row order $a(1), a(2), \ldots, a(n)$. It is the computation of the terms $d(a(i), a(u)) - d(a(i), a(v))$ that constitutes most of the work in computing $DTC_{uv}(\mathbf{a})$, since the assignment $\mathbf{a}$ must constantly be made use of in computing these terms. Also, if a least total cost assignment, say $\mathbf{a}^*$, is found, then for the interchange of locations of *any* two facilities u and v, $DTC_{uv}(\mathbf{a}^*) \leq 0$, so that we have a necessary condition for a least total cost assignment. Unfortunately, the necessary condition is not in general sufficient, since there is no guarantee that an assignment which satisfies the necessary condition cannot be improved upon by considering, for example, three-way interchanges or four-way interchanges. Another point to be made is that many computational procedures that attempt to find a least-cost assignment do so by making pairwise interchanges, so that it is often convenient to use the expression for $DTC_{uv}(\mathbf{a})$ in these procedures.

Procedures that have been developed to attempt to find least-cost solutions to the quadratic assignment problem are of two types: exact and heuristic. Exact procedures do in fact find least total cost assignments. With the exception of total enumeration, all the exact procedures of interest developed thus far are branch and bound, or implicit enumeration, procedures. Branch and bound procedures to date have not proved to be generally computationally satisfactory; it has been estimated that they are unable to solve problems where n is greater than 15. The branch and bound procedures are of considerable theoretical interest, however. It seems likely that if a computationally useful exact solution procedure is developed, the theory of the procedure will be related to, or extend that of, the branch and bound procedures. One exact solution procedure, total enumeration, may be ruled out immediately. Total enumeration requires the explicit consideration of all $n!$ assignments. When $n = 12$, $n! \doteq 4.79(10^8)$; it has been estimated that the time required to solve a problem with $n = 12$ by total enumeration on an IBM 7090 computer would be about 3 years [10]. Even if a later-generation computer were used, it is evident that the use of total enumeration would be infeasible.

While total enumeration is out of the question, and branch and bound methods are of limited value in finding least-cost assignments, there is one notion often used with branch and bound methods that is computationally quite simple and can be useful in developing heuristic procedures. Recall the expression for the total cost of an assignment **a**:

$$TC(\mathbf{a}) = \sum_{1 \leq i < j \leq n} w_{ij}\, d(a(i), a(j))$$

The problem we would like to solve is that of finding an assignment which would minimize the total cost expression, thus giving a minimum value of the total cost. If this problem cannot be solved, it may still be useful to have a lower bound on the minimum value of the total cost, for if upper bounds on the minimum value of the total cost can be found that are not much greater than the lower bound, then a good deal of information is available about the minimum value of the total cost. Upper bounds are very easily obtained, since the total cost of any assignment is certainly an upper bound on the minimum value of the total cost.

A lower bound on the minimum value of the total cost may be obtained as follows. Order the entries of the **W** matrix above the diagonal in nonincreasing order to obtain a vector $\mathbf{w}^*$, whose first entry is the largest entry of **W** above the diagonal, whose second entry is the second largest, and so on, giving $\mathbf{w}^* = (w_1^*, w_2^*, \ldots, w_r^*)$, where $r = n(n-1)/2$. Order the entries of the **D** matrix above the diagonal in nondecreasing order to obtain a vector $\mathbf{d}^*$, whose first entry is the smallest entry of **D** above the diagonal, whose second entry is the second smallest entry of **D** above the diagonal, and so on, giving $\mathbf{d}^* = (d_1^*, d_2^*, \ldots, d_r^*)$. Then a lower bound on the minimum value of the total cost is just the inner product of $\mathbf{w}^*$ with $\mathbf{d}^*$:

$$\sum_{i=1}^{r} w_i^*\, d_i^*$$

That the inner product of $\mathbf{w}^*$ with $\mathbf{d}^*$ is a lower bound is easily motivated, since the expression $TC(\mathbf{a})$ is the inner product of a vector consisting of all entries of $\mathbf{W}$ above the diagonal with a vector consisting of all entries of $\mathbf{D}$ above the diagonal. We have seen previously in Chapters 6 and 7 that to minimize such an inner product exactly the prescribed procedure is followed. For the example problem we have been considering, with $\mathbf{D}$ and $\mathbf{W}$ matrices as given in Figure 8.2,

$$\mathbf{w}^* = (10, 8, 6, 6, 6, 4, 4, 4, 4, 2, 2, 2, 2, 2, 2)$$

and

$$\mathbf{d}^* = (1, 1, 1, 1, 1, 1, 1, 2, 2, 2, 2, 2, 2, 3, 3)$$

and

$$\sum_{i=1}^{15} w_i^* d_i^* = 88$$

so that a lower bound on the minimum total cost is 88. Subsequently, we shall find an upper bound on the minimum total cost of 92 and an assignment with a total cost of 92. Thus, for this example, the minimum total cost lies between 88 and 92. We emphasize that in general the minimum total cost may be strictly greater than the lower bound obtained; there is no guarantee that the lower bound and the minimum total cost will be equal. Of course, the lower bound and the minimum total cost will be equal if an upper bound can be found that is equal to the lower bound.

Because exact solution procedures have not yet been developed to the point where they are of significant practical value, heuristic solution procedures have received considerable attention. For the problem being considered, a heuristic procedure may be characterized as one that has intuitive appeal and seems reasonable; such a procedure might be called a "commonsense" procedure. A number of heuristic procedures have been developed for solving the problem being considered, and several will be discussed in the sequel. Although the heuristic procedures have been found to be useful, they have a major shortcoming, which is inherent in their very nature; it is never possible to tell, except by comparison with the answers provided by exact procedures, whether or not the final answers heuristic procedures provide are least-cost solutions. Consequently, heuristic procedures are commonly evaluated by being compared with other heuristic procedures, with the two major evaluative criteria being the total cost of the final solution provided and the computational effort required to obtain the final answer. The heuristic procedures that provide the better answers often involve more computational effort. While, by design, heuristic procedures are computationally feasible, it is in terms of computer usage that they are truly computationally feasible. Only very small problems can be readily solved by hand using heuristic procedures. Due to the fact that the use of heuristic procedures is inextricably linked with the use of the computer, substantial emphasis will be placed on the computational mechanics of the procedures in the discussion to follow.

8.3 The Steepest-Descent Pairwise-Interchange Procedure

One heuristic computational procedure for attempting to find a least-cost assignment is suggested quite naturally by the expression $DTC_{uv}(\mathbf{a})$. Given an assignment **a**, a distance matrix **D**, and a weight matrix **W**, among all pairwise interchanges of facility locations [there are $n(n-1)/2$ possible pairwise interchanges] find one that causes the greatest decrease in the total cost, make this interchange, and revise the assignment, and then repeat the process until the total cost can be no further decreased. The procedure may be characterized as a steepest-descent procedure because it makes that interchange which results in the greatest decrease in the total cost. This procedure constitutes the basis of CRAFT, a computerized layout program discussed in Chapter 3. We choose to discuss the procedure separately from the CRAFT program since (1) the procedure provides useful insight into other procedures, (2) the procedure is simple to state explicitly in algorithmic form, and (3) the CRAFT program incorporates features that the basic procedure does not have.

The steepest-descent procedure will now be stated explicitly, in algorithmic form. You may find it useful to refer to the flow chart in Figure 8.3 in going through the procedure. The procedure assumes we are given an initial assignment **a** and the matrices **D** and **W**.

0. Given: **a, D, W**.
1. Compute $TC(\mathbf{a})$.
2. Set e to 0. [e is the maximum of 0 and the greatest decrease in $TC(\mathbf{a})$ found so far for the given assignment.]
3. Set i to 1 and j to 2.
4. Compute $DTC_{ij}(\mathbf{a})$.
5. If $DTC_{ij}(\mathbf{a})$ is greater than e, go to step 6; otherwise, go to step 7.
6. Set e to $DTC_{ij}(\mathbf{a})$, set u to i, and set v to j.
7. If $j = n$, go to step 8; otherwise, go to step 9.
8. If $i = n - 1$, go to step 11; otherwise, go to step 10.
9. Increment j by 1 and go to step 4.
10. Increment i by 1, let $j = i + 1$, and go to step 4.
11. If e is positive, go to step 12; otherwise, go to step 14.
12. Replace $TC(\mathbf{a})$ by $TC(\mathbf{a}) - e$.
13. Revise the assignment **a** by interchanging the location of facilities u and v and go to step 2.
14. Stop.

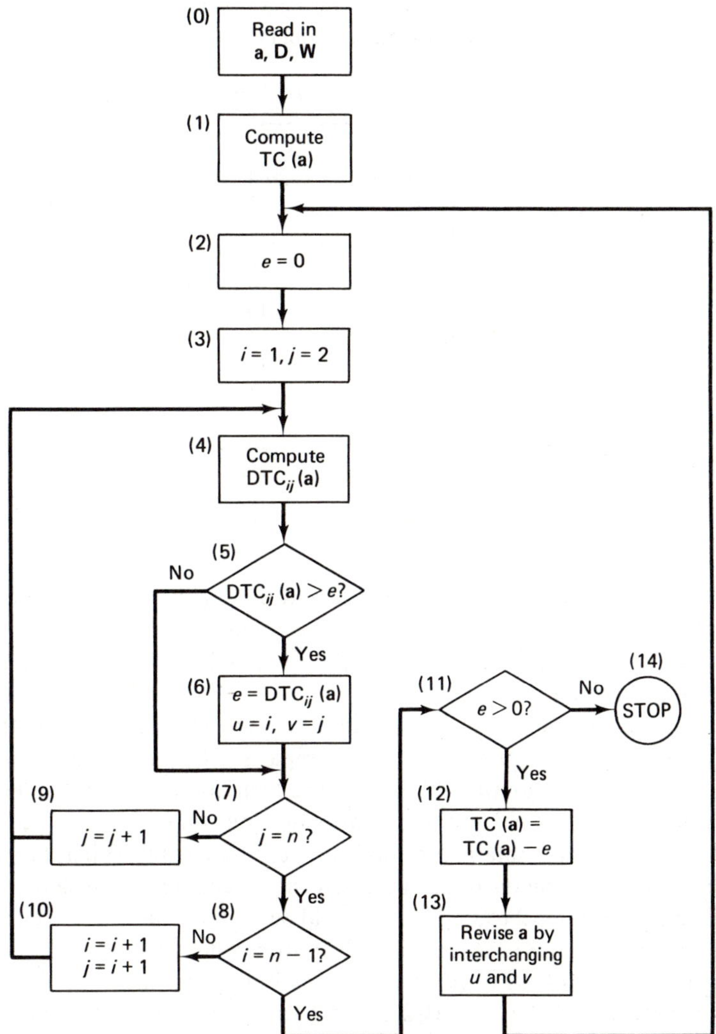

Figure 8.3. Flow chart of steepest-descent pairwise-interchange procedure.

When the procedure is applied to the example problem, given the assignment (2, 4, 5, 3, 1, 6), the values shown in Table 8.1 are computed. The successive values of e are 0, 8, 12, and 16; the final value of u is 2, and of v is 6. Thus the locations of facilities 2 and 6 are interchanged, resulting in the assignment (2, 6, 5, 3, 1, 4) with a corresponding total cost of $114 - 16 = 98$.

Table 8.1. COMPUTATIONAL RESULTS FOR THE EXAMPLE PROBLEM

(i, j)	$DTC_{ij}(\mathbf{a})$
(1, 2)	0
(1, 3)	−4
(1, 4)	−2
(1, 5)	8
(1, 6)	12
(2, 3)	0
(2, 4)	10
(2, 5)	−4
(2, 6)	16
(3, 4)	−4
(3, 5)	8
(3, 6)	10
(4, 5)	16
(4, 6)	−4
(5, 6)	6

When the procedure is applied a second time, given the assignment $\mathbf{a} = (2, 6, 5, 3, 1, 4)$, e remains zero and so the procedure stops; facility 1 is in location 2, 2 is in 6, 3 is in 5, 4 is in 3, 5 is in 1, and 6 is in 4; the assignment is shown in Figure 8.4.

A few concluding remarks about the steepest-descent pairwise-interchange procedure are appropriate. The comparisons made to date [7] indicate that the procedure is among the best of the heuristic procedures in terms of the total cost of the final solution it provides, but that more computational effort is required to obtain the final answer than many other procedures require. If an iteration of the steepest-descent procedure is considered to consist of going from step 2 to step 13, it should be evident why the computational effort required by the procedure is substantial, since, during each iteration, $n(n - 1)/2$ values of $DTC_{ij}(\mathbf{a})$ must be computed. On the other hand, it is the computation of *all* $n(n - 1)/2$ of these values that gives the procedure its

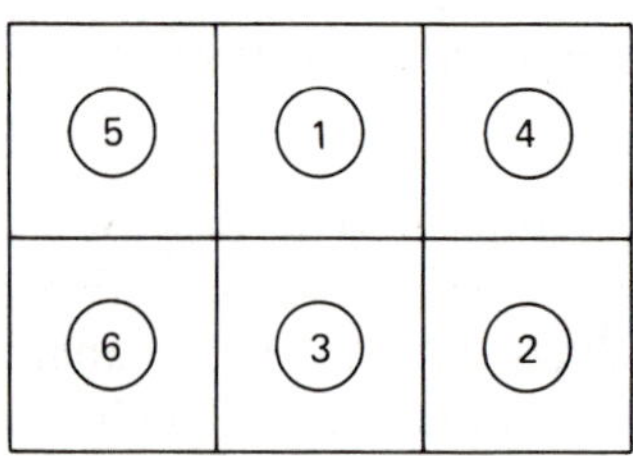

Figure 8.4. Final assignment of facilities to locations resulting from applying pairwise-interchange procedure.

steepest-descent aspect. Several other heuristic procedures that require less computational effort will be considered in the sequel.

8.4 The Vollmann, Nugent, Zartler Procedure

As indicated previously, one of the shortcomings of the steepest-descent pairwise-interchange procedure is the amount of computational effort it requires. In an attempt to find a satisfactory procedure that requires less computational effort, Vollmann, Nugent, and Zartler have devised an alternative procedure(which will subsequently be referred to as the VNZ procedure). They state, based on a common set of test problems given in [12], in which a number of procedures were compared, that the procedure "produces results that are not different with statistical significance from . . . CRAFT, has less storage needs than any procedure examined, and has computation times of one half to one third of the fastest procedure examined. . . ."

The VNZ procedure consists of two phases, which will subsequently be referred to as phase 1 and phase 2. Given an initial assignment, along with the matrices **D** and **W**, in phase 1 two facilities are identified, say M_1 and M_2, that have highest and second highest total costs. A list of other facilities is then established consisting of all facilities that, when interchanged with facility M_1, would cause the total cost of the assignment to decrease. From the list, that facility is chosen which will cause the greatest decrease in the total cost; facility M_1 is then interchanged with the chosen facility, which is then deleted from the list. A second facility is chosen from the list with which to interchange facility M_1, provided the total cost will be decreased, the interchange is made, and the facility is deleted from the list; if an interchange will not reduce the total cost, the facility is deleted from the list, and the process of interchanging facility M_1 with remaining facilities on the list is continued until the list is depleted. Then a similar list is established for facility M_2, and the process repeats itself for facility M_2 until the list is depleted. Then total costs are again computed for each facility, two more facilities, M_1 and M_2(not necessarily the same as the first two), are chosen as previously, and the procedure of phase 1 repeats itself until two facilities M_1 and M_2 are found such that neither can be interchanged with any facility on their respective lists and the total cost will decrease. At this point phase 1 stops.

Notice that in phase 1 only two facilities are chosen, and phase 1 concentrates on interchanging the locations of these two facilities with others; this approach contrasts markedly with the steepest-descent pairwise-interchange procedure in which all facilities are considered for interchanges. The choice of the two facilities is motivated by the fact that each has a high total cost,

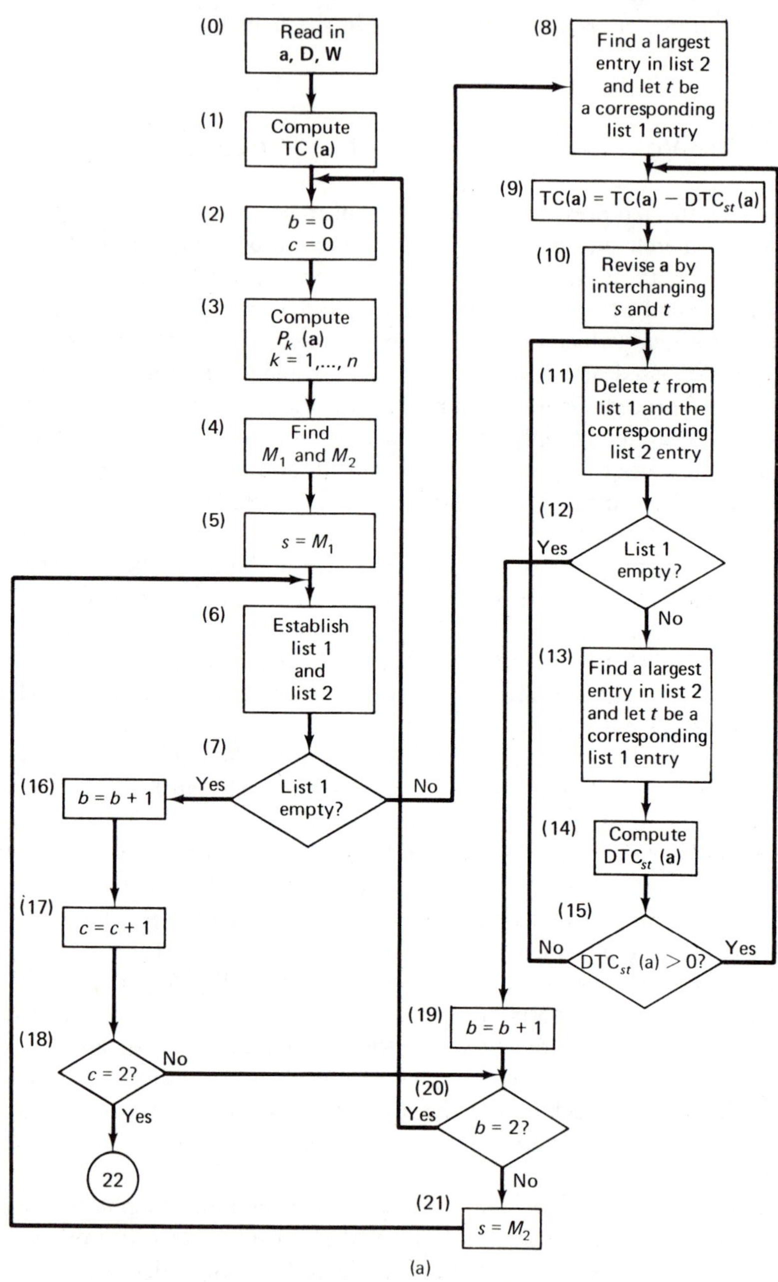

Figure 8.5(a). Flow chart of phase 1 of the VNZ procedure.

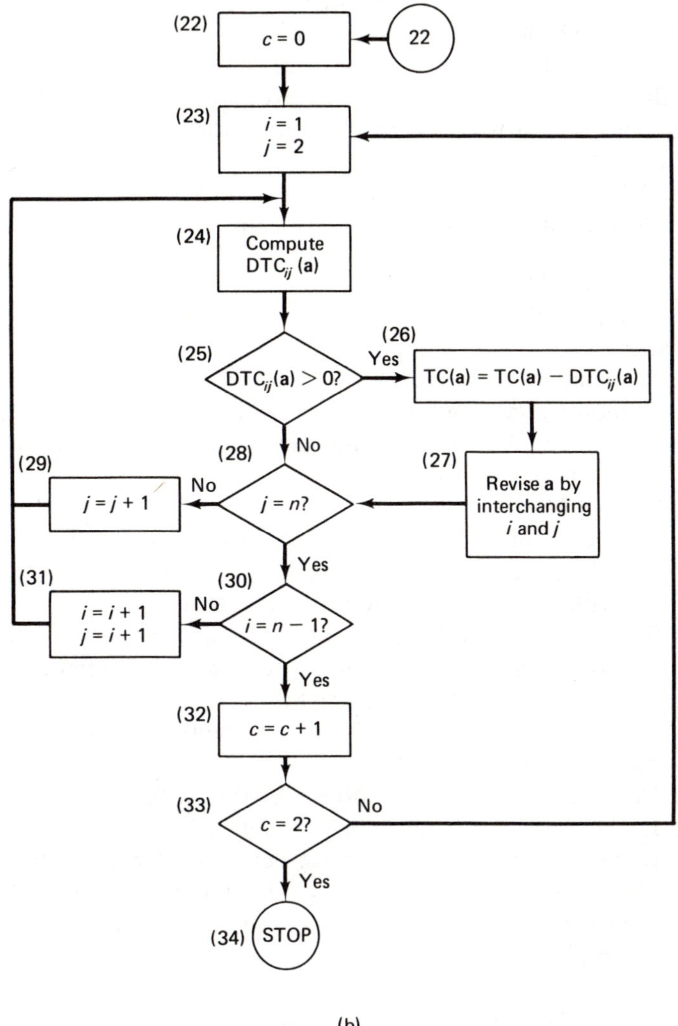

Figure 8.5(b). Flow chart of phase 2 of the VNZ procedure.

which suggests that interchanging these two facilities with others will lead to a greater reduction in total cost than that obtained by most other choices of two facilities.

Phase 2 of the procedure is essentially a double check on phase 1. In phase 2 all pairwise interchanges of facilities are checked twice, and interchanges are made when the total cost is reduced; no attempt is made to use the steepest-descent approach. Since all facilities are considered when making interchanges in phase 2, it is possible that facilities may be interchanged during phase 2 which were not considered for an interchange during phase 1. All pairwise interchanges are checked twice simply to prevent the computational effort in phase 2 from becoming excessive; there appears to be no other obvious reason why two checks, rather than some other number of checks, is made.

The VNZ procedure will now be stated in algorithmic form; a flow chart of the procedure is given in Figures 8.5(a) and (b).

0. Given: $\mathbf{a}$, $\mathbf{D}$, $\mathbf{W}$.
1. Compute $TC(\mathbf{a})$.

Phase 1

2. Set counters b and c to 0.
3. Compute $p_k(\mathbf{a})$, the total cost for facility k, for $k = 1, \ldots, n$.
4. Let M_1 and M_2 be facility numbers with largest and second largest costs computed in step 3, respectively.
5. Set s to M_1.
6. (a) Determine list 1: $\{j: DTC_{sj}(\mathbf{a}) > 0\}$. List 1 is the list of numbers of all facilities that, for the given assignment, if interchanged with facility s will reduce the total cost.
   (b) Determine list 2: $\{DTC_{sj}(\mathbf{a}): DTC_{sj}(\mathbf{a}) > 0\}$. List 2 is the list of cost reductions corresponding to the facility numbers in list 1.
7. If list 1 is not empty, go to step 8; otherwise, go to step 16.
8. Find a largest entry in list 2, and let t be the corresponding facility number in list 1.
9. Compute $DTC_{st}(\mathbf{a})$ and $TC(\mathbf{a}) = TC(\mathbf{a}) - DTC_{st}(\mathbf{a})$.
10. Revise $\mathbf{a}$ by interchanging the locations of facilities s and t.
11. Delete t from list 1 and the corresponding entry from list 2.
12. If list 1 is not empty, go to step 13; otherwise, go to step 19.
13. Find a largest entry in list 2, and let t be the corresponding facility number in list 1.
14. Compute $DTC_{st}(\mathbf{a})$.
15. If $DTC_{st}(\mathbf{a}) > 0$, go to step 9; otherwise, to go step 11.
16. Increment b by 1.
17. Increment c by 1.
18. If $c = 2$, go to step 22; otherwise, go to step 20.
19. Increment b by 1.
20. If $b = 2$, go to step 2; otherwise, go to step 21.
21. Set s to M_2 and go to step 6.

Phase 2

22. Set counter c to 0.
23. Set i to 1 and j to 2.
24. Compute $DTC_{ij}(\mathbf{a})$.

25. If $DTC_{ij}(\mathbf{a}) > 0$, go to step 26; otherwise, go to step 28.
26. $TC(\mathbf{a}) = TC(\mathbf{a}) - DTC_{ij}(\mathbf{a})$.
27. Revise $\mathbf{a}$ by interchanging the locations of i and j.
28. If $j = n$, go to step 30; otherwise, go to step 29.
29. Increment j by 1 and go to step 24.
30. If $i = n - 1$, go to step 32; otherwise, go to step 31.
31. Increment i by 1, let $j = i + 1$, and go to step 24.
32. Increment c by 1.
33. If $c = 2$, go to step 34; otherwise, go to step 23.
34. Stop.

To illustrate the procedure, consider again the example for which the data and the assignment are given in Figures 8.1 and 8.2. Recall that the assignment is given by $\mathbf{a} = (2, 4, 5, 3, 1, 6)$. Computations establish that $TC(\mathbf{a}) = 114$, and that $p_1(\mathbf{a}) = 28$, $p_2(\mathbf{a}) = 36$, $p_3(\mathbf{a}) = 24$, $p_4(\mathbf{a}) = 26$, $p_5(\mathbf{a}) = 52$, and $p_6(\mathbf{a}) = 62$. Since the two largest costs are 62 and 52, $M_1 = 6$, $M_2 = 5$, and $s = 6$. List 1 and list 2 for this example may be obtained from Table 8.1, and are as follows:

List 1	List 2
1	12
2	16
3	10
5	6

Since the largest entry in list 2 is 16, $t = 2$, and the locations of facilities $s = 6$ and $t = 2$ are interchanged, giving the new assignment $\mathbf{a} = (2, 6, 5, 3, 1, 4)$, and $TC(\mathbf{a}) = 114 - 16 = 98$. Facility 2 is now deleted from list 1, and 16 is deleted from list 2. For the new assignment, the terms $DTC_{16}(\mathbf{a})$, $DTC_{36}(\mathbf{a})$, and $DTC_{56}(\mathbf{a})$ are all nonpositive, and so list 1 and list 2 become empty. The counter b takes on the value 1 and $s = 5$. The terms $DTC_{15}(\mathbf{a})$, $DTC_{25}(\mathbf{a})$, $DTC_{35}(\mathbf{a})$, $DTC_{45}(\mathbf{a})$, and $DTC_{56}(\mathbf{a})$ are all nonpositive, so that list 1 is empty. The counter b takes on the value 2, and the counter c takes on the value 1. Since $b = 2$, counters b and c are set to 0, and computations establish, for the assignment $\mathbf{a} = (2, 6, 5, 3, 1, 4)$, that $p_1(\mathbf{a}) = 28$, $p_2(\mathbf{a}) = 36$, $p_3(\mathbf{a}) = 24$, $p_4(\mathbf{a}) = 26$, $p_5(\mathbf{a}) = 36$, and $p_6(\mathbf{a}) = 46$. Thus $M_1 = 6$, $M_2 = 2$ (note that there are two possible choices for M_2), and $s = 6$. Computations establish that the terms $DTC_{16}(\mathbf{a})$, $DTC_{26}(\mathbf{a})$, $DTC_{36}(\mathbf{a})$, $DTC_{46}(\mathbf{a})$, and $DTC_{56}(\mathbf{a})$ are all nonpositive, and so lists 1 and 2 are both empty. Thus counters b and c are incremented by 1, and s is set to 2. Next, computations establish that the terms $DTC_{12}(\mathbf{a})$, $DTC_{23}(\mathbf{a})$, $DTC_{25}(\mathbf{a})$, and $DTC_{26}(\mathbf{a})$ are all nonpositive, and so again lists 1 and 2 are empty. Counters b and c both take on the value 2, and phase 1 is concluded. During phase 2 of the procedure no further interchanges of facilities that will reduce the total cost are found, and the procedure stops.

For the purposes of exposition, several features of the VNZ procedure as originally given have been deleted from the version given here. The procedure can be modified so that, prior to any run of the procedure, the locations of desired facilities in the initial given assignment are fixed. This feature might be useful if the procedure were being used in conjunction with a designer's judgment; after seeing solutions resulting from several different given initial assignments, the designer might decide on definite locations for certain facilities, but might still want to use the procedure as an aid in determining the locations of other facilities. A second feature that the procedure could be modified to incorporate involves the computation of $p_k(\mathbf{a})$ defined in (8.7); the set of summation indices might be changed so that the sum is over all facilities having a distance at least e from facility k. As a motivation for the incorporation for this second feature, if the parameter e were not used, then the cost $p_k(\mathbf{a})$ might be large due to the cost involving facility k and another facility, say i, quite close to facility k. Subsequently the locations of facilities k and i might be interchanged, but the total cost might be only slightly reduced due to the fact that facilities i and k are close together. Since the costs $p_k(\mathbf{a})$ determine the facilities M_1 and M_2 to be moved, it seems reasonable to expect that greater cost reductions may be obtained by interchanging the locations of these two facilities with ones farther away from the two facilities than with ones nearby. This motivation for the incorporation of the parameter e is an entirely heuristic one, of course; there could very well be cases where better assignments could be obtained by not using the parameter e. Furthermore, the proper choice of the value of e is not an obvious one; the choice would have to be based on judgment, experience, and perhaps experimentation with the procedure using various values of e.

8.5 The Hillier Procedure

Unlike the previous procedures, the Hillier procedure addresses itself to a problem with additional special structure, and, by so doing, is able to achieve some special computational advantages.

Figure 8.6 shows the site arrangement that the Hillier procedure assumes: a rectangular region has been partitioned into $n = MN$ grid squares of equal size. Each grid square represents a site, and the grid squares are numbered sequentially 1 through $n = MN$, as shown in Figure 8.6. Due to the similarity of the site arrangement to a matrix, it will sometimes be convenient subsequently to refer to columns and rows of the site arrangement, with the understanding that row 1 constitutes grid squares 1 through N, row 2 constitutes

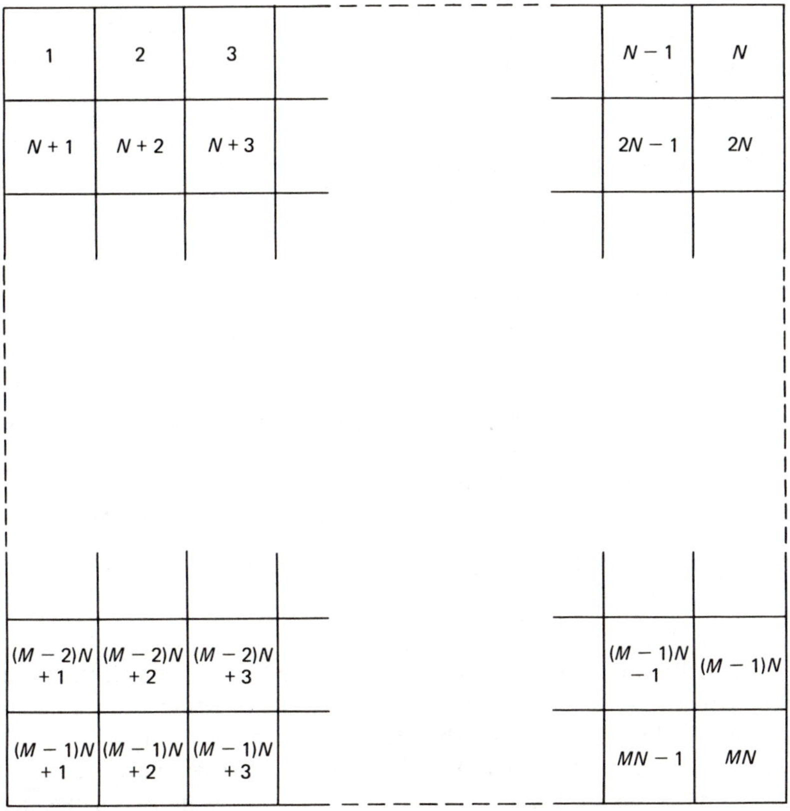

1	2	3		$N-1$	N
$N+1$	$N+2$	$N+3$		$2N-1$	$2N$
$(M-2)N+1$	$(M-2)N+2$	$(M-2)N+3$		$(M-1)N-1$	$(M-1)N$
$(M-1)N+1$	$(M-1)N+2$	$(M-1)N+3$		$MN-1$	MN

Figure 8.6. Site arrangement for the Hillier procedure.

grid squares $N+1$ through $2N$, and so on; that column 1 constitutes grid squares 1, $N+1$, $2N+1, \ldots, (M-1)N+1$, column 2 constitutes columns 2, $N+2$, $2N+2, \ldots, (M-1)N+2$, and so on. There are, of course, M rows and N columns. In the example used to illustrate the previous two procedures, a similar site arrangement has been assumed, but only for convenience; the previous two procedures can be used with any site arrangement.

The distance between sites is considered to be the rectilinear distance between centers of grid squares measured in units of the width of a grid square; thus, for example, $d(1, MN) = (M-1) + (N-1)$. One advantage of the site arrangement, which is substantial if the procedure is computerized, is that it is at least as easy to compute the distances between grid square sites as it is to look up the distances in a table. The ease of computation is due to the fact that the coordinates of a grid square may be readily obtained by knowing only the grid square number, and the values of M and N.

To develop expressions for grid square coordinates, suppose that the grid square site arrangement is located in the first quadrant with its lower left-hand corner at the origin. For any real number r, let $I(r)$ be the largest integer equal to or less than r; thus, for example, $I(3.5) = 3$, $I(4) = 4$; $I[(j-1)/N] = 0$ for $j = 1 \ldots, N$; and $I[(j-1)/N] = 1$ for $j = N+1, \ldots, 2N$. It should then be intuitive that the y coordinate of grid square j is just $M - \{I[(j-1)/N] + 0.5\}$; for example, the y coordinate of grid square j is $M - 0.5$ for $j = 1, \ldots, N$; is $M - 1.5$ for $j = N+1, \ldots, 2N$; and is just $M - [(M-1) + 0.5] = 0.5$ for $j = (M-1)N + 1, \ldots, MN$. Likewise, the x coordinate of grid square j is $j - \{NI[(j-1)/N] + 0.5\}$; thus for $j = 1, \ldots, N$ the x coordinates are $0.5, 1.5, \ldots, N - 0.5$, respectively; for $j = N+1, \ldots, 2N$ the x coordinates are $N + 1 - (N + 0.5) = 0.5$, $N + 2 - (N + 0.5) = 1.5, \ldots,$ and $2N - (N + 0.5) = N - 0.5$. It thus follows that the rectilinear distance between grid squares i and j is given by

$$\begin{aligned} d(i,j) &= |i - \{NI[(i-1)/N] + 0.5\} - j + \{NI[(j-1)/N] + 0.5\}| \\ &\quad + |M - \{I[(i-1)/N] + 0.5\} - M + \{I[(j-1)/N] + 0.5\}| \\ &= |i - NI[(i-1)/N] - j + NI[(j-1)/N]| \\ &\quad + |-I[(i-1)/N] + I[(j-1)/N]| \end{aligned}$$

With the foregoing aspects concerning the site arrangements for the Hillier procedure established, an overview of the procedure can be stated. The basis of the procedure is what Hillier calls a p-step move desirability number. Given an assignment of facilities to sites, a p-step move desirability number for a given facility may be any one of the following: the change in the total cost incurred by moving the given facility p "steps," or grid squares, either to the right, the left, up, down, or along either of the two diagonal lines running through opposite corners of the grid square in which the given facility is initially located. For example, with reference to Figure 8.1(b), the value of the one-step move desirability number obtained by moving facility 6 one step to the left is $(4 + 8 + 6 + 10) - 2 = 26$; the value of the total cost is decreased by $4 + 8 + 6 + 10$, since facility 6 is one unit closer to facilities 1, 2, 3, and 5, and is increased by 2, since facility 6 is one unit farther from facility 4; the net decrease is 26. Thus, if facilities 5 and 6 could both share the same location (and of course they cannot), it would be desirable to move facility 6 to the same site as facility 5, since the total cost could be decreased. The move desirability numbers, which will subsequently be referred to just as move numbers, for purposes of brevity, are recorded in a move desirability table (MDT); the table does not include move numbers for diagonal moves.

The procedure begins, given an assignment, by letting p be the maximum of $M - 1$ and $N - 1$, so that p represents the maximum number of steps a facility can be moved. All p-step move numbers are computed, and the location of the facility with the largest move number is interchanged with the location of the other appropriate facility, provided the total cost is decreased. The move numbers are then updated as necessary, and p-step interchanges are continued until the total cost can no longer be reduced. Then p is decremented by 1, and the procedure repeats itself until finally all one-step interchanges are made that will reduce total cost. When no further reductions in total cost can be obtained by making one-step moves, a "last pass" is made by resetting p to the maximum of $M - 1$ and $N - 1$, and the procedure then considers interchanges until at last all one-step interchanges have again been considered. The procedure then stops.

Before stating the Hillier procedure in detail, it is useful to develop some relationships that will be employed in the procedure. A general move number expression will be developed first. Given an assignment **a**, the change in the total cost of **a** incurred when the location of facility v is changed so that it is the same as the location of facility u will be denoted by $M(v/u:\mathbf{a})$. To develop an expression for $M(v/u:\mathbf{a})$, recall that all costs involving facilities v and u in $TC(\mathbf{a})$ are given by

$$\sum_{\substack{i=1\\ \neq v}}^{n} w_{iu}\, d(a(i), a(u)) + \sum_{\substack{i=1\\ \neq u}}^{n} w_{iv}\, d(a(i), a(v)) + w_{uv}\, d(a(u), a(v)) \tag{8.13}$$

When the location of facility v is changed so that it is the same as that of facility u, all terms involving facilities v and u in the new total cost expression are given in the following double sum:

$$\sum_{\substack{i=1\\ \neq v}}^{n} w_{iu}\, d(a(i), a(u)) + \sum_{\substack{i=1\\ \neq u}}^{n} w_{iv}\, d(a(i), a(u)) + w_{uv}\, d(a(u), a(u)) \tag{8.14}$$

The move number $M(v/u:\mathbf{a})$ is now obtained by subtracting (8.14) from (8.13), so that

$$M(v/u: a) = \sum_{i=1}^{n} w_{iv}[d(a(i), a(v)) - d(a(i), a(u))] \tag{8.15}$$

Now since v and u were arbitrary facility numbers, it follows that

$$M(u/v: a) = \sum_{i=1}^{n} w_{iu}[d(a(i), a(u)) - d(a(i), a(v))] \tag{8.16}$$

A useful expression relating the move numbers defined by (8.15) and (8.16) to the change in total cost resulting from interchanging the locations of facilities u and v is now easily obtained. Recalling expression (8.12) of DTC_{uv}

(**a**), it follows directly from (8.15) and (8.16) that

$$DTC_{uv}(\mathbf{a}) = M(u/v : \mathbf{a}) + M(v/u : \mathbf{a}) - 2w_{uv}\, d(a(u), a(v)) \tag{8.17}$$

Relationship (8.17) is useful because it enables changes in total cost to be computed directly from the move numbers, which are available in a move desirability table.

Another useful relationship is developed in the following remark; the remark will be useful when considering diagonal facility interchanges.

Remark 1: Given facilities s, t, u, and v located at corners of a block of grid squares, where the block has dimensions $p + 1$ by $p + 1$, as illustrated in Figure 8.7, it is true, for $i = 1, \ldots, n$, that

$$d(a(i), a(s)) + d(a(i), a(t)) = d(a(i), a(u)) + d(a(i), a(v)) \tag{8.18}$$

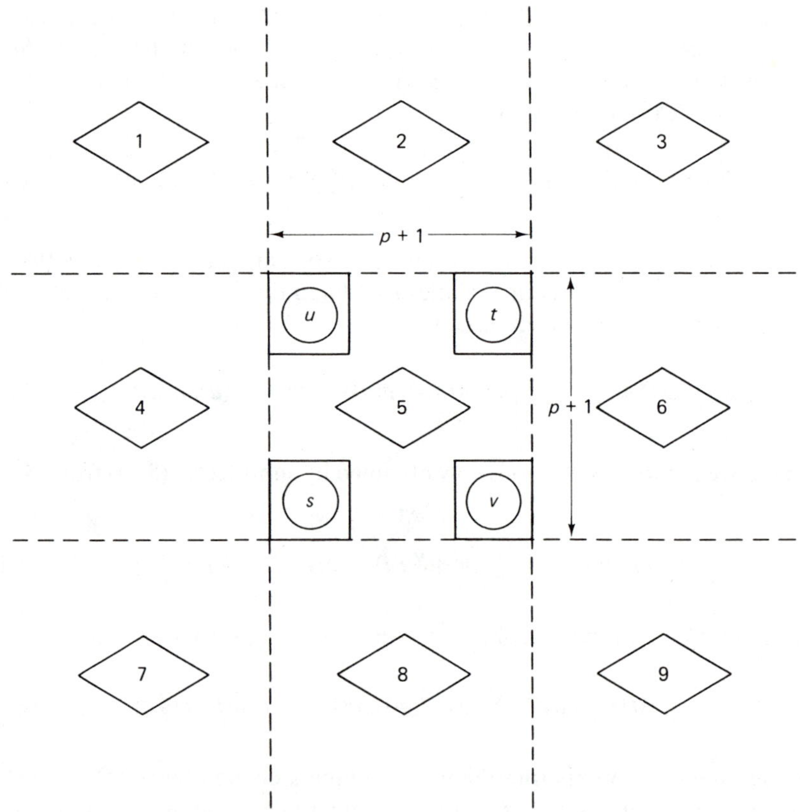

Figure 8.7. Sketch for diagonal moves.

Equation (8.18) is established simply by considering all the possible locations of facility i, as indicated in Figure 8.7. If $i = s$, t, u, or v, then, with reference to Figure 8.7, in each case (8.18) becomes $2p = 2p$, and so (8.18) is certainly true. Next consider a grid square $a(i)$ in any one of the regions 1, 2, or 3 shown in Figure 8.7; for these regions,

$$d(a(i), a(t)) + p = d(a(i), a(v))$$

and

$$p = d(a(i), a(s)) - d(a(i), a(u))$$

so that (8.18) follows from these two equations. It is left as an exercise to show that (8.18) is also true if $a(i)$ is in any one of the remaining regions 4 through 9 shown in Figure 8.7.

The foregoing remark provides the basis for relating move numbers with the change in cost expression for diagonal interchanges of facility locations.

Remark 2: Given facilities s, t, u, and v located at corners of a block of grid squares, where the block has dimensions $p + 1$ by $p + 1$, as illustrated in Figure 8.7, it is true that

$$\begin{aligned} DTC_{uv}(\mathbf{a}) = {} & M(u/t:\mathbf{a}) + M(u/s:\mathbf{a}) + M(v/s:\mathbf{a}) + M(v/t:\mathbf{a}) \\ & - 2w_{uv}\, d(a(u), a(v)) \end{aligned} \tag{8.19}$$

and that

$$\begin{aligned} DTC_{st}(\mathbf{a}) = {} & M(s/v:\mathbf{a}) + M(s/u:\mathbf{a}) + M(t/u:\mathbf{a}) + M(t/v:\mathbf{a}) \\ & - 2w_{st}\, d(a(s), a(t)) \end{aligned} \tag{8.20}$$

To establish (8.19), it is enough to show that

$$\sum_{i=1}^{n} w_{iu}[d(a(i), a(u)) - d(a(i), a(v))] = M(u/t:\mathbf{a}) + M(u/s:\mathbf{a}) \tag{8.21}$$

and

$$\sum_{i=1}^{n} w_{iv}\,[d(a(i), a(v)) - d(a(i), a(u))] = M(v/s:\mathbf{a}) + M(v/t):\mathbf{a} \tag{8.22}$$

since the sum of the terms on the left sides of (8.21) and (8.22) is

$$DTC_{uv}(\mathbf{a}) + 2w_{uv}\; d(a(u), a(v))$$

Since

$$M(u/t : \mathbf{a}) + M(u/s : \mathbf{a}) = \sum_{i=1}^{n} w_{iu}[d(a(i), a(u)) - d(a(i), a(t)) + d(a(i), a(u)) - d(a(i), a(s))]$$

(8.21) will hold if, for $i = 1, \ldots, n$,

$$-d(a(i), a(v)) = -d(a(i), a(t)) + d(a(i), a(u)) - d(a(i), a(s)) \qquad (8.23)$$

But (8.23) is just (8.18) rearranged, and so is true; (8.21) is thus established. It is left as an exercise to derive (8.20); the approach exactly parallels the derivation of (8.19).

With reference to Figure 8.7, the interchange of the locations of facilities u and v, or of s and t, will be called a p-step diagonal interchange. The interchange of the locations of facilities s and u, or of v and t, will be called a p-step up–down interchange. The interchange of the locations of facilities t and u, or of v and s, will be called a p-step right–left interchange.

The Hillier procedure will now be stated; a flow chart of the procedure is given in Figure 8.8.

0. Given: **a**, **D**, **W**; M, N.
1. Compute $TC(\mathbf{a})$.
2. Set c to 1.
3. Compute $p = \max(M - 1, N - 1)$.
4. Set up the p-step move table (MDT).
5. Find the largest positive entry, say $M(q/r : \mathbf{a})$, in the MDT.
6. If $DTC_{qr}(\mathbf{a}) > 0$, go to step 7; otherwise, go to step 12.
7. Set $d = DTC_{qr}(\mathbf{a})$.
8. Set $v = r$.
9. Compute $TC(\mathbf{a}) = TC(\mathbf{a}) - d$.
10. Revise **a** by interchanging the location of facilities q and v.
11. Recompute the entries in the MDT, and go to step 5.
12. Determine whether facility q would move Left, Right, Up, or Down (L, R, U, or D) in order to share the same location as facility r.
13. Given the direction of the move for facility q determined in step 12, find the (at most) two corresponding facilities, say s and t, with which facility q can make a p-step diagonal interchange. (For example, if the move is to the right, facilities s and t would be to the right of facility q.)
14. If $DTC_{qs}(\mathbf{a}) > 0$, go to step 15; otherwise, go to step 17.
15. Set $d = DTC_{qs}(\mathbf{a})$.

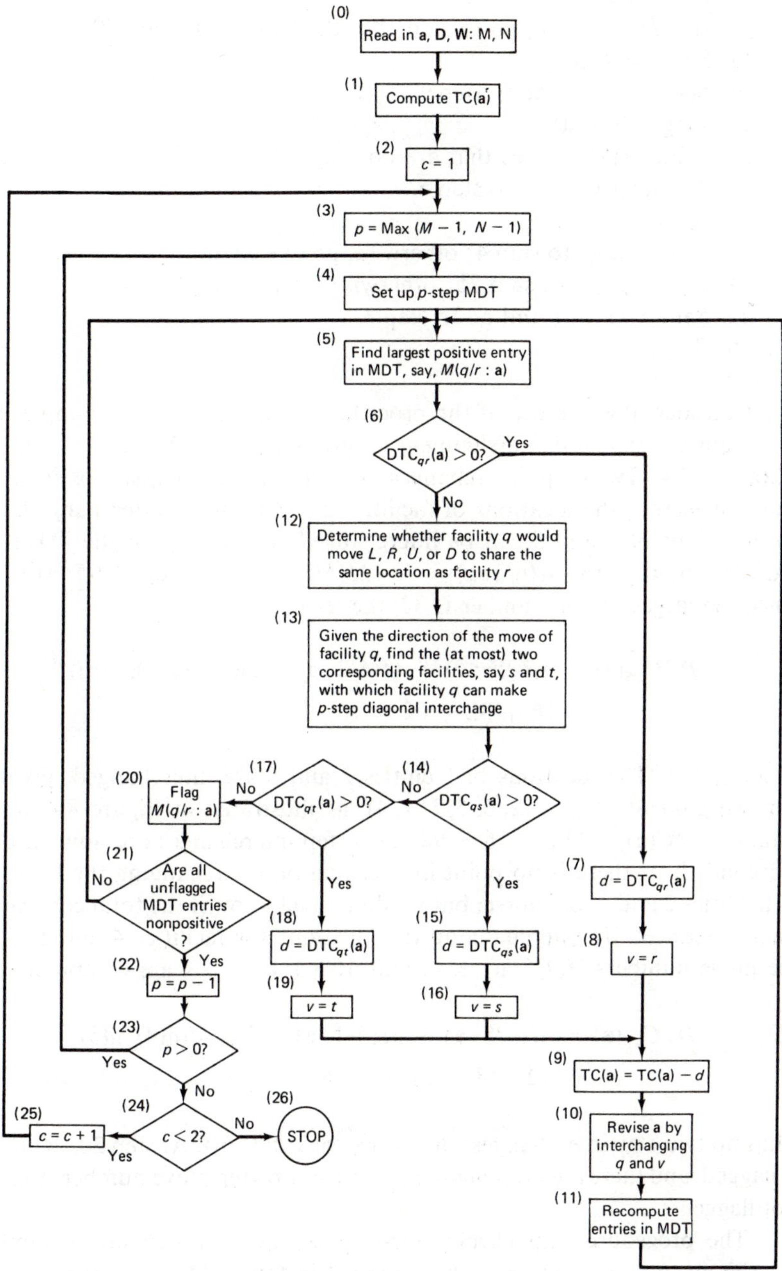

Figure 8.8. Hillier procedure.

16. Set $v = s$, and go to step 9.
17. If $DTC_{qt}(\mathbf{a}) > 0$, go to step 18; otherwise, go to step 20.
18. Set $d = DTC_{qt}(\mathbf{a})$.
19. Set $v = t$, and go to step 9.
20. Flag $M(q/r\colon \mathbf{a})$.
21. If all MDT entries that are not flagged are nonpositive, go to step 22; otherwise, go to step 5.
22. Set $p = p - 1$.
23. If $p > 0$, go to step 4; otherwise, go to step 24.
24. If $c < 2$, go to step 25; otherwise, go to step 26.
25. Set $c = c + 1$ and go to step 3.
26. Stop.

Consider now the use of the procedure as applied to the example. For the example, the initial assignment is given by $\mathbf{a} = (2, 4, 5, 3, 1, 6)$, and $TC(\mathbf{a}) = 114$. Two-step interchanges, which are the largest possible, are first considered; the locations of facilities 2 and 6 can be interchanged, and the locations of 4 and 5 can be interchanged. The entries in the MDT are thus $M(2/6\colon \mathbf{a}) = 16$, $M(6/2\colon \mathbf{a}) = 32$, $M(5/4\colon \mathbf{a}) = 28$ and $M(4/5\colon \mathbf{a}) = 12$. Since the largest move number is 32, the term

$$\begin{aligned} DTC_{26}(\mathbf{a}) &= M(2/6 : \mathbf{a}) + M(6/2 : \mathbf{a}) - 2w_{26}\, d(a(2), a(6)) \\ &= 16 + 32 - 32 = 16 \end{aligned}$$

is computed. The locations of facilities 2 and 6 are interchanged, giving a new assignment of $\mathbf{a} = (2, 6, 5, 3, 1, 4)$, as illustrated in Figure 8.4, and a total cost $TC(\mathbf{a}) = 114 - 16 = 98$. Two-step interchanges are now checked a second time; there is no point in checking on interchanging the locations of facilities 2 and 6, of course, but we do not yet know if the total cost can be further reduced by interchanging the locations for facilities 4 and 5. Thus the move numbers $M(4/5\colon \mathbf{a}) = 12$ and $M(5/4\colon \mathbf{a}) = -4$ are computed, and

$$\begin{aligned} DTC_{45}(\mathbf{a}) &= M(4/5 : \mathbf{a}) + M(5/4 : \mathbf{a}) - 2w_{45}\, d(a(4), a(5)) \\ &= 12 - 4 - 24 = -16 \end{aligned}$$

Thus no two-step interchanges will reduce the total cost. At step 20, $M(4/5\colon \mathbf{a})$ is flagged, and there are no remaining positive two-step move numbers that are not flagged.

The procedure next checks one-step moves. One-step move numbers for the given assignment may be arranged in four tables, giving right, left, up, and down move numbers, as shown in Figure 8.9. In Figure 8.9, the

Right Move Numbers

	1	2	3	4	5	6
1				−8		
2						
3		−8				
4						
5	4					
6			10			

Left Move Numbers

	1	2	3	4	5	6
1					−4	
2			16			
3						−4
4	10					
5						
6						

Up Move Numbers

	1	2	3	4	5	6
1						
2				−4		
3	0					
4						
5						
6					2	

Down Move Numbers

	1	2	3	4	5	6
1			8			
2						
3						
4		−2				
5						4
6						

Figure 8.9. Move numbers for $\mathbf{a} = (2, 6, 5, 3, 1, 4)$.

right move table gives the numbers $M(1/4\colon \mathbf{a}) = -8$, $M(3/2\colon \mathbf{a}) = -8$, $M(5/1\colon \mathbf{a}) = 4$, and $M(6/3\colon \mathbf{a}) = 10$. The other tables in Figure 8.9 are read in a similar manner. Figure 8.10 represents an alternative, more compact display of the move numbers.

Since the largest move number is $M(2/3\colon \mathbf{a}) = 16$, the move (2/3) will be checked; the number

$$DTC_{23}(\mathbf{a}) = M(2/3:\mathbf{a}) + M(3/2:\mathbf{a}) - 2w_{23}\, d(a(2), a(3)) = 16 - 8 - 8 = 0$$

is computed. Since (2/3) is a left move, the left one-step diagonal moves are now checked; there is one such move, (2/1), that is possible. The number

$$\begin{aligned} DTC_{12}(\mathbf{a}) = {} & M(1/4:\mathbf{a}) + M(1/3:\mathbf{a}) + M(2/3:\mathbf{a}) + M(2/4:\mathbf{a}) \\ & - 2w_{12}\, d(a(1), a(2)) = -14 \end{aligned}$$

and so the interchange is not made. The number $M(2/3\colon \mathbf{a})$ is flagged.

		Right	Left	Up	Down
Facility No.	1	−8	−4		8
	2		16	−4	
	3	−8	−4	0	
	4		10		−2
	5	4			4
	6	10		2	

Figure 8.10. Alternative display of move numbers for $\mathbf{a} = (2, 6, 5, 3, 1, 4)$.

Next the left move (4/1) is checked, since $M(4/1: \mathbf{a}) = 10$. Since $DTC_{41}(\mathbf{a}) = -2$, the move (4/1) is not made. The one possible diagonal move results in $DTC_{43}(\mathbf{a}) = -8$, and so is not made. The move number $M(4/1: \mathbf{a})$ is flagged.

We summarize the remaining one-step move computations at this point as follows:

$$M(1/3: \mathbf{a}) = 8 \text{ (a down move)}$$
$$DTC_{13}(\mathbf{a}) = -4$$
$$DTC_{16}(\mathbf{a}) = 0$$
$$DTC_{12}(\mathbf{a}) = -14$$
$$M(5/6: \mathbf{a}) = 4 \text{ (a down move)}$$
$$DTC_{56}(\mathbf{a}) = -14$$
$$DTC_{53}(\mathbf{a}) = -6$$
$$M(5/1: \mathbf{a}) = 4 \text{ (a right move)}$$
$$DTC_{15}(\mathbf{a}) = -8$$
$$DTC_{53}(\mathbf{a}) = -6$$
$$M(6/5: \mathbf{a}) = 2 \text{ (an up move)}$$
$$DTC_{56}(\mathbf{a}) = -14$$
$$DTC_{61}(\mathbf{a}) = 0$$

There are no remaining unflagged positive move numbers at this point, and so all one-step moves have been checked.

At this point the Hillier procedure goes through a "last pass," beginning by checking two-step moves for the assignment $\mathbf{a} = (2, 6, 5, 3, 1, 4)$. Note that two-step moves for this assignment were checked in the first pass, so that computations of the last pass will be identical to those of the previous pass after the assignment $\mathbf{a} = (2, 6, 5, 3, 1, 4)$ has been determined. No further interchanges will be made, and the procedure stops.

Nugent, Vollmann, and Ruml [10] have compared the Hillier procedure with the pairwise-interchange steepest-descent procedure on problems where the number of facilities, n, is 5, 6, 7, 8, 12, 15, 20, and 30. They found that the latter procedure appears to produce assignments of slightly lower total costs, although the differences in total costs were not statistically significant. They also found, as might be expected, that the Hillier algorithm is much better in terms of computation time; with the Hillier procedure computation time increases with about the second power of n, whereas with the other procedure computation time increases with about the third power of n. As has been mentioned previously, the VNZ procedure appears at least as attractive as the other two in terms of total cost of solutions, computation time, and computer storage requirements. It should be evident, however, that each procedure has its own special points of interest.

As a final note on heuristic procedures, it is interesting, for the example problem, to observe that all three heuristic procedures presented yielded the same final assignment of $\mathbf{a} = (2, 6, 5, 3, 1, 4)$, with a total cost of 98. In such a case, one might conclude that the solution obtained is an optimum solution. Unfortunately, such is not the case, as the total cost of 92 for the assignment $\mathbf{a} = (3, 5, 6, 4, 1, 2)$ indicates.

8.6 The Gilmore-Lawler Exact Procedure

In this section an exact procedure, based on the work of Gilmore [4] and Lawler [8], is developed for solving a more general version of the problem considered in this chapter. The procedure is a branch and bound procedure and, as mentioned previously, becomes computationally infeasible if n, the number of facilities to be located, is much more than 15. However, the procedure can be quite useful, even for large values of n, for finding lower bounds on the minimum total cost; all these lower bounds will be at least as large as the lower bound found by the method of Section 8.1, and become particularly useful when combined with "close" upper bounds on the minimum total cost.

We first *redefine* the cost, $TC(\mathbf{a})$, for an assignment $\mathbf{a}$. In addition to the data given previously, it is assumed that a matrix of costs,

$$\mathbf{C} = (c(i, k))$$

is given, where $c(i, k)$ is a cost incurred if facility i is located at site k. Thus, if facility i is located at site $a(i)$, the total cost for an assignment is given by

$$TC(\mathbf{a}) = \sum_{i=1}^{n} c(i, a(i)) + \tfrac{1}{2}\sum_{i=1}^{n}\sum_{\substack{j=1\\ j\neq i}}^{n} w_{ij}\, d(a(i), a(j)) \tag{8.24}$$

Note that (8.24) reduces to the previous total cost expression (8.2) or, equivalently, (8.1), when all $c(i, k) = 0$. Since it will be necessary to make repeated reference to the term subsequently, the minimum value of $TC(\mathbf{a})$ will be denoted by $\overline{TC}$; as previously, our interest will be in finding assignments with a total cost of $\overline{TC}$; also we shall want to find "good" upper and lower bounds on $\overline{TC}$.

We first develop a procedure for finding a lower bound on $\overline{TC}$; this means of obtaining a lower bound provides the basic idea used in developing a computational procedure for finding a least-cost assignment. A statement of the procedure will first be given, and then a justification. The procedure as a whole will be referred to as Rule A, and the lower bound on $\overline{TC}$ obtained by using Rule A will be denoted by $v(0)$.

Rule A

1. Let $\mathbf{w}(i)$ be the row vector obtained from row i of the matrix $\mathbf{W} = (w_{ij})$ by deleting the element in column i of row i of $\mathbf{W}$. Then let $\bar{\mathbf{w}}(i)$ be the row vector obtained by ordering the elements of $\mathbf{w}(i)$ so that they are nonincreasing.

2. Let $\mathbf{d}(k)$ be the row vector obtained from row k of the matrix $\mathbf{D} = (d(k, h))$ by deleting the element in column k of row k of $\mathbf{D}$. Then let $\bar{\mathbf{d}}(k)$ be the row vector obtained by ordering the elements of $\mathbf{d}(k)$ so that they are nondecreasing.

3. For $i = 1, \ldots, n$ and $k = 1, \ldots, n$, let

$$e(i, k) = (\bar{\mathbf{w}}(i))(\bar{\mathbf{d}}(k))'$$

and

$$f(i, k) = c(i, k) + \tfrac{1}{2}e(i, k)$$

4. Solve the linear assignment problem having the cost matrix $\mathbf{F} = (f(i, k))$, and let $v(0)$ be the minimum value of the objective function for this linear assignment problem; $v(0)$ is a lower bound on $\overline{TC}$.

To establish that $v(0)$ is a lower bound on $\overline{TC}$, first note that

$$\sum_{\substack{j=1\\ j\neq i}}^{n} w_{ij}\, d(a(i), a(j))$$

is the inner product of $\mathbf{w}(i)$ with $\mathbf{d}(a(i))$, after ordering the elements of $\mathbf{d}(a(i))$, so that the same approach as used to establish Property 4 in Section 6.6 guarantees that

$$\sum_{\substack{j=1 \\ j \neq i}}^{n} w_{ij}\, d(a(i), a(j)) \geq \bar{\mathbf{w}}(i)(\bar{\mathbf{d}}(a(i)))' = e(i, a(i)) \tag{8.25}$$

It now follows from (8.24) and (8.25) that

$$TC(\mathbf{a}) \geq \sum_{i=1}^{n} c(i, a(i)) + \tfrac{1}{2} \sum_{i=1}^{n} e(i, a(i))$$

and so the definition of $f(i, k)$ in step 3 of Rule A implies that

$$TC(\mathbf{a}) \geq \sum_{i=1}^{n} f(i, a(i)) \tag{8.26}$$

Now (8.26) gives a lower bound on $TC(\mathbf{a})$ for a particular assignment. If we could find the minimum of the right side of (8.26) over *all* assignments, we would have a lower bound on $TC(\mathbf{a})$ for all $\mathbf{a}$, and thus a lower bound on $\overline{TC}$. But minimizing the right side of (8.26) is the same as solving the problem in step 4 of Rule A, and so $v(0)$, the minimum of the right side of (8.26) over all assignments, is a lower bound on $\overline{TC}$.

An alternative approach to finding a lower bound on TC would be to use the method of the first section of this chapter to find a lower bound on the double sum in (8.24), and solve a linear assignment problem with cost matrix $\mathbf{C} = (c(i, k))$ to obtain a lower bound on the first sum in (8.24); the sum of these lower bounds would then be a lower bound on $\overline{TC}$. This approach involves less effort than the procedure of Rule A but, as might be expected, the bound obtained will never be larger than the bound $v(0)$.

At this point it is useful to consider an example, which involves assigning four facilities to four sites. Naturally, such a small problem can be easily solved by enumerating all possible assignments and simply choosing the one with the smallest cost. In practice, one would not bother to apply the procedures developed in the foregoing discussion to such a small problem; the example is intended to illustrate the procedures of the discussion with a minimum of burdensome numerical details.

The following matrices are given for the example:

$$\mathbf{C} = \begin{pmatrix} 1 & 3 & 2 & 1 \\ 2 & 1 & 4 & 2 \\ 4 & 2 & 4 & 4 \\ 3 & 1 & 2 & 2 \end{pmatrix}, \quad \mathbf{D} = \begin{pmatrix} 0 & 1 & 2 & 3 \\ 1 & 0 & 1 & 2 \\ 2 & 1 & 0 & 1 \\ 3 & 2 & 1 & 0 \end{pmatrix}, \quad \mathbf{W} = \begin{pmatrix} 0 & 1 & 3 & 4 \\ 1 & 0 & 2 & 1 \\ 3 & 2 & 0 & 3 \\ 4 & 1 & 3 & 0 \end{pmatrix}$$

Thus

$$\begin{aligned} \mathbf{d}(1) &= (1, 2, 3), & \bar{\mathbf{d}}(1) &= (1, 2, 3) \\ \mathbf{d}(2) &= (1, 1, 2), & \bar{\mathbf{d}}(2) &= (1, 1, 2) \\ \mathbf{d}(3) &= (2, 1, 1), & \bar{\mathbf{d}}(3) &= (1, 1, 2) \\ \mathbf{d}(4) &= (3, 2, 1), & \bar{\mathbf{d}}(4) &= (1, 2, 3) \end{aligned}$$

and

$$\begin{aligned} \mathbf{w}(1) &= (1, 3, 4), & \bar{\mathbf{w}}(1) &= (4, 3, 1) \\ \mathbf{w}(2) &= (1, 2, 1), & \bar{\mathbf{w}}(2) &= (2, 1, 1) \\ \mathbf{w}(3) &= (3, 2, 3), & \bar{\mathbf{w}}(3) &= (3, 3, 2) \\ \mathbf{w}(4) &= (4, 1, 3), & \bar{\mathbf{w}}(4) &= (4, 3, 1) \end{aligned}$$

Since $e(i, k) = \bar{\mathbf{w}}(i)\bar{\mathbf{d}}(k)'$, we have, for example, $e(1, 1) = 13$, $e(1, 2) = 9$, and, in general, for the matrix $\mathbf{E} = (e(i, j))$,

$$\mathbf{E} = \begin{pmatrix} 13 & 9 & 9 & 13 \\ 7 & 5 & 5 & 7 \\ 15 & 10 & 10 & 15 \\ 13 & 9 & 9 & 13 \end{pmatrix}$$

Since $f(i, k) = c(i, k) + e(i, k)/2$, the matrix $\mathbf{F} = (f(i, k))$ is given by

$$\mathbf{F} = \begin{pmatrix} 7.5 & 7.5 & 6.5 & 7.5 \\ 5.5 & 3.5 & 6.5 & 5.5 \\ 11.5 & 7 & 9 & 11.5 \\ 9.5 & 5.5 & 6.5 & 8.5 \end{pmatrix}$$

Solving the assignment problem having the cost matrix **F** gives a least-cost of 26.5, so that $v(0) = 26.5$ is a lower bound on $\overline{TC}$ for this example. Now at this point we observe that, for this example, the value of $TC(\mathbf{a})$ for any assignment **a** must be an integer, since

$$TC(\mathbf{a}) = \sum_{i=1}^{n} c(i, a(i)) + \sum_{1 \le i < j \le n} w_{ij}\, d(a(i), a(j))$$

the entries of the matrices **C**, **D**, and **W** are all integers, and the preceding expression for $TC(\mathbf{a})$ indicates that $TC(\mathbf{a})$ is a sum of products of integers and so is an integer. Thus a lower bound on $\overline{TC}$ is 27. Next, we "pull a rabbit out of a hat" and observe that the assignment $\mathbf{a} = (4, 1, 2, 3)$ has a total cost of 27, and so is a least-cost assignment to this quadratic assignment problem. The assignment $\mathbf{a} = (4, 1, 2, 3)$ might have been obtained from one of the heuristic procedures considered previously; we shall see alternative means of obtaining it subsequently.

Returning now from the example, recall that we defined an assignment to be a permutation of the integers 1, 2, 3, . . . , n. It is useful now to introduce several additional definitions. By a *partial assignment of size q* we mean a permutation of a subset consisting of q of the integers 1, 2, . . ., n; for example, if $n = 7$, $\mathbf{a}^* = (1, 4, 3, 6)$ would be a partial assignment of size 4. When it is unnecessary to specify the size of a partial assignment, we shall simply speak of a partial assignment. Also, for reasons to become evident, when the branch and bound procedure is given in complete form, we shall use the word *node* as an alternative name for a partial assignment. By a *completion* of a partial assignment (or node) of size q we shall mean an assignment for which the first q integers are precisely those of the partial assignment; for example, when $n = 7$ the assignment $\mathbf{a} = (1, 4, 3, 6, 7, 5, 2)$ would be one possible completion of the partial assignment $\mathbf{a}^* = (1, 4, 3, 6)$. By the *value of a partial assignment* we shall mean a lower bound on the total cost of all possible completions of the partial assignment, where it is understood the lower bound is found in a precisely specified manner; the *value of a node* will have the same meaning as the value of a partial assignment, and will be denoted by $v(a^*)$.

We are interested in partial assignments and their values because they are essential to the branch and bound procedure to be developed. We shall see that the idea of finding the value of a partial assignment is essentially the same as that involved in finding a lower bound on the minimum value of $TC(\mathbf{a})$; only the notation becomes a bit more awkward. Suppose that the partial assignment $\mathbf{a}^* = (a^*(1), a^*(2), \ldots, a^*(q))$ is given; then the total cost of any completion, say $\mathbf{a}_q^* = (a^*(1), \ldots, a^*(q), a(q+1), \ldots, a(n))$ is given by

$$\begin{aligned} TC(\mathbf{a}_q^*) = {} & \sum_{i=1}^{q} c(i, a^*(i)) + \sum_{i=q+1}^{n} c(i, a(i)) + \sum_{1 \le i < j \le q} w_{ij}\, d(a^*(i), a^*(j)) \\ & + \sum_{j=q+1}^{n} w_{1j}\, d(a^*(1), a(j)) + \sum_{j=q+1}^{n} w_{2j}\, d(a^*(2), a(j)) \\ & + \cdots + \sum_{j=q+1}^{n} w_{qj}\, d(a^*(q), a(j)) + \sum_{q+1 \le i < j \le n} w_{ij}\, d(a(i), a(j)) \end{aligned}$$

On rearranging terms, $TC(\mathbf{a}_q^*)$ is the sum of terms (8.27), (8.28), and (8.29):

$$\left[\sum_{i=1}^{q} c(i, a^*(i)) + \sum_{i \le i < j \le q} w_{ij}\, d(a^*(i), a^*(j))\right] \tag{8.27}$$

$$\sum_{j=q+1}^{n} \left[c(j, a(j)) + \sum_{i=1}^{q} w_{ij}\, d(a^*(i), a(j))\right] \tag{8.28}$$

$$\tfrac{1}{2}\left[\sum_{i=q+1}^{n} \sum_{\substack{j=q+1 \\ j \ne i}}^{n} w_{ij}\, d(a(i), a(j))\right] \tag{8.29}$$

Notice that (8.27) involves only known, constant terms; (8.28) involves terms for a linear assignment problem, where $a(q+1), \ldots, a(n)$ must be found, suggesting that we may wish to solve a linear assignment problem of size $n-q$ by $n-q$. The terms in (8.29) are the most involved, but are quite similar to the terms in the double sum in the expression for $TC(\mathbf{a})$, suggesting that we may be able to modify the procedure of Rule A to obtain a linear assignment problem giving a lower bound on (8.29). For purposes of notational simplicity, the minimum value of $TC(\mathbf{a}_q^*)$ will be denoted by $\overline{TC}_q^*$. The procedure as a whole for finding a lower bound on $\overline{TC}_q^*$ will be referred to as Rule B, and the lower bound on $\overline{TC}_q^*$ found by using Rule B will be denoted by $v(\mathbf{a}^*)$.

Rule B

1-1. When $q = n-1$, there is only one possible completion of $\mathbf{a}_q^*$, so let $v(\mathbf{a}^*)$ be the total cost of that completion.

2-1. When $q = n-2$, we may assume that $\mathbf{a}^*$ consists of all the integers $1, 2, \ldots, n$ except the integers r and s, so that the only completions of $\mathbf{a}^*$ are $(\mathbf{a}^*, r, s)$ and $(\mathbf{a}^*, s, r)$. In either case (8.29) remains unchanged, since $d(r, s) = d(s, r)$.

2-2. Compute the value of (8.28) for the case when $a(n-1) = r$ and $a(n) = s$, and for the case when $a(n-1) = s$ and $a(n) = r$. Then let b^* be the minimum of these two values (note that we have simply solved a 2 by 2 linear assignment problem).

2-3. Let $v(\mathbf{a}^*)$ be the sum of (8.27), b^*, and $w_{n-1,n}\, d(r, s)$; $v(\mathbf{a}^*)$ is a lower bound on $\overline{TC}_q^*$.

3-1. When $q < n-2$, for $i = q+1, \ldots, n$, let $w_q(i)$ be the row vector obtained from row i of the matrix $\mathbf{W}$ by deleting the entries in columns 1 through q of row i, as well as deleting the entry in column i of row i. Let $\bar{\mathbf{w}}_q(i)$ be the row vector obtained by ordering the elements of $\mathbf{w}_q(i)$ so that they are nonincreasing.

3-2. For all values of k obtained by deleting from the integers $1, 2, \ldots, n$ those specified by $\mathbf{a}^*$ [e.g., if $n = 6$ and $\mathbf{a}^* = (3, 2, 5)$, then $k = 1, 4,$ and 6], let $\mathbf{d}_q^*(k)$ be the row vector obtained from row k of the matrix $\mathbf{D}$ by deleting the entries in columns $a^*(1)$ through $a^*(q)$ of row k, as well as deleting the entry in column k of row k. Then let $\bar{\mathbf{d}}_q^*(k)$ be the row vector obtained by ordering the elements of $\mathbf{d}_q^*(k)$ so that they are nondecreasing. We emphasize the fact that the entries in $\bar{\mathbf{d}}_q^*(k)$ depend on $\mathbf{a}^*$, while the entries in $\bar{\mathbf{w}}_q(i)$ do not.

3-3. Let I be the set of $n-q$ integers obtained by deleting from the integers $1, 2, \ldots, n$ the integers specified by $\mathbf{a}^*$; let I^* be the set of integers in I arranged in increasing order. For $i = q+1, \ldots, n$ and all k in I^*, compute

$$e_q^*(i, k) = (\bar{\mathbf{w}}_q(i)(\bar{\mathbf{d}}_q^*(k))'$$

and

$$f_q^*(i, k) = c(i, k) + \sum_{j=1}^{q} w_{ij}\, d(a^*(j), k) + \tfrac{1}{2}e_q^*(i, k)$$

3-4. Solve the linear assignment problem with cost matrix $(f_q^*(i, k))$, and let f_q^* be the minimum value of the objective function for this linear assignment problem.

3-5. Compute the term defined by (8.27) and call it $k(\mathbf{a}^*)$. Then let $v(\mathbf{a}^*) = k(\mathbf{a}^*) + f_q^*$; $v(\mathbf{a}^*)$ is a lower bound on $\overline{TC}_q^*$.

It should be evident in cases 1 and 2 of Rule B that $v(\mathbf{a}^*)$ is a lower bound on $\overline{TC}_q^*$ [furthermore in case 2-1, $v(\mathbf{a}^*) = \min\,[TC(\mathbf{a}^*, r, s),\ TC(\mathbf{a}^*, s, r)]$; to establish that it is a lower bound in case 3, first note that

$$\sum_{\substack{j=q+1 \\ j \neq i}}^{n} w_{ij}\, d(a(i), a(j)) \geq e_q^*(i, a(i))$$

so the fact that $TC(\mathbf{a}_q^*)$ is the sum of the terms (8.27), (8.28), and (8.29), together with the definition of $f_q^*(i, k)$, implies that

$$TC(\mathbf{a}_q^*) \geq \sum_{i=q+1}^{n} f_q^*(i, a(i)) + k(\mathbf{a}^*) \tag{8.30}$$

Now (8.30) gives a lower bound on $TC(\mathbf{a}_q^*)$ for a particular assignment. If we could find the minimum of the right side of (8.30) over *all* completions of $\mathbf{a}^*$, then we would have a lower bound on the minimum value of $TC(\mathbf{a}_q^*)$; that is, we would have a lower bound on $\overline{TC}_q^*$. But minimizing the right side of (8.30) is equivalent to steps 3-4 and 3-5 of Rule B, and so $v(\mathbf{a}^*)$ is a lower bound on $\overline{TC}_q^*$.

With reference to the example of this section, let us find a lower bound on the partial assignment $\mathbf{a}^* = (a^*(1))$, using Rule B. In particular, when $a^*(1) = 1$, then $k(\mathbf{a}^*) = c(1, a^*(1)) = c(1, 1)$, and $I^* = \{2, 3, 4\}$.

To compute the numbers

$$e_q^*(i, k) = \bar{\mathbf{w}}_q(i)\, \bar{\mathbf{d}}_q^*(k)'$$

we note that

$$\bar{\mathbf{w}}_1(2) = (2, 1), \qquad \bar{\mathbf{d}}_1^*(2) = (1, 2)$$

$$\bar{\mathbf{w}}_1(3) = (3, 2), \qquad \bar{\mathbf{d}}_1^*(3) = (1, 1)$$

$$\bar{\mathbf{w}}_1(4) = (3, 1), \qquad \bar{\mathbf{d}}_1^*(4) = (1, 2)$$

so that

$$\frac{1}{2}\begin{pmatrix} e_{22} & e_{23} & e_{24} \\ e_{32} & e_{33} & e_{34} \\ e_{42} & e_{43} & e_{44} \end{pmatrix} = \begin{pmatrix} 2 & 1.5 & 2 \\ 3.5 & 2.5 & 3.5 \\ 2.5 & 2 & 2.5 \end{pmatrix} \tag{8.31}$$

The matrix $(c(i, k))$ is given by

$$\begin{pmatrix} c_{22} & c_{23} & c_{24} \\ c_{32} & c_{33} & c_{34} \\ c_{42} & c_{43} & c_{44} \end{pmatrix} = \begin{pmatrix} 1 & 4 & 2 \\ 2 & 4 & 4 \\ 1 & 2 & 2 \end{pmatrix} \tag{8.32}$$

while the matrix $(w_{1j}\, d(1, k))$ is given by

$$\begin{pmatrix} w_{12}\, d(1,2) & w_{12}\, d(1,3) & w_{12}\, d(1,4) \\ w_{13}\, d(1,2) & w_{13}\, d(1,3) & w_{13}\, d(1,4) \\ w_{14}\, d(1,2) & w_{14}\, d(1,3) & w_{14}\, d(1,4) \end{pmatrix} = \begin{pmatrix} 1 & 2 & 3 \\ 3 & 6 & 9 \\ 4 & 8 & 12 \end{pmatrix} \tag{8.33}$$

Using step 3-3 of Rule B, the matrix $(f_q^*(i, k))$ is simply the sum of the three matrices (8.31), (8.32), and (8.33), and is thus

$$\begin{pmatrix} f_1^*(2,2) & f_1^*(2,3) & f_1^*(2,4) \\ f_1^*(3,2) & f_1^*(3,3) & f_1^*(3,4) \\ f_1^*(4,2) & f_1^*(4,3) & f_1^*(4,4) \end{pmatrix} = \begin{pmatrix} 4 & 7.5 & 7 \\ 8.5 & 12.5 & 16.5 \\ 7.5 & 12 & 16.5 \end{pmatrix}$$

Solving the linear assignment problem with this matrix results in a minimum objective function value of $7 + 12.5 + 7.5 = 27$, so that $v(\mathbf{a}^*) = v(1) = c(1, 1) + 27 = 28$, and results in the completion of $\mathbf{a}^*$ given by $(a^*(1), a(2), a(3), a(4)) = (1, 4, 3, 2)$.

Proceeding in the same fashion, when $a^*(1) = 2$, the matrix $(f_1^*(i, k))$ is given by

$$\begin{pmatrix} f_1^*(2,1) & f_1^*(2,3) & f_1^*(2,4) \\ f_1^*(3,1) & f_1^*(3,3) & f_1^*(3,4) \\ f_1^*(4,1) & f_1^*(4,2) & f_1^*(4,4) \end{pmatrix} = \begin{pmatrix} 6.5 & 7 & 6.5 \\ 13 & 10.5 & 14.5 \\ 11.5 & 8.5 & 13 \end{pmatrix} \tag{8.34}$$

When $a^*(1) = 3$, the matrix $(f_1^*(i, k))$ is given by

$$\begin{pmatrix} f_1^*(2,1) & f_1^*(2,2) & f_1^*(2,4) \\ f_1^*(3,1) & f_1^*(3,2) & f_1^*(3,4) \\ f_1^*(4,1) & f_1^*(4,2) & f_1^*(4,4) \end{pmatrix} = \begin{pmatrix} 6.5 & 4 & 6.5 \\ 14.5 & 8.5 & 13 \\ 14 & 7.5 & 10.5 \end{pmatrix} \tag{8.35}$$

and when $a^*(1) = 4$ the matrix $(f_1^*(i, k))$ is given by

$$\begin{pmatrix} f_1^*(2,1) & f_1^*(2,2) & f_1^*(2,3) \\ f_1^*(3,1) & f_1^*(3,2) & f_1^*(3,3) \\ f_1^*(4,1) & f_1^*(4,2) & f_1^*(4,3) \end{pmatrix} = \begin{pmatrix} 7 & 4.5 & 7 \\ 16.5 & 10.5 & 10.5 \\ 17.5 & 11 & 8.5 \end{pmatrix} \tag{8.36}$$

Now when the linear assignment problems with matrices given by (8.34), (8.35), and (8.36) are solved, the corresponding minimum objective function values are 28, 25.5, and 26, respectively, so that $v(2) = c(1, 2) + 28 = 31$, $v(3) = c(1, 3) + 25.5 = 27.5$, and $v(4) = c(1, 4) + 26 = 27$; the corresponding completions are given by $\mathbf{a} = (2, 4, 1, 3)$, $\mathbf{a} = (3, 1, 2, 4)$, and $\mathbf{a} = (4, 1, 2, 3)$.

For the assignments $\mathbf{a} = (1, 4, 3, 2)$, $\mathbf{a} = (2, 4, 1, 3)$, $\mathbf{a} = (3, 1, 2, 4)$, and $\mathbf{a} = (4, 1, 2, 3)$, we have $TC(\mathbf{a}) = 28$, $TC(\mathbf{a}) = 33$, $TC(\mathbf{a}) = 28$, and $TC(\mathbf{a}) = 27$, respectively; each of these numbers is of course an upper bound on the minimum value of $TC(\mathbf{a})$, and the best of the upper bounds is, of course, 27.

Furthermore, we also note that the smallest of the numbers $v(1)$, $v(2)$, $v(3)$, and $v(4)$ is a lower bound on $\overline{TC}$, the minimum value of $TC(\mathbf{a})$, since $v(k)$ is a lower bound on the total cost of all assignments for which facility 1 is at site k, $k = 1, 2, 3, 4$, and facility 1 must be at one of the sites 1, 2, 3, or 4. Thus 27, the minimum of the numbers 28, 31, 27.5, and 27, is a lower bound on $\overline{TC}$. Thus, for this example, we may stop, for since 27 is an upper bound as well as a lower bound on $\overline{TC}$, we know that $\overline{TC} = 27$, and that the assignment (4, 1, 2, 3), which has a cost of 27, is a least-cost assignment. For this example, we also note that each of the numbers $v(1)$ through $v(4)$ is at least as large as $v(0)$, as one would expect, since $v(1)$ through $v(4)$ are, roughly speaking, lower bounds on the minimum cost of smaller quadratic assignment problems (3 by 3 in this example) than $v(0)$ is a bound for, and thus may be considered to be bounds for a more constrained version of the original quadratic assignment problem. However, there is no guarantee that the bounds will not decrease when smaller quadratic assignment problems are considered.

To illustrate other aspects of Rule B, as well as to lay the groundwork for other aspects of the branch and bound procedure to be discussed, let us suppose that we do not yet know that an upper bound on $\overline{TC}$ is 27; we would not know this if, for example, we had not computed the total cost for the completions corresponding to $v(1)$ through $v(4)$. Suppose, however, that we had computed $TC(\mathbf{a})$ for the assignment $\mathbf{a} = (1, 4, 2, 3)$, which is an assignment resulting in the lower bound $v(0)$; for this assignment, $TC(\mathbf{a}) = 29$, so that 29 is an upper bound on $\overline{TC}$. We can now make the observation that no assignment having facility 1 in location 2 can be a least total cost assignment; the reason for this observation is as follows: since $v(2) = 31$, any assignment with facility 1 in location 2 will have a total cost of at least 31, and we know that an upper bound on $\overline{TC}$ is 29, which is, of course, less than 31.

Since we are supposing we know only that $26.5 \leq \overline{TC} \leq 29$, and we know we do not need to consider assigning facility 1 to location 2, and since, moreover, $v(4)$ is the smallest of $v(1)$ through $v(4)$, let us consider further the problem of assigning facility 1 to location 4. If facility 1 is assigned to location 4, then facility 2 must be assigned to one of the locations 1, 2, or 3, suggesting that we should compute values for the nodes (4, 1), (4, 2), and (4, 3); in other words, we should compute a lower bound on the total costs of all completions of each of the following partial assignments: (4, 1), (4, 2), and (4, 3).

Consider first computing TC_2^* for the partial assignment $\mathbf{a}^* = (4, 1)$. Since $n = 4$ and there are $q = 2$ entries in $\mathbf{a}^*$, $q = n - 2$, and we use the steps in case 2 of Rule B. Substituting into (8.27) gives $c(1, 4), + c(2, 1) + w_{12}\, d(4, 1) = 6$; substituting into (8.29) gives $w_{34}\, d(2, 3) = 3$; substituting into (8.28) when $a(3) = 2$ and $a(4) = 3$ gives $c(3, 2) + c(4, 3) + w_{13}\, d(4, 2) + w_{23}\, d(1, 2) + w_{14}\, d(4, 3) + w_{24}\, d(1, 3) = 18$, while substituting into (8.28) when $a(3) = 3$ and $a(4) = 2$ gives $c(3, 3) + c(4, 2) + w_{13}\, d(4, 3) + w_{23}\, d(1, 3) + w_{14}\, d(4, 2) + w_{24}\, d(1, 2) = 21$, so that $b^* = 18$, the minimum of 18 and 21. Thus $v(4, 1) = 6 + 18 + 3 = 27$. Equivalently, we have simply computed $TC(4, 1, 2, 3) = 27$ and $TC(4, 1, 3, 2) = 32$, and let $v(4, 1)$ be the smallest of the two numbers. In a similar fashion, we find that $v(4, 2) = 32$ and $v(4, 3) = 36$; furthermore, $v(4, 2) = TC(4, 2, 1, 3)$, and $v(4, 3) = TC(4, 3, 1, 2) = TC(4, 3, 2, 1)$.

Now the smallest of the numbers $v(1)$, $v(3)$, $v(4, 1)$, $v(4, 2)$, and $v(4, 3)$, which is 27, is a lower bound on $\overline{TC}$: in any least-cost assignment, facility 1 must be located at site 1, 3, or 4, and if new facility 1 is located at site 4, then new facility 2 must be located at site 1, 2, or 3. Since we are assured that the assignment (4, 1, 2, 3) has a total cost of 27, we know that the assignment (4, 1, 2, 3) is a least-cost assignment.

The computations we have gone through in the example to illustrate the use of Rules A and B may be readily visualized through the use of several figures. Figure 8.11(a) illustrates the computation of $v(0)$, $v(1)$, $v(2)$, $v(3)$, and $v(4)$. Figure 8.11(b) illustrates the fact that a least-cost assignment cannot have facility 1 located at site 2, and the computation of $v(4, 1)$, $v(4, 2)$, and $v(4, 3)$. The circled numbers in the figures are referred to as nodes, and the lines connecting the nodes are referred to as arcs; the figures should motivate the definitions of nodes and values of nodes, given previously. Nodes in such figures that have no arcs branching from them to the right are sometimes referred to as "leaf nodes," due to the rough similarity of the figures to trees.

We are now in a position to state a complete branch and bound procedure, once a few more definitions are made. Given any node $\mathbf{a}^* = (a^*(1), \ldots, a^*(q))$ of size q, *branching from the node* $\mathbf{a}^*$ will mean computing values for all nodes of size $q + 1$ for which the first q entries are precisely those of $\mathbf{a}^*$; Figure 8.11(b) illustrates branching from node 4. The symbols L and U will represent lower and upper bounds respectively on $\overline{TC}$, the minimum of the total costs of all assignments. The expression *update* U will mean take advantage of available data to find a value of U no larger than the previous value. The precise means by which U is updated will not be specified; we have previously illustrated several means: taking U to be the smallest of the node values of nodes of size $n - 2$, and taking U to be the total cost of completions of partial assignments found in computing values of nodes of size less than $n - 2$. The expression the *assignment* **a** *associated with* U will mean the assignment **a** with a total cost equal to U. The expression *trim the set of leaf nodes* will mean do not consider subsequently nodes with a value greater than U. The abbreviation LCA will stand for least-cost assignment, that is, an assignment with a total cost of $\overline{TC}$.

One possible algorithm for solving the quadratic assignment problem may now be stated. It should be emphasized that a simple version of the

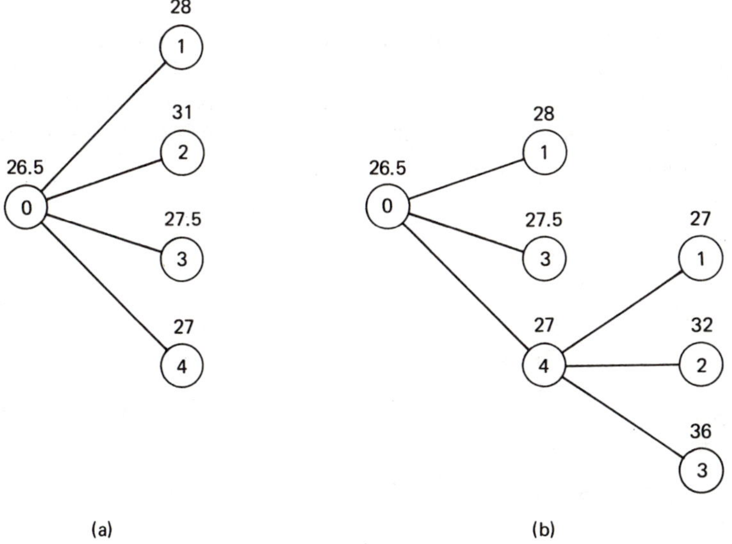

Figure 8.11. Branch and bound tree representation.

algorithm is being stated with the intention of emphasizing fundamental ideas, and that an effort has not been made to state a version of the algorithm in which every attempt has been made to achieve computational efficiencies.

A Branch and Bound Algorithm

0. Use Rule A to compute $v(0)$, let $L = v(0)$, let **a** be the assignment found in step 4 of Rule A, and let $U = TC(\mathbf{a})$. If $L = U$, then stop; **a** is a LCA. If $L < U$, go to the next step.
1. Use Rule B to compute node values $v(1), v(2), \ldots, v(n)$; obtain nodes $1, 2, \ldots, n$; delete node 0. Let L' be the smallest of the set of leaf node values $v(1), v(2), \ldots, v(n)$. If $L' > L$, let $L = L'$; otherwise, do not change L. Update U and trim the set of leaf nodes. If $L = U$, then stop; the assignment **a** associated with U is a LCA. If $L < U$, go to the next step.
2. Branch from any leaf node having a value of L' to create a new collection of leaf nodes and delete the node branched from. Let L' be the smallest of the values of all leaf nodes. If $L' > L$, let $L = L'$; otherwise, do not change L. Update U and trim all leaf nodes, both the new leaf nodes and those previously obtained. If $L = U$, then stop; the assignment **a** associated with U is a LCA. If $L < U$, then repeat this step.

We illustrate the algorithm through the use of the previous example. In the illustration it is convenient to divide steps 1 and 2 into two parts (a) and (b). Part (a) is all of each step up to the sentence "Update U . . . "; step 1b is the rest of step 1, while step 2b is the rest of step 2. In the illustration we choose not to update U in step 1.

Step	*Nodes and Values*	*L'*	*L*	*U*
0.	0 26.5		26.5	29
1a.	1 2 3 4 28, 31, 27.5, 27			
1b.	1 3 4 28, 27.5, 27	27	27	29
2a.	1 3 (4, 1) (4, 2) (4, 3) 28, 27.5, 27, 32, 36	27	27	29
2b.	(4, 1) 27		27	27

Some concluding comments about the branch and bound algorithm appear to be in order. As can be seen, the algorithm develops a sequence of nondecreasing lower bounds on $\overline{TC}$ and a sequence of nonincreasing upper bounds on $\overline{TC}$; when an upper bound is equal to a lower bound, the algorithm stops. While the means of obtaining lower bounds is well specified, the means of obtaining upper bounds has been intentionally left unspecified; the question of how best to obtain upper bounds is not as yet well answered. We suspect, however, that it would be best to obtain first an upper bound by applying one of the heuristic procedures to the quadratic assignment problem, and then not to update the upper bounds further until values are computed for partial assignments of size $n - 2$; each such value is an upper bound on $\overline{TC}$ and is computed as an intrinsic part of the algorithm. The alternative approach previously mentioned for finding upper bounds, computing the total costs for completions of partial assignments, may involve a great deal of work if carried out for each node.

It should also be pointed out that it is important to have a "good" upper bound, that is, one not much larger than $\overline{TC}$, as soon as possible when using the algorithm. The availability of a good upper bound reduces, by virtue of the trimming process, the number of leaf nodes that need to be examined. The fact that it is important to have a good upper bound quickly when using the algorithm strengthens the case for finding an initial bound by using one of the heuristic procedures.

Another point to emphasize is the fact that this algorithm becomes computationally unwieldy very quickly as n increases. We view the primary usefulness of the algorithm to be in providing "good" bounds on $\overline{TC}$, not in providing a least-cost assignment.

Finally, it seems appropriate to remark that the branch and bound procedure we have presented is only one of a number which are applicable to the problem, as Pierce and Crowston [11] point out. Indeed, for virtually any problem that can be solved by branch and bound, a number of different branch and bound procedures are conceivable. We have hardly skimmed the surface of the subject of branch and bound; the amount of literature on the subject is large, and you are referred to references such as [3], [5], and [11] for more material on the topic.

8.7 Summary

In this chapter we have addressed a facility location problem that occurs in a wide variety of contexts, including the plant layout problem. Both heuristic and exact solution procedures for the quadratic assignment problem were presented. In a sense the quadratic assignment problem can be considered as a discrete solution-space formulation of the multifacility location problems addressed in Chapter 5. Our presentation concentrated on those situations in which all new facilities were either points or had, basically, the same area.

Although our discussion of solution procedures concentrated on those costs which are proportional to distances between facilities, the procedures can be modified to incorporate a variety of costs (see Problems 8.13 and 8.14). Consequently, the formulation of the problem can approximate quite closely the actual location problem.

Except for relatively small sized problems, an exact solution to the quadratic assignment problem cannot be obtained at a reasonable computational cost. Therefore, heuristic solution procedures are generally used to obtain "good" solutions to the quadratic assignment problem.

The distinction was made in Chapter 3 between construction procedures and improvement procedures for solving the quadratic assignment problem. One of the first construction procedures for solving the quadratic assignment problem was that suggested by Wimmert [13], and subsequently clarified by Conway and Maxwell [2]. Even though a number of construction procedures have been developed, to date no construction procedure has been shown to be clearly superior to the best improvement procedures [10]. For this reason our discussion in this chapter concentrated on improvement procedures.

REFERENCES

1. ARMOUR, G. C., and E. S. BUFFA, "A Heuristic Algorithm and Simulation Approach to the Relative Location of Facilities," *Management Science*, Vol. 9, No. 2, 1963, pp. 294–309.

2. CONWAY, R. W., and W. L. MAXWELL, "A Note on the Assignment of Facility Locations," *The Journal of Industrial Engineering*, Vol. 12, No. 1, 1961, pp. 34–36.

3. GAVETT, J. W., and N. V. PLYTER, "The Optimal Assignment of Facilities to Locations by Branch and Bound," *Operations Research*, Vol. 14, No. 2, 1966, pp. 210–232.

4. GILMORE, P. C., "Optimal and Suboptimal Algorithms for the Quadratic Assignment Problem," *SIAM Journal*, Vol. 10, No. 2, 1962, pp. 305–313.

5. HANAN, M., and J. KURTZBERG, "A Review of the Placement and Quadratic Assignment Problems," *SIAM Review*, Vol. 14, 1972, pp. 324–342.

6. HILLIER, F. S., "Quantitative Tools for Plant Layout Analysis," *The Journal of Industrial Engineering*, Vol. 14, No. 1, 1963, pp. 33–40.

7. HILLIER, F. S., and M. M. CONNORS, "Quadratic Assignment Problem Algorithms and the Location of Indivisible Facilities," *Management Science*, Vol. 13, No. 1, 1966, pp. 42–57.

8. LAWLER, E. L., "The Quadratic Assignment Problem," *Management Science*, Vol. 9, No. 4, 1963, pp. 586–599.

9. LAWLER, E. L., and D. E. WOOD, "Branch and Bound Methods: A Survey," *Operations Research*, Vol. 14, 1966, pp. 699–719.

10. NUGENT, C. E., T. E. VOLLMANN, and J. RUML, "An Experimental Comparison of Techniques for the Assignment of Facilities to Locations," *Operations Research*, Vol. 16, No. 1, 1968, pp. 150–173.

11. PIERCE, J. F., and W. B. CROWSTON, "Tree-Search Algorithms for Quadratic Assignment Problems," *Naval Research Logistics Quarterly*, Vol. 18, No. 1, 1971, pp. 1–36.

12. VOLLMANN, T. E., C. E. NUGENT, and R. L. Zartler, "A Computerized Model for Office Layout," *The Journal of Industrial Engineering*, Vol. 19, No. 7, 1968, pp. 321–329.

13. WIMMERT, R. J., "A Mathematical Method of Equipment Location," *The Journal of Industrial Engineering*, Vol. 9, No. 6, 1958, pp. 498–505.

PROBLEMS

8.1. Write a computer program for the
   (a) Steepest-descent pairwise-interchange procedure.
   (b) Vollmann, Nugent, Zartler procedure.
   (c) Hillier procedure.
   (d) Gilmore–Lawler procedure.

8.2. Compute the values of $DTC_{ij}(\mathbf{a})$ shown in Table 8.1, and show all work.

8.3. Given an assignment of facilities to sites, if $p_k(\mathbf{a})$ is the total cost for facility k, why isn't the total cost of the assignment just $\sum_{k=1}^{n} p_k(\mathbf{a})$?

8.4. Suppose that p facilities are to be assigned to n sites, and that p is less than n. A weight matrix with entries w_{ij} is given for $i = 1, \ldots, p$ and $j = 1, \ldots, p$, and a weight matrix with entries w_{ij} is defined for $i = 1, \ldots, n$ and $j = 1, \ldots, n$ by using the given weights for $i = 1, \ldots, p$ and $j = 1, \ldots, p$, and by letting all other weights be zero. With "dummy facilities" $p + 1, p + 2, \ldots, n$ created to take up the unused sites, explain why the problem of assigning n facilities to n sites using the n by n weight matrix is equivalent to the problem of assigning p facilities to p sites using the p by p weight matrix.

8.5. Verify, by actually computing the total cost each way, that expressions (8.1) and (8.2) give the same total cost for the example given in Figures 8.1 and 8.2.

8.6. Given the assignment (2, 4, 5, 3, 1, 6) and the **D** and **W** matrices of Figure 8.2, find interchanges of facility locations that will result in total costs of 118, 116, and 108. Given reasons for your answers.

8.7. Given the assignment (2, 3, 4, 1) and the matrices

$$\mathbf{D} = \begin{pmatrix} 0 & 1 & 1 & 2 \\ 1 & 0 & 2 & 1 \\ 1 & 2 & 0 & 1 \\ 2 & 1 & 1 & 0 \end{pmatrix}, \qquad \mathbf{W} = \begin{pmatrix} 0 & 12 & 4 & 2 \\ 12 & 0 & 2 & 3 \\ 4 & 2 & 0 & 12 \\ 2 & 3 & 12 & 0 \end{pmatrix}$$

(a) Compute the total cost of the given assignment.
(b) Find a least-cost assignment by total enumeration. Show all work.
(c) Go through one iteration of the steepest-descent pairwise-interchange procedure. Show all work. (An iteration consists of going through steps 2 through 13 of the procedure.)
(d) Compute a lower bound on the minimum total cost using the method discussed in Section 8.2.
(e) Compute a lower bound on the minimum total cost using the method discussed in Section 8.6.
(f) Find an assignment with a cost equal to one of the lower bounds.

8.8. The following **D** and **W** matrices are given:

$$\mathbf{D} = \begin{pmatrix} 0 & 1 & 1 & 2 & 3 \\ 1 & 0 & 2 & 1 & 2 \\ 1 & 2 & 0 & 1 & 2 \\ 2 & 1 & 1 & 0 & 1 \\ 3 & 2 & 2 & 1 & 0 \end{pmatrix}, \qquad \mathbf{W} = \begin{pmatrix} 0 & 5 & 2 & 4 & 1 \\ 5 & 0 & 3 & 0 & 2 \\ 2 & 3 & 0 & 0 & 0 \\ 4 & 0 & 0 & 0 & 5 \\ 1 & 2 & 0 & 5 & 0 \end{pmatrix}$$

(a) Compute a lower bound on the minimum total cost of an assignment using the method discussed in Section 8.2.
(b) Given the initial assignment $\mathbf{a} = (1, 4, 3, 5, 2)$, apply the steepest-descent pairwise-interchange procedure to the problem until two or more interchanges have been made, and compute the total cost for each assignment obtained.
(c) Repeat part (b), except use the VNZ procedure.
(d) Repeat part (b), except use the Hillier procedure. (Sites 1 and 2 are in columns 1 and 2 of row 1, and sites 3, 4, and 5 are in columns 3, 4, and 5 of row 2.)
(e) Compute a lower bound on the minimum total cost of an assignment using the method discussed in Section 8.6.
(f) Find an assignment with a minimum total cost.

8.9. The following **D** and **W** matrices are given:

$$\mathbf{D} = \begin{pmatrix} 0 & 1 & 2 & 1 & 2 & 3 \\ 1 & 0 & 1 & 2 & 1 & 2 \\ 2 & 1 & 0 & 3 & 2 & 1 \\ 1 & 2 & 3 & 0 & 1 & 2 \\ 2 & 1 & 2 & 1 & 0 & 1 \\ 3 & 2 & 1 & 2 & 1 & 0 \end{pmatrix}, \qquad \mathbf{W} = \begin{pmatrix} 0 & 5 & 2 & 4 & 1 & 0 \\ 5 & 0 & 3 & 0 & 2 & 2 \\ 2 & 3 & 0 & 0 & 0 & 0 \\ 4 & 0 & 0 & 0 & 5 & 2 \\ 1 & 2 & 0 & 5 & 0 & 10 \\ 0 & 2 & 0 & 2 & 10 & 0 \end{pmatrix}$$

(a) Compute a lower bound on the minimum total cost of an assignment using the method discussed in Section 8.2.

(b) Given the assignment $\mathbf{a} = (2, 4, 3, 5, 6, 1)$, apply the steepest-descent pairwise-interchange procedure to the problem until two or more interchanges have been made, and compute the total cost for each assignment obtained.

(c) Repeat part (b), except use the VNZ procedure.

(d) Repeat part (b), except use the Hillier procedure, with $M = 2$ and $N = 3$.

(e) Compute a lower bound on the minimum total cost using the method discussed in Section 8.6.

8.10. Derive Equation (8.18) when $a(i)$ is in any one of the regions 4 through 9 shown in Figure 8.7.

8.11. Derive Equation (8.20) of Remark 2.

8.12. For the Hillier procedure, let an assignment $\mathbf{a}$ be given, and suppose that a new assignment $\mathbf{a}'$ is obtained by interchanging the locations of facilities u and v.

(a) If the interchange is either a L–R or a R–L p-step interchange, show that $M(q/r: \mathbf{a}) = M(q/r: \mathbf{a}')$ if both facilities q and r are located in site columns either to the left of or to the right of all site columns between and including those site columns in which facilities u and v are located.

(b) If the interchange is either an U–D or a D–U p-step interchange, show that $M(q/r: \mathbf{a}) = M(q/r: \mathbf{a}')$ if both facilities q and r are located in site rows either above or below all site rows between and including those site rows in which facilities u and v are located.

(c) If the interchange is a p-step diagonal interchange, the site region is partitioned into four disjoint "corner" sets of grid squares after deleting all columns between and including those site columns in which facilities u and v are located, and all rows between and including those site rows in which facilities u and v are located. Show that $M(q/r: \mathbf{a}) = M(q/r: \mathbf{a}')$ if both facilities q and r are located in any one of the four sets.

8.13. Suppose, for the quadratic assignment problem considered in Section 8.2, that a cost $c(i, k)$ is incurred if facility i is located at site k. Given an assignment $\mathbf{a}$, the cost of the assignment is then given by

$$C_1(\mathbf{a}) = \sum_{i=1}^{n} c(i, a(i)) + \sum_{1 \le i < j \le n} w_{ij}\, d(a(i), a(j))$$

Suppose that a new assignment $\mathbf{a}'$ is obtained by interchanging the locations of facilities u and v, and let $DC_{uv}(\mathbf{a}) = C_1(\mathbf{a}) - C_1(\mathbf{a}')$.

(a) Show that

$$DC_{uv}(\mathbf{a}) = [c(u, a(u)) - c(u, a(v))] + [c(v, a(v)) - c(v, a(u))] + DTC_{uv}(\mathbf{a})$$

where $DTC_{uv}(\mathbf{a})$ is defined by (8.12).

(b) Discuss what modifications would be needed to the heuristic procedures considered previously to incorporate the preceding total cost function.

(c) Develop examples for which the total cost function would be more realistic than the one considered in the chapter.

8.14. Suppose that the total cost function, given an assignment $\mathbf{a}$, is given by

$$K(\mathbf{a}) = C_1(\mathbf{a}) + C_2(\mathbf{a})$$

where $C_1(\mathbf{a})$ is as defined in Problem 8.13 and $C_2(\mathbf{a})$ consists of all remaining costs that result from the assignment $\mathbf{a}$. An example of a cost that would be included in the term $C_2(\mathbf{a})$ is the fixed cost of installing a special conveyor connecting two facilities. Included in $C_2(\mathbf{a})$ are costs not proportional to the distances between facilities, but dependent on the locations of combinations of two or more facilities. Suppose that a new assignment $\mathbf{a}'$ is obtained by interchanging the locations of facilities u and v, and let $DK_{uv}(\mathbf{a}) = K(\mathbf{a}) - K(\mathbf{a}') = DC_{uv}(\mathbf{a}) + C_2(\mathbf{a}) - C_2(\mathbf{a}')$, where $DC_{uv}(\mathbf{a})$ is as defined in Problem 8.13.

(a) Discuss the modifications that must be made to the heuristic procedures considered in order to incorporate the total cost function $K(\mathbf{a})$.

(b) Develop examples for which the total cost function $K(\mathbf{a})$ is more realistic than that considered in the chapter.

8.15. Compute, including complete detail, the matrices given by Equations (8.34), (8.35), and (8.36).

8.16. (a) Use Rule A to obtain a lower bound on the minimum total cost for the example problem given in Section 8.2.

(b) Go through step 1 of the branch and bound algorithm for the example given in Section 8.2.

8.17. Use the branch and bound algorithm to solve the quadratic assignment problem for which the data are

$$\mathbf{C} = \begin{pmatrix} 1 & 2 & 1 \\ 3 & 2 & 2 \\ 1 & 3 & 4 \end{pmatrix}, \qquad \mathbf{D} = \begin{pmatrix} 0 & 2 & 3 \\ 2 & 0 & 1 \\ 3 & 1 & 0 \end{pmatrix}, \qquad \mathbf{W} = \begin{pmatrix} 0 & 1 & 3 \\ 1 & 0 & 4 \\ 3 & 4 & 0 \end{pmatrix}$$

8.18. Using Rule A, find a lower bound for the example problem given in Section 8.2.

8.19. Given the assignment (1, 3, 4, 2) and the matrices

$$\mathbf{D} = \begin{pmatrix} 0 & 1 & 2 & 3 \\ 1 & 0 & 1 & 2 \\ 2 & 1 & 0 & 1 \\ 3 & 2 & 1 & 0 \end{pmatrix}, \qquad \mathbf{W} = \begin{pmatrix} 0 & 12 & 4 & 2 \\ 12 & 0 & 2 & 3 \\ 4 & 2 & 0 & 12 \\ 2 & 3 & 12 & 0 \end{pmatrix}$$

(a) Find a least-cost assignment by total enumeration. Show all work.
(b) Solve the problem using the VNZ procedure.
(c) Solve the problem using the Hillier procedure where $M = 1$ and $N = 4$.

8.20. Solve the quadratic assignment problems, Problems 3.1, 3.2, and 3.3, using the
(a) Steepest-descent pairwise-interchange procedure.
(b) VNZ procedure.
(c) Hillier procedure.

8.21. List 10 applications of the quadratic assignment problem formulation given either in the chapter, in Problem 8.13, or in Problem 8.14. Clearly identify the facilities, the sites, and the costs for each application cited.

8.22. Use the computer program obtained in Problem 8.1 (a) to solve the quadratic assignment problem for which the data are

$$\mathbf{D} = \begin{pmatrix} 0 & 1 & 2 & 3 & 1 & 2 & 3 & 4 \\ 1 & 0 & 1 & 2 & 2 & 1 & 2 & 3 \\ 2 & 1 & 0 & 1 & 3 & 2 & 1 & 2 \\ 3 & 2 & 1 & 0 & 4 & 3 & 2 & 1 \\ 1 & 2 & 3 & 4 & 0 & 1 & 2 & 3 \\ 2 & 1 & 2 & 3 & 1 & 0 & 1 & 2 \\ 3 & 2 & 1 & 2 & 2 & 1 & 0 & 1 \\ 4 & 3 & 2 & 1 & 3 & 2 & 1 & 0 \end{pmatrix} \qquad \mathbf{W} = \begin{pmatrix} 0 & 5 & 2 & 4 & 1 & 0 & 0 & 6 \\ 5 & 0 & 3 & 0 & 2 & 2 & 2 & 0 \\ 2 & 3 & 0 & 0 & 0 & 0 & 0 & 5 \\ 4 & 0 & 0 & 0 & 5 & 2 & 2 & 10 \\ 1 & 2 & 0 & 5 & 0 & 10 & 0 & 0 \\ 0 & 2 & 0 & 2 & 10 & 0 & 5 & 1 \\ 0 & 2 & 0 & 2 & 0 & 5 & 0 & 10 \\ 6 & 0 & 5 & 10 & 0 & 1 & 10 & 0 \end{pmatrix}$$

The sites for this problem are arranged as follows:

1	2	3	4
5	6	7	8

The optimum solution is known to have a total cost of 107. Note the effect of the initial assignment on the final solution by employing a number of different initial assignments. (Anyone who obtains an optimum assignment using the initial assignment selected first automatically becomes a member of the Quadratic Assignment Problem Society, also known as the QUAP Society.)

8.23. Five manufacturing departments are to be assigned among the six sites shown below. Four products are to be processed through the five departments according to the processing sequences shown, with the indicated frequencies of movement between departments.

Product	*Processing Sequence*	*Material-Handling Volume (loads/day)*
1	A, B, C, D, E	20
2	A, C, B, C, D, E	30
3	A, D, E	15
4	A, C, D, B, E	40

1	2	3
4	5	6

(a) Develop the **W** and **D** matrices, assuming rectilinear travel between the centroids of the sites.

(b) Specify an initial assignment and employ the three heuristic algorithms to solve the problem.

8.24. Solve Problem 8.23 assuming the six sites are arranged in a row along the x axis.

8.25. O. R. Mann is designing his house so that distance traveled is minimized. Using standard room modules of size 12 by 12 feet an eight-room house (including the garage) is to be designed. Based on the **W** matrix and site arrangement given below, develop a house design using each of the heuristic algorithms. Evaluate each on the basis of practical limitations.

W =

	L.R.	B.R. 1	B.R. 2	B.R. 3	Den	D.R.	Kit.	Gar.
L.R.	0	2	1	1	5	4	3	2
B.R. 1	2	0	2	2	6	3	4	1
B.R. 2	1	2	0	3	5	4	4	0
B.R. 3	1	2	3	0	4	3	5	0
Den	5	6	5	4	0	4	6	10
D.R.	4	3	4	3	4	0	8	0
Kit.	3	4	4	5	6	8	0	12
Gar.	2	1	0	0	10	0	12	0

1	2	3	4
5	6	7	8
9	10	11	12
13	14	15	16

Site Arrangement

(Note: The number of sites is greater than the number of rooms!)

8.26. Pierce and Crowston [11] list a variety of applications of the quadratic assignment formulation. Discuss pros and cons of each application cited by them.

8.27. A variation of the steepest-descent pair-wise interchange procedure is the natural selection procedure in which the first pair-wise interchange yielding a decrease in total cost is used to obtain an improved solution. A form of the natural selection procedure is used in phase 2 of the VNZ procedure. Write a computer program for the natural selection procedure and solve Problem 8.22 using the program.

MINIMAX LAYOUT AND LOCATION PROBLEMS

9.1 Introduction

In previous discussions about facility layout and location we have set as an objective the minimization of some appropriate *total* cost function. To be more specific, to date we have been concerned with the *minimization* of the *sum* of costs that results from a particular solution. Thus we have been concerned with "minisum" layout and location problems.

As emphasized in Chapter 1, a "minisum" objective might be inappropriate for a number of real-world layout and location problems. It was suggested that an alternative, and perhaps more appropriate, objective might be to obtain a solution which *minimizes* the *maximum* cost resulting from a given location solution. Such a situation was labeled a minimax layout and location problem. In this chapter we consider a number of minimax formulations of facility layout and location problems.

As an illustration of a location problem that might be formulated with a minimax objective, consider the problem of locating health outreach clinics in rural areas. In such a situation, clinics might be located so that the maximum distance a patient must travel to a clinic is minimized. A similar example concerns the location of storerooms in a manufacturing plant that

9

minimizes the maximum distance employees must travel to reach a storeroom. Yet another example concerns the placement of fire stations in a large, metropolitan area such that the maximum distance between any location within the city and the nearest fire station is minimized. As can be seen, the number and variety of location problems that can be formulated appropriately as minimax problems are sizable.

Although no substantive effort has been exerted to verify the belief, a minimax objective appears to reflect accurately many managers' preferences in locating facilities. In a sense, a minimax solution can be interpreted as a "grease the squeaky wheel" solution, since one wishes to minimize the effects of the worst situation, that is, maximum cost. In solving a minimax location problem, costs other than the maximum cost are not considered.

As an overview of the chapter, the discussion of minimax problems begins with a consideration of rectilinear minimax location problems in Section 9.2. Both single and multiple-facility problems are considered. In Section 9.3 we consider some Euclidean minimax facility location problems. A minimax facility layout problem is treated in Section 9.4. Our treatment of minimax problems concludes in Section 9.5 with a discussion of a number of related minimax facility layout and location problems.

9.2 Rectilinear Location Problems

In this section we consider minimax facility location problems under the assumption of rectilinear distances. Both single and multiple-facility location problems are treated. Thus the problems considered are minimax analogs of the minimum total cost (minisum) location problems considered in Sections 4.2 and 5.2.

9.2.1 *Single-facility rectilinear minimax location problems*

As a first example of a minimax location problem, suppose that existing facilities are located at points $(a_1, b_1), \ldots, (a_m, b_m)$ and that a new facility is to be located at the point (x, y). The maximum rectilinear distance between the new facility and any existing facility is denoted by $f(x, y)$, and is given by

$$f(x, y) = \max_{1 \leq i \leq m} (|x - a_i| + |y - b_i|) \tag{9.1}$$

The minimax location problem is then to find a location (x^*, y^*) for the new facility that makes $f(x, y)$ as small as possible. In other words, we want to minimize the maximum distance between the new facility and any existing facility; hence the use of the term "minimax".

It is important to recognize that minimax location problems occur in different physical contexts than do minimum total cost location problems. As an example, suppose that particularly dangerous fires may break out in a plant at any of the known points $(a_1, b_1), \ldots, (a_m, b_m)$, and a single expensive piece of apparatus needed for fighting these fires is to be located at a point (x, y) to be determined. In such a situation it is arguably unreasonable to minimize the sum of the travel distances between the piece of apparatus and all the points (a_i, b_i), since a single fire may conceivably put the plant out of business; the piece of apparatus, if used, would very likely never move from its location to all the potential fire points, and so its travel distance would never be the sum of all the travel distances. However, the travel distance of the piece of apparatus could very well be the distance to any one of the points (a_i, b_i), and since we want the piece of apparatus to respond as quickly as possible, we shall require that it be located so as to minimize the maximum travel distance from the piece of apparatus to any one of the points (a_i, b_i). To generalize from this example, one might take a minimax approach to a location problem when it is more important to provide quick service, or convenient access, than it is to minimize long-term total costs.

We now state a procedure for finding all points $(x,^* y^*)$ that minimize the function defined by (9.1); a justification for the procedure will be given subsequent to its statement. In the expressions for c_1 through c_4 that follow, set all g_i to zero, and compute

$$c_1 = \min_{1 \le i \le m} (a_i + b_i - g_i), \qquad c_2 = \max_{1 \le i \le m} (a_i + b_i + g_i)$$

$$c_3 = \min_{1 \le i \le m} (-a_i + b_i - g_i), \qquad c_4 = \max_{1 \le i \le m} (-a_i + b_i + g_i)$$

$$c_5 = \max (c_2 - c_1, c_4 - c_3)$$

Any point (x^*, y^*) on the line segment joining the points

$$\tfrac{1}{2}(c_1 - c_3, c_1 + c_3 + c_5) \tag{9.2}$$

and

$$\tfrac{1}{2}(c_2 - c_4, c_2 + c_4 - c_5) \tag{9.3}$$

is a minimax location, that is, minimizes the function defined by (9.1); the minimum value of the function is $c_5/2$.

As an example of the use of the procedure, suppose that $m = 5$ and the points $(a_1, b_1), \ldots, (a_5, b_5)$ are given by (2, 10), (7, 9), (7, 12), (3, 15), and (1, 14), respectively; then $c_1 = 12$, $c_2 = 19$, $c_3 = 2$, $c_4 = 13$, and $c_5 = 11$. Any point on the line segment joining (5, 12.5) and (3, 10.5) is a minimax location, and the minimum value of (9.1) for this example is 5.5. The points $(a_1, b_1), \ldots, (a_5, b_5)$ are plotted in Figure 9.1, as well as the line segment joining the points (5, 12.5) and (3, 10.5). The rectangle in the figure is a contour line of (9.1) for this example of value 8; a method for constructing contour lines will be discussed later in the section.

A justification for the solution procedure will now be given. We shall, in fact, develop a procedure for solving the more general problem of minimizing the function $f(x, y)$ defined by

$$f(x, y) = \max_{1 \le i \le m} (|x - a_i| + |y - b_i| + g_i) \tag{9.4}$$

Note that (9.4) becomes (9.1) when all g_i are zero, which is why all g_i are zero in the procedure for minimizing (9.1). As a motivation for including the term g_i, (x, y) might be the location of an ambulance, and g_i the travel distance from the point (a_i, b_i) to the nearest hospital. Thus a minimax solution to (9.4) would minimize the maximum travel distance for an ambulance to respond to an emergency at any point (a_i, b_i) and then go to the nearest hospital. As will be seen, the procedure for finding all points which minimize

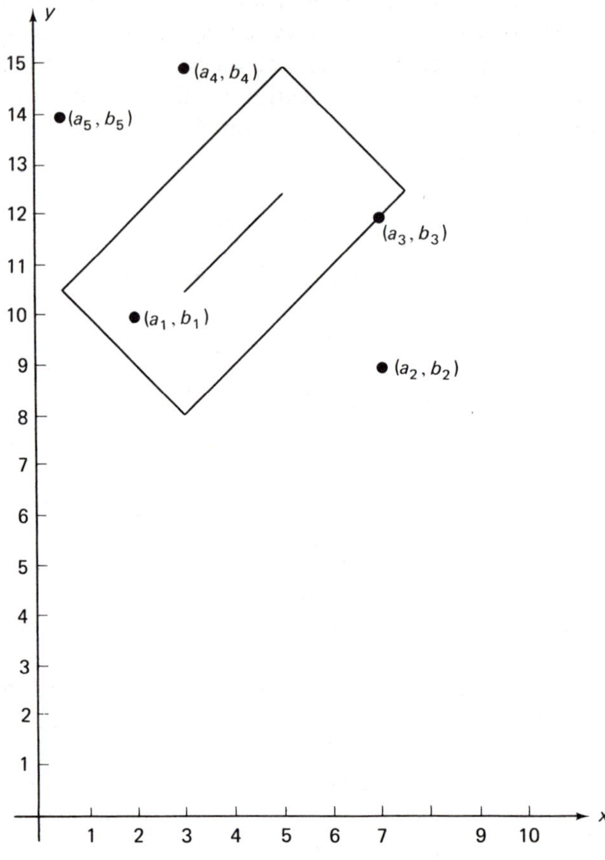

Figure 9.1. Example relation to Eq. (9.1).

(9.4) is identical to that for finding all points which minimize (9.1), with the single exception that the g_i are not set to zero, but take on their actual values.

To solve the problem of minimizing (9.4), we construct the following equivalent problem:

minimize z
subject to

$$|x - a_i| + |y - b_i| + g_i \leq z, \quad i = 1, \ldots, m$$

To motivate the equivalence of the two problems, note that since z is at least as large as each term $|x - a_i| + |y - b_i| + g_i$, it will be at least as large as the maximum of these; furthermore, since z is to be as small as possible, it will be equal to the maximum of the terms and, since z is being minimized, the maximum of the terms will be as small as possible.

We now convert the preceding problem to a linear-programming problem. Note that each constraint may be written

$$|x - a_i| \leq z - g_i - |y - b_i|$$

or, equivalently,

$$-z + g_i + |y - b_i| \leq x - a_i \leq z - g_i - |y - b_i|$$

These two inequalities may be written

$$|y - b_i| \leq x - a_i + z - g_i$$
$$|y - b_i| \leq -x + a_i + z - g_i$$

which are, in turn, equivalent to

$$-x + a_i - z + g_i \leq y - b_i \leq x - a_i + z - g_i$$
$$+x - a_i - z + g_i \leq y - b_i \leq -x + a_i + z - g_i$$

We thus have, after rearranging the four inequalities, the following linear-programming problem:

minimize z
subject to

$$x + y - z \leq a_i + b_i - g_i, \quad i = 1, \ldots, m \qquad (9.5)$$
$$x + y + z \geq a_i + b_i + g_i, \quad i = 1, \ldots, m \qquad (9.6)$$
$$-x + y - z \leq -a_i + b_i - g_i, \quad i = 1, \ldots, m \qquad (9.7)$$
$$-x + y + z \geq -a_i + b_i + g_i, \quad i = 1, \ldots. m \qquad (9.8)$$

Now this linear-programming problem is equivalent to an even simpler one. Consider the first m inequalities; since $x + y - z$ is no greater than each $a_i + b_i - g_i$, it is no greater than the smallest $a_i + b_i - g_i$, which is just c_1. Thus the first m inequalities are equivalent to the inequality $x + y - z \leq c_1$. In a similar manner, each of the other three sets of m inequalities is equivalent to just one, resulting in the following linear-programming problem:

minimize z
subject to

$$\begin{aligned} x + y - z &\leq c_1 \\ x + y + z &\geq c_2 \\ -x + y - z &\leq c_3 \\ -x + y + z &\geq c_4 \end{aligned} \qquad (9.9)$$

Problem (9.9) is a particularly simple one, and optimum solutions to it are readily obtainable. Multiplying the first constraint by -1 and adding it to the second gives $z \geq \frac{1}{2}(c_2 - c_1)$, while multiplying the third constraint by -1 and adding it to the fourth gives $z \geq \frac{1}{2}(c_4 - c_3)$, so that $z \geq \max [\frac{1}{2}(c_2 - c_1), \frac{1}{2}(c_4 - c_3)] = \frac{1}{2}c_5$. Thus $\frac{1}{2}c_5$ is a lower bound on the minimum value of the objective function, so any feasible solution (x, y, z) with an objective function value of $\frac{1}{2}c_5$ is an optimum solution. It is left as an exercise to verify that the points $(x_1, y_1, z_1) = \frac{1}{2}(c_1 - c_3, c_1 + c_3 + c_5, c_5)$ and $(x_2, y_2, z_2) = \frac{1}{2}(c_2 - c_4, c_2 + c_4 - c_5, c_5)$ both satisfy all the constraints of the linear-programming problem, and so are optimum. Furthermore, since convex combinations of optimum solutions are also optimum solutions, taking convex combinations of (x_1, y_1, z_1) and (x_2, y_2, z_2) gives as minimax locations all points on the line segment joining the points (9.2) and (9.3), which completes the justification of the solution procedure.

Next we consider the function $f(x, y)$ defined by

$$f(x, y) = \max_{1 \leq i \leq m} [w_i(|x - a_i| + |y - b_i|) + g_i] \tag{9.10}$$

Note that (9.10) is a generalization of (9.4), where the "weights" w_i may be any positive numbers. As an example of a possible use of (9.10), suppose that (x, y) is to be the location of a "convenience" center, and "users" of the center are located at the points $(a_i, b_i), \ldots, (a_m, b_m)$. User i requires a time of g_i to prepare to go to the center, and then travels to the center at a time per unit distance rate of w_i, so that $w_i(|x - a_i| + |y - b_i|) + g_i$ is the total time to prepare to go to the center and then go there. The center is to be located so that the maximum total time for any user will be minimized.

Prior to stating a procedure for minimizing (9.10), we give a procedure for constructing contour lines of (9.10); the construction of contour lines of (9.10) is of interest for the same reasons as given in Chapter 4 for constructing contour lines. The procedure, of course, also may be used to construct contour lines of (9.1) and (9.4).

A contour line of (9.10) consists of all points (x, y) for which $f(x, y) = k$, where k is a chosen constant. The procedure for constructing contour lines is to develop a simple characterization of the set $\mathbf{S}(k) = \{(x, y): f(x, y) \leq k\}$, that is, the set of all points (x, y) for which $f(x, y) \leq k$. Let M be the collection of integers i between 1 and m for which w_i is positive, and $\bar{M}$ be the integers between 1 and m for which w_i is zero. (We allow here the possibility that $w_i = 0$.) Define the following terms:

$$\bar{g} = \max_{i \in \bar{M}} (g_i)$$

$$c_1(k) = \min_{i \in M} \frac{k - g_i}{w_i} + a_i + b_i$$

$$c_2(k) = \max_{i \in M} \frac{-k + g_i}{w_i} + a_i + b_i$$

$$c_3(k) = \min_{i \in M} \frac{k - g_i}{w_i} - a_i + b_i$$

$$c_4(k) = \max_{i \in M} \frac{-k + g_i}{w_i} - a_i + b_i$$

If $k < \bar{g}$, there are no points in the set $\mathbf{S}(k)$, and thus no contour lines of (9.10). If $\bar{g} \leq k$, then

$$\mathbf{S}(k) = \{(x, y) \colon c_2(k) \leq x + y \leq c_1(k), c_4(k) \leq -x + y \leq c_3(k)\}$$

It should be evident that an equivalent condition for there to be points in $\mathbf{S}(k)$ is that $c_2(k) \leq c_1(k)$ and that $c_4(k) \leq c_3(k)$; when the latter two inequalities hold, a contour line of (9.10) of value k is just the boundary of the set $\mathbf{S}(k)$. Furthermore, the set $\mathbf{S}(k)$ is just a rectangle with two parallel sides making a 45-degree angle with the x axis, and the other two parallel sides making a -45-degree angle with the x axis; when $c_2(k) \leq c_1(k)$ and $c_4(k) \leq c_3(k)$, the coordinates of the corners of the rectangle, beginning with the top corner and proceeding clockwise, are given respectively by $\frac{1}{2}(c_1(k) - c_3(k), c_1(k) + c_3(k))$, $\frac{1}{2}(c_1(k) - c_4(k), c_1(k) + c_4(k))$, $\frac{1}{2}(c_2(k) - c_4(k), c_2(k) + c_4(k))$, and $\frac{1}{2}(c_2(k) - c_3(k), c_2(k) + c_3(k))$. It is left as an exercise to show that the corners of $\mathbf{S}(k)$ are in fact as given. The derivation of the procedure for constructing contour lines will not be given in any detail. The procedure is based on the fact that, for $f(x, y)$ given by (9.10),

$$f(x, y) = \max \{\max_{i \in M} [w_i(|x - a_i| + |y - b_i|) + g_i], \max_{i \in \bar{M}} (g_i)\}$$

and that, when $\bar{g} \leq k$, the inequality

$$\max_{i \in M} [w_i(|x - a_i| + |y - b_i|) + g_i] \leq k$$

is equivalent to inequalities (9.5) through (9.8) with g_i deleted, z replaced by $(k - g_i)/w_i$, and the condition $i = 1, \ldots, m$ replaced by $i \in M$.

As an example of the construction of contour lines, consider the example with $m = 5$ given previously, so that all $w_i = 1$ and all $g_i = 0$. When $k = 5.5$, the set $\mathbf{S}(k)$ should be the line joining the points (3, 10.5) and (5, 12.5), since k is the smallest value of the function. A computation establishes that $c_1(5.5) = 17.5$, $c_2(5.5) = 13.5$, $c_3(5.5) = 7.5$, $c_4(5.5) = 7.5$, and the coordinates of the four corners of the set $\mathbf{S}(5.5)$, beginning with the top corner and proceeding clockwise, are given, respectively, by (5, 12.5), (5, 12.5), (3, 10.5), and (3, 10.5); thus $\mathbf{S}(5.5)$ is a degenerate rectangle—a line—and agrees with what we know to be the answer. When $k = 8$, $c_1(8) = 20$, $c_2(8) = 11$, $c_3(8) = 10$,

$c_4(8) = 5$, and the coordinates of the four corners of **S**(8) are (5, 15), (7.5, 12.5), (3, 8), and (0.5, 10.5), so that **S**(8) is in this case a nondegenerate rectangle, the boundary of which consists of all points for which the value of (9.10) is 8. The contour line is illustrated in Figure 9.1.

Naturally, it is also of interest to find all points which minimize (9.10); that is, to find all minimax locations, as well as to compute contour lines. We now state a procedure for finding minimax locations, and shall provide a motivation for the procedure subsequently. It is first useful to define the linear transformations T and T^{-1} of points in the plane by

$$T(x, y) = (x + y, -x + y) \tag{9.11}$$

$$T^{-1}(r, s) = \tfrac{1}{2}(r - s, r + s) \tag{9.12}$$

As a notational convenience, $T(a_i, b_i)$ will be represented by $(a'_i, b'_i) = (a_i + b_i, -a_i + b_i)$. Also, for all $1 \le i < j \le m$, define the numbers α_{ij} and β_{ij} by

$$\alpha_{ij} = \max\left(\frac{w_i w_j |a'_i - a'_j| + w_i g_j + w_j g_i}{(w_i + w_j)}, g_i, g_j\right)$$

$$\beta_{ij} = \max\left(\frac{w_i w_j |b'_i - b'_j| + w_i g_j + w_j g_i}{(w_i + w_j)}, g_i, g_j\right)$$

We shall now give a procedure for minimizing (9.10), then give an example, and then motivate the procedure. The problem of minimizing (9.10) may be transformed into an equivalent linear-programming problem, using basically the same approach as we used for finding a linear-programming problem (9.9) equivalent to the problem of minimizing (9.4). However, the resultant linear-programming problem is not nearly as simple as the problem (9.9), and it is necessary to know more about linear programming than we are assuming you know in order to understand the derivation of the solution procedure via linear programming. Therefore, we are going to be content with providing a motivation for the solution procedure instead of a complete derivation; if you want all the details we refer you to [2]. The solution procedure is as follows. Let p_1 and p_2 be indices for which

$$z_1 = \max_{1 \le i < j \le m} (\alpha_{ij}) = \alpha_{p_1 p_2}$$

and, when $a'_{p_1} \le a'_{p_2}$, let

$$r^* = \frac{w_{p_1} a'_{p_1} + w_{p_2} a'_{p_2} - g_{p_1} + g_{p_2}}{w_{p_1} + w_{p_2}}$$

When $a'_{p_1} > a'_{p_2}$, let

$$r^* = \frac{w_{p_1} a'_{p_1} + w_{p_2} a'_{p_2} + g_{p_1} - g_{p_2}}{w_{p_1} + w_{p_2}}$$

Let q_1, q_2, be indices for which

$$z_2 = \max_{1 \leq i < j \leq m} (\beta_{ij}) = \beta_{q_1 q_2}$$

and, when $b'_{q_1} \leq b'_{q_2}$, let

$$s^* = \frac{w_{q_1} b'_{q_1} + w_{q_2} b'_{q_2} - g_{q_1} + g_{q_2}}{w_{q_1} + w_{q_2}}$$

When $b'_{q_1} > b'_{q_2}$, let

$$s^* = \frac{w_{q_1} b'_{q_1} + w_{q_2} b'_{q_2} + g_{q_1} - g_{q_2}}{w_{q_1} + w_{q_2}}$$

Then
$$z_0 = \max (z_1, z_2)$$

is the minimum value of (9.10), and $T^{-1}(r^*, s^*)$ is a minimax location. To find all possible minimax locations, three cases must be considered:

Case 1: $z_0 = z_1 = z_2$. $T^{-1}(r^*, s^*)$ is the only minimax location.

Case 2: $z_0 = z_1 > z_2$. Compute

$$s_1 = \max_{1 \leq i \leq m} b'_i - \frac{(z_0 - g_i)}{w_i}, \qquad s_2 = \min_{1 \leq i \leq m} b'_i + \frac{(z_0 - g_i)}{w_i}$$

Any point on the line segment joining the points $T^{-1}(r^*, s_1)$ and $T^{-1}(r^*, s_2)$ is a minimax location.

Case 3: $z_0 = z_2 > z_1$. Compute

$$r_1 = \max_{1 \leq i \leq m} a'_i - \frac{(z_0 - g_i)}{w_i}, \qquad r_2 = \min_{1 \leq i \leq m} a'_i + \frac{(z_0 - g_i)}{w_i}$$

Any point on the line segment joining the points $T^{-1}(r_1, s^*)$ and $T^{-1}(r_2, s^*)$ is a minimax location.

At this point it may be helpful to consider an example. Suppose that $m = 4$, the points (a_1, b_1) through (a_4, b_4) are given by (3, 3), (3, 6), (6, 3), and (7, 8), respectively, that $w_1 = 2$, $w_2 = 3$, $w_3 = 4$, $w_4 = 2$, and that $g_1 = 1$, $g_2 = g_3 = g_4 = 0$. Then computations establish that $\alpha_{12} = \frac{21}{5}$, $\alpha_{13} = \frac{28}{6}$, $\alpha_{14} = \frac{38}{4}$, $\alpha_{23} = 0$, $\alpha_{24} = \frac{36}{5}$, $\alpha_{34} = \frac{48}{6}$, and that $\beta_{12} = \frac{21}{5}$, $\beta_{13} = \frac{28}{6}$, $\beta_{14} = \frac{6}{4}$, $\beta_{23} = \frac{72}{7}$, $\beta_{24} = \frac{12}{5}$, $\beta_{34} = \frac{32}{6}$. Thus

$$z_1 = \max_{1 \leq i < j \leq 4} (\alpha_{ij}) = \alpha_{14} = \tfrac{38}{4}$$

$$z_2 = \max_{1 \leq i < j \leq 4} (\beta_{ij}) = \beta_{23} = \tfrac{72}{7}$$

so that $(p_1, p_2) = (1, 4)$, $(q_1, q_2) = (2, 3)$, and $z_0 = z_2 > z_1$. Also, $r^* = \frac{41}{4}$ and $s^* = -\frac{3}{7}$. Since we have the Case 3 situation, we compute $r_1 = \frac{69}{7}$, $r_2 = \frac{149}{14}$, and find that $T^{-1}(r_1, s^*) = (\frac{36}{7}, \frac{33}{7})$ and $T^{-1}(r_2, s^*) = (\frac{155}{28}, \frac{143}{28})$. Any point on the line segment joining $T^{-1}(r_1, s^*)$ and $T^{-1}(r_2, s^*)$ is a minimax location. It is left as an exercise to verify that the computation of the four corners of the set $S(z_0)$ for this example also gives the line segment joining $T^{-1}(r_1, s^*)$ and $T^{-1}(r_2, s^*)$.

We now develop a motivation of the solution procedure just stated. It is convenient first to state two remarks explicitly. The details of the derivation of the first remark are left as a homework problem.

Remark 1: Given any points (x, y) and (a_i, b_i), if $(r, s) = T(x, y)$ and $(a_i', b_i') = T(a_i, b_i)$, then

$$|x - a_i| + |y - b_i| = \max(|r - a_i'|, |s - b_i'|)$$

Remark 2: Given the conditions of Remark 1, if $f(x, y)$ is the maximum cost function defined by (9.10), then

$$f(x, y) = g(r, s)$$

where $$g(r, s) = \max[f_1(r), f_2(s)]$$

with $$f_1(r) = \max_{1 \le i \le m} (w_i|r - a_i'| + g_i)$$

$$f_2(s) = \max_{1 \le i \le m} (w_i|s - b_i'| + g_i)$$

To prove Remark 2, we first notice that Remark 1 implies that

$$w_i(|x - a_i| + |y - b_i|) + g_i = w_i \max(|r - a_i'|, |s - b_i'|) + g_i$$

which is the same as

$$\max(w_i|r - a_i'|, w_i|s - b_i'|) + g_i = \max(w_i|r - a_i'| + g_i, w_i|s - b_i'| + g_i)$$

Thus $$f(x, y) = \max_{1 \le i \le m} [\max(w_i|r - a_i'| + g_i, w_i|s - b_i'| + g_i)]$$

so that interchanging the order of maximization gives

$$\begin{aligned} f(x, y) &= \max[\max_{1 \le i \le m} (w_i|r - a_i'| + g_i), \max_{1 \le i \le m} (w_i|s - b_i'| + g_i)] \\ &= \max[f_1(r), f_2(s)] = g(r, s) \end{aligned}$$

The importance of Remark 2 is that it allows the problem of minimizing $f(x, y)$ to be separated into two problems: if $f_1(r)$ is minimized and $f_2(s)$ is minimized, then certainly $g(r, s)$, and thus $f(x, y)$, will be minimized. Furthermore, if $(\bar{r}, \bar{s})$ is any point that minimizes $g(r, s)$, then Remark 2 guarantees that $(\bar{x}, \bar{y}) = T^{-1}(\bar{r}, \bar{s})$ minimizes $f(x, y)$.

Assuming for a minute that r^* and s^* as given minimize $f_1(r)$ and $f_2(s)$, respectively, it should be intuitive that (r^*, s^*) minimizes $g(r, s)$, since $g(r, s)$ is the maximum of $f_1(r)$ and $f_2(s)$, and this maximum is as small as possible. Thus with $z_1 = f_1(r^*)$, $z_2 = f_2(s^*)$, and $z_0 = \max(z_1, z_2) = g(r^*, s^*)$, z_0 is the minimum value of $g(r, s)$, and is the minimum value of $f(x, y)$.

To motivate the solution procedure specified in Cases 2 and 3, it is enough to consider the case where $z_0 = z_1 > z_2$. In this case, for any $\bar{s}$ such that $z_0 \geq f_2(\bar{s})$ we still have $z_0 = z_1$, and so $(r^*, \bar{s})$ also minimizes $g(r, s)$, and thus $T^{-1}(r^*, \bar{s})$ minimizes $f(x, y)$. Now the inequality $z_0 \geq f_2(\bar{s})$ is equivalent to $w_i|\bar{s} - a_i'| + g_i \leq z_0$, $i = 1, \ldots, m$, which in turn can be shown to be equivalent to $s_1 \leq \bar{s} \leq s_2$; the procedure for establishing the equivalence is outlined in a chapter problem.

To complete the motivation of the solution procedure, if we could establish that $\alpha_{p_1p_2}$ is a lower bound on the minimum value of $f_1(r)$, and that $\alpha_{p_1p_2} = f_1(r^*)$, this would guarantee that r^* minimizes $f_1(r)$. Similarly, if we could establish that $\beta_{q_1q_2}$ is a lower bound on $f_2(s)$ and that $\beta_{q_1q_2} = f_2(s^*)$, this would guarantee that s^* minimizes $f_2(s)$.

In a chapter problem we outline the procedure for showing that $\alpha_{p_1p_2}$ is a lower bound on the minimum value of $f_1(r)$; the procedure for showing that $\beta_{q_1q_2}$ is a lower bound on the minimum value of $f_2(s)$ is exactly analogous. To show that $\alpha_{p_1p_2} = f_1(r^*)$ and $\beta_{q_1q_2} = f_2(s^*)$ requires a substantial discussion, and we therefore refer you to [2] instead; to provide some reassurance, however, we shall return to the example, compute $f_1(r^*)$ and $f_2(s^*)$ directly, and verify that $f_1(r^*) = \alpha_{p_1p_2} = \alpha_{14} = \frac{38}{4}$, and that $f_2(s^*) = \beta_{q_1q_2} = \beta_{23} = \frac{72}{7}$:

$$\begin{aligned} f_1(r^*) &= \max(2|r^* - 6| + 1, 3|r^* - 9|, 4|r^* - 9|, 2|r^* - 15|) \\ &= \max[2(4\tfrac{1}{4}) + 1, 4(1\tfrac{1}{4}), 2(4\tfrac{3}{4})] = 9\tfrac{1}{2} = \tfrac{38}{4} \\ f_2(s^*) &= \max[2|s^* - 0| + 1, 3|s^* - 3|, 4|s^* - (-3)|, 2|s^* - 1|] \\ &= \max[2(\tfrac{3}{7}) + 1, 3(3\tfrac{3}{7}), 4(2\tfrac{4}{7}), 2(1\tfrac{3}{7})] = 10\tfrac{2}{7} = \tfrac{72}{7} \end{aligned}$$

We also strongly recommend that you graph $f_1(r)$ and $f_2(s)$ and verify graphically that the minimum and minimum value for each function are as given. [In fact, graphing $f_1(r)$ and $f_2(s)$ should be a required problem.]

9.2.2 *Multifacility rectilinear minimax location problems*

We now consider a multifacility minimax location problem. We shall see that, as with one of the previous problems, the problem can be decomposed into two separate problems. Each of the two resultant problems may then be set up as a linear-programming problem, and solved either directly or via duality. Once optimum solutions $r_1^*, \ldots, r_n^*$ and $s_1^*, \ldots, s_n^*$ are obtained to the separate problems, $(x_j^*, y_j^*) = T^{-1}(r_j^*, s_j^*)$ is an optimum location for new facility j, $j = 1, \ldots, n$.

To formulate the problem, suppose that m existing facilities are located at known points $(a_1, b_1), \ldots, (a_m, b_m)$, and n new facilities are to be located at points $(x_1, y_1), \ldots, (x_n, y_n)$. The cost $w_{ij}(|x_j - a_i| + |y_j - b_i|) + g_{ij}$ is incurred due to travel between new facility j and existing facility i (where w_{ij} is nonnegative) for all i and j, and the cost $v_{jk}(|x_j - x_k| + |y_j - y_k|) + h_{jk}$ is incurred due to travel between new facilities j and k(where v_{jk} is nonnegative) for all $j < k$. Thus the maximum cost incurred due to movement between new facilities is given by

$$f^*((x_1, y_1), \ldots, (x_n, y_n)) = \max_{1 \leq j < k \leq n} [v_{jk}(|x_j - x_k| + |y_j - y_k|) + h_{jk}]$$

and the maximum cost due to movement between new and existing facilities is given by

$$f^{**}((x_1, y_1), \ldots, (x_n, y_n)) = \max_{\substack{1 \leq i \leq m \\ 1 \leq j \leq n}} [w_{ij}(|x_j - a_i| + |y_j - b_i|) + g_{ij}]$$

The maximum cost due to movement between facilities is thus given by

$$\begin{aligned} &f((x_1, y_1), \ldots, (x_n, y_n)) \\ &\quad = \max [f^*(x_1, y_1), \ldots, (x_n, y_n)), f^{**}((x_1, y_1), \ldots, (x_n, y_n))] \end{aligned} \tag{9.13}$$

In addition, we suppose there exist upper bounds c_{ij} upon the rectilinear distance apart that new facility j and existing facility i can be, and upper bounds d_{jk} upon the rectilinear distance apart that new facilities j and k can be. Thus the problem of interest may be stated as

minimize $f((x_1, y_1), \ldots, (x_n, y_n))$
subject to

$$|x_j - a_i| + |y_j - b_i| \leq c_{ij}, \quad 1 \leq i \leq m, 1 \leq j \leq n \tag{9.14}$$

$$|x_j - x_k| + |y_j - y_k| \leq d_{jk}, \quad 1 \leq j \leq k \leq n \tag{9.15}$$

That is, new facilities are to be located so as to minimize the maximum cost, subject to upper bound constraints on distances among facilities.

Several remarks about the problem are in order. First, it is useful to make the chaining assumption, just as in Chapter 4, to help assure a well-formulated problem. Second, in cases where it is unreasonable to have any one of the constraints (9.14) or (9.15), the constraint may either be deleted, or c_{ij} or d_{jk} may be chosen sufficiently large that the constraint is not active. Third, it is possible to choose the c_{ij} and d_{jk} in such a way that the constraints (9.14) and (9.15) would not be consistent, that is, there would be no new facility locations satisfying (9.14) and (9.15), so that some care is required in the choice of the c_{ij} and d_{jk}. Finally, as concerns applications of the problem, they occur in contexts similar to those considered previously in this chapter, with the evident exception that more than one new facility is to be located. In fact, it should be evident that this problem includes the rectilinear minimax problems considered previously as special cases. The reasons for considering a sequence of increasingly more general problems are that insight may be gained by considering simpler problems first, and that special advantage can be taken of the structure of simpler problems to develop specialized solution procedures.

We now convert the multifacility minimax problem into an equivalent constrained problem. The motivation for the conversion is identical to that given for converting the problem of minimizing (9.4) into a constrained problem. One equivalent constrained version of the multifacility problem is

$$\begin{aligned}
&\text{minimize} \quad z \\
&\text{subject to} \\
&w_{ij}(|x_j - a_i| + |y_j - b_i|) + g_{ij} \le z, \quad 1 \le i \le m, 1 \le j \le n \\
&v_{jk}(|x_j - x_k| + |y_j - y_k|) + h_{jk} \le z, \quad 1 \le j < k \le n \\
&|x_j - a_i| + |y_j - b_i| \le c_{ij}, \quad 1 \le i \le m, 1 \le j \le n \\
&|x_j - x_k| + |y_j - y_k| \le d_{jk}, \quad 1 \le j < k \le n
\end{aligned}$$

This problem will now be converted further. Recalling definition (9.11) of the transformation T, we let $(a_i', b_i') = T(a_i, b_i)$, and $(r_j, s_j) = T(x_j, y_j)$. Then, by means of the same approach as that used to prove Remark 2, the problem may be rewritten as

$$\begin{aligned}
&\text{minimize} \quad z \\
&\text{subject to} \\
&\max(w_{ij}|r_j - a_i'| + g_{ij},\ w_{ij}|s_j - b_i'| + g_{ij}) \le z, \quad 1 \le i \le m, 1 \le j \le n \\
&\max(v_{jk}|r_j - r_k| + h_{jk},\ v_{jk}|s_j - s_k| + h_{jk}) \le z, \quad 1 \le j < k \le n \\
&\max(|r_j - a_i'|, |s_j - b_i'|) \le c_{ij}, \quad 1 \le i \le m, 1 \le j \le n \\
&\max(|r_j - r_k|, |s_i - s_k|) \le d_{jk}, \quad 1 \le j < k \le n
\end{aligned}$$

Denote this above problem by P0. Problem P0 may be separated into two problems, which we shall label P1 and P2. Problem P1 may be stated as

$$\begin{aligned}
&\text{minimize} \quad z_1 \\
&\text{subject to}
\end{aligned}$$

$$w_{ij}|r_j - a_i'| + g_{ij} \le z_1, \quad 1 \le i \le m, 1 \le j \le n \tag{9.16}$$

$$v_{jk}|r_j - r_k| + h_{jk} \le z_1, \quad 1 \le j < k \le n \tag{9.17}$$

$$|r_j - a_i'| \le c_{ij}, \quad 1 \le i \le m, 1 \le j \le n \tag{9.18}$$

$$|r_j - r_k| \le d_{jk}, \quad 1 \le j < k \le n \tag{9.19}$$

Problem P2 is obtained by replacing z_1, a_i', r_j, and r_k in P1 by z_2, b_i', s_j, and s_k, respectively. The motivation for the conversion of the problem into two problems follows from the observation that the inequality $\max(|r_j - a_i'|, |s_j - b_i'|) \leq c_{ij}$ is equivalent to the two inequalities $|r_j - a_i'| \leq c_{ij}$ and $|s_j - b_i'| \leq c_{ij}$; the other inequalities are converted in an exactly analogous manner. The symbol z is replaced by z_1 and z_2 in problems P1 and P2 respectively in order to be able to distinguish between the objective functions of the two problems.

Problems P0, P1, and P2 are related in the following sense:

Remark 3: If $(r_1^*, \ldots, r_n^*, z_1^*)$ and $(s_1^*, \ldots, s_n^*, z_2^*)$ are optimum solutions to P1 and P2 respectively, and if $z_0^* = \max(z_1^*, z_2^*)$, then $(r_1^*, \ldots, r_n^*, s_1^*, \ldots, s_n^*, z_0^*)$ is an optimum solution to P0.

Since problem P0 is equivalent to the multifacility location problem of interest, an optimum solution to the minimax location problem may be obtained as follows:

Remark 4: Given the assumptions of Remark 3, if $(x_j^*, y_j^*) = T^{-1}(r_j^*, s_j^*)$ for $1 \leq j \leq n$, then (x_j^*, y_j^*) is a minimax location for new facility j, and $z_0^* = f((x_1^*, y_1^*), \ldots, (x_n^*, y_n^*))$.

The justification for these two remarks is essentially the same as that given for the procedure for minimizing (9.10). Now if we can find a satisfactory way of solving P1 (and thus P2 as well), Remark 4 specifies how P0 may be solved. It should not be too surprising at this point to find that problem P1 may be written as an equivalent linear program, and thus solved by solving the equivalent linear program. To write the linear program, we assume all the weights are positive; the modification to the linear program needed when all weights are nonnegative and some are zero will be indicated subsequently.

The equivalent linear program, which we denote by P1′, is

minimize z_1
subject to

$$-r_j + \frac{z_1}{w_{ij}} \geq -a_i' + \frac{g_{ij}}{w_{ij}}, \quad 1 \leq i \leq m, 1 \leq j \leq n \tag{9.20}$$

$$r_j + \frac{z_1}{w_{ij}} \geq a_i' + \frac{g_{ij}}{w_{ij}}, \quad 1 \leq i \leq m, 1 \leq j \leq n \tag{9.21}$$

$$r_j - r_k + \frac{z_1}{v_{jk}} \geq \frac{h_{jk}}{v_{jk}}, \quad 1 \leq j < k \leq n \tag{9.22}$$

$$-r_j + r_n + \frac{z_1}{v_{jk}} \geq \frac{h_{jk}}{v_{jk}}, \qquad 1 \leq j < k \leq n \tag{9.23}$$

$$r_j \geq d_j, \qquad 1 \leq j \leq n \tag{9.24}$$

$$-r_j \geq -D_j, \qquad 1 \leq j \leq n \tag{9.25}$$

$$r_j - r_k \geq -d_{jk}, \qquad 1 \leq j < k \leq n \tag{9.26}$$

$$-r_j + r_k \geq -d_{jk}, \qquad 1 \leq j < k \leq n \tag{9.27}$$

In this problem the numbers d_j and D_j, for $1 \leq j \leq n$, are defined as

$$d_j = \max_{1 \leq i \leq m} (a'_i - c_{ij})$$

$$D_j = \min_{1 \leq i \leq m} (a'_i + c_{ij})$$

Linear program P1′ is obtained from P1 by manipulating the absolute value inequalities: for example, the inequality $|r_j - r_k| \leq d_{jk}$, (9.19), is equivalent to inequalities $-d_{jk} \leq r_j - r_k \leq d_{jk}$, giving constraints (9.26) and (9.27). It is left as an exercise to show that (9.16), (9.17), and (9.18) are equivalent to (9.20) and (9.21), (9.22) and (9.23), and (9.24) and (9.25), respectively. In program P1, when any w_{ij} is zero, the corresponding constraints in (9.20) and in (9.21) are replaced by the single constraint $g_{ij} \leq z_1$; when any v_{jk} is zero, the corresponding constraint in (9.22) and in (9.23) is replaced by $h_{jk} \leq z_1$. Furthermore, if in the original problem constraints (9.14) and (9.15) were deleted, then constraints (9.24) through (9.27) would be deleted. We note that problem P1′ may now be solved directly as a linear-programming problem, or the problem may be solved more efficiently by solving its dual. The reader is referred to [3] for a particularly efficient means of solving the dual of P1′.

Note that problem P1′ does not include nonnegativity constraints as written. Nonnegativity constraints may be included in both problem P1′ and P2′ if the x-y axes are chosen appropriately for the original problem, P0. One proper choice of the axes is such that all existing facilities are located in the first quadrant such that $b_s \geq a_L$, where b_s is the smallest of the b_i, and a_L is the largest of the a_i. The condition $b_s \geq a_L$ implies that $b_i \geq a_i$, so all existing facilities will lie on or above the line $y = x$. To see that the specified choice of axes guarantees nonnegativity in problems P1′ and P2′, consider the smallest rectangle, denoted by **R**, with each side parallel to an x or y axis, enclosing the existing facility locations. If a_s is the smallest of the a_i, and b_L is the largest of the b_i, then $\mathbf{R} = \{(x, y) = a_s \leq x \leq a_L, b_s \leq y \leq b_L\}$. It is geometrically evident that the optimum solution to P0 will have every new-facility location (x_i, y_j) lying on or in the rectangle, so that $0 \leq a_s \leq x_j \leq a_L$, and $0 \leq b_s \leq y_j \leq b_L$. But recalling the use of the transformation T

defined by (9.11), $(r_j, s_j) = (x_j + y_j, -x_j + y_j)$, and $r_j = x_j + y_j \geq a_s + b_s \geq 0$, while $s_j = -x_j + y_j \geq -a_L + b_s \geq 0$, since $b_s \geq a_L$.

To summarize matters, nonnegativity constraints may be included in problem P1′ thus allowing the problem to be solved by any version of the simplex method, if a choice of axes is made so that the set **R** lies in the first quadrant on or above the line $y = x$.

Recall that to solve problem P0 problem P2 must also be solved. Problem P2 may be solved by solving the equivalent linear program P2′ obtained by replacing each r_j in P1′ by s_j, replacing z_1 by z_2, replacing each a'_i by b'_i, and replacing a'_i in the definitions of d_j and D_j by b'_i. Once both problems P1′ and P2′ are solved, Remark 4 provides the means for solving the original minimax location problem. Two comments about the use of the solutions to P1′ and P2′ to solve the minimax problem are in order. First, there may be alternative minimax locations besides those obtained using Remark 4; it is necessary to consider various cases just as when minimizing (9.10); you are referred to [3] for a means of finding all alternative minimax locations. Second, as mentioned earlier, constraints (9.14) and (9.15) of the minimax problem may be inconsistent; if this is so, then the constraints of P1′ or P2′ will be found to be inconsistent when solving the problems, and no minimax locations will exist. Conversely, if constraints of P1′ or P2′ are inconsistent, then constraints (9.14) or (9.15), or both, will be inconsistent, and no minimax location will exist.

As an example problem, we suppose that two new facilities are to be located, and three existing facilities are located at $(a_1, b_1) = (0, 9)$, $(a_2, b_2) = (4, 8)$, and $(a_3, b_3) = (7, 7)$. Note that all existing facilities are located in the first quadrant, $a_L = 7$, and $b_s = 7$, so that nonnegativity constraints may be included in problems P1′ and P2′ without affecting the answers. The weights for the problem are given by $v_{12} = 3$ and

$$\mathbf{W} = (w_{ij}) = \begin{pmatrix} 3 & 1 \\ 1 & 1 \\ 1 & 3 \end{pmatrix}$$

All g_{ij} and h_{ij} are zero. The upper bound on the distance between new facilities 1 and 2 is given by $d_{12} = 3$, and the upper bounds on distances between existing and new facilities are given by

$$\mathbf{C} = (c_{ij}) = \begin{pmatrix} 4 & 6 \\ 2 & 2 \\ 8 & 4 \end{pmatrix}$$

To determine the data for problem P1′, we compute $a'_1 = 9$, $a'_2 = 12$, $a'_3 = 14$, use the equation for d_j to compute $d_1 = \max(9 - 4, 12 - 2, 14 - 8)$

$= 10$, $d_2 = \max(9 - 6, 12 - 2, 14 - 4) = 10$, and use the equation for D_j to compute $D_1 = \min(9 + 4, 12 + 2, 14 + 8) = 13$, $D_2 = \min(9 + 6, 12 + 2, 14 + 4) = 14$. We now have all the data for problem P1′. The problem is too large to write out conveniently in the text; however, an optimum solution is given by $(r_1^*, r_2^*, z_1^*) = (\frac{32}{3}, \frac{37}{3}, 5)$.

To determine the data for problem P2′, we compute $b_1' = 9$, $b_2' = 4$, and $b_3' = 0$. We now have $d_j = \max_{1 \leq i \leq m}(b_i' - c_{ij})$ and $D_j = \min_{1 \leq i \leq m}(b_i' + c_{ij})$, so that $d_1 = \max(9 - 4, 4 - 2, 0 - 8) = 5$, $d_2 = \max(9 - 6, 4 - 2, 0 - 4) = 3$, $D_1 = \min(9 + 4, 4 + 2, 0 + 8) = 6$, and $D_2 = \min(9 + 6, 4 + 2, 0 + 4) = 4$. Recall that problem P2′ is obtained by replacing r_j, r_k, a_i', and z_1 by s_j, s_k, b_i', and z_2, respectively, and using the indicated expressions for d_j and D_j. When problem P2′, for which we have all the needed data, is solved, an optimum solution is given by $(s_1^*, s_2^*, z_2^*) = (6, 3, 9)$. Since optimum solutions to P1′ and P2′ are optimum solutions to P1 and P2, respectively, we can now use Remarks 3 and 4 to obtain an optimum solution to P0, and thus minimax new-facility locations. We have $z_0^* = \max(z_1^*, z_2^*) = 9$, $(x_1^*, y_1^*) = T^{-1}(r_1^*, s_1^*) = (\frac{7}{3}, \frac{25}{3})$, and $(x_2^*, y_2^*) = T^{-1}(r_2^*, s_2^*) = (\frac{14}{3}, \frac{23}{3})$. A plot of existing facility locations and optimum new-facility locations is given in Figure 9.2.

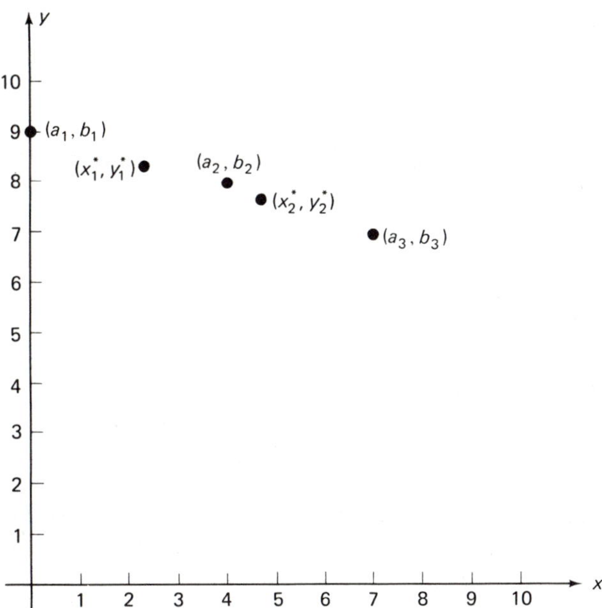

Figure 9.2. Plot of existing facility and optimum new-facility locations for a minimax multifacility problem involving rectilinear distances.

9.3 Some Euclidean Minimax Facility Location Problems

In this section we discuss briefly some minimax facility location problems involving Euclidean distances. These problems, due to their inherent nonlinearity, are not all as well solved as those involving rectilinear distances. The Euclidean problems may all be formulated as general convex programming problems [13, 18], a class of nonlinear programming problems. However, since procedures for solving the Euclidean problems that take advantage of their special structure are still being developed, it seems best to state the problems, give, or refer to, efficient solution procedures where such exist, and warn the reader that, if he wishes to be assured he is using the most efficient means for solving such problems, he will have to keep in touch with the research literature. If finding a particularly efficient solution procedure for solving the problems is not of interest to the reader, then he can solve them by using any algorithm that solves convex programming problems; the question of which algorithm will best solve a particular problem is still unanswered at this point in time.

We begin by considering the problem of minimizing the function

$$f(x, y) = \max_{1 \leq i \leq m} [(x - a_i)^2 + (y - b_i)^2]^{1/2} \tag{9.28}$$

where, as previously, $(a_1, b_1), \ldots, (a_m, b_m)$ are existing facility locations and (x, y) is a new facility to be located in such a way that (9.28) is minimized. A problem equivalent to minimizing (9.28) is to find x, y, and z so that we

minimize z
subject to

$$[(x - a_i)^2 + (y - b_i)^2]^{1/2} \leq z, \quad 1 \leq i \leq m \tag{9.29}$$

The problem (9.29) has a useful geometrical interpretation; the constraints of (9.29) simply state that each existing facility location must lie in a circle with center (x, y) and radius z, so that the geometrical problem is to find the smallest circle which will enclose all the existing facility locations. Nair and Chandrasekaran [15] point out that when one wishes to locate a radar station, a radio transmitter, or some similar type of broadcasting device at a point (x, y), and the points to be scanned or broadcast to are the "existing facilities," a good case can be made for using the model (9.28), since the required power of the station or transmitter is proportional to the radius of the circle enclosing the existing facilities. One would want to locate the station or transmitter so that the power is minimized. Alternatively, one could think of (x, y) as the location of an emergency care center, which treats cases that arrive from the "existing facilities" by helicopters; the helicopter would be located at the emergency care center and return there after responding to an emergency.

The following algorithm for minimizing (9.28), that is, for finding a minimax location, is due to Elzinga and Hearn [4], and appears to be the most efficient one extant. In the algorithm, existing facility locations will be referred to as points.

Step 1. Choose any two points and go to step 2.

Step 2. Let the two points define the diameter of a circle. If this circle contains all the points, then the center of the circle is a minimax location, so stop. Otherwise, choose some point outside the circle, together with the two points defining the circle, and go to step 3.

Step 3. (a) If the three points define a right triangle or an obtuse triangle, return to step 2 with the two points opposite the angle equal to or greater than 90 degrees.

(b) If the three points, called the defining points, define an acute triangle, construct a circle through the three defining points. If the circle contains *all* the points, then the center of the circle is a minimax location, so stop. Otherwise, go to step 4.

Step 4. Choose some point, call it D, outside the circle, and label as A a point from among the three defining points that is farthest from D. Pass a line through the center of the current circle and point A to divide the plane into two half-planes. Of the two remaining defining points, label as B the point in the same half-plane with D, and label as C the other point. With points A, C, and D, go to step 3.

Several comments about the algorithm are in order. Note that in step 3 the three points may lie on a straight line; in this case the point lying on the line between the other two points is deleted. To construct a circle through any three points, say P_1, P_2, and P_3, forming corners of an acute triangle, as must be done in step 3(b), you may recall from geometry that the procedure is as follows: Construct perpendicular bisectors of any two legs of the triangle; the intersection of the two perpendicular bisectors is the center of a circle passing through P_1, P_2, and P_3. In step 4 the point D may be chosen which is farthest from the center of the circle; doing so tends to speed up the algorithm, but at the cost of the computational effort needed to find the point D. Finally, although the algorithm has been presented in geometric terms, it should be evident that all the steps can be readily done algebraically as well, so that the algorithm may be programmed for a computer. In fact, the algorithm is quite efficient when programmed, and the reported average central processing unit time to solve problems with 100 existing facilities on an IBM 7094 is less than 0.5 second.

It is perhaps simplest to motivate the algorithm by considering an example. Suppose that there are nine existing facilities with locations P_1 through P_9, given by $P_1 = (0.0, 1.75)$, $P_2 = (5.75, 3.50)$, $P_3 = (6.75, 8.00)$, $P_4 = (8.50, 1.50)$, $P_5 = (8.00, 0.75)$, $P_6 = (3.50, 9.75)$, $P_7 = (9.50, 8.50)$, $P_8 = (9.00, 0.50)$, and $P_9 = (5.25, 1.00)$. The nine points as well as the minimax location (x^*, y^*) are shown in Figure 9.3. The steps the algorithm goes through for

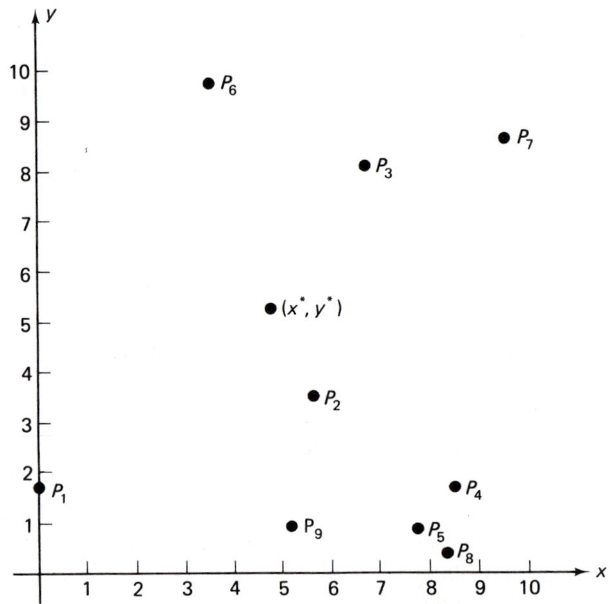

Figure 9.3. Plot of example solved using the Elzinga-Hearn algorithm.

this example are shown in Table 9.1. In step 1, points, P_1 and P_9 define a circle with a radius of 2.652. In step 2, point P_2 lies outside this circle, so the algorithm determines in step 3(a) that P_1, P_2, and P_9 form an obtuse triangle. The point P_9 is dropped and the algorithm returns to step 2 with P_1 and P_2 defining a circle with radius of 3.005. The algorithm goes through steps 3(a) and 2 again with different defining points until the points P_1, P_3, and P_4 are determined to form an acute triangle in step 3(b). A circle is constructed through the three points with a radius of 4.803. Not all the points lie within this circle, so the algorithm proceeds to step 4, chooses P_6 as point D, P_4 as point A, P_1 as point C, and P_3 as point B. The iterations continue until finally

Table 9.1. EXAMPLE OF THE ELZINGA–HEARN ALGORITHM

Step Number	*Defining Points*	*Outside Point*	*Radius**
1	P_1, P_9		2.652
2	P_1, P_9	P_2	
3(a)	P_1, P_9, P_2		
2	P_1, P_2	P_3	3.005
3(a)	P_1, P_2, P_3		
2	P_1, P_3	P_4	4.600
3(b)	P_1, P_3, P_4		4.803
4	P_1, P_3, P_4	P_6	
3(b)	P_1, P_4, P_6		5.200
4	P_1, P_4, P_6	P_5	
3(b)	P_1, P_6, P_5		5.247
4	P_1, P_6, P_5	P_7	
3(a)	P_1, P_5, P_7		
2	P_1, P_7	P_8	5.827
3(b)	P_1, P_7, P_8		5.844

* Rounded to three significant places.

the points P_1, P_7, and P_8 define a circle, with radius of 5.844 that encloses all the points. The center of the circle, which is the minimax location, is given by $(x^*, y^*) = (5.006, 4.765)$.

A motivation for the algorithm may be given as follows. At each step in the algorithm a circle is defined by either two points or three points. Each time a new circle is defined it has a radius strictly greater than that of the previous circle, so that the algorithm never examines the same combination of two points or three points more than once. Since the number of combinations of two points and three points is finite, the algorithm eventually terminates, and it terminates in such a way that the most recently defined circle contains all the points. Table 9.1 illustrates all the aspects of the algorithm, and we recommend that you use the algorithm to solve completely the example on which the table is based. If you are interested in a proof that the algorithm finds a minimax new-facility location, you are referred to [4]; the proof is concerned with showing that the radius of the defining circles increases strictly.

In [4], Elzinga and Hearn also give a geometrical procedure for minimizing the function

$$f(x, y) = \max_{1 \le i \le m} \{[(x - a_i)^2 + (y - b_i)^2]^{1/2} + g_i\}$$

as well as geometrical derivations of the solution procedures presented in the previous section for minimizing functions (9.1) and (9.4), involving rectilinear distances.

Finally, we consider the problem identical to minimizing (9.13) subject to (9.14) and (9.15), with the exception that the distances are Euclidean instead of rectilinear, and all g_{ij} and h_{jk} are zero. Converting the resultant problem to an equivalent constrained problem gives

$$\begin{aligned}
&\text{minimize} \quad z \\
&\text{subject to} \\
&\quad w_{ij}[(x_j - a_i)^2 + (y_j - b_i)^2]^{1/2} \le z, \quad 1 \le i \le m, 1 \le j \le n \\
&\quad v_{jk}[(x_j - x_k)^2 + (y_j - y_k)^2]^{1/2} \le z,. \quad 1 \le j < k \le n \\
&\quad [(x_j - a_i)^2 + (y_j - b_i)^2]^{1/2} \le c_{ij}, \quad 1 \le i \le m, 1 \le j \le n \\
&\quad [(x_j - x_k)^2 + (y_j - y_k)^2]^{1/2} \le d_{jk}, \quad 1 \le j < k \le n
\end{aligned}$$

This problem is a convex programming problem, but one that is a bit awkward to work with due to the square-root terms. Therefore, squaring each side of each constraint and replacing z^2 by $\hat{z}$ gives the following equivalent problem, which is also a convex programming problem:

$$\begin{aligned}
&\text{minimize} \quad \hat{z} \\
&\text{subject to} \\
&w_{ij}^2[x_j^2 + y_j^2 - 2(a_i x_j + b_i y_j)] - \hat{z} + w_{ij}^2(a_i^2 + b_i^2) \\
&\qquad\qquad\qquad \le 0, \quad 1 \le i \le m, 1 \le j \le n \\
&v_{jk}^2[x_j^2 + x_k^2 + y_j^2 + y_k^2 - 2(x_j x_k + y_j y_k)] - \hat{z} \le 0, \quad 1 \le j < k \le n \\
&x_j^2 + y_j^2 - 2(a_i x_j + b_i y_j) + a_i^2 + b_i^2 - c_{ij}^2 \le 0, \quad 1 \le i \le m, 1 \le j \le n \\
&x_j^2 + x_k^2 + y_j^2 + y_k^2 - 2(x_j x_k + y_j y_k) - d_{jk}^2 \le 0, \quad 1 \le j < k \le n
\end{aligned}$$

As an example of a solution to this problem with the last two types of constraints deleted, we give one due to Love, Wesolowsky, and Kraemer [14]. The existing facilities are five cities in Wisconsin: Columbus, Horican, Marshfield, Monroe, and Wausau. The cities have coordinate locations given by $(a_1, b_1) = (39.12, 28.11)$, $(a_2, b_2) = (39.50, 28.28)$, $(a_3, b_3) = (37.88, 29.87)$, $(a_4, b_4) = (38.59, 27.03)$, and $(a_5, b_5) = (38.38, 30.28)$, respectively. The term $v_{12} = 1$, and the w_{ij} are given by

$$W = (w_{ij}) = \begin{pmatrix} 1 & 4 \\ 4 & 1 \\ 4 & 1 \\ 4 & 1 \\ 1 & 4 \end{pmatrix}$$

The minimax location for new facility 1 is given by $(x_1^*, y_1^*) = (38.2964,$

28.4653), and for new facility 2 by $(x_2^*, y_2^*) = (38.7266, 29.1613)$; the minimum value of $\hat{z}$ is 34.3724.

A few concluding remarks about the multifacility minimax problem considered in this section are in order. To date no one has developed an algorithm that takes advantage of the special structure of the problem (other than the fact that it is a convex programming problem) to solve it (we refer the reader to our discussion in the first paragraph concerning means for solving convex programming problems). Also, it should be evident that the multifacility problem includes the one considered by Elzinga and Hearn as a special case, and that more general versions of the one-facility minimax problem involving weights other than 1 may be formulated as a special case of the multifacility problem. To date no one really understands very well why not having all weights the same (which is equivalent to not having all weights with a value of 1) in the one-facility minimax problem (much less the multifacility problem) makes the problem so much more difficult to solve in a manner that takes advantage of the special structure of the problem; hopefully, further research will enlighten us.

9.4 A Minimax Facility Configuration Problem

To visualize the problem we consider in this section, imagine an enormous checkerboard with an infinite number of rows and columns, and suppose that a unit of measurement is being used so that the side length of each square on the board is 1. Suppose also that each square represents a possible site for a department in a facility, and that any n squares, or departments, constitute a facility configuration. Alternatively, instead of being departments, the squares may represent modular construction units, which constitute a facility configuration. One possible configuration made up of nine squares is shown in Figure 9.4.

Now to state the problem we are considering more precisely, note the dots shown in Figure 9.4. Each dot represents a lattice point, that is, a point in the plane with integer coordinates, such as (1, 1), (3, 12), and so on. For the configuration shown in the figure, each square in the configuration has a lattice point as its center. In fact, it is true in general that each lattice point in the plane is the center of a square of unit dimensions, and that the squares with lattice points as centers do not overlap, so that the specification of n unit squares is equivalent to the specification of n lattice points. We therefore define a *facility configuration of size n* to be a set of n distinct lattice points, denote a facility configuration of size n by $\mathbf{S}_n$, and denote the collection of *all possible* facility configurations of size n by $\mathbf{C}_n$. Figure 9.4 shows one possible facility configuration of size 9.

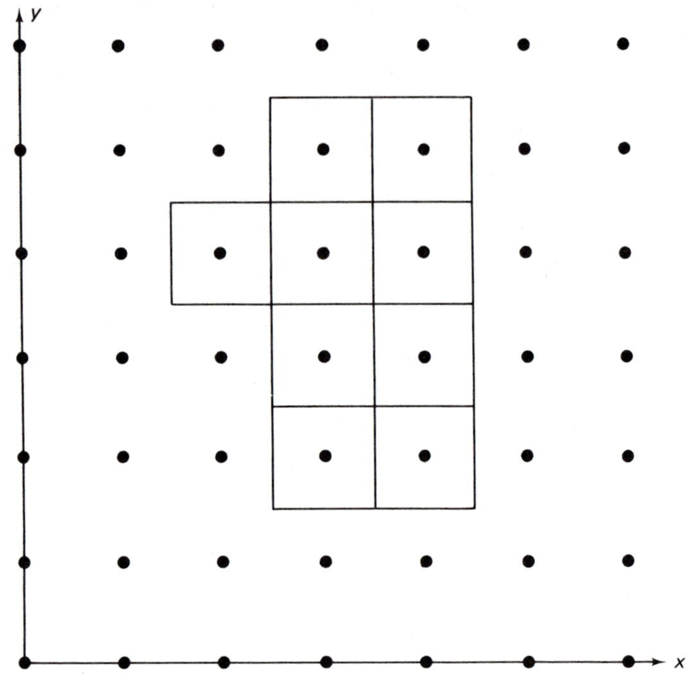

Figure 9.4. Facility configuration of size 9 and diameter 4.

As the section title indicates, we are interested in minimax configurations, and so we have to define a minimax configuration. For any configuration of size n, $\mathbf{S}_n$, define the rectilinear diameter of $\mathbf{S}_n$ to be the maximum of the rectilinear distances between all pairs of lattice points in the configuration, and denote the rectilinear diameter of $\mathbf{S}_n$ by $d(\mathbf{S}_n)$. The configuration of size 9 in Figure 9.4 has a rectilinear diameter of 4. In terms of the facility configuration problem, the diameter of a configuration is just the distance between the centers of any two departments in the configuration that are farthest apart in a rectilinear distance sense. We can now state the problem we wish to consider: for a fixed size n, we wish to find all configurations of minimum rectilinear diameter. In other words, for a fixed size n, we wish to find all configurations for which the maximum rectilinear distance between centers of departments is minimized; that is, we wish to find all *minimax* configurations of size n. Again with reference to Figure 9.4, it is evident that there exist facility configurations of size 9 having diameters greater than 4; if every configuration of size 9 has a diameter of at least 4, then the figure shows a minimax facility configuration of size 9. To find out if the configuration in Figure 9.4 is a minimax configuration of size 9, you will have to complete the section; you may gain some insight however, by trying to construct a facility configuration of size 9 that has a diameter less than 4 (so that the diameter would have to be 3 or less.)

In terms of the physical problem being considered, we are simply trying to choose sites for departments in a minimax fashion, assuming that the travel distance between any two departments may be approximated by the rectilinear distance between the centers of departments. One might consider taking such an approach when it is infeasible to consider the additional step of assigning activities to sites; roughly speaking, we are making no distinction between different departments. In cases where a new facility is being built and nothing is known about the flow of material between departments, or when the facility is expected to be versatile and to be able to serve a variety of different purposes, so that data on the flow of materials might change in an unknown manner over time, there is something to be said for a minimax approach. We wish to emphasize again, however, that the minimax facility configurations we shall obtain should not be considered as final solutions, but as *design aids;* they simply give optimum answers to the problem we have formulated, and it is evident that our formulation is an idealized one. We have seen in Chapter 2 that facility design problems often involve many more aspects than the one we are considering. We should also point out that the problem being considered is somewhat akin to the one which the Hillier algorithm of Chapter 8 addresses, when all the entries of the **W** matrix are taken to be 1; there, however, the objective function involved a sum of distances, whereas here it involves a maximum of distances. Also, the solutions we shall obtain are optimum ones, whereas optimum solutions to the quadratic assignment problem may be computationally infeasible to obtain.

We now begin to develop the notions we need to find minimax configurations. In many cases we shall simply motivate results that are intuitively believable, rather than prove them; the reader interested in a more rigorous treatment is referred to [6].

Given any point $\mathbf{P} = (a, b)$ and any nonnegative number t, we shall define a *diamond* with *center* **P** and *radius* t, denoted by $\mathbf{D}(\mathbf{P}, t)$, to be the set of all points (x, y) for which $|x - a| + |y - b| \leq t$; that is, $\mathbf{D}(\mathbf{P}, t) = \{(x, y): |x - a| + |y - b| \leq t\}$. Thus the set $\mathbf{D}(P, t)$ simply consists of all points having a rectilinear distance from **P** of at most t. The *diameter* of $\mathbf{D}(\mathbf{P}, t)$ is defined to be twice the radius. It should be evident that a diamond is defined in manner analogous to that of a circle, and has similar properties; for example, the rectilinear distance between any two points in a diamond is no greater than the diameter, and the rectilinear distance between any two points on opposite edges of a diamond is equal to the diameter. Graphs of the boundaries of diamonds of diameter 4 are shown in Figures 9.5 and 9.6. We shall call a *Type I diamond* a diamond having exactly two vertices coincident with lattice points and having as its diameter an odd positive integer. We shall call a *Type IIa diamond* a diamond having all four vertices coinci-

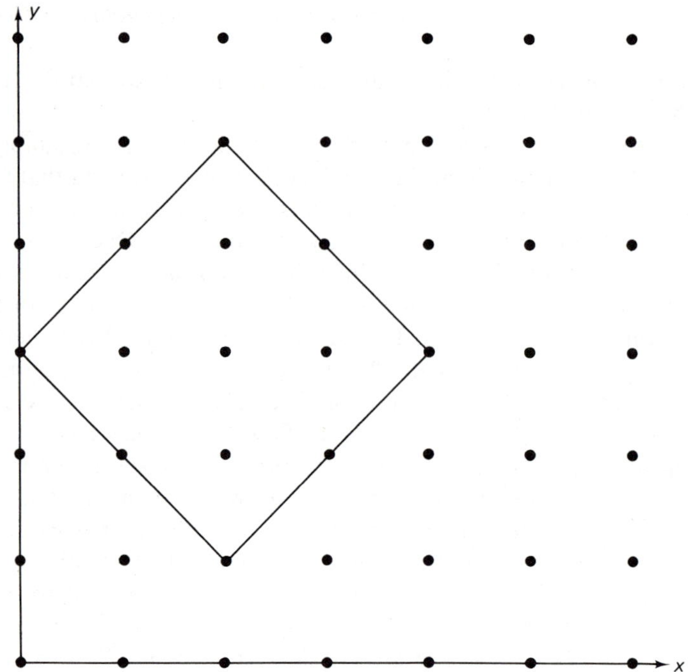

Figure 9.5. Type IIa diamond of diameter 4.

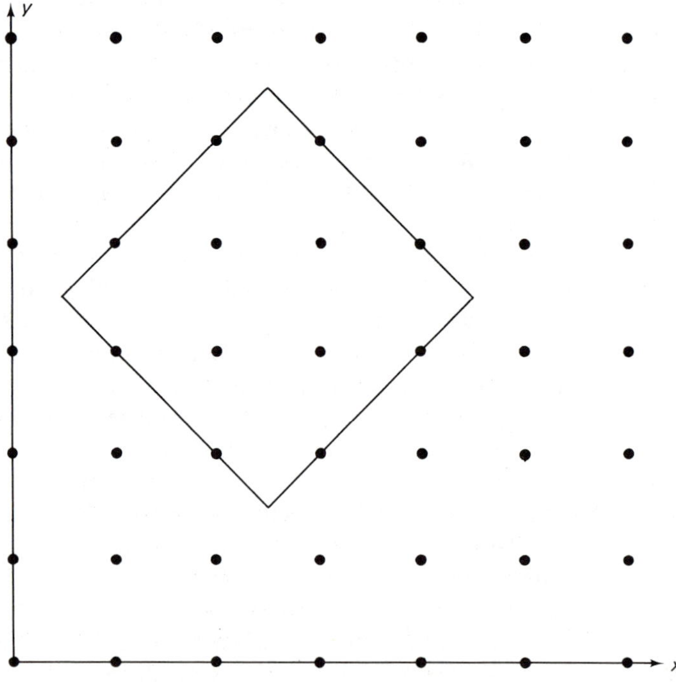

Figure 9.6. Type IIb diamond of diameter 4.

dent with lattice points and having as its diameter a nonnegative even integer. Finally, we shall call a *Type IIb diamond* a diamond such that each coordinate of its center is an integer plus 0.5; for example, (2.5, 3.5) is the center of the Type IIb diamond shown in Figure 9.6. Illustrations of Type IIa and IIb diamonds are given in Figures 9.5 and 9.6, respectively. Given any configuration $\mathbf{S}_n$ consisting of n distinct lattice points, $\mathbf{D}(\mathbf{P}, t)$ will be called a smallest diamond containing $\mathbf{S}_n$ if $\mathbf{D}(\mathbf{P}, t)$ contains $\mathbf{S}_n$ and the diameter of any other diamond containing $\mathbf{S}_n$ is at least as large as $2t$, the diameter of $\mathbf{D}(\mathbf{P}, t)$.

The reader who is uninterested in proofs or motivation, or who wishes to obtain an overview of the main results first, is advised at this point to refer to the definition of the function $g(i)$ in Remark 2, then skip directly to the statement of Result 1, skip the proof of the result, and go through the discussion immediately following the proof of Result 1.

The preliminary results we need may now be stated; all references to remarks or results are to remarks or results of this section. The first remark is as follows:

Remark 1: For any configuration $\mathbf{S}_n$ in $\mathbf{C}_n$, $d(\mathbf{S}_n)$ is equal to the diameter of any smallest diamond containing $\mathbf{S}_n$.

Remark 1 may be motivated by considering, for example, any fixed configuration of nine lattice points in Figure 9.5; note that the diamond containing the configuration is such that at least two lattice points in the configuration lie on opposite edges of the diamond, and thus the configuration and the diamond have the same diameter. (It should also be evident that the choice of a smallest containing diamond may not be unique.)

Remark 2: For every nonnegative integer i, define the function $g(i)$ as follows:

$$g(i) = \frac{i^2 + 2i + 1}{2}, \quad \text{if } i \text{ is an odd positive integer}$$

$$g(i) = \frac{i^2 + 2i + 2}{2}, \quad \text{if } i \text{ is an even nonnegative integer}$$

A Type I or a Type IIa diamond of diameter i contains exactly $g(i)$ lattice points. A Type IIb diamond of diameter i contains exactly $g(i) - 1$ lattice points.

Remark 2 may be motivated by considering the Type IIa and IIb diamonds of diameter 4 in Figures 9.5 and 9.6, respectively. The diamonds have diameter 4, $g(4) = 13$, and the Type IIa diamond contains 13 lattice points, while the Type IIb diamond contains $g(4) - 1 = 12$ lattice points. In general, it is straightforward, simply by counting and using the expression for summing an arithmetic progression, to establish that Remark 2 is true.

It is left as an exercise to verify that $g(i) = g(i - 1) + i$ if i is an odd positive integer, and that $g(i) = g(i - 1) + i + 1$ if 1 is an even positive integer. Since $g(0) = 1$, it follows from the expressions relating $g(i)$ and $g(i - 1)$ that the function $g(i)$ is strictly increasing and integer valued.

Next we have

Remark 3: Let $\mathbf{D}(\mathbf{P}, t)$ be any diamond of integral diameter; there exists a Type I or Type II diamond of the same diameter such that every lattice point in $\mathbf{D}(\mathbf{P}, t)$ is in the Type I or Type II diamond.

To motivate Remark 3, consider Figure 9.7. The lines in Figure 9.7 passing through lattice points that make a 45-degree angle with the x axis will be called 45-degree lines, while the lines passing through lattice points that make a -45-degree angle with the x axis will be called -45-degree lines. Consider a diamond of diameter 3, shown in the figure in dotted lines, and suppose that the diamond is translated in a northeast direction, with its upper left edge remaining parallel to the 45-degree lines, until its upper right edge first coincides with a -45-degree line; note that every lattice point in the original diamond is in the translated diamond. Next, translate the diamond just obtained in a northwest direction, with its upper right edge remaining

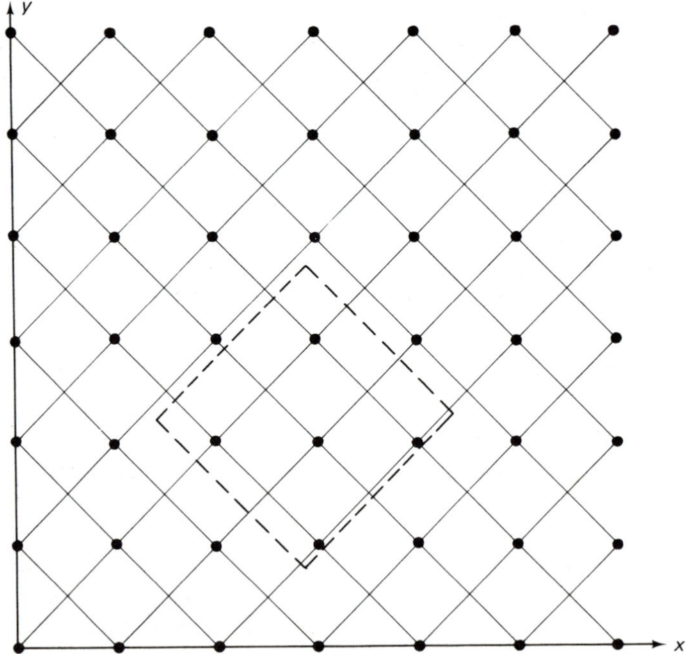

Figure 9.7. Diamond to illustrate Remark 3.

coincident with the −45-degree line, until its upper left edge first coincides with a 45-degree line; the diamond so obtained is a Type I diamond, and every lattice point in the first translated diamond is in the second translated diamond, the Type I diamond. Thus every lattice point in the original diamond is in the Type I diamond. When the same procedure is used with a diamond of even diameter, such as one of diameter 2, the second translated diamond will be either a Type IIa or IIb diamond. The ideas embodied in this approach can be made rigorous by using an algebraic approach; you are referred to [6] for an algebraic approach.

An immediate consequence of Remarks 2 and 3, the proof of which is left as an exercise, is

Remark 4: A diamond of positive integral diameter i contains at most $g(i)$ lattice points.

We now have all the ammunition needed to establish our main results. The result below is the most important one.

Result 1: Let $f(n)$ be the minimum rectilinear diameter of all facility configurations of size n. For any positive integer i,

$$f(n) = i, \quad \text{for } n = g(i-1)+1, \ldots, g(i)-1, g(i)$$

Proof: Let n take on any one of the values $g(i-1)+1, \ldots, g(i)-1$, $g(i)$, and let $\mathbf{S}_n$ be a configuration of size n of minimum diameter, say j, so that $j = f(n)$. Let $\mathbf{D}$ be any smallest diamond containing $\mathbf{S}_n$, and let p be the number of lattice points in $\mathbf{D}$. By Remark 1, the diameter of $\mathbf{D}$ is j, and so, by Remark 4, $p \leq g(j)$. Since $\mathbf{S}_n$ is contained in $\mathbf{D}$, $n \leq p$, and so $n \leq g(j)$. Furthermore, $n \geq g(i-1)+1$, so $n > g(i-1)$, and thus $g(j) > g(i-1)$. Since $g(i)$ is a strictly increasing function, $g(j) > g(i-1)$ implies that $j > i-1$; so, since $f(n) = j$ is an integer, $f(n) \geq i$. Making use of Remark 2, let $\mathbf{D}^*$ be a diamond of diameter i containing $g(i)$ lattice points, let n take on any one of the values $g(i-1)+1, \ldots, g(i)-1$, $g(i)$, and let $\mathbf{S}_n$ be any collection of n lattice points in $\mathbf{D}^*$. Since $\mathbf{S}_n$ is contained in $\mathbf{D}$, $d(\mathbf{S}_n) \leq i$. By the definition of $f(n)$, $f(n) \leq d(\mathbf{S}_n)$, and so $f(n) \leq i$. Since $f(n) \geq i$ as well, we must have $f(n) = i$.

The result is illustrated in Table 9.2. With reference to the table, minimax configurations of size 2 have diameter 1, minimax configurations of size 3 through 5 have diameter 2, minimax configurations of size 6 through 8 have diameter 3; minimax configurations of size 201 through 221 have diameter 20. Again with reference to the table, a facility configuration of size 9 having a rectilinear diameter of 4 would be a minimax facility configuration of size 9; one such minimax facility configuration is illustrated in Figure 9.4. Thus we see that Table 9.2 shows the minimum rectilinear diameter of facili-

Table 9.2. DIAMETERS OF MINIMAX CONFIGURATIONS FOR SIZES 2 THROUGH 221

i	$g(i-1)+1$	$g(i)$	i	$g(i-1)+1$	$g(i)$
1	2	2	11	62	72
2	3	5	12	73	85
3	6	8	13	86	98
4	9	13	14	99	113
5	14	18	15	114	128
6	19	25	16	129	145
7	26	32	17	146	162
8	33	41	18	163	181
9	42	50	19	182	200
10	51	61	20	201	221

ty configurations of sizes 2 through 221; minimum diameters of facility configurations of even larger size may be readily computed using Result 1.

There is a useful corollary of Result 1, which we now state. It follows from the fact that the function $g(i)$ is strictly increasing and integer valued that the statement "n takes on one of the values $g(i-1)+1, \ldots, g(i)-1, g(i)$" is equivalent to the statement "i is the smallest positive integer for which $n \leq g(i)$." The equivalence of these two statements is used in the proof of the corollary; we leave the proof of the corollary as an exercise.

Result 2: Suppose $\mathbf{S}_n$ is a configuration of size n. An equivalent condition for $\mathbf{S}_n$ to be a minimax configuration of size n is that the rectilinear diameter of $\mathbf{S}_n$ is i, where i is the smallest positive integer for which $n \leq g(i)$.

The usefulness of Result 2 lies in the fact that it gives a very direct means of checking whether or not a configuration of a given size is a minimax configuration. For example, the configuration of size 9 of diameter 4 is a minimax configuration, since 4 is the smallest positive integer i for which $9 \leq g(i)$, as $g(3) = 8$ and $g(4) = 13$.

As well as finding the minimum diameter for configurations of a given size, it is also of interest to be able to construct minimax configurations. The following results allow us to construct all minimax facility configurations of interest.

Result 3: Let n be a positive integer at least 2, let i be the smallest positive integer for which $n \leq g(i)$, and suppose that i is an odd integer. If $\mathbf{S}_n^*$ is a minimax configuration, then $d(\mathbf{S}_n^*) = i$ and $\mathbf{S}_n^*$ is contained in a Type I diamond of diameter i, say $\mathbf{D}(i)$. Furthermore, for any configuration $\mathbf{S}_n$ of size n such that $\mathbf{S}_n$ is contained in $\mathbf{D}(i)$, we have $d(\mathbf{S}_n) = i$, so that $\mathbf{S}_n$ is a minimax configuration, and there are $\binom{g(i)}{n}$ such configurations.

At this point we shall prove Result 3. If you really understand the material of this section you should be able to prove Result 3 by yourself; however, if you wish to check your proof, do not wish to try to prove the result, or do not understand the material, then you should go over the proof, which is as follows. We have $d(\mathbf{S}_n^*) = i$ by Result 2, and $\mathbf{S}_n^*$ is contained in a Type I diamond by Remark 3. Since $n \leq g(i)$, and $g(i)$ is the number of lattice points in $\mathbf{D}(i)$, we can choose any n lattice points in $\mathbf{D}(i)$ to constitute a configuration, say $\mathbf{S}_n$. Since $\mathbf{S}_n$ is contained in $\mathbf{D}(i)$, we have $d(\mathbf{S}_n) \leq i$; $d(\mathbf{S}_n) \geq i$ by Result 2, and so $d(\mathbf{S}_n) = i$ and $\mathbf{S}_n$ is thus a minimax configuration. Since the number of combinations of $g(i)$ lattice points taken n at a time is $\binom{g(i)}{n}$, there will be $\binom{g(i)}{n}$ choices of n lattice points in $\mathbf{D}(i)$.

Notice that there is no physical reason to distinguish between two configurations of the same size if one configuration can be obtained from another by a sequence of translations and/or rotations. Thus, for a given n such that $i = f(n)$ is an odd integer, Result 3 implies that it is enough to consider only configurations of size n in a *single* Type I diamond of diameter i when constructing minimax configurations of size n. For example, any choice of n lattice points in a Type I diamond of diameter 5 will be a minimax configuration for any value of n between 14 and 18.

It still remains to consider the case where $f(n) = i$ is an even positive integer when constructing minimax configurations; we do so in the following result.

Result 4: Let n be a positive integer at least 2, let i be the smallest positive integer for which $n \leq g(i)$, and suppose that i is an even integer. If $\mathbf{S}_n^*$ is a minimax configuration, $\mathbf{S}_n^*$ is contained in a Type IIa diamond of diameter i, say $\mathbf{D}^*(i)$, or in a Type IIb diamond of diameter i, say $\mathbf{D}^{**}(i)$. Furthermore, for any configuration $\mathbf{S}_n$ of size n such that $\mathbf{S}_n$ is contained in $\mathbf{D}^*(i)$, we have $d(\mathbf{S}_n) = i$, so that $\mathbf{S}_n$ is a minimax configuration, and there are $\binom{g(i)}{n}$ such configurations; for any configuration $\mathbf{S}_n$ of size n such that $\mathbf{S}_n$ is contained in $\mathbf{D}^{**}(i)$, we also have $d(\mathbf{S}_n) = i$, so that $\mathbf{S}_n$ is a minimax configuration, and there are $\binom{g(i) - 1}{n}$ such configurations.

The proof of Result 4 is entirely analogous to the proof of Result 3, and is left as an exercise.

Due to Result 4, when $f(n) = i$ is an even positive integer, it is only necessary to consider a single Type IIa diamond and a single Type IIb diamond when constructing minimax configurations of size n. Notice that if $n = g(i)$ then there is a unique configuration of size n, which consists of the set of all lattice points in a Type IIa diamond.

Let us define two configurations of the same size to be equivalent if one configuration can be obtained from the other by a sequence of translations and/or rotations. Then when $g(i-1)+1 \leq n < g(i)$ there may be a number of configurations that are not equivalent. For example, consider the Type IIa and IIb diamonds shown in Figures 9.5 and 9.6 respectively. Each diamond has a diameter of 4; minimax configurations of size 9 through 13 have a diameter of 4. By an inspection of the figures it is easy to see that there are, for example, a number of minimax configurations in the Type IIa diamond which are not equivalent to any minimax configuration of size 9 in the Type IIb diamond. Thus it is not enough to consider only Type IIa diamonds or IIb diamonds when constructing minimax configurations; both types must be considered in order to find all possible minimax configurations of a given size that are not equivalent. On the other hand, there are cases when Type IIa and IIb diamonds of the same diameter contain equivalent minimax configurations, as can easily be seen by constructing Type IIa and IIb diamonds of diameter 2 and considering minimax configurations of size 3.

At this point it is perhaps useful to summarize the procedure for constructing minimax configurations of size n. For $n \leq 221$, look up the number i in Table 9.2 for which $g(i-1)+1 \leq n \leq g(i)$; for $n > 221$, let i be the smallest integer i for which $n \leq g(i)$. When i is an odd positive integer, construct a Type I diamond of diameter i; any set of n lattice points in the Type I diamond constitutes a minimax configuration of diameter i. When i is an even positive integer, any set of n lattice points in a Type IIa or IIb diamond of diameter i constitutes a minimax configuration of diameter i [for the case when $n = g(i)$ the only minimax configuration of diameter i is the set of n lattice points in the Type IIa diamond]. Generally, a number of choices of minimax configurations will be available, and the choice of the final configuration will be up to the analyst.

As an illustration of the solution procedure, in a context different from a facility design context, consider a developer of a planned community who wishes to design a central business district which will consist of 25 blocks. The developer wishes to arrange the central business district in such a way that the maximum travel time between any two businesses is minimized. The streets will be arranged in a rectilinear manner, and the rectilinear distance between businesses in any two blocks will be approximated by the rectilinear distances between the centers of the blocks. All blocks will be considered to be unit squares.

From Table 9.2 it is found that the minimum rectilinear diameter is 6 for a configuration of size 25. A minimax configuration of 25 blocks is shown in Figure 9.8(a) for this example. If we assume that only 10 blocks are desired, alternative minimax configurations of diameter 4 are available; a minimax configuration constructed using a Type IIa diamond of diameter 4 is shown

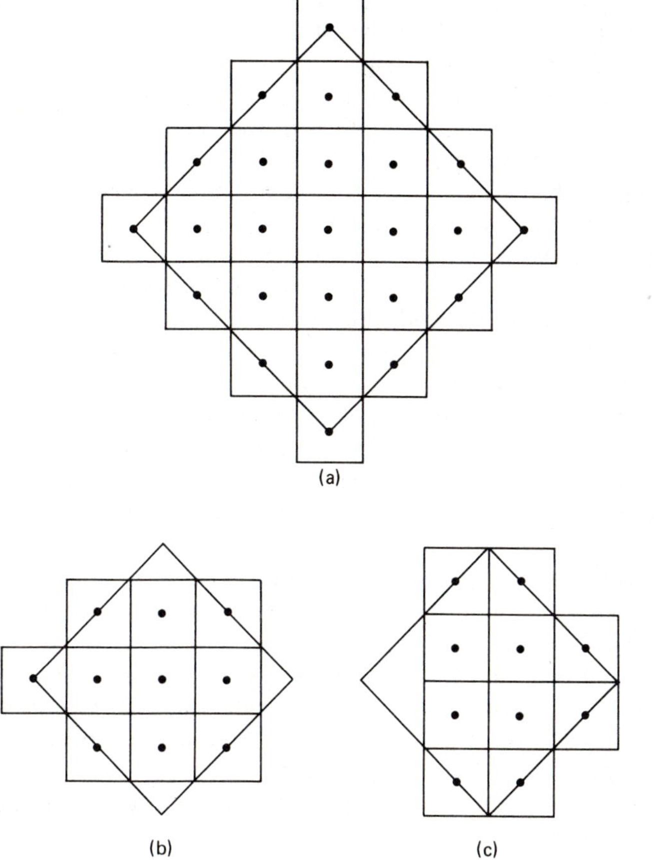

Figure 9.8. Example business district designs of sizes 25 and 10.

in Figure 9.8(b), while a minimax configuration involving a Type IIb diamond is shown in Figure 9.8(c).

9.5 Other Minimax Location and Layout Problems

Depending on what you thought of Sections 9.2, 9.3, 9.4, you may be either happy, or disappointed, to learn that there are a number of other minimax location and layout problems we still have not treated. If you are happy, we regret to inform you that our treatment will be rather brief; if you are disappointed, the brevity of our treatment should make you happy.

There are a number of reasons for the decision to make our treatment relatively brief; one reason, which is unfortunately rather obvious, is that we are running out of space. Another, more cogent reason is that the structure of the models of the problems we shall consider parallels rather closely the structure of the models of Chapters 6 and 7, with the important exception that the objective functions are different. A third reason is that the minimax problems we shall discuss are not yet as well studied as the minimum total cost models we have discussed previously, and thus require more care in their use, so that a really exhaustive discussion is beyond the scope of a first text on layout and location problems. Finally, we feel it is important for you to recognize that at some point in the learning process it is necessary to stop relying on books and to turn, instead, to the research literature published in journals; in the field which this text treats, we believe that journals are the most important general means by which new knowledge is first disseminated, and we would be doing you a disservice if we left you with any other impression. Therefore, we strongly encourage you to refer to the original journal papers referenced in the discussion to follow. Indeed, if you wish to solve either of the first two types of problems we shall discuss, you will have no choice in the matter, since the algorithms needed for solving the problems are stated only in the references.

After the preceding six-sentence apologia, rationale, and sermon we shall now consider the minimax problems that are supposed to be the subject of this section.

The simplest minimax problem we shall consider is usually referred to in the literature as the bottleneck assignment problem [9], due to an early application involving a bottleneck production facility; in keeping with the terminology we have been using in this chapter, however, we shall refer to the problem as the minimax assignment problem. The constraints of the minimax assignment problem are identical to those of the assignment problem of Chapter 6, but, for each assignment, the objective function is defined to be the maximum of the c_{ij} for which the corresponding $x_{ij} = 1$ [where $\mathbf{C} = (c_{ij})$ is a nonnegative "cost matrix"]; for example, for a 3 by 3 problem, if $x_{12} = 1$, $x_{21} = 1$, $x_{33} = 1$, thus specifying an assignment, then the value of the objective function for the assignment is the maximum of c_{12}, c_{21}, and c_{33}. In the minimax assignment problem, one wants to find an assignment that will minimize the value of the objective function; to date, the best algorithm for solving the problem appears to be the one due to Garfinkel [9]. As a possible example in a location context of the minimax assignment problem, suppose that n emergency service facilities are to be assigned to n sites; as usual, when there are more sites than facilities, we can take up the unused sites by "dummy facilities" with costs of zero. For the nonnegative cost matrix $\mathbf{C} = (c_{ij})$, c_{ij} would be the cost of assigning facility i to site j, and might be, for example, the maximum cost to service any emergency that might be handled by facility i if located at site j. One would want to find an assignment of facilities to sites that would minimize the maximum service time over all possible assignments. If you wish actually to solve any minimax assignment problems, you will have to refer to the literature [9].

A next logical step beyond the minimax assignment problem is what we shall call the minimax transportation problem, a problem variously referred to in the literature as the time transportation problem [1, 16] and the bottleneck transportation problem [10]. (At this point we assume you are familiar with the regular transportation problem.) The constraints for the minimax transportation problem are identical to those of the regular transportation problem, but for a given feasible solution the objective function is defined to be the maximum of the c_{ij} for which the corresponding x_{ij} in the feasible solution are positive. In the minimax transportation problem one wants to find a feasible solution that will minimize the value of the objective function. As a possible example in a layout context of the minimax transportation problem, suppose that we again consider the layout problem we solved as a generalized assignment problem in Chapter 6, but that we now have a different cost matrix $\mathbf{C} = (c_{ij})$. Since the generalized assignment problem constraints are a special case of the transportation problem constraints, we may consider a minimax generalized assignment problem, and suppose that a basic feasible solution is given to the latter problem, so that each x_{ij} is either zero or one. Recall that i is the index of an item and j is the index of a grid square, so that $\max \{c_{ij}: x_{ij} = 1; i = 1, \ldots, m; j = 1, \ldots, n\}$ is the maximum cost incurred for the assignment of items to grid squares.

It takes a bit of effort to find the c_{ij} for the layout application of the minimax generalized assignment problem, so we are going to devote a paragraph to one possible means of specifying the c_{ij}. As in Chapter 6, suppose again that there are p docks, that k is the index of a dock, and that w_{ik} is the number of pallet loads per time period of item i moving to and from storage by means of dock k. Given a basic feasible solution, $\max (d_{kj}: x_{ij} = 1; j = 1, \ldots, n)$ is the maximum distance between grid squares taken up by item i and dock k. Under the pessimistic, but possible, assumption that all pallet loads of item i moving to and from dock k move to and from those grid squares taken up by item i that are farthest from dock k, $w_{ik} \max (d_{kj}: x_{ij} = 1; j = 1, \ldots, n)$ is the total distance traveled by item i to and from dock k per time period. Thus $\max [w_{ik} \max (d_{kj}: x_{ij} = 1, j = 1, \ldots, n): k = 1, \ldots, p]$ is the maximum distance item i travels to and from any dock, while $\max [w_{ik} \max (d_{kj}: x_{ij} = 1; j = 1, \ldots, n): k = 1, \ldots, p; i = 1, \ldots, m]$ is the maximum distance traveled by any item to and from any dock. This latter maximum, after rearranging the order of maximization and defining c_{ij} by

$$c_{ij} = \max (w_{ik} d_{kj}: k = 1, \ldots, p) \tag{9.30}$$

becomes $\max \{c_{ij}: x_{ij} = 1; i = 1, \ldots, m; j = 1, \ldots, n\}$, which is the objective function we are seeking. Thus the minimax layout problem may be solved by using cost matrix (9.30) and any algorithm suitable for solving the minimax generalized assignment problem; the most efficient algorithm proposed to date appears to be the threshold algorithm of Garfinkel and Rao [10]. Other relevant work on the minimax transportation problem has been done by Hammer [11, 12], and Szwarc [16]. It is evident, we hope, that one would use a minimax approach to a warehouse-layout problem when it is much more important to give quick service than it is to minimize the total cost of item movement. One other point is perhaps worth mentioning concerning the use of the minimax generalized assignment problem to solve layout problems. Particularly when the d_{kj} are rectilinear distances, a number of alternative minimax solutions may exist, some of which may not exhibit the symmetry one would hope for. Hammer [12] gives a useful method for choosing from among the alternative minimax solutions one with the least *total* cost. Hammer's procedure is as follows. Let c^* be the minimum value of the objective function for the minimax generalized assignment problem, let M be a number much larger than c^*, and define a new "cost" matrix $\mathbf{C}^* = (c^*_{ij})$ as follows: $c^*_{ij} = M$ if $c_{ij} > c^*$, and $c^*_{ij} = c_{ij}$ if $c_{ij} \leq c^*$. Solve the *total cost* generalized assignment problem having the cost matrix $\mathbf{C}^*$ and the same constraints as the minimax generalized assignment problem; any least total cost solution will also be a minimax solution. Limited computational evidence indicates that layouts so obtained exhibit a high degree of symmetry.

Finally, we shall present a "continuous" problem that is a minimax analog of the continuous problems considered in Chapter 7. To motivate the formulation of the problem, we shall present it first in a very specific physical context, and then indicate subsequently other contexts in which the problem occurs. To begin with, suppose that we consider a stadium design problem, where the word "stadium" is used in a generic sense, and might just as well represent a theater or a coliseum. Suppose that a region $\mathbf{L}$ in the plane is given in which the stadium will be located. There will be n different classes of spectators, where the word "spectator" is also used in a generic sense, and might just as well represent a theatergoer or a ticket holder. The total area each spectator class will be allotted is known, and we denote the area for class i by A_i for $i = 1, \ldots, m$. Let the set of points in the region $\mathbf{L}$ that class i may take up be denoted by $\mathbf{S}_i$, so that $\mathbf{S}_i$ has an area of A_i; let $\{\mathbf{S}_1, \mathbf{S}_2, \ldots, \mathbf{S}_m\}$ represent a stadium design (where we assume that $\mathbf{S}_i$ and $\mathbf{S}_j$ do not overlap for all $i \neq j$), and denote the collection of all such stadium designs by $H_m(\mathbf{L}:A)$. Let spectators in class 1 pay the most for tickets, spectators in class 2 the next most, and so on.

We now make the following assumptions.

A1: There exists a "view function" $f(\mathbf{X})$ with the following properties: for any two spectators D and E at any two points $\mathbf{Y}$ and $\mathbf{Z}$, respectively, in $\mathbf{L}$, spectators D and E have the same view if $f(\mathbf{Y}) = f(\mathbf{Z})$, and spectator D

has a better view than does spectator E if $f(\mathbf{Y}) < f(\mathbf{Z})$; that is, the better the view the lower the value of $f(\ )$.

A2: For each spectator class i there exists a "dissatisfaction function" g_i with the following properties. For any two spectators D and E in class i at locations $\mathbf{Y}$ and $\mathbf{Z}$, respectively, spectator D will be at least as dissatisfied with his location as spectator E will be if the view of spectator E is at least as good as that of spectator D; that is, if $f(\mathbf{Y}) \geq f(\mathbf{Z})$, then $g_i[f(Y)] \geq g_i[f(\mathbf{Z})]$. For any two spectators D and E in classes i and $i+1$, respectively, and at locations $\mathbf{Y}$ and $\mathbf{Z}$, respectively, if the view of spectator E is at least as good as that of spectator D, then spectator D will be at least as dissatisfied with his view as will spectator E; that is, if $f(\mathbf{Y}) \geq f(\mathbf{Z})$, then $g_i[f(\mathbf{Y})] \geq g_{i+1}[f(\mathbf{Z})]$.

Now for a given stadium design $\{\mathbf{S}_1, \ldots, \mathbf{S}_m\}$, due to assumption A2,

$$G_i(\mathbf{S}_i) = \max \{g_i[f(\mathbf{X})]: \mathbf{X} \in \mathbf{S}_i\}$$

will be the maximum dissatisfaction of any spectator in class i, while

$$G(\mathbf{S}_1, \ldots, \mathbf{S}_m) = \max \{G_i(\mathbf{S}_i): i = 1, \ldots, m\}$$

will be the maximum dissatisfaction of any spectator. The stadium design problem is to find a stadium design $\{\mathbf{S}_1^*, \ldots, \mathbf{S}_m^*\}$ that will minimize the maximum dissatisfaction of any spectator. The following theorem states conditions which, if satisfied by a stadium design, guarantee that the stadium design will minimize the maximum dissatisfaction of any spectator; for a proof of the theorem you are referred to [7] or [8].

Theorem: Given assumptions A1 and A2, suppose that there exist constants $k_1 < k_2 < \ldots < k_m$ such that $\{\mathbf{S}_1^*, \ldots, \mathbf{S}_m^*\}$ is in $H_m(\mathbf{L}:A)$, where

$$\mathbf{S}_1^* = \{\mathbf{X} \in \mathbf{L}: f(\mathbf{X}) \leq k_1\}$$
$$\mathbf{S}_i^* = \{\mathbf{X} \in \mathbf{L}: k_{i-1} \leq f(\mathbf{X}) \leq k_i\}, \quad i = 2, \ldots, m$$

Then $G(\mathbf{S}_1^*, \ldots, \mathbf{S}_m^*) \leq G(\mathbf{S}_1, \ldots, \mathbf{S}_m)$ for any $\{\mathbf{S}_1, \ldots, \mathbf{S}_m\}$ in $H_m(\mathbf{L}:A)$.

Several remarks about the theorem and the two assumptions appear appropriate. Note that the theorem states that spectators in class i, who paid more for their tickets than spectators in class $i+1$, will have a view at least as good as spectators in class $i+1$; this seems reassuring. Furthermore, it is not necessary to know what the dissatisfaction functions are in order to use the theorem; we only need to know that the functions exist and satisfy assumptions A2, and we feel the assumptions A2 are not unreasonable. We do need to know what the "view function" $f(\mathbf{X})$ is if we are to use the theorem, however, and it is here that difficulties lurk, since different spectators might evaluate the view of a given location differently; that is, the view function might not be the same for all spectators. There are simple cases, however, where the determination of a view function does not seem an insurmountable task. For example, in an outdoor theater design problem, $f(\mathbf{X})$ might simply be the Euclidean distance from the point **X** to the center of the stage. If the axis is chosen so that the center of the stage is the origin, and **L** is the union of the first and fourth quadrants, thus forcing all spectators to sit in front of the stage, then $\mathbf{S}_1^*$ is just a semicircle of area A_1, $\mathbf{S}_2^*$ is just a "band" around $\mathbf{S}_1^*$ of area A_2, and so on, so that $\mathbf{S}_1^* \cup \mathbf{S}_2^* \cup \ldots \cup \mathbf{S}_i^*$ is a semicircle of area $A_1 + A_2 + \ldots + A_i$ for $i = 2, \ldots, m$. Such theater designs have existed for several thousand years, with Roman theaters being classic examples [1]. We do not know if the Romans intended to design minimax theaters, but, in the sense of the theorem, they certainly did!

Believe it or not, we have more comments to make about the theorem. If you have taken a course in advanced calculus, you will recognize that we should require the function $f(\mathbf{X})$ to be continuous on **L**, and each set $\mathbf{S}_j$ in any spectator design $\{\mathbf{S}_1, \ldots, \mathbf{S}_m\}$ to be a closed and bounded set in order for all the maxima we have been computing to exist; if you have not had such a course, we simply suggest that you ignore these requirements, which are rather technical niceties one has to worry about only in the last analysis. A more interesting comment is that the theorem is still true when assumptions A1 and A2 are replaced by assumptions A1′ and A2′, which are given next, $H_m(\mathbf{L}:A)$ is defined as in Chapter 7, and $G_j(\mathbf{S}_j)$ and $G(\mathbf{S}_1, \ldots, \mathbf{S}_m)$ are defined as previously (and the requirements just specified are still imposed):

A1′: The function $f(\mathbf{X})$ is continuous on **L**.

A2′: Each function g_i is nondecreasing, and $g_i(y) \geq g_{i+1}(y)$ for all y and $i = 1, 2, \ldots, m - 1$.

The usefulness of the version of the theorem based on assumptions A1′ and A2′ is that it has applications to facility design problems which are not stadium design problems. As an example, suppose that we consider the minimax analog of the warehouse-layout problem discussed in Section 7.3. Docks are located at points $\mathbf{P}_1, \ldots, \mathbf{P}_n$, and the matrix $\mathbf{W} = (w_{ij})$ is given, where w_{ij} is the number of pallet loads per time period of item i going in and out of storage via dock j. Then, if we assume that all movement of item i for the time period is to and from the points in the storage region farthest from dock j,

$$w_{ij}\,(\max_{X \in S_i} |\mathbf{X} - \mathbf{P}_j|)$$

is the total distance item i moves in and out of storage via dock j, and the maximum total distance over all items and docks, denoted by $G(\mathbf{S}_1, \ldots, \mathbf{S}_m)$, is given by

$$\max_{\substack{1 \leq i \leq m \\ 1 \leq j \leq n}} \{w_{ij}\,(\max_{\mathbf{X} \in \mathbf{S}_i} |\mathbf{X} - \mathbf{P}_j|)\} \tag{9.31}$$

Expression (9.31), we should point out, is completely analogous to the maximum expression, developed just prior to the definition (9.30), of the costs for the minimax generalized assignment problem, and hence compares to the minimax generalized assignment problem, just as the material of Chapter 7 compares with that of Chapter 6. Also, just as in Chapter 7, it is necessary to assume that the matrix $\mathbf{W} = (w_{ij})$ factors in order to get expression (9.31) into a form where the theorem can be applied. With w_j and c_i defined as usual, it can be shown that (9.31) may be rewritten as

$$G(\mathbf{S}_1, \ldots, \mathbf{S}_m) = \max_{1 \leq i \leq m} \{\max_{X \in S_i} c_i[f(\mathbf{X})]\}$$

where
$$f(\mathbf{X}) = \max_{1 \leq j \leq n} (w_j |\mathbf{X} - \mathbf{P}_j|) \tag{9.32}$$

Thus, if $g_i(y)$ is defined by $g_i(y) = c_i y$, we have

$$G(\mathbf{S}_1, \ldots, \mathbf{S}_m) = \max_{1 \leq j \leq n} \{\max_{X \in S_i} g_i[f(\mathbf{X})]\}$$

which is the form we want for the theorem. In order for assumption A2′ to hold, simply number the items so that

$$c_1 \geq c_2 \geq \ldots \geq c_m \tag{9.33}$$

At this point it seems useful to consider an example of the minimax approach to warehouse layout. Suppose that there is only one dock, located at the origin, and that the distance is rectilinear, so that $f(\mathbf{X})$ is just the rectilinear distance between X and the origin. Suppose that $\mathbf{L}$ is the union of the first and fourth quadrants. Then, with $B_0 = 0$ and $B_i = A_1 + \ldots + A_i$ for $i = 1, \ldots, m$, if we find a minimax layout using the theorem, the same approach as in Chapter 7 gives $\mathbf{S}_i^* = \{\mathbf{X} \in \mathbf{L}: B_{i-1}^{1/2} \leq f(\mathbf{X}) \leq B_i^{1/2}\}$ for $i = 1, \ldots, m$. Thus, with reference to (9.33), the item with the largest "turnover," c_1, is closest to the dock, the item with the second largest "turnover," c_2, is next closest, and so forth. We therefore see that inequalities (9.33) specify the relative closeness of items to the docks, so that we obtain a ranking rule different from the c_i/A_i rule of Chapter 7. To some, the ranking rule (9.33) is more acceptable initially, on an intuitive basis, than the c_i/A_i rule, thus at least suggesting the possibility that some analysts, on an intuitive basis, prefer a conservative minimax approach to a minimum total cost approach. Except in cases where it is essential to provide quick service, the minimum total cost approach of Chapters 6 and 7 seems preferable, and it is a bit mysterious to us as to why the ranking rule (9.33) seems so intuitively acceptable.

A few more comments will conclude this section. Notice that to use the theorem it is necessary to be able to construct, graphically, sets defined by $\{\mathbf{X} \in \mathbf{L}: f(\mathbf{X}) \leq k\}$, where the function $f(\mathbf{X})$ is given by (9.32); when distances are rectilinear, we have seen a method for constructing such sets in Section 9.2. Such sets are represented by the set $\mathbf{S}(k)$ defined in Section 9.2. When distances are Eulidean and $f(X)$ is defined by (9.32), it is not difficult to show that $\{\mathbf{X} \in \mathbf{E}_2 : f(\mathbf{X}) \leq k\}$ is the intersection of n disks, with disk j having a center at $\mathbf{P}_j$ and a radius of k/w_j, for $j = 1, \ldots, n$; when n is small it is not too difficult to construct these intersections by graphical means. To obtain $\{\mathbf{X} \in \mathbf{L} : f(\mathbf{L}) \leq k\}$ one then simply intersects $\mathbf{L}$ with the intersection of the n disks.

9.6 Summary

We have considered a number of minimax analogs of minisum facility layout and location problems considered earlier in Chapters 4, 5, 6, and 7. Additionally, the minimax facility configuration problem treated in Section 9.3 is analogous to a special class of quadratic assignment problems.

It should again be emphasized that the minimax results obtained should be interpreted as design aids, rather than final solutions. As with minisum solutions, the minimax solutions obtained can serve as benchmarks against which other solutions are compared. An *ideals* approach is still recommended.

The interest in minimax formulations of facility layout and location problems has developed only recently, when compared with the interest in minisum facility layout and location problems. Consequently, efficient solution procedures that take advantage of the special structure of the problems have not been fully developed. However, as research continues efficient solution procedures will undoubtedly be developed. Therefore, it is highly desirable that you consult the facility layout and location literature to keep abreast of such developments.

REFERENCES

1. Bieber, M., *The History of the Greek and Roman Theatre*, Princeton University Press, Princeton, N.J., 1961.
2. Dearing, P. M., *On Some Minimax Location Problems Using Rectilinear Distances*, unpublished Ph.D. dissertation, University of Florida, Gainesville, Fla., 1972.
3. Dearing, P. M., and R. L. Francis, "A Network Flow Solution to a Multi-facility Minimax Location Problem Involving Rectilinear Distances," *Transportation Science*, Vol. 9, 1974.
4. Elzinga, J., and D. W. Hearn, "Geometrical Solutions for Some Minimax Location Problems," *Transportation Science*, Vol. 6, No. 4, 1971, pp. 379–394.

5. FRANCIS, R. L., "A Geometrical Solution Procedure for a Rectilinear Distance Minimax Location Problem," *AIIE Transactions*, Vol. 4, No. 4, 1971, pp. 328–332.

6. FRANCIS, R. L., "A Minimax Facility Configuration Problem Involving Lattice Points," *Operations Research*, Vol. 21, No. 1, 1973, pp. 101–111.

7. FRANCIS, R. L., "On Some Optimum Facility Design Problems," unpublished Ph.D. dissertation, Northwestern University, Evanston, Ill., 1967; also available as order no. 67–15,232, University Microfilms, P.O. Box 1346, Ann Arbor, Mich. 48106.

8. FRANCIS, R. L., "Sufficient Conditions for Some Optimum-Property Facility Designs," *Operations Research*, Vol. 15, No. 3, 1967, pp. 448–456.

9. GARFINKEL, R. S., "An Improved Algorithm for the Bottleneck Assignment Problem," *Operations Research*, Vol. 19, No. 7, 1971, pp. 1747–1750.

10. GARFINKEL, R. S., and M. R. RAO, "The Bottleneck Transportation Problem," *Naval Research Logistics Quarterly*, Vol. 18, No. 4, 1971, pp. 465–472.

11. HAMMER, P. L., "Communication on 'The Bottleneck Transportation Problem' and 'Some Remarks on the Time Transportation Problem'," *Naval Research Logistics Quarterly*, Vol. 18, No. 4, 1971, pp. 487–490.

12. HAMMER, P. L., "Time Minimizing Transportation Problems," *Naval Research Logistics Quarterly*, Vol. 16, No. 3, 1969, pp. 345–357.

13. HIMMELBLAU, D. M., *Applied Nonlinear Programming*, McGraw-Hill Book Company, New York, 1972.

14. LOVE, R. F., G. O. WESOLOWSKY, and S. A. KRAEMER, "A Multifacility Minimax Location Method for Euclidean Distances," *International Journal of Production Research*, Vol. 11, No. 1, 1973, pp. 37–45.

15. NAIR, K. P. K., and R. CHANDRASEKARAN, "Optimal Location of a Single Service Center of Certain Types," *Naval Research Logistics Quarterly*, Vol. 18, No. 4, 1971, pp. 503–510.

16. SZWARC, W., "Some Remarks on the Time Transportation Problem," *Naval Research Logistics Quarterly*, Vol. 18, No. 4, 1971, pp. 473–487.

17. WESOLOWSKY, G. O., "Rectangular Distance Location Under the Minimax Optimality Criterion," *Transportation Science*, Vol. 6, No. 2, 1972, pp. 103–113.

18. ZANGWILL, W., *Nonlinear Programming: A Unified Approach*, Prentice-Hall, Inc., Englewood Cliffs, N.J., 1969.

PROBLEMS

9.1. The points $(a_1, b_1), \ldots, (a_{12}, b_{12})$ are given by (0, 3), (2, 7), (3, 7), (3, 10), (4, 7), (8, 7), (9, 6), (8, 5), (7, 6), (5, 6), (4, 3), and (2, 1), respectively. The boundary of a region **S** may be obtained by connecting points (a_i, b_i) and (a_{i+1}, b_{i+1}) with a straight line, for $i = 1, \ldots, 11$, and by connecting points

(a_{12}, b_{12}) and (a_1, b_1) with a straight line. The region represents a region within a city that must be served by a fire station. Assuming that travel distance between any two points in the region is rectilinear, at what points can the fire station be located so as to minimize its maximum travel distance to any fire within the region?

9.2. For the example location problem in Section 9.2 for which $z_0 = \frac{72}{7}$, compute the four corner points of the rectangle $\mathbf{S}(z_0)$ and verify that $\mathbf{S}(z_0)$ gives the same minimax locations as obtained in the section for this example.

Given z_0, notice that it is almost as easy to compute the corner points of $\mathbf{S}(z_0)$, thus obtaining minimax locations, as it is to compute minimax locations using the procedure developed in Section 9.2.1. Since this is the case, what is the point in using the procedure developed in Section 9.2.1?

9.3. If $(r, s) = T(x, y) = (x + y, -x + y)$, and $(a', b') = T(a, b) = (a + b, -a + b)$, show that

$$|x - a| + |y - b| = \max(|r - a'|, |s - b'|)$$

[*Hint:* Use the following facts: (1) If c is any number, $|c| = \max(c, -c)$. (2) If c and d are any numbers, $|c| + |d| = \max(c, -c) + \max(d, -d) = \max(c + d, c - d, -c + d, -c - d) = \max[\max(c + d, -c - d), \max(c - d, -c + d)] = \max(|c + d|, |c - d|)$.]

9.4. Set up the problem of minimizing

$$f_1(r) = \max_{1 \le i \le m} (w_i|r - a_i'| + g_i)$$

as an equivalent linear-programming problem, and write the dual of the linear-programming problem. Note that the dual problem has only two constraints.

9.5. Show that $\alpha_{pq} \le f_1(r)$ for any value of r, and thus α_{pq} is a lower bound on the minimum value of $f_1(r)$, where α_{pq} is the maximum of the α_{ij}, $i < j$. [*Hint:* Use the triangle inequality and the definition of $f_1(r)$ to show that $(w_p w_q|a_p' - a_q'| + w_p g_q + w_q g_p)/(w_p + w_q) \le w_q(w_p|r - a_p'| + g_p)/(w_p + w_q) + w_p(w_q|r - a_q'| + g_q)/(w_p + w_q) \le w_q f_1(r)/(w_p + w_q) + w_p f_1(r)/(w_p + w_q) \le f_1(r)$. Next, show that $g_p \le f_1(r)$ and $g_q \le f_1(r)$. It will then follow (why?) that $\alpha_{pq} \le f_1(r)$.]

9.6. (a) Graph the function

$$f_1(r) = \max(2|r - 6| + 1, 4|r - 9|, 2|r - 15|)$$

Verify that $r^* = 10\frac{1}{4}$ minimizes $f_1(r)$, and that the minimum value of $f_1(r)$ is $9\frac{1}{2}$. [*Hint:* To graph $f_1(r)$, first graph the three functions $2|r - 6| + 1$, $4|r - 9|$, and $2|r - 15|$; the graph of $f_1(r)$ is then the maximum of these three graphs. Furthermore, r^* is obtained by computing the value of r where the graphs of $2|r - 6| + 1$ and $2|r - 15|$ intersect.]

(b) Graph the function

$$f_2(s) = \max(2|s - 0| + 1, 3|s - 3|, 4|s - (-3)|, 2|s - 1|)$$

Verify that $s^* = \frac{3}{7}$ minimizes $f_2(s)$, and that the minimum value of $f_2(s)$ is $10\frac{2}{7}$.

9.7. Show that the inequalities $w_i|r - a_i'| + g_i \leq z_0, i = 1, \ldots, m$, are equivalent to $r_1 \leq r \leq r_2$. [*Hint:* First show that the inequalities are equivalent to $-(z_0 - g_i)/w_i \leq r - a_i' \leq (z_0 - g_i)/w_i, i = 1, \ldots, m$.]

9.8. Show, by computing them, that the four corners of the set $S(k)$ defined in Section 9.2.1 are in fact as given in the text when $\bar{g} \leq k$.

9.9. Show that the vectors $\frac{1}{2}(c_1 - c_3, c_1 + c_3 + c_5, c_5)$ and $\frac{1}{2}(c_2 - c_4, c_2 + c_4 - c_5, c_5)$ are feasible solutions to the linear-programming problem (9.9).

9.10. Solve, geometrically, the problem given in Section 9.3 illustrating the Elzinga–Hearn algorithm.

9.11. Write a computer program for solving problems with the Elzinga–Hearn algorithm.

9.12. Solve the problem for which the existing facilities are given by (2, 10), (7, 9), (7, 12), (3, 15), and (1, 14), respectively, using the Elzinga–Hearn algorithm. (Notice that the existing facility locations are the same as those given for the rectilinear minimax problem in Section 9.2.1.)

9.13. Construct a Type I diamond of diameter 3, and draw minimax configurations of sizes 6, 7, and 8.

9.14. For the function $g(i)$ defined in Remark 2, show that $g(i) - g(i - 1) = i$ if i is an odd positive integer, and that $g(i) - g(i - 1) = i + 1$ if i is an even nonnegative integer.

9.15. Construct three different minimax configurations of size 9 that are not equivalent.

9.16. Prove Remark 4 of Section 9.4.

9.17. Prove that the statement "n takes on one of the values $g(i - 1) + 1, \ldots, g(i) - 1, g(i)$" is equivalent to the statement "i is the smallest positive integer for which $n \leq g(i)$."

9.18. Prove Result 2 of Section 9.4.

9.19. Prove Result 4 of Section 9.4.

9.20. A problem different from that considered in Section 9.4, but of some interest, is to find a configuration of given diameter, say i, having maximum size. Prove that any configuration of diameter i has at most $g(i)$ lattice points, and characterize completely configurations of diameter i that consist of $g(i)$ lattice points.

9.21. (a) Suppose that i is an odd positive integer. Prove that any configuration of diameter i of maximum size consists of the set of all lattice points in a Type I diamond of diameter i.

(b) Suppose that i is an even positive integer. Prove that any configuration of diameter i of maximum size consists of the set of all lattice points in a Type IIa diamond of diameter i.

9.22. Prepare a descriptive handout suitable for class use of Garfinkel's algorithm [9] for solving the bottleneck assignment problem.

9.23. Prepare a descriptive handout suitable for class use of the Garfinkel–Rao threshold algorithm [10] for solving the bottleneck transportation problem.

9.24. (a) Make up what you consider to be a realistic location problem to be solved using a minimax assignment algorithm.
(b) Solve the problem you have made up.

9.25. A warehouse is subdivided into eight grid squares of unit dimensions, is 2 grid squares wide and 4 grid squares high, and has two docks. The distance matrix **D** and the "weight" matrix **W** are given by

$$\mathbf{D} = (d_{kj}) = \begin{pmatrix} 1 & 2 & 1 & 2 & 3 & 4 & 5 & 6 \\ 5 & 6 & 3 & 4 & 1 & 2 & 1 & 2 \end{pmatrix}, \qquad \mathbf{W} = \begin{pmatrix} 2 & 2 \\ 4 & 4 \end{pmatrix}$$

(a) With the matrix $\mathbf{C} = (c_{ij})$ defined by (9.30), show that

$$\mathbf{C} = \begin{pmatrix} 10 & 12 & 6 & 8 & 6 & 8 & 10 & 12 \\ 20 & 24 & 12 & 16 & 12 & 16 & 20 & 24 \end{pmatrix}$$

(b) Suppose that two items are to go in the warehouse, and that $A_1 = A_2 = 4$. Explain why, when item 2 takes up grid squares 3, 4, 5, and 6 while item 1 takes up the remaining grid squares, a minimax layout is obtained.
(c) Make a sketch of a warehouse of the indicated matrix that would have the given distance matrix, and indicate which grid squares would be taken up by item 1 and which by item 2.

9.26. An alternative means of representing the maximum distance for a layout problem that can be solved as a minimax generalized assignment problem is given by

$$\max_{i,k} [w_{ik}(\max_{j \in S_i} d_{kj})]$$

where $\mathbf{S}_i$ is the collection of numbers of grid squares taken up by item i. When the matrix $\mathbf{W} = (w_{ik})$ factors, so that $w_{ik} = c_i w_k$, the expression may be rewritten, on interchanging the order of maximization, as

$$\max_{i} [c_i(\max_{j \in \mathbf{S}_i} f_j)]$$

where

$$f_j = \max_{k} (w_k d_{kj})$$

It can then be shown that a minimax layout can be obtained as follows. Number the items so that $c_1 \geq c_2 \geq \ldots \geq c_m$. Let $j_1, j_2, \ldots, j_n$ be a permutation of the grid square numbers so that $f_{j_1} \leq f_{j_2} \leq \ldots \leq f_{j_n}$. A minimax layout $\{\mathbf{S}_1^*, \mathbf{S}_2^*, \ldots, \mathbf{S}_m^*\}$ is now given by $\mathbf{S}_1^* = \{j_1, \ldots, j_{B_1}\}$, $\mathbf{S}_i^* = \{j_{B_{i-1}+1}, \ldots, j_{B_i}\}$, $i = 2, \ldots, m$, where $B_i = A_1 + \ldots + A_i$, $i = 1, \ldots, m$.

Solve the layout problem in Problem 9.25 using this procedure, and verify that the same minimax layout is obtained.

Notice that this ordering procedure is a discrete analog of the procedure for solving the continuous minimax layout problems, and is a more efficient approach than the use of a minimax transportation algorithm. The ordering procedure can only be used, however, when the matrix $\mathbf{W} = (w_{ik})$ factors.

9.27. (a) Design a minimax baseball stadium for two spectator classes when the view function $f(\mathbf{X})$ is defined to be the distance to the base which is farthest from the point $\mathbf{X}$ in the Euclidean distance sense.
(b) What arguments, pro and con, can you give for the use of the suggested view function?

9.28. For the stadium design problem, notice that we never add dissatisfactions, but simply compare relative values. Thus we are assuming dissatisfactions are measured using an ordinal scale, rather than an interval or ratio scale. What arguments, pro and con, can you give for measuring dissatisfactions using an ordinal scale?

9.29. An alternative experimental approach to finding a view function is to determine a number of different seats to which a person would be indifferent, sketch a line through these locations, and consider the line to be an indifference contour line of a view function. This procedure may be repeated, beginning with different seats, to determine a family of contour lines that then, in effect, define the view function.
(a) Construct such contour lines for a stadium design problem of your choice.
(b) As a class project, construct such contour lines for a mutually agreeable stadium design problem, pooling individual indifference contour lines to obtain contour lines generally acceptable to the class.

9.30. In reference [8] an example is given of a two-dock two-item minimax warehouse layout. Develop, in detail, the algebra that justifies this example.

9.31. Solve Problem 4.4 using a minimax objective.

9.32. Solve Problem 4.5 using a minimax objective based on Euclidean distances and assuming the central heating facility is to be located to minimize the maximum length of pipe between the central heating facility and the buildings it services.

9.33. Consider Problem 4.6 in which a central baggage delivery point is to be determined. Based on rectilinear distances between the baggage receiving areas and the baggage pickup point, determine the location for the baggage pickup that minimizes the maximum of the weighted distances traveled.

9.34. Solve Problem 4.8 using a minimax objective.

9.35. Determine the minimax solution to Problem 4.11.

9.36. Given existing facilities at points $P_1 = (0, 0)$ and $P_2 = (6, 0)$, with $w_1 = w_2 = 1$, plot several contour lines for the single-facility minimax location problem based on rectilinear distances.

9.37. Show that the minimax solution to the Euclidean problem is also a minimax solution to the squared Euclidean problem.

9.38. Obtain a minimax solution to Problem 4.23 assuming (a) rectilinear distance, (b) Euclidean distance, and (c) squared Euclidean distance.

9.39. Obtain a minimax solution to Problem 4.25 assuming rectilinear distances.

9.40. Solve Problem 4.26 based on a minimax formulation.

9.41. Solve Problem 4.31 based on a minimax formulation.

9.42. Solve Problem 4.32 based on a minimax formulation.

9.43. Solve Problem 5.4 based on a minimax formulation.

9.44. Solve Problem 5.6 based on a minimax formulation.

9.45. Solve Problem 5.7 based on a minimax formulation.

9.46. Solve Problem 5.8 based on a minimax formulation.

9.47. Solve Problem 5.29 based on a minimax formulation.

9.48. Consider the following minimax optimization problem:

$$\begin{aligned} \text{minimize} \quad & f(\mathbf{X}) = \max\,(c_{ij}x_{ij}\colon \; i = 1, \ldots, n, j = 1, \ldots, n) \\ \text{subject to} \quad & \sum_{i=1}^{n} x_{ij} = 1, \quad j = 1, \ldots, n \\ & \sum_{j=1}^{n} x_{ij} = 1, \quad i = 1, \ldots, n \\ & x_{ij} \geq 0 \end{aligned}$$

Convert the minimax problem to an equivalent linear-programming problem. Is the optimum solution to the equivalent linear-programming problem guaranteed to be a 0, 1 solution? Justify your answer.

9.49. Consider the following version of the minimax transportation problem of minimizing the function

$$f(\mathbf{x}) = \max_{i,j} c_{ij}x_{ij}$$

subject to the constraints

$$\begin{aligned} & \sum_{j=1}^{n} x_{ij} = a_i, \quad i = 1, \ldots, n \\ & \sum_{i=1}^{n} x_{ij} = b_j, \quad j = 1, \ldots, m \\ & x_{ij} \geq 0, \quad \text{for all } i, j \end{aligned}$$

Formulate the problem as an equivalent linear-programming problem.

9.50. Consider the following alternative minimax formulation of the multifacility location problem:

$$\text{minimize} \quad \phi = \sum_{j=1}^{n} \alpha_j z_j$$

where α_j is an appropriately chosen weighting factor and

$$z_j = \text{maximum}\,[v_{jk}\, d(\mathbf{X}_j, \mathbf{X}_k),\, w_{ji}\, d(\mathbf{X}_j, \mathbf{P}_i)\colon 1 \leq k \leq n,\, 1 \leq i \leq m]$$

with $d(\mathbf{X}_j, \mathbf{X}_k)$ and $d(\mathbf{X}_j, \mathbf{P}_i)$ the rectilinear distances between the points $\mathbf{X}_j$ and $\mathbf{X}_k$ and the points $\mathbf{X}_j$ and $\mathbf{P}_i$, respectively. Formulate the problem as an equivalent linear-programming problem. Give a physical interpretation to the objective function.

DISCRETE PLANT LOCATION AND COVERING PROBLEMS

10.1 Introduction

In this the last chapter we undertake a discussion of discrete plant location and covering problems. Due to the structure of the mathematical formulations presented, the subject matter of this chapter could logically be referred to as integer programming formulations of facility location problems.

In Chapter 5 the continuous location-allocation problem was considered and the promise made to address its discrete counterpart in this chapter. We do so, but call the problem a *discrete plant location problem*, since the problem occurs frequently in the context of locating industrial plants. (An alternative name commonly given to the problem under consideration is the *warehouse location* problem.) You may recall from an earlier discussion of the location-allocation problem that the decision variables are not only the number and location of new facilities, but also their sizes (capacities).

A problem related to the discrete plant location problem is the covering problem. In a facility location context, the problem of determining the number and location of, say, warehouses such that a warehouse is within 100 miles of each customer can be formulated as a covering problem. In

10

addition to the location of plants and warehouses, a number of other facility location problems can be formulated as covering problems. As an illustration, one can formulate as a covering problem the problem of determining the number and location of community colleges within a state such that a student will be within a 1-hour drive of a college. Additionally, the problem of locating fire stations within, say, 15-minute driving time from all points in a city can be formulated as a covering problem. Similarly, many of the problems of determining the number and locations of public schools, police stations, libraries, hospitals, public buildings, post offices, parks, civil-defense units, military bases, radar installations, branch banks, shopping centers, and waste-disposal facilities can be formulated as covering problems.

As an overview of the chapter, due to the similarity in the mathematical structure of the discrete plant location problems considered and covering problems, both subjects are treated in this chapter. Furthermore, since many location problems other than discrete plant location problems can be formulated as covering problems, a distinction is made between the two topics in the title and organization of the chapter.

The discussion begins with a brief treatment of the discrete plant location problem. The formulation due to Efroymson and Ray [24] is presented, followed by a discussion of related research. Next, two variations of the covering problem are considered: the *total cover* problem and the *partial cover* problem. The total cover problem is solved using a technique suggested by Toregas, Swain, ReVelle, and Bergman [62]. The partial cover problem is solved using a heuristic solution procedure developed by Ignizio [36].

10.2 Discrete Plant Location Problems

In this section the discrete plant location problem is examined. To be determined are the number, locations, and sizes of plants (warehouses) required to supply products to customers. The locations of a finite number of potential plant sites and customers are assumed to be known. We begin by describing the zero–one problem formulation and solution procedure developed by Efroymson and Ray.

10.2.1 *Efroymson and Ray (E/R) formulations*

Efroymson and Ray [24] formulate the discrete plant location problem as a mixed integer programming problem and employ branch and bound methods in solving the problem. Before presenting the E/R formulation, the following notation is introduced:

m = number of customers

n = number of possible plant sites

y_{ij} = fraction or portion of the demand of customer i which is satisfied by a plant located at site j; $i = 1, \dots, m$, $j = 1, \dots, n$

$x_j = \begin{cases} 1, & \text{if a plant is located at site } j \\ 0, & \text{otherwise} \end{cases}$

c_{ij} = cost of supplying the entire demand of customer i from a plant located at site j

f_j = fixed cost resulting from locating a plant at site j

Based on this notation, the following discrete plant location problem, called P0, is formulated:

P0.

$$\text{minimize} \quad z = \sum_{i=1}^{m} \sum_{j=1}^{n} c_{ij} y_{ij} + \sum_{j=1}^{n} f_j x_j \tag{10.1}$$

$$\text{subject to} \quad \sum_{i=1}^{m} y_{ij} \leq m x_j, \quad j = 1, \dots, n \tag{10.2}$$

$$\sum_{j=1}^{n} y_{ij} = 1, \quad i = 1, \ldots, m \tag{10.3}$$

$$y_{ij} \geq 0, \text{ for all } i, j \tag{10.4}$$

$$x_j = (0, 1), \text{ for all } j \tag{10.5}$$

The objective function (10.1) gives the cost when $\sum_{j=1}^{n} x_j$ plants are to be located at those sites corresponding to positive-valued x_j. Constraints (10.2) indicate that the total fraction of customer demand supplied by a plant at site j either equals zero when x_j equals zero or cannot exceed the number of customers when x_j equals one. By (10.3), all demands for customer i must be met by some combination of plants. Nonnegativity and integer restrictions on the decision variables y_{ij} and x_j are given by (10.4) and (10.5), respectively. Typically, the c_{ij} are transportation costs. If there were no fixed costs, the optimum solution to the problem would be to build a plant at every site. On the other hand, if there were no transportation costs, the optimum solution to the problem would be to build one plant at the site with the smallest fixed cost. Thus we see that in the general problem a balance must be achieved between transportation costs and fixed costs.

The solution procedure employed by Efroymson and Ray is to first solve P0 as a linear-programming problem, ignoring the integer restrictions on x_j. Obviously, if all the x_j values obtained by solving the linear program are integer valued, then the problem has been solved. However, if some x_j value is not integer valued, then a branch and bound approach is employed.*

The branch and bound algorithm begins by establishing a lower bound on the optimum value of z. The lower bound, z_0, is the objective function value obtained by solving P0 without integer restrictions on x_j. If for some index k, x_k is noninteger, then two problems are solved:

1. P0 is solved after x_k is set equal to zero and an objective function value of z_{k0} is obtained.
2. P0 is solved after x_k is set equal to one and an objective function value of z_{k1} is obtained. (Notice that $z_{k0} \geq z_0$ and $z_{k1} \geq z_0$.)

A lower bound of $\bar{z}_k = \min(z_{k0}, z_{k1})$ is defined, since the value of z^*, the optimum value of P0, must be at least as great as $\bar{z}_k$. Additionally, two nodes can be added to a branch and bound tree; one node corresponds to each solution yielding values of z_{k0} and z_{k1}, respectively.

The branch and bound process continues by branching from a node having a fractional x_j. Two new branches are formed, one with x_j set equal

*It is assumed that you are familiar with the branch and bound procedure; otherwise, you might want to skip to Section 10.2.2.

to zero and the other with x_j set equal to one. If a node is reached where all the x_j values are integers, an upper bound is obtained on the optimum value of z. A node where all the x_j values are integers will be called a *terminal* node, since no further branching from the node occurs. A *nonterminal* node involves at least one fractional x_j value.

Branching occurs from nonterminal nodes having objective function values less than the current upper bound. If no such node can be found, the branch and bound algorithm terminates. The current upper bound is then the optimum solution.

A variety of node selection rules and branching rules can be used to achieve computational efficiencies in solving the plant location problem. For a discussion of several rules which have been found to perform well in solving P0, see Khumawala [38]. In our subsequent branch and bound solution to an example problem, we state the node selection and branching rules we employ.

At each node in the branch and bound tree a linear-programming problem is solved. However, since an unlimited plant capacity is assumed, the linear-programming problem is easily solved. To establish the claim, three sets K_0, K_1, and K_2 are defined:

$$K_0 = \{j : x_j = 0\}$$

$$K_1 = \{j : x_j = 1\}$$

$$K_2 = \{j : x_j \text{ unassigned}\}$$

and

$$K_0 \cup K_1 \cup K_2 = \{1, \ldots, n\}$$

A new problem, P1, is defined:

P1. minimize $z = \sum_{i=1}^{m} \sum_{j=1}^{n} c_{ij} y_{ij} + \sum_{j=1}^{n} f_j x_j$

subject to

$$\sum_{j=1}^{n} y_{ij} = 1, \quad i = 1, \ldots, m \tag{10.6}$$

$$\sum_{i=1}^{m} y_{ij} \leq 0, \quad j \in K_0 \tag{10.7}$$

$$\sum_{i=1}^{m} y_{ij} \leq m, \quad j \in K_1 \tag{10.8}$$

$$\sum_{i=1}^{m} y_{ij} \leq m x_j, \quad j \in K_2 \tag{10.9}$$

$$y_{ij} \geq 0, \quad \text{for all } i, j$$

$$x_j \geq 0, \quad j \in K_2$$

Since $y_{ij} \geq 0$, from (10.7) we set $y_{ij} = 0$ for all values of i and those values of j belonging to the set K_0. Furthermore, on summing the first m constraints given by (10.6),

$$\sum_{i=1}^{m}\sum_{j=1}^{n} y_{ij} = \sum_{i=1}^{m} 1 = m$$

Therefore, (10.8) is redundant since

$$\sum_{i=1}^{m} y_{ij} \leq \sum_{i=1}^{m}\sum_{j=1}^{n} y_{ij} = \sum_{i=1}^{m} 1 = m$$

Finally, $$\sum_{j=1}^{n} f_j x_j = \sum_{j\in K_0} f_j x_j + \sum_{j\in K_1} f_j x_j + \sum_{j\in K_2} f_j x_j$$

or, from the definitions of K_0, K_1, and K_2,

$$\sum_{j=1}^{n} f_j x_j = \sum_{j\in K_1} f_j + \sum_{j\in K_2} f_j x_j$$

Since $y_{ij} = 0$ for $j \in K_0$,

$$\sum_{i=1}^{m}\sum_{j=1}^{n} c_{ij} y_{ij} = \sum_{i=1}^{m} \sum_{j\in K_1\cup K_2} c_{ij} y_{ij}$$

while $$\sum_{j=1}^{n} y_{ij} = \sum_{j\in K_1\cup K_2} y_{ij}$$

At this point, P1 can be simplified to obtain P2, where $z = \hat{z} + \sum_{j\in K_1} f_j$:

P2. minimize $\hat{z} = \sum_{i=1}^{m} \sum_{j\in K_1\cup K_2} c_{ij} y_{ij} + \sum_{j\in K_2} f_j x_j$

subject to $\sum_{j\in K_1\cup K_2} y_{ij} = 1, \quad i = 1, \ldots, m$

$$\sum_{i=1}^{m} y_{ij} \leq m x_j, \quad j \in K_2$$

$$y_{ij} \geq 0, \quad j \in K_1 \bigcup K_2, \quad i = 1, \ldots, m$$

$$x_j \geq 0, \quad j \in K_2$$

Continuing our "simplified" approach, we shall simplify P2. To achieve the desired simplification, the following remark is stated:

Remark 1: If y_{ij}^*, x_j^* is a minimum feasible solution to P2 and if all $f_j > 0$, then

$$\sum_{i=1}^{m} y_{ij}^* = m x_j^*, \quad j \in K_2 \tag{10.10}$$

$$x_j^* \leq 1, \quad j \in K_2 \tag{10.11}$$

Proof: To prove (10.10), notice if, for some $j \in K_2$, $\sum_{i=1}^{m} y_{ij} < mx_j$, then by letting x_j equal $(1/m)\sum_{i=1}^{m} y_{ij}$ the value of $\hat{z}$ in P2 can be reduced. Consequently, for y_{ij}^* and x_j^* to be optimum values, (10.10) must hold. To prove (10.11), observe that

$$x_j^* = \frac{1}{m}\sum_{i=1}^{m} y_{ij}^* \leq \frac{1}{m}\sum_{i=1}^{m}\sum_{j\in K_1\cup K_2} y_{ij}^* = \frac{1}{m}\sum_{i=1}^{m} 1 = 1$$

Applying Remark 1, P2 can be solved by solving P3:

P3.

$$\text{minimize} \quad \hat{z} = \sum_{i=1}^{m}\sum_{j\in K_1\cup K_2} c_{ij}y_{ij} + \sum_{j\in K_2} f_j x_j$$

$$\text{subject to} \quad \sum_{j\in K_1\cup K_2} y_{ij} = 1, \quad i = 1,\ldots,m \tag{10.12}$$

$$\sum_{i=1}^{m} y_{ij} = mx_j, \quad j \in K_2 \tag{10.13}$$

$$y_{ij} \geq 0, \quad j \in K_1 \cup K_2, \quad i = 1,\ldots,m$$

$$x_j \geq 0, \quad j \in K_2$$

A further simplification of the optimization problem can be achieved by making use of the following remark:

Remark 2: Define a new term g_j, where

$$g_j = \begin{cases} 0, & \text{if } j \in K_1 \\ f_j, & \text{if } j \in K_2 \end{cases}$$

Given any feasible solution y_{ij}, x_j to P3,

$$\sum_{j\in K_2} f_j x_j = \sum_{i=1}^{m}\sum_{j\in K_1\cup K_2} \frac{g_j}{m} y_{ij}$$

Proof: To prove Remark 2, let y_{ij} and x_j be feasible solutions to P3. Consequently, from (10.13),

$$\begin{aligned}\sum_{j\in K_2} f_j x_j &= \sum_{j\in K_2} f_j\left(\frac{1}{m}\sum_{i=1}^{m} y_{ij}\right)\\ &= \sum_{j\in K_1\cup K_2} g_j\left(\frac{1}{m}\sum_{i=1}^{m} y_{ij}\right)\\ &= \sum_{i=1}^{m}\sum_{j\in K_1\cup K_2} \frac{g_j}{m} y_{ij}\end{aligned}$$

Thus P3 simplifies to P4:

P4. minimize $\hat{z} = \sum_{i=1}^{m} \sum_{j \in K_1 \cup K_2} \left(c_{ij} + \frac{g_j}{m}\right) y_{ij}$

subject to $\sum_{j \in K_1 \cup K_2} y_{ij} = 1, \quad i = 1, \ldots, m$

$y_{ij} \geq 0, \quad j \in K_1 \cup K_2$
$\quad i = 1, \ldots, m$

Notice that P4 is essentially an assignment problem with one set of constraints omitted. Consequently, P4 can be solved almost by inspection. To facilitate the determination, consider the following two remarks (which we ask you to prove as an exercise):

Remark 3: For $i = 1, \ldots, m$, let

$$\alpha_i = \underset{j \in K_1 \cup K_2}{\text{minimum}} \left(c_{ij} + \frac{g_j}{m}\right) \tag{10.14}$$

A lower bound on the value of the objective function of P4 is $\hat{z} = \sum_{i=1}^{m} \alpha_i$.

Remark 4: For $i = 1, \ldots, m$, let s be an index dependent upon i such that

$$c_{is} + \frac{g_s}{m} = \alpha_i$$

Define, for $i = 1, \ldots, m$, the following feasible solution to P4:

$$y_{is}^* = 1$$
$$y_{ij}^* = 0, \quad j \neq s, j \in K_1 \cup K_2$$

The feasible solution so defined is a minimum feasible solution to P4.

From Remark 3, a minimum feasible solution to P4 is available if a feasible solution can be found having an objective function value equal to the lower bound. From Remark 4, y_{is}^* is a feasible solution to P4 resulting in a value of the objective function equal to the lower bound. Consequently, y_{is}^* is a minimum feasible solution to P4.

For each node in the branch and bound tree, we have a means of solving the optimization problem. That is, P4 is solved as indicated by Remark 4 and the value of x_j^* computed by

$$x_j^* = \frac{1}{m} \sum_{i=1}^{m} y_{ij}^*, \quad j \in K_2 \tag{10.15}$$

To obtain the value of the objective function for P0, recall that $z = \hat{z} + \sum_{j \in K_1} f_j$.

To illustrate the E/R solution procedure, consider a plant location problem involving five customers and three plant sites. Let the cost data for the problem be

$$\mathbf{C} = (c_{ij}) = \begin{pmatrix} 4 & 6 & 8 \\ 3 & 4 & 7 \\ 10 & 5 & 7 \\ 12 & 8 & 6 \\ 8 & 4 & 6 \end{pmatrix}$$

and
$$\mathbf{f} = (5 \quad 4 \quad 6)$$

To begin, let $K_2 = (1, 2, 3)$. Therefore, $g_j = f_j$ for $j = 1, 2, 3$. Solving for α_1,

$$\alpha_1 = \text{minimum}\,(4 + \tfrac{5}{5}, 6 + \tfrac{4}{5}, 8 + \tfrac{6}{5}) = 5, \quad \text{for } s = 1$$

Since $s = 1$, $y_{11}^* = 1$, $y_{12}^* = y_{13}^* = 0$. In a similar fashion, the remaining values of α_i and y_{ij}^* given in Table 10.1 are obtained. In particular, the remaining computations of the α_i values are given as follows:

$$\begin{aligned} \alpha_2 &= \text{minimum}\,(3 + \tfrac{5}{5}, 4 + \tfrac{4}{5}, 7 + \tfrac{6}{5}) = 4, && \text{for } s = 1 \\ \alpha_3 &= \text{minimum}\,(10 + \tfrac{5}{5}, 5 + \tfrac{4}{5}, 7 + \tfrac{6}{5}) = 5.8, && \text{for } s = 2 \\ \alpha_4 &= \text{minimum}\,(12 + \tfrac{5}{5}, 8 + \tfrac{4}{5}, 6 + \tfrac{6}{5}) = 7.2, && \text{for } s = 3 \end{aligned}$$

and
$$\alpha_5 = \text{minimum}\,(8 + \tfrac{5}{5}, 4 + \tfrac{4}{5}, 6 + \tfrac{6}{5}) = 4.8, \quad \text{for } s = 2$$

Solving for x_j^* gives $x_1^* = 0.4$, $x_2^* = 0.4$, $x_3^* = 0.2$, and $z_0 = 26.8$. All x_j are not integer valued; therefore P0 is not yet solved.

At this point we could choose to branch from either $j = 1, 2, 3$, since all x_j have fractional values at node zero. A simple branching rule will be used, a lexicographic selection rule. Thus we branch on site 1 by first letting $x_1 = 0$, followed by $x_1 = 1$. Letting $x_1 = 0$ gives $K_0 = \{1\}$ and $K_2 = \{2, 3\}$. Therefore, $g_1 = 0$, $g_2 = 4$, and $g_3 = 6$. From Remarks 3 and 4, the values of α_i, y_{ij}^*, x_j^*, and z_{10} given in Table 10.1 are obtained. As a sample calculation, the value of α_1 is obtained as follows:

$$\alpha_1 = \text{minimum}\,(6 + \tfrac{4}{5}, 8 + \tfrac{6}{5}) = 6.8, \quad \text{for } s = 2$$

Therefore, $y_{12}^* = 1$ and $y_{13}^* = 0$. Of course, since $x_1 = 0$, then $y_{11}^* = 0$.

Table 10.1. SUMMARY OF BRANCH AND BOUND CALCULATIONS

Node Number	*Node Value*	i	α_i	y_{i1}^*	y_{i2}^*	y_{i3}^*
0	(x_j unrestricted)	1	5.00	1	0	0
	$x_1^* = 0.4$	2	4.00	1	0	0
	$x_2^* = 0.4$	3	5.80	0	1	0
	$x_3^* = 0.2$	4	7.20	0	0	1
	$z_0 = 26.8$	5	4.80	0	1	0
1	($x_1^* = 0$)	1	6.80	0	1	0
	$x_2^* = 0.8$	2	4.80	0	1	0
	$x_3^* = 0.2$	3	5.80	0	1	0
	$z_{10} = 29.4$	4	7.20	0	0	1
		5	4.80	0	1	0
2	($x_1^* = 1$)	1	4.00	1	0	0
	$x_2^* = 0.67$	2	3.00	1	0	0
	$x_3^* = 0.33$	3	6.33	0	1	0
	$z_{11} = 31.67$	4	8.00	0	0	1
		5	5.33	0	1	0
3	($x_1^* = 0, x_2^* = 0$)	1	8.00	0	0	1
	$x_3^* = 1$	2	7.00	0	0	1
	$z_{20} = 40.0$ (U.B.)	3	7.00	0	0	1
		4	6.00	0	0	1
		5	6.00	0	0	1
4	($x_1^* = 0, x_2^* = 1$)	1	6.00	0	1	0
	$x_3^* = 0$	2	4.00	0	1	0
	$z_{21} = 31.0$ (U.B.)	3	5.00	0	1	0
		4	8.00	0	1	0
		5	4.00	0	1	0

Branching next by letting $x_1 = 1$ results in $K_1 = \{1\}$, $K_2 = \{2, 3\}$, $g_1 = 0$, $g_2 = 4$, and $g_3 = 6$. Furthermore, from the cost data, if a plant is located at site number 1, customers 1 and 2 will be serviced by the plant, since $\min(c_{11}, c_{12}, c_{13}) = c_{11}$ and $\min(c_{21}, c_{22}, c_{23}) = c_{21}$, and if a plant at any site other than at site 1 serves customers 1 and 2, an additional fixed cost will be incurred. Consequently, the location problem reduces to one involving three customers ($i = 3, 4, 5$) and two sites ($j = 2, 3$). The reduced cost matrix is

$$\mathbf{C} = (c_{ij}) = \begin{pmatrix} 10 & 5 & 7 \\ 12 & 8 & 6 \\ 8 & 4 & 6 \end{pmatrix}$$

and the value of m reduces to 3. Solving for α_i results in the following calculations:

$$\alpha_3 = \text{minimum}\,(10,\ 5 + \tfrac{4}{3},\ 7 + \tfrac{6}{3}) = 6.33, \quad \text{for } s = 2$$
$$\alpha_4 = \text{minimum}\,(12,\ 8 + \tfrac{4}{3},\ 6 + \tfrac{6}{3}) = 8.00, \quad \text{for } s = 3$$
$$\alpha_5 = \text{minimum}\,(8,\ 4 + \tfrac{4}{3},\ 6 + \tfrac{6}{3}) = 5.33, \quad \text{for } s = 2$$

The resulting values of y_{ij}^*, x_j^*, and z_{11} are given in Table 10.1. A lower bound of min (29.4, 31.67) = 29.4 is obtained. The node selection rule we use is to branch from the node corresponding to the current lower bound. Thus the next branching is made from the node $x_1 = 0$, as shown in Figure 10.1.

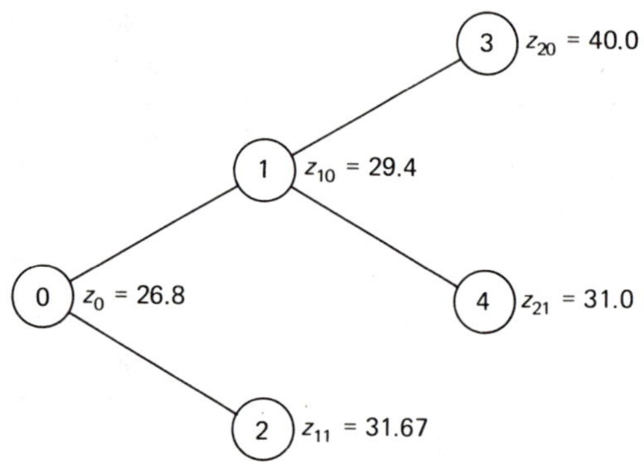

Figure 10.1. Branch and bound tree for the Efroymson-Ray example problem.

Letting $x_1 = 0$ and $x_2 = 0$ gives $K_0 = \{1, 2\}$ and $K_2 = \{3\}$. However, since only one plant is available for location, it must provide service to the customers. Consequently, $x_3^* = 1$ with the values of y_{ij}^* and z_{20} given in Table 10.1 for node 3. Since node 3 is a feasible solution, an upper bound of $z_{20} = 40.00$ is established.

Next, we branch from node 1 by letting $x_1^* = 0$ and $x_2^* = 1$. Since no plant is to be assigned to site 1, the cost matrix reduces to

$$\mathbf{C} = (c_{ij}) = \begin{pmatrix} 6 & 8 \\ 4 & 7 \\ 5 & 7 \\ 8 & 6 \\ 4 & 6 \end{pmatrix}$$

Furthermore, since a plant is to be located at site 2, customers 1, 2, 3, and 5 will choose to be served by the plant at site 2. Thus, the cost matrix reduces further to

$$\mathbf{C} = (c_{4j}) = (8 \quad 6)$$

Additionally, $K_0 = \{1\}$, $K_1 = \{2\}$, $K_2 = \{3\}$, $g_2 = 0$, and $g_3 = 6$, with m set equal to one. The value of α_4 is obtained as follows:

$$\alpha_4 = \text{minimum}\,(8, \quad 6 + 6) = 8$$

With s equal to 2, $x_1^* = x_3^* = 0$, $x_2^* = 1$, and $z_{21} = 31.0$. Since the solution is an integer solution and $z_{21} = 31.0$ is less than the old upper bound of 40.0, a new upper bound of $z_{21} = 31.0$ is established and node 3 is eliminated from further consideration. Furthermore, since the current upper bound is less than the lower bound at node 2, node 2 is also pruned from the branch and bound tree.

As can be seen from Figure 10.1, by pruning nodes 2 and 3, only one node remains, and it is a terminal node; consequently, an optimum solution to P0 is obtained with $x_1^* = x_3^* = 0$, $x_2^* = 1$, $y_{i2}^* = 1$, for $i = 1, \ldots, 5$, and $z^* = 31.0$. Thus only one plant is required, and it is to be located at site 2. The plant must have sufficient capacity to meet the demands for all customers.

Efroymson and Ray present a number of simplifications that can be used to reduce the number of evaluations required in solving P0. Additionally, Khumawala [38] suggests a number of efficient branching rules, as well as a more efficient method for solving the linear-programming problem at each node.

In reporting their computational experience, Efroymson and Ray indicate a number of 50-plant 200-customer problems were solved. The average solution time was reported to be about 10 minutes on an IBM 7094 computer. Khumawala solved 16 test problems of size 25 by 50 and reported average solution times of approximately 10 seconds on a CDC 6500 computer.

10.2.2 *Related research*

A sizeable amount of research has been devoted to the discrete plant location problem and a number of persons are involved currently in the study of the problem. Due to the magnitude of the research, it is not possible to treat the subject in depth in a single chapter. Consequently, we chose to give a detailed discussion of the Efroymson and Ray problem and then to briefly cite a number of alternative approaches to the discrete plant location problem.

In addition to the Efroymson and Ray procedure, a number of other exact solution procedures have been developed for a variety of plant location problems. Among these are the procedures of Curry and Skeith [16], Davis and Ray [19], Ellwein and Gray [26], Gray [35], ReVelle and Swain [51], Sa [55], and Spielberg [60].

Depending on the structure of the problem and the availability of an efficient procedure that will yield an exact solution, heuristic procedure methods are often used. One of the best known heuristic procedures for solving the warehouse location problem is that of Kuehn and Hamburger [41]. (The distinction is made here between a warehouse location problem and a plant location problem, because of the various cost terms included in the Kuehn–Hamburger model.)

The Kuehn–Hamburger heuristic program consists of two parts: the main program, which locates warehouses one at a time until no additional warehouses can be added without increasing total costs; and the "bump and shift" routine, which is entered after processing in the main program is completed. The bump and shift routine "attempts to modify solutions arrived at in the main program by evaluating the profit implications of dropping individual warehouses or of shifting them from one location to another" [41, p. 645].

Feldman, Lehrer, and Ray [29] develop a heuristic solution procedure that involves a "drop" heuristic, rather than the "add" heuristic employed by Kuehn and Hamburger. Thus the solution procedure begins by assuming plants are assigned to all sites. Plants are "dropped" one at a time from the list of plant sites assigned a plant until no plant can be dropped without increasing total cost.

A number of other heuristic procedures have been developed to solve plant location problems. Among these are the procedures of Balinski and Mills [6], Ballou [9], Baumol and Wolfe [10], Drysdale and Sandiford [21], Khumawala and Kelly [39], Manne [46], Sa [55], and Shannon and Ignizio [58].

10.3 Covering Problems

In this section a class of zero–one programming problems is considered briefly. The problems are often referred to as *covering problems.* To motivate the discussion, consider the following zero–one programming problem:

P5. minimize $z = \sum_{j=1}^{n} c_j x_j$

subject to $\sum_{j=1}^{n} a_{ij} x_j \geq 1, \quad i = 1, \ldots, m$

$x_j = (0, 1), \quad j = 1, \ldots, n$

The a_{ij} values in P5 are referred to as covering coefficients and take on the value of one if customer i is *covered* by site j; otherwise, a_{ij} equals zero. Likewise, x_j is set equal to one if a facility is assigned to site j and to zero, otherwise. By the constraint in P5 it is required that each of the m customers be *covered* by at least one of the n facilities. The objective, then, is to cover the customers at minimum cost, where c_j is the cost of assigning a facility to site j.

As an illustration of the meaning of the term "cover," consider the customers to be residences in a community, the facilities to be fire stations, and let residence i be covered if there is a fire station located within a 5-minute drive of the residence. As another example, let the facilities be plants, and let customer i be covered or served by a plant if it is assigned to either sites 1, 2, or 3. Thus $a_{i1} = a_{i2} = a_{i3} = 1$ and all other a_{ik} are zero for $k \neq 1, 2, 3$.

Since P5 is also an integer linear-programming problem, any appropriate integer programming solution procedure can be used to solve P5. However, due to the special structure of P5, a number of algorithms have been developed especially for covering problems. As an illustration, Bellmore and Ratliff [13], Edmonds [23], Garfinkel and Nemhauser [30], Lawler [43], Norman and Rabin [48], and Pierce [49], among others, have developed algorithms which take advantage of the special structure of variations of P5.

Covering problems arise in a variety of contexts including product delivery [7], switching circuit design [4], network defense [14], network attack [13], truck dispatching [17], information retrieval [20], political districting [31], airline crew scheduling [44], central facilities location [51], assembly-line balancing [56], warehouse location [58], and emergency service facility location [62]. A number of additional applications are provided in the examples and problems. For a general review of the covering problem and its applications, see Balinski [4], Garfinkel [30], and Nemhauser and Garfinkel [47], as well as the work of Fejes To'th [28] and Rogers [52].

Generally speaking there are four broad approaches reported in the literature for solving covering problems. The first of these is an implicit enumeration approach, such as branch and bound [42, 44, 49]. A second approach is to use cutting-plane methods and solve iteratively a number of linear-programming problems [15]. The third approach is to employ reduction techniques [4, 42, 54, 63] with the fourth approach involving the use of heuristic methods [36].

In treating the subject of covering problems, we present the cutting-plane method used by Toregas, Swain, ReVelle, and Bergman [62] and the heuristic algorithm developed by Ignizio [36]. However, as suggested in Chapter 9, you should consult the literature when faced with a particular problem in order to benefit from advances in solution methods.

10.3.1 *Total cover problem*

A common formulation of a covering problem involves the determination of the minimum number of facilities required to cover a set of customers. In such a situation P5 reduces to the total cover problem, P6:

P6.
$$\text{minimize} \quad z = \sum_{j=1}^{n} x_j$$
$$\text{subject to} \quad \sum_{j=1}^{n} a_{ij} x_j \geq 1, \quad i = 1, \ldots, m$$
$$x_j = (0, 1), \quad j = 1, \ldots, n$$

where, as in P5,

$$x_j = \begin{cases} 1, & \text{if a facility is located at site } j \\ 0, & \text{otherwise} \end{cases}$$

$$a_{ij} = \begin{cases} 1, & \text{if customer } i \text{ is covered by site } j \\ 0, & \text{otherwise} \end{cases}$$

As an illustration of the total cover problem, suppose that it is desired to determine the minimum number of fire extinguishers in a building and still provide an extinguisher within, say, 150 feet (running distance!) of each department in the building. If potential site j is within 150 feet of department i, then a_{ij} equals one; otherwise, a_{ij} equals zero. A number of emergency service facility location problems can be formulated as total cover problems.

In modeling the problem of locating emergency facilities Toregas, Swain, ReVelle, and Bergman [62] employ the total cover problem formulation, P6. They further propose a method for solving P6 that is quite simple and powerful. To motivate the discussion, consider the following linear-programming problem:

P7.
$$\text{minimize} \quad f = \sum_{j=1}^{n} x_j$$
$$\text{subject to} \quad \sum_{j=1}^{n} a_{ij} x_j \geq 1, \quad i = 1, \ldots, m$$
$$x_j \geq 0, \quad j = 1, \ldots, n$$

Let f^* be the minimum value of f obtained by solving P7. Obviously, if some optimum solution to P7 is a zero–one solution, it is an optimum solution to P6; otherwise, the following linear-programming problem is solved:

P8.
$$\text{minimize} \quad g = \sum_{j=1}^{n} x_j$$

$$\text{subject to} \quad \sum_{j=1}^{n} a_{ij}x_j \geq 1, \quad i = 1, \ldots, m$$

$$\sum_{j=1}^{n} x_j \geq [f^*] + 1$$

$$x_j \geq 0, \quad j = 1, \ldots, n$$

where $[f^*]$ is the integer portion of f^* obtained by solving P7.

The additional constraint added to P8 is referred to as a single cut constraint and is justified by reasoning that since the optimum value of z in P6, say z^*, must be integer valued then $z^* \geq [f^*] + 1$. Consequently, $\sum_{j=1}^{n} x_j \geq [f^*] + 1$. Of course, it is not necessary that there exist a zero–one optimum solution to P8. However, Toregas et al. report that considerable computational experience failed to produce a situation where an optimum solution to P6 was not obtained either directly by solving P7 or by solving P8.

As an illustration of the approach to be taken, consider the problem of locating branch banks in a city that consists of five major subdivisions. The president of the bank indicates he would like to see a branch bank within 5 miles of the center of each subdivision. After searching for potential branch-bank locations, four sites were identified and distances to the centers of the subdivision were calculated. The covering coefficients were found to be

$$\mathbf{A} = (a_{ij}) = \begin{pmatrix} 1 & 0 & 1 & 0 \\ 1 & 1 & 1 & 0 \\ 0 & 0 & 0 & 1 \\ 0 & 0 & 1 & 0 \\ 1 & 1 & 0 & 1 \end{pmatrix}$$

For this example, P7 can be given as

$$\begin{aligned} \text{minimize} \quad & f = x_1 + x_2 + x_3 + x_4 \\ \text{subject to} \quad & x_1 + x_3 \geq 1 \\ & x_1 + x_2 + x_3 \geq 1 \\ & x_4 \geq 1 \\ & x_3 \geq 1 \\ & x_1 + x_2 + x_4 \geq 1 \\ & x_j \geq 0, \quad j = 1, \ldots, 4 \end{aligned}$$

By inspection, $x_3^* = x_4^* = 1$ and $x_1^* = x_2^* = 0$. Since a zero–one solution is obtained, P6 is solved and branch banks will be placed at sites 3 and 4.

As a second illustration of the total cover problem, consider the example problem presented by Toregas et al. Thirty major areas have been identified

Table 10.2. DISTANCES BETWEEN THIRTY CITIES IN NEW YORK STATE

	1	2	3	4	5	6	7	8	9	10	11	12	13	14	15	16	17	18	19	20	21	22	23	24	25	26	27	28	29	30
1	0	244	140	128	281	196	181	51	248	167	338	54	203	146	295	211	295	78	169	38	167	112	71	220	157	16	135	7	90	165
2	244	0	158	359	37	111	66	268	60	112	101	278	272	328	51	222	77	200	106	281	332	263	294	33	284	233	109	248	161	164
3	140	158	0	202	194	56	92	170	117	46	215	137	256	170	209	206	160	62	114	177	279	105	136	136	239	129	78	144	92	148
4	128	359	202	0	395	258	294	179	319	248	416	90	331	61	410	339	361	176	290	100	295	106	70	337	285	143	254	133	211	293
5	281	37	194	395	0	145	102	305	92	148	69	317	309	366	19	259	74	236	143	318	369	299	330	70	321	272	146	285	198	201
6	196	111	56	258	145	0	60	229	61	34	159	189	269	226	162	219	104	118	112	233	315	161	192	100	274	185	91	200	128	161
7	181	66	92	294	102	60	0	208	67	47	157	220	225	262	117	175	114	134	59	218	279	197	228	46	237	170	49	185	101	117
8	51	268	170	179	305	229	208	0	275	195	366	105	152	197	319	180	322	108	186	81	116	163	124	242	106	41	159	48	107	175
9	248	60	117	319	92	61	67	275	0	87	111	254	292	287	111	242	56	175	126	285	346	222	253	60	304	237	116	252	168	184
10	167	112	46	248	148	34	47	195	87	0	185	179	235	216	163	185	130	93	79	204	281	151	182	90	240	156	57	171	94	127
11	338	101	215	416	69	159	157	366	111	185	0	348	373	381	88	323	55	273	207	375	433	316	351	134	385	327	206	342	258	265
12	54	278	137	90	317	189	220	105	254	179	348	0	257	95	329	250	293	86	205	65	221	58	20	254	211	69	169	61	126	204
13	203	272	256	331	309	269	225	152	292	235	373	257	0	349	323	66	339	236	168	233	60	315	274	239	46	193	178	200	171	108
14	146	328	170	61	366	226	262	197	287	216	381	95	349	0	379	343	326	179	284	144	313	69	75	306	303	161	248	151	219	297
15	295	51	209	410	19	162	117	319	111	163	88	329	323	379	0	273	93	251	157	332	383	314	345	84	335	284	160	299	212	215
16	211	222	206	339	259	219	175	180	242	185	323	250	66	343	273	0	289	192	118	248	126	293	270	189	92	198	128	213	129	58
17	295	77	160	361	74	104	114	322	56	130	55	293	339	326	93	289	0	218	173	332	393	261	296	106	351	284	163	299	215	231
18	78	200	62	176	236	118	134	108	175	93	273	86	236	179	251	192	218	0	130	118	221	124	106	178	194	67	94	82	64	146
19	169	106	114	290	143	112	59	186	126	79	207	205	168	284	157	118	173	130	0	206	228	219	225	73	180	156	36	171	79	60
20	38	281	177	100	318	233	218	81	285	204	375	65	233	144	332	248	332	118	206	0	191	123	76	257	187	52	172	37	127	202
21	167	332	279	295	369	315	279	116	346	281	433	221	60	313	383	126	393	221	228	191	0	279	238	299	52	157	230	162	187	168
22	112	263	105	106	299	161	197	163	222	151	316	58	315	69	314	293	261	124	219	123	279	0	60	241	269	127	183	119	164	247
23	71	294	136	70	330	192	228	124	253	182	351	20	274	75	345	270	296	106	225	76	238	60	0	272	228	86	189	76	146	224
24	220	33	136	337	70	100	46	242	60	90	134	254	239	306	84	189	106	178	73	257	299	241	272	0	251	209	85	224	135	131
25	157	284	239	285	321	274	237	106	304	240	385	211	46	303	335	92	351	194	180	187	52	269	228	251	0	147	188	154	147	120
26	16	233	129	143	272	185	170	41	237	156	327	69	193	161	284	198	284	67	156	52	157	127	86	209	147	0	124	15	77	152
27	135	109	78	254	146	91	49	159	116	57	206	169	178	248	160	128	163	94	36	172	230	183	189	85	188	124	0	139	52	70
28	7	248	144	133	285	200	185	48	252	171	342	61	200	151	299	213	299	82	171	37	162	119	76	224	154	15	139	0	92	167
29	90	161	92	211	198	128	101	107	168	94	258	126	171	219	212	129	215	64	79	127	187	164	146	135	147	77	52	92	0	83
30	165	164	148	293	201	161	117	175	184	127	265	204	108	297	215	58	231	146	60	202	168	247	224	131	120	152	70	167	83	0

From *Operations Research*, Vol. 19, No. 6, 1971, p. 1367.

Table 10.3. COVER COEFFICIENTS FOR THE TOTAL COVER EXAMPLE PROBLEM

	1	2	3	4	5	6	7	8	9	10	11	12	13	14	15	16	17	18	19	20	21	22	23	24	25	26	27	28	29	30
1	1	0	0	0	0	0	0	1	0	0	0	1	0	0	0	0	0	0	0	1	0	0	0	0	0	1	0	1	0	0
2	0	1	0	0	1	0	1	0	1	0	0	0	0	0	1	0	0	0	0	0	0	0	0	1	0	0	0	0	0	0
3	0	0	1	0	0	1	0	0	0	1	0	0	0	0	0	0	0	1	0	0	0	0	0	0	0	0	0	0	0	0
4	0	0	0	1	0	0	0	0	0	0	0	0	0	1	0	0	0	0	0	0	0	0	0	0	0	0	0	0	0	0
5	0	1	0	0	1	0	0	0	0	0	1	0	0	0	1	0	0	0	0	0	0	0	0	0	0	0	0	0	0	0
6	0	0	1	0	0	1	1	0	1	1	0	0	0	0	0	0	0	0	0	0	0	0	0	0	0	0	0	0	0	0
7	0	1	0	0	0	1	1	0	1	1	0	0	0	0	0	0	0	0	1	0	0	0	0	1	0	0	1	0	0	0
8	1	0	0	0	0	0	0	1	0	0	0	0	0	0	0	0	0	0	0	0	0	0	0	0	0	1	0	1	0	0
9	0	1	0	0	0	1	1	0	1	1	0	0	0	0	0	0	1	0	0	0	0	0	0	1	0	0	0	0	0	0
10	0	0	1	0	0	1	1	0	1	1	0	0	0	0	0	0	1	0	0	0	0	0	0	0	0	0	1	0	0	0
11	0	0	0	0	1	0	0	0	0	0	1	0	0	0	0	0	0	0	0	0	0	0	0	0	0	0	0	0	0	0
12	1	0	0	0	0	0	0	0	0	0	0	1	0	0	0	0	0	0	0	1	0	1	1	0	0	1	0	1	0	0
13	0	0	0	0	0	0	0	0	0	0	0	0	1	0	0	1	0	0	0	0	1	0	0	0	1	0	0	0	0	0
14	0	0	0	1	0	0	0	0	0	0	0	0	0	1	0	0	0	0	0	0	0	1	0	0	0	0	0	0	0	0
15	0	1	0	0	1	0	0	0	0	0	0	0	0	0	1	0	0	0	0	0	0	0	0	0	0	0	0	0	0	0
16	0	0	0	0	0	0	0	0	0	0	0	0	1	0	0	1	0	0	0	0	0	0	0	0	0	0	0	0	0	1
17	0	0	0	0	0	0	0	0	1	1	0	0	0	0	0	0	1	0	0	0	0	0	0	0	0	0	0	0	0	0
18	0	0	1	0	0	0	0	0	0	0	0	0	0	0	0	0	0	1	0	0	0	0	0	0	0	1	0	0	1	0
19	0	0	0	0	0	0	1	0	0	0	0	0	0	0	0	0	0	0	1	0	0	0	0	0	0	0	1	0	0	1
20	1	0	0	0	0	0	0	0	0	0	0	1	0	0	0	0	0	0	0	1	0	0	0	0	0	1	0	1	0	0
21	0	0	0	0	0	0	0	0	0	0	0	0	1	0	0	0	0	0	0	0	1	0	0	0	1	0	0	0	0	0
22	0	0	0	0	0	0	0	0	0	0	0	1	0	1	0	0	0	0	0	0	0	1	1	0	0	0	0	0	0	0
23	0	0	0	0	0	0	0	0	0	0	0	1	0	0	0	0	0	0	0	0	0	1	1	0	0	0	0	0	0	0
24	0	1	0	0	0	0	1	0	1	0	0	0	0	0	0	0	0	0	0	0	0	0	0	1	0	0	0	0	0	0
25	0	0	0	0	0	0	0	0	0	0	0	0	1	0	0	0	0	0	0	0	1	0	0	0	1	0	0	0	0	0
26	1	0	0	0	0	0	0	1	0	0	0	1	0	0	0	0	0	1	0	1	0	0	0	0	0	1	0	1	0	0
27	0	0	0	0	0	0	1	0	0	1	0	0	0	0	0	0	0	0	1	0	0	0	0	0	0	0	1	0	1	0
28	1	0	0	0	0	0	0	1	0	0	0	1	0	0	0	0	0	0	0	1	0	0	0	0	0	1	0	1	0	0
29	0	0	0	0	0	0	0	0	0	0	0	0	0	0	0	0	0	1	0	0	0	0	0	0	0	0	1	0	1	0
30	0	0	0	0	0	0	0	0	0	0	0	0	0	0	0	1	0	0	1	0	0	0	0	0	0	0	0	0	0	1

in New York State with distances between major areas given in Table 10.2. It is desired to locate facilities among the areas such that no area is more than, say, 69 miles from a facility. The resulting matrix of covering coefficients is given in Table 10.3. Solving P7 yields the solution $f^* = 8.5$, $x_4^* = x_{12}^* = x_{13}^* = x_{26}^* = 1$, $x_2^* = x_3^* = x_5^* = x_9^* = x_{11}^* = x_{16}^* = x_{18}^* = x_{19}^* = x_{27}^* = 0.5$, and all other x_j^* equal zero. Adding the cut constraint $\sum_{j=1}^{30} x_j \geq [8.5] + 1 = 9.0$ yields a solution to P8 of $g^* = 9.0$, $x_2^* = x_4^* = x_7^* = x_{12}^* = x_{13}^* = x_{16}^* = x_{17}^* = x_{18}^* = x_{26}^* = 1$ and all other x_j^* equal zero. Since the optimum solution to P8 is a zero–one solution, an optimum solution to P6 is obtained and facilities will be placed in areas 2, 4, 7, 12, 13, 16, 17, 18, and 26. A graphical representation of the optimum solution is given in Figure 10.2.

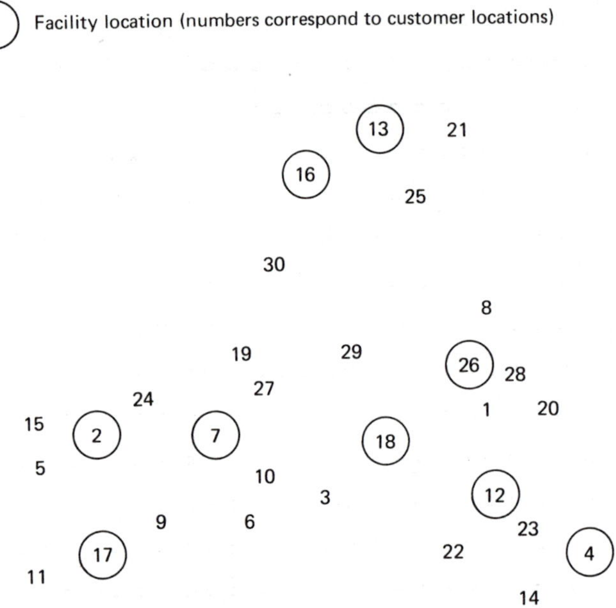

Figure 10.2. Integer solution to the total cover example problem.

10.3.2 *Partial cover problem*

A zero–one programming formulation related to the *total cover* problem is the *partial cover* problem. Recall that the total cover problem involves a determination of the minimum number and locations of facilities such that all customers are covered. In certain cases it is not possible to provide the number of facilities required to "totally cover" all customers; rather, the number of facilities available for location might only be sufficient to "partially cover" the set of customers. In such a situation it is desired that the k available facilities be assigned to sites in such a way that the maximum number of customers are covered. Mathematically, the partial cover problem can be formulated as

P9.

$$\text{maximize} \quad \tilde{z} = \sum_{i=1}^{m} \max_{j} a_{ij} x_j$$

$$\text{subject to} \quad \sum_{j=1}^{n} x_j \leq k$$

$$x_j = (0, 1), \quad \text{for all } j$$

where k is the maximum number of facilities available for assignment to sites.

As indicated by the term $\max_j a_{ij} x_j$ in the objective function of P9, if a particular customer is "covered" by more than one of the facilities that have been assigned to sites, only the maximum a_{ij} value is included in the computation of $\tilde{z}$. (Note that when each a_{ij} is zero or one $\tilde{z}$ is just the number of covered customers.) The constraint in P9 indicates at most k facilities are to be assigned to sites.

As an illustration of a partial cover problem, suppose that there are nine suburban areas which are to be served by two high schools. A suburb will be said to be "covered" if a high school is located within 10-minute driving time of the centroid of the suburb. Thus, if suburb i is covered by a high school located at site j, then a_{ij} equals one. For this example, suppose that the matrix of covering coefficients is

$$\mathbf{A} = (a_{ij}) = \overset{j}{\begin{pmatrix} 1 & 1 & 0 & 0 & 1 & 0 & 0 & 0 & 0 \\ 1 & 1 & 0 & 0 & 0 & 0 & 0 & 0 & 0 \\ 0 & 0 & 1 & 0 & 0 & 1 & 0 & 0 & 0 \\ 0 & 1 & 0 & 1 & 1 & 0 & 0 & 1 & 0 \\ 0 & 0 & 0 & 0 & 1 & 1 & 0 & 0 & 0 \\ 0 & 0 & 0 & 0 & 0 & 1 & 0 & 0 & 0 \\ 0 & 0 & 0 & 0 & 0 & 0 & 1 & 1 & 0 \\ 0 & 0 & 1 & 0 & 0 & 0 & 0 & 1 & 0 \\ 1 & 0 & 0 & 0 & 0 & 0 & 0 & 0 & 1 \end{pmatrix}} \, i$$

Since no site covers more than three suburbs (no more than three 1's appear in any column of **A**), with two high schools a maximum of six suburbs can be covered. By inspection, six suburbs are covered if high schools are placed at either sites 1 and 6, sites 1 and 8, sites 2 and 6, or sites 6 and 8. Thus there are four optimum solutions to this particular partial cover problem.

In the example problem, suppose that three high schools are available for assignment to suburbs. By the preceding reasoning, an upper bound on the number of suburbs that can be covered is nine. However, since there are only nine suburbs, if a solution can be found to P9 such that $\tilde{z} = 9$, the solution obtained will also be an optimum solution to the total cover problem. Again, by inspection, placing high schools at sites 1, 6, and 8 will cover all nine suburbs. Thus, since two high schools only produce a partial cover and three high schools produce a total cover, an optimum solution to the total cover problem, P6, is $x_1^* = x_6^* = x_8^* = 1$ and $z^* = 3$. With $k = 3$, the optimum solution to the partial cover problem, P9, is $x_1^* = x_6^* = x_8^* = 1$ and $\tilde{z}^* = 9$.

Before presenting a method for solving P9, notice that P9 is a more general formulation than indicated by the previous discussion. As an illustration, consider a minimization version of P9 [36]:

P10.

$$\text{minimize} \quad \tilde{z} = \sum_{i=1}^{m} \min_{j \in \theta(x)} a_{ij}$$

$$\text{subject to} \quad \sum_{j=1}^{n} x_j \leq k$$

$$x_j = (0, 1), \quad \text{for all } j$$

where $\theta(x) = \{j : x_j = 1\}$; θ must be non-empty. In P10 let a_{ij} be the distance between customer i and site j. Thus the objective is to locate at most k facilities such that the total distance traveled between customer locations and facility locations is minimized. The term $\min_{j \in \theta(x)} a_{ij}$ indicates that if a customer can be served by more than one assigned facility he will choose that facility which is closest. The set $\theta(x)$ contains the indices for the sites that are assigned facilities. Both the central facilities location problem studied by ReVelle and Swain [51] and the facility location problem studied by Curry and Skeith [16] and Shannon and Ignizio [58] are illustrations of P10.

We choose to refer to P9 and P10 as partial cover problems when the a_{ij} are zero–one cover coefficients. However, when the a_{ij} are not restricted to be zero–one valued, for lack of a better label, we refer to P9 and P10 as *generalized partial cover* problems, since the latter allow a more general definition of a_{ij}.

As an illustration of a generalized partial cover problem, consider a facility location problem involving five customers and four potential sites for locating at most three facilities. The distances between customer locations and potential sites are given as follows:

$$\mathbf{D} = (d_{ij}) = \begin{pmatrix} 1 & 9 & 17 & 24 \\ 10 & 2 & 8 & 15 \\ 16 & 8 & 2 & 11 \\ 20 & 12 & 4 & 5 \\ 24 & 16 & 10 & 1 \end{pmatrix}$$

The number of trips made per month between a facility and customer i, a_i, equals 75, 171, 153, 137, and 805 for $i = 1, \ldots, 5$, respectively. The objective is to allocate facilities to sites and assign customers to sites in such a way that the total distance traveled per month is minimized. A maximum of three facilities is available for allocation to sites.

Based on the data for the illustration, the facility location problem can be formulated using P10. The matrix of a_{ij} values is obtained by letting a_{ij} equal $a_i\, d_{ij}$. Thus,

$$\mathbf{A} = (a_{ij}) = \begin{pmatrix} 75 & 675 & 1{,}275 & 1{,}800 \\ 1{,}710 & 342 & 1{,}368 & 2{,}565 \\ 2{,}448 & 1{,}224 & 306 & 1{,}683 \\ 2{,}740 & 1{,}644 & 548 & 685 \\ 19{,}320 & 12{,}880 & 8{,}050 & 805 \end{pmatrix}$$

Subsequently, it will be found that the optimum solution to the generalized partial cover problem is to assign facilities to sites 2, 3, and 4, assign customers 1 and 2 to the facility at site 2, assign customers 3 and 4 to the facility at site 3, and assign customer 5 to the facility at site 4.

Notice that in the generalized partial cover problem formulation an individual customer interacts with only one facility, whereas a given facility might interact with a number of customers. Furthermore, in the present illustration a customer interacts with the closest facility.

A number of different solution procedures are available for solving P9 and P10. Specifically, either dynamic programming [16], branch and bound [51], or heuristic methods [36] can be used to solve P9 and P10. Of these, we shall present a heuristic procedure developed by Ignizio [36].

The heuristic algorithm consists of two parts:

1. The main program, which picks facility locations one at a time until either the maximum number of locations has been chosen or until the addition of another facility will not decrease the total distance traveled.
2. The improvement check and elimination subroutine, which is designed to remove from solution those locations selected earlier that become uneconomical in combination with subsequent selections.

To motivate the steps of the heuristic algorithm, suppose that a number of facilities have been assigned to sites. Let $\theta(x)$ be the set of indices for those sites assigned a facility and let

$$a_i^* = \min_{j \in \theta(x)} a_{ij}$$

Define the column vector $\mathbf{a}^* = (a_i^*)$. Notice that $\sum_{i=1}^{m} a_i^*$ is the total cost of the current assignment. Site j, for $j \notin \theta(x)$, will only be assigned a facility if such an assignment will reduce the total cost. Thus, if site t is not currently assigned a facility, then assigning a facility to site t will decrease total cost by the amount DTC_t, where

$$DTC_t = \sum_{i=1}^{m} \max(a_i^* - a_{it}, 0) \tag{10.16}$$

Using a steepest-descent approach, the main program assigns a facility to that site which maximizes DTC_j, for $j \notin \theta(x)$, given some $DTC_j > 0$; otherwise, no additional facility is justified.

Since facilities are assigned to sites one at a time, it is possible for some combination of assigned sites to eliminate the need for a facility at some site previously assigned a facility. To determine the effect of removing a facility from a site, recall that the current assignment of facilities, $\theta(x)$, has a total cost of

$$TC(\theta(x)) = \sum_{i=1}^{m} \min_{j \in \theta(x)} a_{ij} = \sum_{i=1}^{m} a_i^* \tag{10.17}$$

If site t is removed from $\theta(x)$ to give a new set, $\theta'(x)$, the total cost of the new assignment will be

$$TC(\theta'(x)) = \sum_{i=1}^{m} \min_{j \in \theta'(x)} a_{ij} \tag{10.18}$$

Therefore, if the facility that is assigned to site t is eliminated, ΔTC_t, the change in total cost, is obtained by subtracting (10.17) from (10.18):

$$\Delta TC_t = \sum_{i=1}^{m} \left(\min_{\substack{p \in \theta(x) \\ p \neq t}} a_{ip} - a_i^* \right) \tag{10.19}$$

If the site having the minimum value of ΔTC_t is the site most recently assigned a facility, it should not be eliminated from the set $\theta(x)$. Otherwise, the site would be assigned a facility on the next iteration. Thus, if the minimum value of ΔTC_t corresponds to any site, other than the site most recently assigned a facility, then the site is removed from $\theta(x)$.

Based on the preceding justification, the steps of the heuristic algorithm for solving P10 will be presented. Concurrently, the algorithm will be illustrated using the example problem involving five customers and four sites. The steps of the algorithm are as follows:

Step 1 (*First site selection*). Let the cover matrix **A** consist of n column vectors $\mathbf{a}_1, \ldots, \mathbf{a}_n$. Calculate $c_j = \sum_{i=1}^{m} a_{ij}$ for $j = 1, \ldots, n$. Let t correspond to the site index having minimum c_j. Let $\mathbf{a}^* = \mathbf{a}_t$, set $x_t = 1$, and place t in the *ordered* set $\theta(x)$, where $\theta(x) = \{t : x_t = 1\}$. If $k = 1$, go to step 7; otherwise go to step 2 (see Table 10.4).

Table 10.4. STEP 1 CALCULATIONS

i \ j	a_{ij} 1	2	3	4	$\mathbf{a}^*$
1	75	675	1,275	1,800	1,800
2	1,710	342	1,368	2,565	2,565
3	2,448	1,224	306	1,683	1,683
4	2,740	1,644	548	685	685
5	19,320	12,880	8,050	805	805
c_j	26,293	16,765	11,547	7,538 min	$\theta(x) = \{4\}$

Step 2 (*Selection of next site*). For each $j \notin \theta(x)$, calculate

$$DTC_j = \sum_{i=1}^{m} \max\,(a_i^* - a_{ij}, 0)$$

where $\mathbf{a}^* = (a_i^*)$. If all $DTC_j = 0$, go to step 4; otherwise, let t correspond to the index j having maximum DTC_j. Set $x_t = 1$, place t in the next position in $\theta(x)$, and go to step 3 (see Table 10.5).

Table 10.5. STEP 2 CALCULATIONS

i \ j	a_{ij} 1	2	3	$\mathbf{a}^*$
1	75	675	1,275	1,800
2	1,710	342	1,368	2,565
3	2,448	1,224	306	1,683
4	2,740	1,644	548	685
5	19,320	12,880	8,050	805
DTC_j	2,580	3,807 max	3,236	$\theta(x) = \{4, 2\}$

Step 3 (*Formation of best combination*). Let $\mathbf{a}^* = (a_i^*)$, where, for $i = 1, \ldots, m$,

$$a_i^* = \min_{t \in \theta(x)} a_{it}$$

If $\sum_{t \in \theta(x)} x_t = 2$ and $k = 2$, go to step 7; if $\sum_{t \in \theta(x)} x_t = 2$ and $k > 2$, go to step 2; otherwise, go to step 4 (see Table 10.6).

Table 10.6. STEP 3 CALCULATIONS, FOLLOWED BY STEP 2 REPEATED

i \ j	a_{ij}: 1	a_{ij}: 3	$\mathbf{a}^*$
1	75	1,275	675
2	1,710	1,368	342
3	2,448	306	1,224
4	2,740	548	685
5	19,320	8,050	805
DTC_j	600	1,055 max	$\theta(x) = \{4, 2, 3\}$

Step 4 (*Formation of current assignment*). Let $h = \sum_{t \in \theta(x)} x_t$. Thus $\theta(x) = \{j_1, \ldots, j_h\}$. Let $R = (\mathbf{a}_{j_1}, \mathbf{a}_{j_2}, \ldots, \mathbf{a}_{j_h})$. If step 4 is entered directly from step 2, go to step 7; otherwise, go to step 5.

Step 5 (*Combination improvement and elimination check*). For each column of R, calculate

$$\Delta TC_t = \sum_{i=1}^{m} \left(\min_{\substack{p \in \theta(x) \\ p \neq t}} a_{ip} - a_i^* \right)$$

If $\min \Delta TC_t = \Delta TC_{j_h}$, go to step 6; otherwise, remove from R the $\mathbf{a}_t$ having $\min \Delta TC_t$, remove t from $\theta(x)$, set $x_t = 0$, set $a_i^* = \min_{t \in \theta(x)} a_{it}$, and go to step 2 (see Table 10.7).

Table 10.7. STEP 5 CALCULATIONS

i \ t	a_{it}: 4	a_{it}: 2	a_{it}: 3	$\mathbf{a}^*$
1	1,800	675	1,275	675
2	2,568	342	1,368	342
3	1,683	1,224	306	306
4	685	1,644	548	548
5	805	12,880	8,050	805
ΔTC_t	7,245	1,626	1,055 min	

$\min \Delta TC_t = \Delta TC_{j_h}$; $h = k$; go to step 7

Step 6 (*Check*). If $h = \sum_{t \in \theta(x)} x_t = k$, go to step 7; otherwise, go to step 2.

Table 10.8. STEP 7 (FINAL SOLUTION)

t	a_{it}		
i	4	2	3
1	1,800	(675)	1,275
2	2,568	(342)	1,368
3	1,683	1,224	(306)
4	685	1,644	(548)
5	(805)	12,880	8,050

$\bar{z} = 2{,}676$

Step 7 (*Assignment*). From matrix R, for each value of i find the index t having $\min_{t \in \theta(x)} a_{it}$. Assign customer i to a facility at site t for those i and t corresponding to each min a_{it} (see Table 10.8).

As indicated previously, the solution obtained is to assign facilities to sites 2, 3, and 4; customers 1 and 2 will be served by a facility at site 2; customers 3 and 4 will be served by a facility at site 3; and customer 5 will be served by a facility at site 4. The solution is identical to that obtained by Curry and Skeith [16] using dynamic programming.

As an exercise at the end of the chapter, you are asked to modify the heuristic procedure as required to solve P9, a maximization problem, and P6, the total cover problem. The modifications required are straightforward. In particular, once the procedure for solving P9 is obtained, P6 is solved by noting that coverage achieved must be exactly equal to the number of customers. Thus facilities are assigned to sites one at a time until all customers are covered.

In reporting their computational experience in using the heuristic procedure to solve P10, Shannon and Ignizio [58] solved several hundred problems of different sizes. In 85% of the cases an optimum solution was obtained, and in no case was the error greater than 4%. Three 40 by 100 problems were solved in an average time of 2.3 seconds on a Univac 1108 computer. The optimum solution was obtained in two cases; the solution to the third problem was within 4% of the optimum.

The heuristic procedure is easily applied and does not require a background in mathematical programming or branch and bound procedures. The use of the heuristic procedure in solving total and partial cover problems is explored further in the homework problems.

10.4 Summary

In this chapter we have considered briefly discrete plant location and covering problems. Due to the magnitude of the literature treating the subjects and space limitations for this chapter, a detailed treatment of either subject was not feasible. Consequently, we chose to present the Efroymson and Ray formulation of the discrete plant location problem, the Toregas, Swain, ReVelle, and Bergman solution procedure to the total cover problem, and the heuristic procedure developed by Ignizio for solving the partial cover problem.

Since the discussion in this chapter was necessarily brief, a large number of references are provided at the end of the chapter. Due to the amount of research currently underway on plant location problems, as well as covering problems, it is anticipated that numerous additions to the literature will be made prior to the publication of this material. Consequently, you are advised to consult the recent research literature for significant research contributions.

As in preceding chapters, it is emphasized that the analytical approaches presented in this chapter are design aids; the analytical models are not panaceas for all the discrete plant location and covering problems that exist. Actual problems, with their immense complexities, tend to defy exact reproduction in the form of a mathematical model. Consequently, the results obtained from analysis should be used as an aid or supplement to your intuition, rather than as a replacement for it. Recall, it is recommended that an "ideals approach" be used; the solutions obtained from analysis should be interpreted as benchmark solutions against which alternative solutions are compared.

REFERENCES

1. ATKINS, R. J., and R. H. SHRIVER, "New Approaches to Facilities Location," *Harvard Business Review*, Vol. 46, No. 3, 1968, pp. 70–79.
2. BALAS, E., "An Additive Algorithm for Solving Linear Programs with Zero–One Variables," *Operations Research*, Vol. 13, No. 4, 1965, pp. 517–546.
3. BALAS, E., "Discrete Programming by the Filter-Method," *Operations Research*, Vol. 15, No. 5, 1967, pp. 915–957.
4. BALINSKI, M. L., "Integer Programming: Methods, Uses, Computations," *Management Science*, Vol. 12, No. 3, 1965, pp. 253–313.
5. BALINSKI, M. L., "On Finding Integer Solutions to Linear Programs," *Mathematica*, Princeton, N.J., May 1964.

6. BALINSKI, M. L., and H. MILLS, "A Warehouse Problem," prepared for: Veteran's Administration, *Mathematica*, Princeton, N.J., April 1960.

7. BALINSKI, M. L., and R. E. QUANDT, "On an Integer Program for a Delivery Problem," *Operations Research*, Vol. 12, No. 2, 1964, pp. 300–304.

8. BALLOU, R. H., "Dynamic Warehouse Location Analysis," *Journal of Marketing Research*, Vol. 15, No. 3, 1969, pp. 271–276.

9. BALLOU, R. H., "Locating Warehouses in a Logistics System," *The Logistics Review*, Vol. 4, No. 19, 1968, pp. 23–40.

10. BAUMOL, W. J., and P. WOLFE, "A Warehouse Location Problem," *Operations Research*, Vol. 16, No. 2, 1958, pp. 252–263.

11. BAZARAA, M. S., and J. J. GOOD, "On the Quadratic Set Covering Problem," a working paper, Department of Industrial and Systems Engineering, Georgia Institute of Technology, Atlanta, Ga., 1972.

12. BEALE, E. M. L., "Survey of Integer Programming," *Operational Research Quarterly*, Vol. 16, No. 2, 1965, pp. 219–228.

13. BELLMORE, M., H. GREENBERG, and J. JARVIS, "Multi-Commodity Networks," *Management Science*, Vol. 16, No. 6, 1970, pp. 427–433.

14. BELLMORE, M., and H. D. RATLIFF, "Optimal Defense of Multi-Commodity Networks," *Management Sciences*, Vol. 18, No. 4, 1971, pp. 174–185.

15. BELLMORE, M., and H. D. RATLIFF, "Set Covering and Involutory Bases," *Management Science*, Vol. 18, No. 3, 1971, pp. 194–206.

16. CURRY, G. L., and R. W. SKEITH, "A Dynamic Programming Algorithm for Facility Location and Allocation," *AIIE Transactions*, Vol. 1, No. 2, 1969, pp. 133–138.

17. DANTZIG, G. B., and J. H. RAMSER, "The Truck Dispatching Problem," *Management Science*, Vol. 6, No. 1, 1960, pp. 80–91.

18. DAVIS, R. E., D. A. KENDRICK, and M. WEITZMAN, "A Branch-and-Bound Algorithm for Zero–One Mixed Integer Programming Problems," *Operations Research*, Vol. 19, No. 4, 1971, pp. 1036–1044.

19. DAVIS, P. S., and T. L. RAY, "A Branch-Bound Algorithm for the Capacitated Facilities Location Problem," *Naval Research Logistics Quarterly*, Vol. 16, No. 3, 1969, pp. 331–343.

20. DAY, R. H., "On Optimal Extracting for a Multiple File Data Storage System: An Application of Integer Programming," *Operations Research*, Vol. 13, No. 3, 1965, pp. 482–494.

21. DRYSDALE, J. K., and P. J. SANDIFORD, "Heuristic Warehouse Location—A Case History Using a New Method," *Canadian Operational Research Journal*, Vol. 7, No. 1, 1969, pp. 45–61.

22. EDMONDS, J., "Covers and Packing in a Family of Sets," *Bulletin of the American Mathematical Society*, Vol. 68, No. 5, 1962, pp. 494–499.

23. EDMONDS, J., "Paths, Trees, and Flowers," *Canadian Journal of Mathematics*, Vol. 17, No. 3, 1965, pp. 449–467.

24. EFROYMSON, M. A., and T. L. RAY, "A Branch-Bound Algorithm for Plant Location," *Operations Research*, Vol. 14, No. 3, 1966, pp. 361–368.

25. EILON, S., C. D. T. WATSON-GANDY, and N. CHRISTOFIDES, *Distribution Management: Mathematical Modelling and Practical Analysis*, Hafner Publishing Company, Inc., New York, 1971.

26. ELLWEIN, L. B., and P. GRAY, "Solving Fixed Charge Allocation Problems with Capacity and Configuration Constraints," *AIIE Transactions*, Vol. 3, No. 4, 1971, pp. 290–299.

27. ELSON, D. G., "Site Location via Mixed Integer Programming," *Operational Research Quarterly*, Vol. 23, No. 1, 1972, pp. 31–44.

28. FEJES TO'TH, L., *Regular Figures*, Pergamon Press, Inc., Elmsford, N.Y., 1964.

29. FELDMAN, E., F. A. LEHRER, and T. L. RAY, "Warehouse Locations Under Continuous Economies of Scale," *Management Science*, Vol. 2, No. 9, 1966, pp. 670–684.

30. GARFINKEL, R. S., "Set Covering: A Survey," presented at the XVIII International Conference of the Institute of Management Sciences, London, 1970.

31. GARFINKEL, R. S., and G. L. NEMHAUSER, "The Set Partitioning Problem: Set Covering with Equality Constraints," *Operations Research*, Vol. 17, No. 5, 1969, pp. 840–856.

32. GEOFFRION, A. M., "An Improved Implicit Enumeration Approach for Integer Programming," *Operations Research*, Vol. 17, No. 3, 1969, pp. 437–454.

33. GEOFFRION, A. M., "Integer Programming by Implicit Enumeration and Balas Method," *SIAM Review*, Vol. 9, No. 2, 1967, pp. 178–190.

34. GERSON, M., and R. B. MAFFEI, "Technical Characteristics of Distribution Simulators," *Management Science*, Vol. 10, No. 1, 1963, p. 62.

35. GRAY, P., "Mixed Integer Programming Algorithm for Site Selection and Other Fixed Charge Problems," Technical Report 6, Department of Operations Research, Stanford University, Stanford, Calif. 1967.

36. IGNIZIO, J. P., "A Heuristic Solution to Generalized Covering Problems," unpublished Ph.D. Dissertation, Virginia Polytechnic Institute and State University, Blacksburg, Va., 1971.

37. KHUMAWALA, B. M., "Branch and Bound Algorithms for Locating Emergency Service Facilities," Institute Paper 355, Herman C. Krannert Graduate School of Industrial Administration, Purdue University, Lafayette, Ind., 1972.

38. KHUMAWALA, B. M., "An Efficient Branch and Bound Algorithm for the Warehouse Location Problem," *Management Science*, Vol. 18, No. 12, 1972, pp. 718–731.

39. KHUMAWALA, B. M., and D. L. KELLY, "Warehouse Location with Concave Costs," Institute Paper 360, Herman C. Krannert Graduate School of Industrial Administration, Purdue University, Lafayette, Ind., 1972.

40. KHUMAWALA, B. M., and D. C. WHYBARK, "A Comparison of Some Recent, Warehouse Location Techniques," *The Logistics Review*, Vol. 7, 1971, pp. 3–19.

41. KUEHN, A. A., and M. J. HAMBURGER, "A Heuristic Program for Locating Warehouses," *Management Science*, Vol. 9, No. 4, 1963, pp. 643–666.

42. LAWLER, E. L., "Covering Problems: Duality Relations and a New Method of Solution," *Journal of the SIAM on Applied Mathematics*, Vol. 14, No. 5, 1966, pp. 1115–1132.

43. LAWLER, E. L., and D. E. WOOD, "Branch-and-Bound Methods: A Survey," *Operations Research*, Vol. 14, No. 4, 1966, pp. 699–719.

44. LEMKE, C. E., H. M. SALKIN, and K. SPIELBERG, "Set Covering by Single Branch Enumeration with Linear Programming Subproblems," *Operations Research*, Vol. 19, No. 4, 1971, pp. 998–1022.

45. LEMKE, C. E., and K. SPIELBERG, "Direct Search Algorithms for Zero-One and Mixed Integer Programming," *Operations Research*, Vol. 15, No. 5, 1967, pp. 892–914.

46. MANNE, A. S., "Plant Location Under Economies-of-Scale: Decentralization and Computation," *Management Science*, Vol. 11, No. 2, 1964, pp. 213–235.

47. NEMHAUSER, G., and A. GARFINKEL, *Integer Programming*, John Wiley & Sons, Inc., New York, 1972.

48. NORMAN, R. Z., and M. O. RABIN, "An Algorithm for a Minimum Cover of a Graph," *Proceedings of the American Mathematical Society*, Vol. 10, No. 2, 1959, pp. 315–319.

49. PIERCE, J. F., "Application of Combinatorial Programming to a Class of All Zero-One Programming Problems," *Management Science*, Vol. 15, No. 3, 1968, pp. 191–209.

50. REVELLE, C., D. MARKS, and J. C. LIEBMAN, "An Analysis of Private and Public Sector Location Models," *Management Science*, Vol. 16, No. 11, 1970, pp. 692–707.

51. REVELLE, C., and R. SWAIN, "Central Facilities Location," *Geographical Analysis*, Vol. 2, No. 1, 1970, pp. 30–42.

52. ROGERS, C. A., *Packing and Covering*, Cambridge University Press, New York, 1964.

53. ROJESKI, P., and C. REVELLE, "Central Facilities Location Under an Investment Constraint," *Geographical Analysis*, Vol. 2, No. 4, 1970, pp. 343–360.

54. ROTH, R., "Computer Solutions to Minimum-Cover Problems," *Operations Research*, Vol. 17, No. 3, 1969, pp. 455–465.

55. SA, G., "Branch-and-Bound and Approximate Solutions to the Capacitated Plant Location Problem," *Operations Research*, Vol. 17, No. 6, 1969, pp. 1005–1016.

56. SALVESON, M. E., "The Assembly-Line Balancing Problem," *Transactions ASME*, Vol. 77, No. 8, 1955, pp. 939–947.

57. SCOTT, A. J., "Location-Allocation Systems: A Review," *Geographical Analysis*, Vol. 2, No. 2, 1970, pp. 95–119.

58. SHANNON, R. E., and J. P. IGNIZIO, "A Heuristic Programming Algorithm for Warehouse Location," *AIIE Transactions*, Vol. 2, No. 4, 1970, pp. 334–339.

59. SHYCON, H. N., and R. B. MAFFEI, "Simulation-Tool for Better Distribution," *Harvard Business Review*, Vol. 38, No. 6, 1960, pp. 65–75.

60. SPIELBERG, K., "Algorithm for the Simple Plant-Location Problem with Some Side Conditions," *Operations Research*, Vol. 17, No. 1, 1969, pp. 85–111.

61. SPIELBERG, K., "Plant Location with Generalized Search Origin," *Management Science*, Vol. 16, No. 3, 1969, pp. 165–178.

62. TOREGAS, C., R. SWAIN, C. REVELLE, and L. BERGMAN, "The Location of Emergency Service Facilities," *Operations Research*, Vol. 19, No. 6, 1971, pp. 1363–1373.

63. TOREGAS, C. and C. REVELLE, "Optimal Location under Time or Distance Constraints," *Papers of the Regional Science Association*, Vol. 28, 1972, pp. 133–143.

64. WATSON-GANDY, C. D. T., and S. EILON, "The Depot Siting Problem with Discontinuous Delivery Cost," *Operational Research Quarterly*, Vol. 23, No. 3, 1972, pp. 277–288.

65. WHYBARK, D. C., and B. M. KHUMAWALA, "A Survey of Facility Location Methods," Institute Paper 350, Herman C. Krannert Graduate School of Industrial Administration, Purdue University, Lafayette, Ind., 1972.

PROBLEMS

10.1. Write a computer program for the Efroymson and Ray branch and bound solution procedure.

10.2. Write a computer program for the heuristic procedure given in Section 10.3.1.

10.3. Write a computer program for the heuristic solution procedure given in Section 10.3.2.

10.4. (a) List 10 applications of the total cover problem formulation, P6.
(b) List five applications of the partial cover problem formulation, P9 or P10.

10.5. Prepare a written summary of Khumawala's branch and bound procedure for solving the Efroymson and Ray problem. See [38] for a discussion of Khumawala's procedure.

10.6. Perform a detailed literature survey on the discrete plant location problem.

10.7. Perform a detailed literature survey on the covering problem.

10.8. Prepare a detailed written description of the discrete plant location algorithm developed by Ellwein and Gray [26].

10.9. Using complete enumeration, solve the example problem given in Section 10.2.1.

10.10. Prove Remarks 3 and 4 in Section 10.2.1.

10.11. Solve the example problem in Section 10.2.1. with the following changes in the costs:

(a)
$$\mathbf{C} = \begin{pmatrix} 6 & 8 & 10 \\ 10 & 5 & 7 \\ 12 & 8 & 6 \\ 8 & 4 & 6 \\ 4 & 5 & 8 \end{pmatrix}, \qquad \mathbf{f} = (8 \quad 10 \quad 6)$$

(b)
$$\mathbf{C} = \begin{pmatrix} 0 & 6 & 8 \\ 3 & 4 & 7 \\ 5 & 0 & 7 \\ 6 & 8 & 6 \\ 8 & 4 & 0 \end{pmatrix}, \qquad \mathbf{f} = (10 \quad 8 \quad 12)$$

(c)
$$\mathbf{C} = \begin{pmatrix} 14 & 16 & 18 \\ 13 & 14 & 17 \\ 20 & 15 & 17 \\ 22 & 18 & 16 \\ 18 & 14 & 16 \end{pmatrix}, \qquad \mathbf{f} = (0 \quad 0 \quad 0)$$

10.12. Modify the Ignizio heuristic procedure such that it can be used to solve P6 and P9.

10.13. Solve the total cover problems having the following cover matrices. Use the solution procedure given in Sections 10.3.1 and 10.3.2.

(a)
$$\mathbf{A} = \begin{pmatrix} 1 & 0 & 0 & 1 & 1 \\ 1 & 1 & 0 & 0 & 1 \\ 0 & 0 & 1 & 0 & 0 \\ 0 & 1 & 1 & 1 & 0 \\ 0 & 1 & 0 & 0 & 1 \end{pmatrix}$$

(b)
$$\mathbf{A} = \begin{pmatrix} 1 & 1 & 0 & 0 & 0 \\ 0 & 1 & 0 & 1 & 1 \\ 0 & 0 & 1 & 1 & 0 \\ 1 & 0 & 0 & 1 & 0 \\ 1 & 1 & 0 & 0 & 1 \end{pmatrix}$$

(c)
$$\mathbf{A} = \begin{pmatrix} 1 & 1 & 0 & 0 & 0 & 0 & 0 & 0 & 1 \\ 1 & 1 & 0 & 1 & 0 & 0 & 0 & 0 & 0 \\ 0 & 0 & 1 & 0 & 0 & 0 & 0 & 1 & 0 \\ 0 & 0 & 0 & 1 & 0 & 0 & 0 & 0 & 0 \\ 1 & 0 & 0 & 1 & 1 & 0 & 0 & 0 & 0 \\ 0 & 0 & 1 & 0 & 1 & 1 & 0 & 0 & 0 \\ 0 & 0 & 0 & 0 & 0 & 0 & 1 & 0 & 0 \\ 0 & 0 & 0 & 1 & 0 & 0 & 1 & 1 & 0 \\ 0 & 0 & 0 & 0 & 0 & 0 & 0 & 0 & 1 \end{pmatrix}$$

(d)
$$\mathbf{A} = \begin{pmatrix} 1 & 0 & 1 & 0 \\ 1 & 0 & 0 & 1 \\ 0 & 1 & 1 & 0 \\ 0 & 1 & 0 & 0 \end{pmatrix}$$

(e)
$$\mathbf{A} = \begin{pmatrix} 1 & 0 & 1 & 0 \\ 0 & 1 & 1 & 0 \\ 1 & 0 & 0 & 1 \\ 0 & 0 & 0 & 1 \end{pmatrix}$$

(f)
$$\mathbf{A} = \begin{pmatrix} 1 & 0 & 1 & 0 \\ 1 & 1 & 1 & 0 \\ 0 & 0 & 0 & 1 \\ 0 & 0 & 1 & 0 \\ 1 & 1 & 0 & 1 \end{pmatrix}$$

10.14. A logging company is cutting trees in four different areas of Ashley County, Arkansas. The logs are transported by truck to a site along a railroad where special loading equipment is located for loading the logs on rail cars. The firm has identified five candidate loading sites and has equipment available for at most three loading sites. Based on the volume of material handling between logging areas and loading sites, the mileages traveled per day between combinations of logging areas and loading sites are summarized below. It is the policy of the firm that all loads of logs from a given logging area are transported to the same loading site. Determine the optimum allocation of loading equipment to sites and the assignment of logging areas to loading sites to minimize distance traveled.

	Candidate Loading Sites				
Logging Area	1	2	3	4	5
1	0	100	250	150	400
2	40	50	500	300	30
3	400	200	50	350	160
4	100	150	60	10	250

10.15. In Problem 10.14 suppose that the cost of transporting the equipment to the sites, setting up the equipment, and returning the equipment to the equipment storage yard is given to be

Candidate loading sites	1	2	3	4	5
Set-up and transportation cost ($)	200	100	50	150	200

Furthermore, let the travel cost per mile traveled by logging trucks between the logging areas and loading sites be $0.50. Determine the optimum allocation of loading equipment to sites and the assignment of logging areas to loading sites to minimize total cost over a 10-day planning horizon.

10.16. The community of Snyder, Arkansas wishes to locate municipal parks for use by the public. The residential area in the community has been approximated by the 15 grid squares shown in P10.16. Each grid square has an area of 0.25 square mile. Potential park sites are available in eight of the grid squares. It is desired that a park be located within 1 mile of the center of each grid square. Assume rectilinear travel and that each park site is in the center

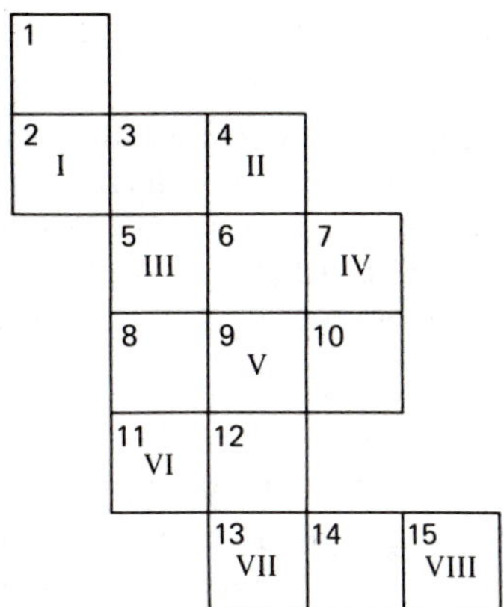

Figure P10.16

of the indicated grid square. Determine the minimum number of park sites necessary to satisfy the desired maximum distance criteria.

10.17. Solve Problem 10.16 when a maximum distance of 1.5 miles is used. Also, determine the locations and allocations of grid squares to parks when a maximum of three parks is allowed.

10.18. In a rural section of North Carolina, health outreach clinics are to be located over an area coincident with 10 magisterial districts. Five potential sites are available for locating health outreach clinics. Distances between the centroids of the districts and the potential sites are tabulated, along with the population of each district. Determine the locations of the clinics that minimize the total distance traveled per unit time, based on a maximum of (a) one clinic, (b) two clinics, and (c) three clinics.

Magisterial District	*Potential Sites* 1	2	3	4	5	*Population*
1	15	10	25	20	30	3,000
2	30	10	20	20	35	4,500
3	30	15	10	15	30	2,500
4	25	0	15	10	25	5,000
5	20	10	5	10	20	2,000
6	30	20	5	10	20	4,000
7	15	25	20	5	10	6,000
8	0	25	30	20	15	7,000
9	20	25	20	10	5	3,000
10	10	30	30	20	10	5,000

10.19. In Problem 10.18 suppose that a magisterial district is "covered" if a health outreach clinic is located within X miles of the centroid of the district. Determine the minimum number of clinics required to cover the districts when X equals (a) 5, (b) 10, (c) 15, and (d) 20 or more.

10.20. A study is being conducted to determine the optimum number and location of fire towers in a large national forest. It is desired that the minimum number of towers be used to provide a coverage of all tracts in the forest. The forest consists of 12 tracts to be included in the surveillance by rangers located in the fire towers. Eight locations have been selected as potential sites for fire towers due to their altitudes and visibility ranges. The coverage matrix is

$$\mathbf{A} = \begin{pmatrix} 1 & 1 & 0 & 0 & 0 & 0 & 0 & 0 \\ 0 & 1 & 0 & 0 & 0 & 0 & 0 & 0 \\ 0 & 1 & 1 & 0 & 0 & 0 & 0 & 0 \\ 1 & 1 & 0 & 1 & 0 & 0 & 0 & 0 \\ 0 & 1 & 1 & 1 & 0 & 1 & 0 & 0 \\ 0 & 0 & 1 & 1 & 1 & 1 & 0 & 1 \\ 1 & 0 & 1 & 1 & 1 & 1 & 0 & 0 \\ 1 & 0 & 0 & 1 & 0 & 0 & 0 & 0 \\ 0 & 0 & 0 & 1 & 1 & 1 & 1 & 1 \\ 1 & 0 & 0 & 0 & 1 & 0 & 1 & 0 \\ 0 & 0 & 0 & 0 & 1 & 1 & 1 & 1 \\ 0 & 0 & 0 & 0 & 1 & 1 & 1 & 1 \end{pmatrix}$$

Solve the problem by inspection and using the heuristic procedure.

10.21. Based on a maximum of three fire towers, determine the location of the towers that maximizes the coverage of tracts in Problem 10.20. Use the Ignizio heuristic procedure. Is the solution obtained a feasible solution to the total cover problem? Why or why not?

10.22. A textile plant has a number of automatic spinning machines that are assigned to operators who patrol the assigned area and repair any breaks which occur in the continuous filament fiber. The plant can be divided into 20 squares contained in a rectangle with a width of 4 and a length of 5. A patrol operator who is "based" in a square can patrol all machines in that square, as well as the eight adjacent squares. Formulate the problem of determining the minimum number of operators required to patrol the entire plant. Write out the corresponding cover matrix **A**.

10.23. The Republic of East Venutia is threatened by its neighbor, The Republic of West Venutia. Radar installations are to be located throughout East Venutia to provide protection against an attack by the West Venutians. Ten key areas of East Venutia have been designated as vital areas. Additionally, ten sites have been designated as feasible sites for the radar installations. Determine the minimum number of radar installations to be located throughout the country to provide radar coverage for the ten vital areas. Letting a_{ij} equal one (zero) if an installation at site i covers (does not cover) vital area

j, the following covering coefficients are provided:

Site	Vital Area 1	2	3	4	5	6	7	8	9	10
1	1	1	0	1	0	0	1	1	0	0
2	1	1	1	0	0	0	0	0	0	0
3	0	1	1	1	1	1	0	0	0	0
4	1	1	1	1	1	1	1	1	0	0
5	0	1	1	1	1	1	1	0	0	0
6	0	0	1	1	1	1	1	0	1	0
7	1	0	0	1	1	1	1	1	1	1
8	1	0	0	1	0	0	0	1	0	1
9	0	0	0	0	0	1	1	0	1	1
10	0	0	0	0	0	1	1	1	1	1

10.24. Consider the covering problem having the cover matrix

$$\mathbf{A} = \begin{pmatrix} 1 & 0 & 1 & 0 & 0 \\ 1 & 0 & 0 & 0 & 1 \\ 1 & 1 & 0 & 0 & 0 \\ 0 & 1 & 0 & 0 & 0 \\ 0 & 1 & 0 & 0 & 0 \\ 0 & 0 & 1 & 0 & 0 \\ 0 & 0 & 1 & 0 & 0 \\ 0 & 0 & 0 & 1 & 1 \\ 0 & 0 & 0 & 1 & 1 \end{pmatrix}$$

(a) Solve the total cover problem by enumerating all possible assignments.
(b) Solve the partial cover problem (P9) for the case of $k = 3$ by enumerating all possible assignments.
(c) Solve the partial cover problem (P9) for the case of $k = 3$ by using the heuristic solution procedure.

10.25. Solve the Toregas et al. example problem in Section 10.3.1 assuming a city is covered if a facility is located within s miles of the city, where s equals (a) 60 miles, (b) 61 miles, (c) 62 miles, (d) 68 miles, and (e) 70 miles.

10.26. Suppose nine facilities are available for assignment to sites and consider the cover matrix given in Table 10.3. Assign the facilities to sites using the Ignizio heuristic to achieve maximum coverage.

10.27. Using the data in Table 10.2 instead of the data in Table 10.3, solve Problem 10.27 by assigning facilities to sites to minimize the sum of the distances between customers and facilities.

10.28. Solve Problems 10.27 and 10.28 when the number of facilities equals (a) 8, (b) 10, and (c) 11.

10.29. Using the cover matrices obtained in Problem 10.26, solve Problem 10.27 when 10 facilities are available for assignment to sites. Use the Ignizio heuristic procedure.

10.30. ReVelle and Swain [51] present an example problem having the following generalized cover coefficients:

	1	2	3	4	5	6	7	8	9	10
1	0	0.875	1.50	2.00	2.875	3.625	3.50	4.25	4.50	4.43
2	1.750	0	2.75	2.250	7.500	7.75	6.50	6.75	8.50	8.36
3	3.750	3.44	0	1.88	6.25	6.07	5.00	7.50	7.50	7.32
4	3.500	1.97	1.31	0	5.70	4.81	3.72	3.94	5.46	5.34
5	3.450	4.50	3.00	3.90	0	2.85	3.81	5.92	4.80	4.94
6	8.350	8.91	5.58	6.32	5.46	0	2.88	6.90	3.74	4.89
7	15.050	13.98	8.60	9.14	13.69	5.38	0	7.53	4.30	4.00
8	12.750	10.11	9.00	6.75	14.80	9.00	5.25	0	6.00	4.50
9	10.800	10.20	7.20	7.51	9.60	3.90	2.40	4.80	0	1.20
10	6.650	6.27	4.39	4.58	6.16	3.19	1.40	2.25	0.75	0

Given four available facilities, use the Ignizio heuristic to solve the central facilities location problem, P10.

INDEX